www.wadsworth.com

www.wadsworth.com is the World Wide Web site for Wadsworth and is your direct source to dozens of online resources.

At *www.wadsworth.com* you can find out about supplements, demonstration software, and student resources. You can also send email to many of our authors and preview new publications and exciting new technologies.

www.wadsworth.com
Changing the way the world learns®

Statistical Analysis
in the Social Sciences

McKEE J. McCLENDON
Case Western Reserve University

Australia • Canada • Mexico • Singapore • Spain
United Kingdom • United States

THOMSON
WADSWORTH

Sociology Editor: *Robert Jucha*
Technology Project Manager: *Dee Dee Zobian*
Assistant Editor: *Stephanie Monzon*
Editorial Assistant: *Melissa Walter*
Marketing Manager: *Matthew Wright*
Marketing Assistant: *Tara Pierson*
Advertising Project Manager: *Linda Yip*
Project Manager, Editorial Production: *Cheri Palmer*
Print/Media Buyer: *Jessica Reed*

Permissions Editor: *Joohee Lee*
Production Service: *Ruth Cottrell*
Copy Editor: *Ruth Cottrell*
Illustrator: *Lotus Art*
Cover Designer: *Belinda Fernandez*
Cover Image: *ImageState*
Compositor: *ATLIS Graphics*
Printer: *Transcontinental Printing/Louiseville*

For more information about our products,
contact us at:
Thomson Learning Academic Resource Center
1-800-423-0563
For permission to use material from this text,
contact us by:
Phone: 1-800-730-2214   Fax: 1-800-730-2215
Web: http://www.thomsonrights.com

Library of Congress Control Number: 2003107839

ISBN 0-534-63783-3

Wadsworth/Thomson Learning
10 Davis Drive
Belmont, CA 94002-3098
USA

Asia
Thomson Learning
5 Shenton Way #01-01
UIC Building
Singapore 068808

Australia/New Zealand
Thomson Learning
102 Dodds Street
Southbank, Victoria 3006
Australia

Canada
Nelson
1120 Birchmount Road
Toronto, Ontario M1K 5G4
Canada

Europe/Middle East/Africa
Thomson Learning
High Holborn House
50/51 Bedford Row
London WC1R 4LR
United Kingdom

Latin America
Thomson Learning
Seneca, 53
Colonia Polanco
11560 Mexico D.F.
Mexico

Spain/Portugal
Paraninfo
Calle/Magallanes, 25
28015 Madrid, Spain

To
Jeanne,
Pat, Barb,
Kara, and Paul

Contents

★The asterisk is used throughout the book to indicate optional material.

Preface

*S*tatistical Analysis in the Social Sciences is an introductory statistics text for upper-
level undergraduates in the social sciences. It embeds basic descriptive and inferential statistical concepts within a unified framework: the linear regression
model. The regression model is used not only for conducting standard regression
analyses but also for other statistical analyses that are ordinarily taught as separate and
distinct techniques, such as correlation, difference of means, analysis of variance, and
certain nonparametric tests. This extensive use of a single statistical model will increase students' depth of understanding and give them a sense of mastery and self-
confidence that they might otherwise fail to achieve in their first statistics course.

The basic statistical tools covered in this text consist of describing or summarizing the observed attributes of a single variable, describing the association between
two or more observed variables, and making estimates of the likely values of each of
these quantities in a population. Although these topics comprise the content of most
introductory statistics texts, the typical text covers this material under a rather complex typology of seemingly disparate and unrelated techniques. For example, measures and tests of associations are often covered under these categories: difference of
means, correlation, regression analysis, analysis of variance, association between ordinal variables (e.g., rho, tau, gamma, *d*), association between ordinal and nominal variables (e.g., Wilcoxon–Mann–Whitney test, Kruskal–Wallis test), and association
between nominal variables (e.g., chi square, phi, Cramér's *V*, lambda).

This book presents the basic statistical tools from a more unified framework. With
the exception of the case of nominal variables, the regression model is the statistical
technique that is used to analyze all associations among variables. The concepts and

notation of the regression model are also consistent with those of two important univariate descriptive statistics, the mean and the standard deviation (or variance); or, to put it another way, the mean and the standard deviation are basic statistical concepts that lie at the very core of regression analysis. The consistent utilization of the regression model provides a basic statistical framework for the cumulative development of statistical concepts and techniques. It also provides a consistent set of statistical terminology and notations. The end result is an increase in students' understanding of basic statistical procedures and less reliance on memorization of rules and formulas.

Another benefit of using a unifying statistical model is that there is less of a cookbook approach to statistical analysis. The typical text teaches the student to first identify the level(s) of measurement of the variables to be analyzed and the general type of analysis to be conducted (central tendency, variation, or association). After this is done, the student can then look up the specific statistical technique with which to conduct the analysis. In this approach, the researcher finds the statistical technique that fits the types of variables and data to be analyzed; in other words, the technique is fitted to the data. Although the use of a single unifying model does not obviate the need to pay careful attention to the levels of measurement of the variables, the emphasis in this text is on fitting the data to the model rather than the other way around. Fitting the data to the model involves transforming the data in such a way that it can be validly used by the regression technique. For example, nominal variables are recoded into dummy variables (Chapter 11), and ordinal variables are transformed into rank-scored variables (Chapter 12).

How can the regression model be justified as a unifying framework for most types of association among variables? Quantitative methodologists know that the most commonly used parametric statistical techniques are based on a general linear model. In particular, tests for differences of means, bivariate and partial correlations, regression analysis, and analysis of variance are all manifestations of the same general linear model (e.g., Fox, 1997, pp. 204–219). Regression is the most general and flexible of these techniques. It can be used to test a difference of means (the t test of a dummy variable slope), to compute and test correlations (the standardized regression slope), and to conduct an analysis of variance (e.g., the F test for a set of dummy variables). In addition, regression analysis is the most frequently used statistical technique in the social sciences. Thus, a more extended treatment of and reliance on regression than it normally receives in introductory texts is fully warranted. Despite widespread knowledge about the general linear model and the versatility of regression analysis, there are no texts suitable for use in an undergraduate introductory statistics course in the social sciences that incorporate a general linear model approach to teaching parametric statistics.

Regression analysis is also used to execute a number of the more commonly used nonparametric techniques that are designed for ordinal dependent variables. The validity of using regression in this capacity is not nearly as well known as its validity for parametric analyses. This method for conducting nonparametric analyses consists of applying a parametric technique, such as regression, to ordinal variables that have been scored for ranks (Conover and Iman, 1981; Conover, 1980). When two ordinal variables, for example, are scored for ranks, the Pearson correlation equals the Spearman correlation. And when an ordinal dependent variable scored for ranks is regressed on a dummy variable, it is equivalent to the Wilcoxon–Mann–Whitney test. The use of regression for conducting nonparametric tests involving ordinal variables is covered in Chapter 12.

The implementation of the proposed framework simply involves giving the students more instruction in regression and omitting some of the other traditional techniques. In this case, however, it is not a matter of substituting depth for breadth, because the increased depth in treatment of regression is designed to give students the tools needed to handle the same range of statistical problems that are covered in a more traditional fashion in other texts.

OVERVIEW OF CHAPTERS

Chapter 1 offers an overview of social research and the role of statistics in this process. Chapters 2 through 6 provide a traditional coverage of univariate descriptive and inferential statistics. Chapter 2 shows how to create meaningful graphical and numerical descriptions of univariate distributions (i.e., frequency distributions). Chapter 3 introduces three ways to describe numerically the central tendency of the distribution, namely, the mode, the median, and the mean. Chapter 4 shows how to describe the amount and the shape of dispersion in a univariate distribution. Chapter 5, which covers random sampling error, establishes the conceptual foundation for inferential statistics, including the normal distribution. Chapter 6 applies the principles developed in the previous chapter to the use of the t distribution to make inferences about population means (i.e., confidence intervals and hypothesis tests).

Chapters 7 through 12 develop the regression model as a unified method for analyzing associations between variables. Although students will learn to use regression techniques to conduct all of their statistical analyses of associations except those involving nominal dependent variables, they will undoubtedly read and hear about alternative statistical techniques when they begin to practice social research and read the social research literature. Therefore, they will be informed about these alternative statistics at appropriate points in the text. Sections marked with an asterisk (*) provide brief descriptions of well-known alternative techniques, such as analysis of variance, Spearman's rho, or the Kruskal–Wallis test. These sections, however, may be skipped without any loss of understanding of the basic concepts and techniques.

Chapter 7 lays the foundation for studying associations between interval-level variables by using scatterplots to display associations visually and by using crossproducts and the covariance as the fundamental numerical measure of linear association. Chapter 8 introduces the least-squares bivariate regression model and two measures of the strength of linear association, namely, the coefficient of determination and the standardized slope. This is the first point in the text that the general linear model becomes apparent as a result of the equivalence between the standardized slope and the Pearson correlation. Chapter 9 covers hypothesis tests for the bivariate regression slope. As such, it is a rather straightforward extension of the material in Chapter 6 on hypothesis tests for the mean. Chapter 10 provides an introduction to multiple regression—that is, regression analysis with two or more independent variables. Many introductory statistics texts do not cover multiple regression. Aside from the fact that multiple regression is an important topic in its own right, however, it is necessary to cover this topic because multiple regression is used to analyze differences of means among three or more groups, a problem for which one-way analysis of variance is commonly used. The F test is also introduced in this chapter. Chapter 11 describes how nominal independent variables can be included in regression analysis by the use of dummy variable coding. When a student is learning how to do a dummy variable regression for a dichotomous nominal variable, the student is also learning to do the t test for a difference of means. This is another instance of the general linear model and is discussed in an optional ★ section. When the nominal independent variable has more than two categories/groups, the F test for the regression equation is equivalent to the F test in a one-way analysis of variance, which is discussed in an optional ★ section. Chapter 12 covers the analysis of relationships involving ordinal dependent variables, which can be accomplished by including rank-scored ordinal variables in a conventional regression analysis. The equivalence of several forms of such regression analyses to well-known nonparametric techniques—namely, the Spearman correlation, the Mann-Whitney-Wilcoxon test, and the Kruskal-Wallis test—is described in optional ★ sections.

Chapter 13 covers associations for nominal dependent variables. This chapter is the only one dealing with associations that does not use regression analysis. Although it would be possible to introduce logistic regression for nominal dependent variables to maintain continuity with the previous chapters, logistic regression is beyond the level of this book. Therefore, this chapter uses the Pearson chi-square test and chi-square-based measures of association for the analysis of crosstabular tables involving nominal variables.

It is desirable, although not absolutely necessary, that the chapters in the book be assigned in the presented order. The material in each chapter depends on an understanding of the concepts contained in the previous chapters. The major exception is the fact that Chapter 13, on associations between nominal variables, does not depend on Chapter 12, which deals with associations between ordinal variables. Furthermore, if the instructor does not want to cover multiple regression, Chapters 10 and 11 could be skipped and it would still be possible for the student to understand the material in Chapter 12 (except the section on polytomous independent variables) and all of the material in Chapter 13.

PEDAGOGICAL TECHNIQUES

Linear regression analysis is a relatively simple statistical technique that beginning students can readily understand if it is presented carefully and used repeatedly. At the core of the regression model is the idea of a linear slope, or the change in Y per unit increase in X. This concept is related to basic algebra and the concept of a straight line. The concept of the linear slope is used throughout all the chapters on association except for the last one dealing with nominal variables. I have found that it is easier for students to understand and interpret a slope than it is to read a simple crosstabulation table.

The text emphasizes an elementary algebraic understanding of statistics. In this regard, there is a relatively strong emphasis on understanding the formulas for statistical concepts. For example, a student should understand the crucial role that deviation scores play in defining the standard deviation as a measure of dispersion. Moreover, it is important for students to understand why the sum of cross-products of the deviation scores of X and Y (a basic statistic used for measuring association in regression analysis) indicates whether there is a positive or a negative relationship between the two variables. The formulas emphasized are those that give the most meaningful and useful insight into the nature of the statistic. That is, only definitional formulas are used. Computational formulas are not presented.

Computers are used for carrying out statistical calculations. A section in each chapter describes and illustrates how to use SPSS software to conduct the statistical analyses covered in the chapter. SPSS is emphasized because of its widespread use in the social sciences. The output of most statistical packages is very much the same for the basic statistics covered in this text. Different software outputs may be organized somewhat differently, but they are quite consistent in content and in the way they are labeled.

Although the book emphasizes a mathematical understanding of statistical concepts, the level of math required is not difficult. The only mathematics needed is the ability to read and understand simple algebraic expressions and functions. Derivations and proofs are not given except for a very few cases in which they significantly increase understanding. Furthermore, statistical concepts receive a great deal of verbal discussion.

Graphics are also given a relatively strong emphasis for several reasons. First, bar charts, histograms, frequency polygons, scatterplots, graphs of linear equations, and other graphics are helpful in summarizing data and presenting research results to others. As such, the student must be familiar with these tools. Second, graphics are valuable for helping students to understand statistical concepts. A histogram of a skewed distribution, for example, provides a means of seeing the concept of skewness that complements and enhances the understanding that is based on frequency distributions, numerical measures, and verbal interpretations. Finally, graphics are also useful for diagnostic purposes. Scatterplots, for example, can be used to detect nonlinear relationships between X and Y. In sum, graphics are useful for presenting the results of statistical analyses, teaching the concepts of statistical analysis, and conducting statistical analyses.

Practice exercises are included at the end of each chapter. Some of them consist of questions about statistical quantities that are given in the exercise. The students may be asked, for example, to interpret the numerical value of a slope, or they may have to make a decision about the null hypothesis or give the probability of a Type 1 error from computer output. Other exercises involve computing statistics with a pocket calculator from small data sets or other statistics. I believe that it is important for a student to stay close to the data when learning new concepts. Calculating statistics is one way to do this. Sometimes this process may involve calculating a statistic, such as the standard deviation, from raw data. In such instances, only a few cases are used to minimize the drudgery of the task. Furthermore, computational formulas are not used because the objective is to give the student hands-on familiarity with the elements and terms in definitional formulas. When the statistic is a more complex one that can be meaningfully defined in terms of more elementary statistics that the student has already learned, calculations are based on the more elementary statistics and not on raw data. For example, because the linear regression slope can be defined as the covariance of two variables divided by the variance of the independent variable, slopes are calculated from the values of covariances and variances rather than from raw data. The incorporation of calculations in the exercises is not inconsistent with the overall emphasis on definitional formulas and on using the computer for calculations.

Additional exercises involve using SPSS to analyze selected variables from the 2000 General Social Survey (GSS). Two independent samples from the 2000 GSS (SPSS files *gss2000a.sav* and *gss2000b.sav*) can be downloaded from this book's companion website at http://sociology.wadsworth.com. These SPSS files contain fewer than 1,500 cases (1,398 cases in *gss2000a* and 1,419 cases in *gss2000b*) and 50 or fewer variables so that they can be used with the student version of SPSS (SPSSâ Student Version 11.0 CD-ROM, Windows), which can be bundled with orders of this text. Each exercise specifies whether *gss2000a, gss2000b,* or both contain the variables to be used in the exercise. A file containing all of the cases in the 2000 GSS (*gss2000.sav*) can also be downloaded for use with the professional version of SPSS. These exercises focus on interpreting the results of analyses and providing the student with a realistic exposure to the conduct of statistical analyses with computer software.

ACKNOWLEDGMENTS

I have been greatly influenced by Judd and McClelland (1989) for showing how the regression model can be used to conduct a wide array of parametric analyses that are often taught as disparate statistical techniques. I am also indebted to Conover and Iman (1981) for explaining how certain nonparametric tests can be conducted with regression analysis. The helpful comments and suggestions of Fred Halley, Ralph McNeal, and several anonymous reviewers were greatly appreciated. I would like to thank Ruth Cottrell for her committed and careful copy editing. I am also grateful to Ted Peacock, Dick Welna, and Janet Tilden for their contributions to this project.

This work would not have been possible without the support and encouragement of my wife Patricia and daughter Kara. Thank you with all my heart.

1

Statistics and Social Science

WHAT IS STATISTICS?

Everyone "knows" that college graduates get better jobs and earn more money than those who lack a college degree. But how much difference does college graduation actually make? Is it worth the financial sacrifice and the years spent in school? And if there is a significant financial payoff, why does it exist? Do college graduates possess skills that make them more productive employees, or does college attendance function as a screening device to make it easier for employers to find satisfactory workers?

Statistical analysis can yield answers to questions like these. Before statistics can help, however, valid data must be collected about these issues. To determine the effect of a college education, for example, we have to select a large number of representative persons and survey them to find out how much education they have, how much money they earn, what kinds of jobs they hold, and even what they studied and how much they learned in school. This body of data, however, is not what is meant by statistics. *Statistics* is a set of mathematical techniques for making sense out of data. It consists of systematic methods for summarizing and describing data. It also consists of techniques for using the data to make estimates about the characteristics of larger aggregates of people for whom we do not have data.

DATA AND DATA ANALYSIS

In general, **data** are individual facts or pieces of information. The word *data* is plural, and it refers to multiple pieces of information about a particular subject. A *datum*, on the other hand, is a single fact. An entire collection of data is a **data set.** Although data are most commonly in the form of numbers, they can also consist of letters, special

1

characters, and combinations of different types of characters. These characters are systematically arranged in a **data matrix.** Table 1.1 shows a small data matrix consisting of seven characteristics of a sample of ten persons 18 or more years of age who were respondents in the 1996 General Social Survey (we will have more to say about this survey later). A matrix is a rectangular array of data composed of multiple rows and columns. The columns in Table 1.1 represent different social, political, and economic characteristics, and the rows correspond to the different persons surveyed. At the intersection of each row and column is a *cell,* and each cell contains a datum. The datum in the cell corresponding to row 1 and column 3 (counting Sex as column 1), for example, indicates that this particular respondent (person 1) is 31 years of age. The datum in row 9 and column 6 states that person 9 is a Democrat. In general, d_{ij} is the datum in row i and column j. Therefore, the two data just described are $d_{13} = 31$ and $d_{96} =$ Democrat.

In more formal statistical terminology, the rows in a data matrix are the **units of analysis** and the columns are **variables.** A general data matrix, with the appropriate datum d_{ij} in each cell, is shown in Table 1.2. We can also refer to the units of analysis as **cases.** Units are cases, and cases are units. Table 1.2 contains 6 units, or 6 cases.

The units of analysis in Table 1.1 are adults, whereas in another table the units might be children. The units of analysis, however, are not always individual persons. Instead, the units may be aggregates of individuals, such as schools, hospitals, clubs, cities, or nations. Table 1.3 shows a data set consisting of seven characteristics (variables) that describe a sample of twenty nations. In this data set, each unit of analysis has a unique name: the name of the country. In Table 1.1, the names of the units of analysis were not shown because the identities of the people taking part in a survey are almost always kept confidential. The datum in row 6 and column 3 of Table 1.3 indicates that Burkina Faso (a country in West Africa) has a literacy rate of 18 percent ($d_{63} = 18$), the lowest in this sample of nations. The datum in row 17 and column 4 shows that the United States has a gross domestic product per capita (the total output in dollars of goods and services produced inside the country) of $23,474, the highest in the sample.

Although we have been pointing to data that describe individual persons and nations, statistical analysis is not usually concerned with describing and understanding each unit of analysis in its own right. Although the data set might contain a large number of rich characteristics about each unit that could be used to develop a unique portrait of each individual or nation, this is rarely the goal of statistical research.

Statistics does not focus on individuals for two reasons. First, the emphasis is usually on the columns in the data matrix. Each column is called a variable because it contains two or more related attributes that distinguish some units from others. The units vary in terms of the attributes that compose the variable. Statistical analysis is concerned with summarizing variables. For example, in Table 1.1, the average age is about 42 and the most frequent political party identification is Republican. In Table 1.3, the average fertility for the twenty nations is 3.6 children per woman, and the average life expectancy of women is 67 years of age. At this point you need not be concerned about how these averages are determined. Statistical analysis is also interested in summarizing how variables are related to one another. For example, Republicans have higher family incomes

Table 1.1 Selected Characteristics of a Sample of 10 Persons from the 1996 General Social Survey

Person	Sex	Marital Status	Age	Siblings	Highest Degree	Political Party	Family Income
1	Male	Married	31	4	High School	Democrat	$35000–39999
2	Female	Widowed	55	18	None	Democrat	$17500–19999
3	Male	Never married	20	2	High School	Independent	$25000–29999
4	Female	Married	33	3	High School	Independent	$35000–39999
5	Male	Married	26	3	Bachelor	Republican	$25000–29999
6	Female	Widowed	72	1	High School	Republican	$35000–39999
7	Male	Never married	33	2	Graduate	Republican	Refused
8	Female	Married	53	2	Bachelor	Republican	$50000–59999
9	Female	Widowed	54	9	High School	Democrat	Less than $1000
10	Male	Divorced	39	2	Bachelor	Republican	$50000–59999

Table 1.2 The General Form of the Data Matrix

Units of Analysis	Variables				
	1	2	3	4	5
1	d_{11}	d_{12}	d_{13}	d_{14}	d_{15}
2	d_{21}	d_{22}	d_{23}	d_{24}	d_{25}
3	d_{31}	d_{32}	d_{33}	d_{34}	d_{35}
4	d_{41}	d_{42}	d_{43}	d_{44}	d_{45}
5	d_{51}	d_{52}	d_{53}	d_{54}	d_{55}
6	d_{61}	d_{62}	d_{63}	d_{64}	d_{65}

than Democrats (Table 1.1), and females have lower life expectancies in countries with higher fertility rates (Table 1.3).

Second, statistics does not focus on individual units of analysis because the units for which the data have been collected have often been carefully selected to represent a larger population of units. We are not interested in the persons or nations in the data set per se but in what they can tell us about the larger group that they represent. Based on the data for a relatively small sample of persons living in different parts of the United States, for example, we may be able to reach some conclusions about the United States in general. For example, we might estimate that it is very likely (i.e., there is a 95-percent probability) that between 15 and 25 percent of the adult population of the United States has never been married.

Table 1.3 Selected Variables for a Sample of 20 Nations, Circa 1995

Country	Population (1,000s)	Urban: Percent Living in Cities	Literacy: Percent Who Read	Gross Domestic Product per Capita ($)	Fertility Rate: Average Number of Children	Average Female Life Expectancy	Average Male Life Expectancy
Afghanistan	20500	18	29	205	6.9	44	45
Armenia	3700	68	98	5000	3.2	75	68
Australia	17800	85	100	16848	1.9	80	74
Bangladesh	125000	16	35	202	4.7	53	53
Bulgaria	8900	68	93	3831	1.8	75	69
Burkina Faso	10000	15	18	357	6.9	50	47
Cambodia	10000	12	35	260	5.8	52	50
Denmark	5200	85	99	18277	1.7	79	73
Gabon	1300	46	61	4283	4.0	58	52
India	911600	26	52	275	4.5	59	58
Italy	58100	69	97	17500	1.3	81	74
Japan	125500	77	99	19860	1.6	82	76
Lebanon	3620	84	80	1429	3.4	71	67
Pakistan	128100	32	35	406	6.4	58	57
South Korea	45000	72	96	6627	1.7	74	68
Switzerland	7000	62	99	22384	1.6	82	75
United States	260800	75	97	23474	2.1	79	73
Ukraine	51800	67	97	2340	1.8	75	65
Uzbekistan	22600	41	97	1350	3.7	72	65
Zambia	9100	42	73	573	6.7	45	44

COMPONENTS OF SOCIAL SCIENCE

The statistical analysis of data is an important component of the social sciences. It is only one part of the research enterprise, however, and it is heavily dependent on the completion of prior work. In particular, before statistical analysis can be executed successfully, we must decide what to investigate, execute a plan for selecting the units to be analyzed, and collect data on the selected units. More formally, these components of research are called theory, sampling, and measurement. Although these components of research must be executed in the above sequence, not all research is carried out in such a systematic fashion. Exploratory research may begin with only fuzzy ideas about who and what to observe. As an exploratory study progresses, more precise decisions about the nature of the topic and the variables to be observed may be made and revised several times. However, the type of statistical analysis that makes inferences about larger populations requires an unbiased method of sampling that cannot be done without a plan—including the specification of the population. Moreover, all statistical analysis requires a standardized measurement design to insure that the same method of making observations is used for each unit and that the same variables are observed for each unit whenever applicable.

Theory

There is much disagreement about the nature of theory. For our purposes, the meaning of the term **theory** is very broad. Before statistical analysis can begin, there must have been prior thinking about the research topic, the concepts or variables relevant to the topic to be investigated, the units to be analyzed, and the population to be studied. This preparatory thinking about topics, concepts, and populations is what we mean by theory, or theoretical thinking. Without it, the systematic data required by statistical analysis would not be available. This notion of theory-based research can accommodate secondary analyses by investigators who were not involved in the collection of the data, because theory was necessary before the data were originally collected and because the secondary analyst must make a decision about whether the variables and the sample are appropriate for his or her research interests. The notion of theory that we are using here, however, does not require formulating hypotheses about how variables are related before performing statistical analysis. The breadth of this definition of theory allows for the existence of exploratory statistical analysis, which is known as *inductive research*. Sometimes, however, the theoretical work does include hypotheses. In this case, the theory is more fully developed. In this type of research, called *deductive research*, the statistical analysis may confirm or fail to confirm the theory.

Sampling

Sampling is the process of selecting a subset of units to be observed from a larger aggregate called the population. The objective is to select units in such a way that no selection bias will occur. **Selection bias** means that some units in the population are more likely to be selected than other units. If selection bias is present, the sample units will not be representative of the population. The objective is to sample in a way that avoids any conscious or unconscious selection biases. An intuitively clear model of an unbiased selection procedure is to write the name of each member of a population on a separate card, put the cards in a box, stir them up thoroughly, and draw out a predetermined number of the cards without looking. If the cards were thoroughly and completely mixed up before the drawing, the names on the selected cards will be a **random sample** of the population. Such random or blind selection means that it is impossible to predict ahead of time which units will be selected. Thus, this process eliminates any selection biases, which means that each unit in the population has an equal chance of being selected. If the sample size is relatively large, there will be a high probability that the characteristics of the units in the sample will closely approximate those of all the units in the population. This means that we can infer what the population is like on the basis of the traits that we have observed in the sample. We can also calculate the probability that our inference about the population is in error. In sum, we can do very sophisticated analyses of our sample observations when that sample is a random sample.

Before we can randomly select a sample of units to be observed, we must have completed our theoretical thinking about the topic, the units of analysis, and the population to be studied. For example, we may know that we want to learn more about the types of school programs and policies that enhance educational achievement, but we need to be much more specific before we can select a sample. Do we want to look at all schools—public, parochial, and private—or perhaps focus only on public schools because those schools are affected the most by state policies? As we make such decisions, we are not only narrowing the topical focus but also defining our population more precisely. Part of this process also involves deciding what our units of analysis will be. For example, will the units be schools or students? Sometimes we may want to incorporate several different units of analysis in the same study. This type of study is called a multilevel analysis.

Once we have theoretically defined our topic, our population, and our units of analysis, we have to make some practical decisions before we can select a sample of units. It is not possible to take a random sample unless we can acquire or construct a list, or a pseudo-list, of all the units in the population. Such a list is called a **sampling frame** because it is from such a list that we select our sample. Sometimes available lists do not include all units in the theoretical population. Phone books, for example, do not include households with unlisted telephone numbers. We can reach unlisted households if we dial telephone numbers randomly within lists of known telephone exchanges (a pseudo-list). We cannot, of course, reach households without telephones by this method. In the end, however, we may decide that even though the theoretical population is all households within a certain geographical area, we will not attempt to conduct interviews in the small percentage of households without telephones (about 5 percent). Thus, the study population, as opposed to the theoretical population, may be defined for practical reasons as those households with telephones. The list of working telephone exchanges will be the sampling frame from which the sample will be selected by random digit dialing. The fact that the list of exchanges does not include all households means that our sample will be somewhat biased because not all households will have an equal chance of being included in the sample (households without telephones have no chance of being selected).

Measurement

Measurement is the process of making and recording observations about the characteristics of the selected sample. An observation is a datum—that is, a fact about a particular variable for a specific unit of analysis. Thus, measurement is the process of data collection. It may take several different forms. Perhaps the most common method involves the use of standardized questionnaires. Respondents to sample surveys answer a series of questions, one for each variable in the data matrix. These questions may be read by the respondents from a questionnaire and the answers recorded by the respondents on the questionnaire (this is called a self-administered questionnaire), or the questions may be read by interviewers and the respondents' answers recorded by the interviewers. This type of data collection involves indirect observation because neither the interviewers nor the investigators directly observe the behaviors and attitudes that are being measured; instead, the researchers rely on self-reports by the selected persons. Measurement, however, may also be conducted through first-hand observations by the researcher. For example, an investigator may observe a group of children playing together and make a record for each child of each friendly interaction with another child and each unfriendly interaction. Measurement by direct observation is more time consuming. Also, researchers may find it difficult to obtain first-hand observations of many types of behavior and attitudes. For these reasons, data collection by direct observation is much less common than indirect observation.

A third type of data is collected from official records, which themselves must have originally been collected by either direct or indirect methods. For example, personnel records at companies may keep track of pay, hours of work, and job performance evaluations by supervisors. Such data may be acquired by direct observations of the employee: the employer gives the employee a pay raise and makes a record of it; the employee punches in and out of work on a time clock, which results in a direct observation of hours of work (if the employee fills in a time card once each week, however, that would be indirect observation); or supervisors record their evaluations of their direct observations of the employees' work.

The quality of the statistical analysis depends to a great extent on the quality of measurement. Measurement quality is a function of the **reliability** and **validity** of the observations. Reliability refers to the dependability or repeatability of the observations. If someone or something is dependable, that means you can count on the same performance time and time again. A reliable bathroom scale, for example, will give the same reading of your weight each time you stepped on the scale during a short period of time (i.e., short enough that your true weight would not change). If your scale is unreliable, the readings will tend to vary. If you weighed yourself on an unreliable scale ten times during a period of a couple of minutes, the average of these ten weights might equal your true weight (approximately). Although some of your measures of weight would be too high and some would be too low, the average is likely to be on target, but the variance of the observations around the true weight would be unpredictable or random. Thus, if you weighed yourself only once on an unreliable scale, you would not know whether the reading was correct, too high, or too low.

In social science measurement we normally make only one observation of an attribute during a limited period of time in which that attribute would reasonably be expected not to change. For example, during a survey interview we would ask about a respondent's weight only once. To the extent that respondents do not monitor their weights on a regular basis, they will be making educated guesses about their current weights. Some of the reported weights will be about right, but some will be too heavy and some too light. The major consequence for statistical analysis is that our ability to predict persons' weights on the basis of another variable such as age, for example, will be impeded by the unreliability of the reported weights. The amount of bias will depend on the degree of unreliability. Therefore, we must try to make our observations as reliable as possible. Perfect reliability, however, is unattainable. There will always be some random error or noise in our data. Thus, it is important to have special studies to determine just how reliable different types of measures are. Once we have estimates of their reliability, we can make some corrections for the biases that would otherwise result.

Measurement validity can be illustrated by returning to the example of the bathroom scale. If the average of ten measurements equals a person's true weight, as stated above, we would say that the scale is valid but not reliable. Imagine another scale on which you get the same weight each time you step on the scale but each weight is two pounds too heavy. We would say that the observations were reliable (they were all equal) but not valid (they were all too high). Consider two different possible reasons for these biased readings. First, you may have failed to notice that the needle was pointing at 2 rather than 0 when you were not standing on the scale. Or, second, perhaps an internal spring has lost some of its tension. In the first case the problem was not with the scale per se; instead, the problem was with the observer who did not use the scale properly. In the second case, however, the problem was inherent in the scale itself.

Analogous problems of invalidity can also arise in a survey interview that asks about the respondents' weights. First, the question might read, *How much do you weigh first thing in the morning?* The interviewers, however, might streamline the question and simply ask, *How much do you weigh?* If the interviewing is done at night in a large majority of cases, as is typical in many surveys, the reported weights might tend to be too high if the respondents are taking into consideration the fact that they recently ate dinner. In this case the biased reports of weights are due to the improper way the observer (the interviewer) used the instrument (the questionnaire). If the question were asked exactly as written, however, the *first thing in the morning* for persons working night shifts might be after they got off work rather than after they awakened. Consequently, their reported weights would be too high. In this case, the biased reports would be due to a problem with the instrument (the wording of the question). In social science measurement an additional validity problem that is not possible in physical measurement may arise. Survey respondents may sometimes answer questions to make themselves look acceptable. We call such a phenomenon a social desirability bias. For example, heavy respondents may underreport their weights, especially in telephone surveys where the interviewer cannot see how heavy they are.

Invalid measurements also create problems for statistical analysis. When the goal is to estimate the average value of a variable (e.g., weight), biased measures mean biased estimates of the average. For example, we may either overestimate or underestimate the weight of the average person, depending on which of the above types of invalidity is most common. Invalidity may also present a problem when we are trying to estimate the relationship between two variables. If the objective is to predict a person's weight based on age, for example, we expect that the older a person is, the heavier he or she will be predicted to be. However, if older people are the ones most likely to underreport their weights (because they are more likely to be especially heavy), that bias will result in an underestimate of the amount by which weight increases as people get older (generally speaking). So, as with unreliability of measures, it behooves us to do whatever is possible to minimize biased or invalid measurements.

An additional aspect of the quality of data is an even more fundamental issue than the reliability and validity of the data. **Missing data** are a problem to varying degrees in almost every study. Missing data take two forms: **unit nonresponse** and **variable nonresponse.** Unit nonresponse means that some of the units of analysis (persons, groups) that were randomly selected for the sample fail to be included in the study. For example, in survey research this may occur because a respondent refuses to participate in an interview or because a respondent is unable to participate due to illness. It also may occur because some of the selected persons cannot be located or contacted. For instance, some potential respondents are never home no matter how many times the interviewer calls or comes to the home. The failure to gain the participation of some of the selected units generally means that the data will be missing on all variables for such cases. Unit nonresponse potentially creates two kinds of problems. First, the sample size (number of cases) may be smaller than was planned. We can cope with this problem by simply selecting a larger sample than is actually needed in anticipation of the fact that there will be an unavoidable amount of unit nonresponse. Second, the units that are missing may be different in some characteristics from those units that respond, which means that the observed (i.e., participating) units will not be a representative sample of the population.

The greater the difference between the missing units and the participating units, and the greater the number of missing units, the greater will be the unrepresentativeness or bias of the observed sample. This is called **nonresponse bias.** Nonresponse bias is very difficult to correct validly. The best way to address this problem is to make every reasonable effort to locate every selected unit and then to gain their participation. By minimizing the nonresponse rate, we hope to minimize the nonresponse bias. There are also methods for giving greater weight in the statistical analysis to observed cases that are believed to be similar to those that are missing, but this method is only partially successful in dealing with the problem.

Variable nonresponse means that although the study has gained the participation of a particular unit, there may be missing data for some of that unit's variables. In a survey, a respondent may refuse to provide the desired information on some questions. This is particularly problematic on questions about sensitive topics such as drug use, sexual behavior, or income. In Table 1.1, for example, person 7 refused to report the family's income. Another reason for variable nonresponse is that a respondent may be unable to answer the question, either because he or she does not have the necessary knowledge (e.g., the spouse's salary is not known) or because he or she can't remember the desired information (e.g., the amount spent on prescription drugs during the past year). There are methods for making estimates of missing values that help to deal with this problem, but they are beyond the scope of this book (Allison 2001).

Statistical Analysis

After data have been collected from a sample of units and organized into a data matrix, **statistical analysis** of that data can begin. Statistical analysis consists of two somewhat distinct tasks. The first task is the description, or summarization, of the observed sample data, and the second is using these sample observations to make estimates about the characteristics of the larger population from which the sample was selected. These statistical tasks are called *descriptive statistics* and *inferential statistics*. Every study that involves a randomly selected sample of cases involves both descriptive and inferential statistics. If no sampling occurs—that is, if the study is a **census** of the entire population of interest—then only descriptive statistics are used. Although it is useful to make a distinction between descriptive and inferential statistics, in practice they go hand in hand because an inferential statistic, such as a confidence interval, indicates the probability that a descriptive statistic, such as an arithmetic average, is within a certain range of values in a population.

Descriptive Statistics

We can use **descriptive statistics** to describe the characteristics of a single variable (univariate statistics), to describe the nature of the relationship between two variables (bivariate statistics), or to describe the relationships among three or more variables at a time (multivariate statistics). A data matrix provides a complete description of the variables in the data set because it gives the observed value of each variable for each and every case (except for missing data). As the number of cases and variables in the matrix increase beyond a relatively small number, however, the data set becomes too large to be comprehended in this form. If, for example, we have only five cases and five variables, we can probably easily recognize the most important features of the observations. But imagine the difficulty of comprehending a data set with hundreds of cases and dozens of variables, a size that is typical of many studies. Some method of simplifying the complexity must be achieved so that we can recognize the most important aspects and patterns.

Descriptive statistics provide a relatively small menu of statistical methods for summarizing the most important aspects of a single variable. Perhaps the most common method is to compute an average value for the variable (see Chapter 3). Actually, there are several different types of averages, which are called measures of central tendency. You will have to learn which are appropriate and which are not for a particular variable, depending on the kind of variable it is (see the section titled "Levels of Measurement"). A second important type of univariate descriptive statistic summarizes how different the cases are from one another. For a particular variable, the cases may be very similar (homogeneous), or the cases may be quite dissimilar from one another (heterogeneous). The descriptive statistics that describe the amount of similarity or dissimilarity among the cases are called *measures of dispersion*. There are several different measures of dispersion (see Chapter 4).

When we consider ways to describe relationships between two or more variables (also called associations between variables), we find that the number of options is much more extensive than it is for describing a single variable. This book, however, significantly reduces the number of methods by emphasizing the use of a powerful and flexible technique called **regression analysis** (Chapters 8–12). It involves finding the formula for the line that best describes the linear relationship between two variables. There are situations involving certain types of variables, however, for which regression analysis is either not possible or overly complex. In these cases we use a method called **crosstabular analysis,** or simply crosstabulation. For both regression analysis and crosstabular analysis, there are several statistics for describing different aspects of the way that two or more variables are associated with one another. Nevertheless, the emphasis on regression analysis reduces considerably the number of techniques that have to be learned for describing associations between variables.

The statistics that describe central tendency, dispersion, and association each provide numerical summaries of these variable attributes. Such descriptions are read and interpreted according to the rules of mathematics. If, for example, we read that 75 percent of the persons in a sample are female, we can interpret this statistic to mean that 75 out of every 100 persons are female or, equivalently, that there are 3 females for each male. In addition to using words and numbers to convey the meaning of statistics, we can also use graphics. A graph can be defined as a mathematical picture. The dot graph in Figure 1.1 conveys the same information as the statistic "75 percent female" without using numbers.

Bar charts, histograms, and scatterplots are often used to depict variables and relationships between variables. Statistics in numerical form are more precise than graphical statistics. Furthermore, certain types of graphs can create biased perceptions. Therefore, graphical displays should be used to supplement, rather than replace, numer-

Male ♂ ♂ ♂ ♂
Female ♀ ♀ ♀ ♀ ♀ ♀ ♀ ♀ ♀ ♀ ♀ ♀

FIGURE 1.1 A Distribution of Males and Females

ical statistics. In modern statistics there is an increasing emphasis on the use and development of good graphing techniques.

Inferential Statistics

In addition to using statistics to describe our sample observations, we often want to make estimates, or inferences, about the values of these statistics in the larger population from which the sample is selected. For example, we might calculate that in our sample survey of Americans 75 percent of the respondents support gun control legislation. But what percentage of all Americans favor such laws? This value is unknown. It is possible, however, to use our sample data to make inferences such as the following: There is a 95-percent probability that between 72 and 78 percent of Americans support gun control legislation. An **inferential statistic** estimates that the value of the population parameter lies within a certain range of values. It also gives the probability that this estimate is true. An inferential statistic such as the one just given is called a **confidence interval.** Two types of imprecision are involved in inferential statistics. First, they do not specify the exact value but instead indicate a range of values (an interval) for the unknown parameter. Second, they give a probability that the value of the statistic in the population is within the specified range. Ideally, the inference would predict that the population statistic has a high probability of being within a narrow range of values. The conditions under which this ideal may be realized will be discussed in some detail in this text.

The preceding example was a *univariate* inferential statistic. Inferential statistics also exist for relationships between variables. The sample survey may also have found that 85 percent of female respondents favored gun control, compared with 65 percent of male respondents. The two variables in this case are gender and opinion on gun control. An inferential statistic might indicate that there is a 95-percent probability that the percentage of females who support gun control is 15 to 25 percentage points higher than the male percentage. The inferential statistic again specifies a range of values for the population statistic (the gender difference in the percentage supporting gun control) and a probability that the population statistic is within this range.

Hypothesis Tests

Hypothesis testing is a second type of inferential statistic, one that is closely related to confidence intervals. In this method the investigator states two rival hypotheses about the value of a population's parameter. Let's assume that we are interested in determining whether a majority of the population favors gun control or opposes gun control. One hypothesis is that 50 percent of the population favors gun control (or 50 percent opposes it); that is, the population is evenly split on the issue. This type of hypothesis is called the **null hypothesis** because usually the objective is to attempt to reject or nullify this hypothesis. The second or **alternative hypothesis** simply states that the percentage favoring gun control is not equal to 50 percent; that is, it is either greater than or less than 50 percent. The logic of hypothesis testing is to determine the likelihood or probability that the null hypothesis is true. If the probability is very small—commonly defined as 5 percent or less—we reject the null hypothesis in favor of its logical alternative. For example, if we find that 75 percent of our sample respondents favor gun control, we use statistical theory to determine the probability that this could occur by chance due to random sampling when the population parameter is equal to 50 percent, the null hypothesis. If there is a probability of only .01 (one in a hundred) that the departure from 50 percent occurred by chance, for example, then we normally reject the null hypothesis and conclude that the majority of the population favors gun control, with the qualification that there is a probability of .01 that this alternative hypothesis is incorrect.

The preceding test is called a **two–alternative hypothesis test** because the alternative hypothesis states that the population parameter can be either of two possibilities: greater than or less than the value specified by the null hypothesis. The two-alternative test is generally used when there is no theory about the value of the population's parameter. If there had been a theory that the majority of the population favored gun control prior to the data collection, we might conduct a **one-alternative test.** In a one-alternative test, for example, the null hypothesis would state that the population percentage favoring gun control is less than or equal to 50 percent and the alternative hypothesis would specify that the percentage is greater than 50 percent. The advantage of using the one-alternative test is that the probability of incorrectly rejecting the null hypothesis is only half as great as it is for a two-alternative test, if it is found that the sample statistic is the one predicted by the alternative hypothesis. For example, if the sample percentage is 75 percent, as it was earlier, we can reject the null hypothesis in favor of the alternative hypothesis with a probability of only .005 that we are incorrect, which is half the probability of .01 in the two-alternative example already given. The disadvantage of the one-alternative test is that we can draw no conclusion about the population's parameter if the sample statistic is found to be in the opposite direction of that predicted by the alternative hypothesis. To summarize, if we are doing exploratory research rather than testing a hypothesis, we want to conduct two-alternative tests. If, on the other hand, we have a preexisting theory about the population's parameter, we may choose to conduct a one-alternative test.

We can also conduct hypothesis tests for associations between variables. Assume once again that we are studying the relationship between the variables "gender" and "opinion on gun control." If we have no theory about the relationship, a two-tailed hypothesis test is appropriate. The **null hypothesis** would state that there is no gender difference in opinions on gun control. The **alternative hypothesis** would state that there is a gender difference; either females are more in favor of gun control than males, or females are less in favor of gun control than males. If 85 percent of the females in our sample favored gun control in comparison with only 65 percent of males, as earlier, we have to determine the probability that this result could occur in our sample observations by chance when the null hypothesis is true (no gender difference in the population). If the probability is very small (e.g., 2 percent) that the gender difference could have occurred by chance, we reject the null hypothesis and conclude that females are more in favor of gun control than males. If, prior to data collection, we had formulated a theory that predicted females are more supportive of gun control than males, we would conduct a one-alternative hypothesis test. The null hypothesis would state that females are equally favorable toward or less favorable toward gun control compared with males. The alternative hypothesis would specify that females are more supportive than males. Assuming that the results were as previously described, we reject the null hypothesis with only a 1-percent probability of being incorrect.

LEVELS OF MEASUREMENT

There are many different descriptive and inferential statistics. Whether a statistic can be used validly with any particular variable, however, depends on the level of measurement of the variable. The **level of measurement** refers to the amount of information that the variable provides about the phenomenon being measured. As the level of measurement increases, we can use more powerful statistical techniques to analyze the variable. Before discussing the different levels of measurement, however, we must describe two properties that all variables must possess.

All variables, no matter what their level of measurement, must have exhaustive and mutually exclusive attributes. The requirement for **exhaustive attributes** is that it must be possible to assign a valid score for the variable for every unit (person) to which the variable applies. The investigator must anticipate and allow for the full range of attributes possessed by the population that is being studied. No unit can be *off the response scale.* If you are using a bathroom scale to weigh college football players and the maximum reading on the scale is 260 pounds, the weight variable will not be exhaustive. You could always add a residual category labeled "260 or more," but you would undoubtedly be classifying together a lot of players who varied considerably in weight. Or, consider the following survey question: Are you currently married, divorced, widowed, or have you never been married? A few respondents might respond "I am legally separated," indicating that your categories are not exhaustive. Some investigators and/or respondents might also argue that the response alternatives for the abortion variable in Table 1.4 are not exhaustive because there should be a middle response category that states *neither agree nor disagree.* Others might maintain that there should also be a don't know alternative. Still others, however, might argue that the research literature on wording questions does not support adding such additional response categories. This lack of agreement indicates that although exhaustiveness is a property of variables for which we should strive, it is not always easy to determine whether we have achieved it in practice.

The requirement that every variable should have **mutually exclusive attributes** means that there should be one and only one attribute that is valid for each unit of analysis. That is, a person cannot have more than one age, cannot have more than one net worth (wealth), and cannot both agree and disagree with legalized abortion. However, if you asked respondents "Which sports do you follow closely?", many might mention two or more sports. The answers to this question would not constitute a single, mutually exclusive variable. One solution is to create a separate variable for each sport, each with two response categories indicating whether or not respondents followed that particular sport closely.

The consequence of creating a variable with exhaustive and mutually exclusive categories is that the units of analysis can be compared with one another to determine the level of measurement of the variable (Table 1.4). The most elementary comparison

Table 1.4 Levels of Measurement

	Nominal	**Ordinal**	**Interval**	**Ratio**
Properties	Mutually exclusive and exhaustive attributes	Same as prior level, plus rank-ordered attributes	Same as prior level, plus known differences between attributes	Same as prior level, plus true zero attribute
Examples	**Marital Status**	**"Abortion Should Be Legal."**	**Wealth**	**Age**
	Married	Strongly disagree	$-$20,000	25
	Divorced	Somewhat disagree	$50,000	18
	Widowed	Somewhat agree	$200,000	40
	Never married	Strongly agree	$5,000,000	75
Valid Questions	Is unit A different from unit B?	Is unit A higher or lower than unit B?	How much higher or lower is unit A than unit B?	How many times higher is unit A than unit B?

between two units is whether they are the same or different—that is, whether they are equal or not equal. Differences between units may involve quality or quantity. If the properties of the variable do not permit us to make quantitative differentiations—that is, if differences are only qualitative—then the variable is measured at the nominal level, or it is a nominal variable. *Merriam Webster's Collegiate Dictionary,* 10th Edition, defines nominal as *in name or form only.* This implies that the **nominal level of measurement** is measurement in name only, that it is not truly measurement. Although this is a strong and controversial position, it is widely agreed that the nominal level of measurement is the lowest or least complex level. Nominal variables consist of named categories that do not allow us to use quantifiers like *higher* or *lower* and *more* or *less* to describe differences between cases. Marital status, occupation, and gender, for example, are nominal variables. Thus, we can't say that married persons are higher than or lower than divorced persons; they are just different. Or, we can't say that one occupation (physician) is higher than another occupation (mechanic), or that one gender is greater than the other is. (Although society often attaches different amounts of status to nominally different roles, it is the *status* of the roles that is quantitatively different, not the roles themselves.) When we analyze nominal variables statistically, we must use statistics that do not involve arithmetic or quantitative operations other than counting. As you will see, many statistics are appropriate for nominal variables, such as percentages for measuring relative frequency, the mode for measuring central tendency, the index of qualitative variation for measuring dispersion, Cramér's V for measuring association, and chi square for conducting hypothesis tests.

In addition to equal and not equal, the **ordinal level of measurement** permits us to make comparisons like *greater than* and *less than.* Now we can classify cases as equal to, less than, or greater than other cases. The abortion variable in Table 1.4 is an ordinal variable because a person who strongly agrees is higher on support for abortion than a person who simply somewhat agrees, a person who somewhat agrees is higher than one who somewhat disagrees, and so on. The academic rank of college students (i.e., freshman, sophomore, junior, and senior) is another example of an ordinal variable. For these variables and others, such as social class identification (e.g., upper class, middle class, working class, and lower class), we can rank order the units of analysis (persons, in these examples) with allowances for ties (persons with the same attribute). (You will learn later how to assign scores that properly reflect the rank order of ordinal variables with ties.) Although we can rank order the attributes of ordinal variables, we cannot perform arithmetic operations because the distances or intervals between attributes are not known. We cannot say how much greater, quantitatively speaking, *strongly agree* is than *agree,* for example, only that it is greater. Consequently, statistics that require addition and subtraction of scores are not valid for ordinal variables, strictly speaking. (You will see, however, that meaningful statistics can be computed by performing arithmetic operations on rank scores.) A number of statistics exploit the ordinal information contained in these variables, such as the median as a measure of central tendency and Spearman's rho as a measure of association.

With the achievement of an **interval level of measurement,** we can employ almost all of the most powerful statistics used in the social sciences. With interval variables, we know the distances or intervals between attributes. If we subtract $50,000 in wealth from $250,000 we get $200,000, the distance or quantitative difference between these two attributes. Notice that wealth does not have a true zero; that is, zero does not represent the smallest amount of wealth because people can be in debt. They can have negative wealth, or negative net worth. The Fahrenheit and Celsius temperature scales are also interval variables because temperatures colder than zero exist. The absence of a true zero does not prevent the arithmetic operations of addition and subtraction. Consequently, the majority of statistics covered in this book can be used for interval variables. It includes all the regression statistics that are the backbone of the measures of association used in this book, plus more elementary statistics such as the mean, standard deviation, and covariance that regression statistics are built upon. Statistics such as the harmonic mean and the coefficient of variation are not valid for interval variables because they require a true zero. These statistics, however, are used rather infrequently. Consequently, it is not necessary, by and large, for us to have variables with true zero points, called **ratio level of measurement.** Nevertheless, ratio variables are described so we can complete our discussion of levels of measurement.

There are a relatively large number of ratio variables. A number of these variables are based on measures of time, such as age (number of years ago that a person was born), education (number of years of school completed), work (number of years of work experience), and hours (length of work week). Because none of these variables can have negative amounts of time, they all have true zeroes. Many other ratio variables are based on counting, such as number of siblings, number of children, number of doctor visits in the past year, and frequency of church attendance. With ratio variables you can not only subtract scores to find, for example, that one person attended church 52 times more often than another person (104 times minus 52 times) but also compute a ratio of their church attendance and say that the first person attended church twice as often as the second person ($104 \div 52 = 2$). It is not valid to compute ratios for interval variables because they do not have a true zero. For example, we cannot compute the ratio of a positive amount of wealth to a negative amount of wealth, such as $50,000 \div -$20,000 $= -2.5$. Nor can we take the ratio of two positive amounts of wealth, such as $200,000/$50,000 = 4$ because $0 is not the smallest amount of wealth. Because ratio variables have all of the properties of interval measurement plus a true zero, all of the regression statistics that are appropriate for interval variables are also valid for ratio variables. Thus, for all practical purposes in the social sciences, ratio and interval variables are equally good levels of measurement.

Parametric and Nonparametric Statistics

The discussion of levels of measurement stressed that different statistics are valid for different levels of measurement. The statistics stressed the most in this book (regression analysis) require at least an interval level of measurement. Traditionally, these statistics have been called **parametric statistics** because when we conduct hypothesis tests, we must make some assumptions about the values of certain population parameters. Although discussing these assumptions is not the point, to the degree that these assumptions are not true, the results of the tests will not be valid. When it appears that the assumptions underlying parametric tests are not valid, it may be necessary to use **nonparametric statistics** that do not make assumptions about population parameters, even though the variables may be at a higher level of measurement than nonparametric statistics require. Nonparametric statistics do not require the interval level of measurement. There are certain nonparametric statistics for nominal variables (Cramér's V and the chi-square hypothesis test) and other nonparametric statistics for ordinal variables (Spearman's rho and its hypothesis test). The way nonparametric statistics are traditionally taught requires students to learn formulas, notation, and concepts that appear to be unrelated to those that they learned for parametric statistics, such as regression analysis. Thus, students feel that they are learning totally new statistical techniques. This book, however, shows how to conduct nonparametric analyses of ordinal variables by applying parametric techniques (regression analysis) to ordinal variables that have been appropriately scored for ranks. Thus, you will not have to learn an entirely different type of statistic with different terminology and notation when you analyze ordinal variables. You will use parametric techniques that you have learned to use for interval variables, but you will apply them to ordinal variables. The end result will be the same as if you had learned to use a new (or seemingly new) nonparametric technique. When you have interval variables, you will use parametric statistics and make the necessary assumptions about population parameters. When you have ordinal variables, you will use the same parametric techniques but apply them to rank-scored variables; you will not, however, have to make assumptions about population parameters. As a consequence, you will have to learn fewer statistical techniques than are traditionally presented in a statistics text. Unfortunately, if you are analyzing nominal variables, you will have to learn appropriate nonparametric techniques. You cannot simply apply the parametric techniques to the nominal variables and get the correct results.

FORMULAS AND COMPUTING

Statistics is a branch of mathematics. Because mathematics involves formulas, you have to learn formulas for statistics. You may not want to learn formulas because memorization tends to be difficult and boring. This book does not emphasize memorization of formulas. Instead, it emphasizes understanding formulas. It is important to understand formulas because the formulas that are stressed, called **definitional formulas,** provide meaningful definitions of the statistics. Thus, when you understand the formula, you will understand the statistic. And when you understand the statistic, you will be able to use and interpret it more intelligently. You may have heard the notion that you can prove anything with statistics. A comparable point of view is conveyed by the title of the book *How to Lie with Statistics* (Huff, 1954). However, if you *understand* statistics, people will not be able to lie to you with statistics. You will not be fooled. Furthermore, understanding the formulas will help you to remember them or deduce them instead of relying on rote memorization.

A related misunderstanding about formulas is that they are important because they specify how to compute statistics. That is, formulas are viewed as recipes for statistics. Although it is true that there are **computational formulas** for computing statistics by hand or with a calculator, that is not the reason you need to learn them. By and large, in this modern computer age, you will not have to calculate statistics. You can simply go to a good statistical software package and pick the appropriate statistic for your research problem from a statistical menu. The computer program will then do all the work with lightning speed. All you have to do is learn how to make the program work, which is usually relatively easy. You can get plenty of statistical results from the computer quickly, but then comes the hard part: You have to make sense of the statistical results, the numbers. Although this text shows you how to use a computer program (SPSS) to get statistical results, the emphasis is on understanding the results rather than how to compute them.

SUMMARY

Statistical analysis is a set of mathematical techniques for making sense out of data. A data matrix is a rectangular array of data in which the rows are the units of analysis and the columns are variables. Units of analysis are persons or groups of persons. A variable consists of two or more related attributes that distinguish some units from others. A datum is an observed attribute for a specific unit.

Statistical analysis is a component of social science research that follows theory, sampling, and measurement. Theory consists, minimally, of decisions about the research topic, the relevant concepts and variables to be investigated, the units to be analyzed, and the population to be studied. Sampling is the process of randomly selecting a subset of units to be observed from a larger aggregate called the population. Measurement consists of making observations, either direct or indirect (e.g., standardized questionnaires), about the attributes of the units of analysis. The quality of measurement is assessed by the reliability (repeatability) and validity (unbiasedness) of the observations.

Statistical analysis is composed of descriptive statistics and inferential statistics. Univariate descriptive statistics describes the central tendency and dispersion of a single variable. Descriptive statistics is also used to describe relationships between two or more variables, such as regression analysis and crosstabular analysis. Graphs may be used to provide a visual description of a variable or of a relationship between variables. Inferential statistics is used to make estimates, or inferences, about the values of descriptive statistics in the larger population from which the sample observations are selected. Inferential statistics consists of confidence intervals around the estimate of the population's parameter and hypothesis tests about the null hypothesis versus the alternative hypothesis.

Statistical analysis depends on the level of measurement of the variable(s). The measured attributes of a variable must be exhaustive and mutually exclusive. Because the nominal level of measurement specifies only whether one case has the same or a different attribute in comparison with another case, it provides the least information. The ordinal level of measurement provides more information than the nominal level because it includes a ranking of the attributes from highest to lowest. The interval level of measurement gives even more information because it also states the interval, or difference, between any pair of attributes. The ratio level of measurement provides the most information because it has an absolute zero point that makes it possible to compute the ratio of one attribute divided by another.

Statistical analysis also consists of parametric and nonparametric statistics. Parametric statistics, which requires at least the interval level of measurement, makes certain assumptions about population parameters. Nonparametric statistics, which is designed for the nominal and ordinal levels of measurement, does does not make such assumptions. This text emphasizes parametric statistics, especially regression analysis.

Statistical analysis can be done better when the researcher knows certain statistical formulas. It is important to know formulas, not because they are needed to compute statistics, but because understanding the formulas will help the researcher to better interpret the statistical results provided by a computer.

KEY TERMS

data	statistical analysis	mutually exclusive attributes
data matrix	descriptive statistics	nominal level of measurement
unit of analysis	regression analysis	ordinal level of measurement
variables	inferential statistics	interval level of measurement
theory	confidence intervals	
sampling	hypothesis test	
random sample	null hypothesis	parametric statistics
measurement	alternative hypothesis	nonparametric statistics
reliability	exhaustive attributes	definitional formulas
validity		

EXERCISES

1. The following table lists selected characteristics of a sample of persons.
   a. Is "Hours TV Watched per Day" a unit of analysis or a variable?
   b. Is person 5 a unit of analysis or a variable?
   c. What is the value of d_{64}?

	Self-Rated Health	Age at Birth of First Child	Religious Preference	Income Change in the Past Year	Attitude toward Premarital Sex	Race	Hours TV Watched per Day
1	Good	23	Protestant	$9876	Not wrong at all	White	0
2	Poor	18	Jewish	$432	Sometimes wrong	White	5
3	Excellent	33	Catholic	$1822	Always wrong	Black	3
4	Good	24	Protestant	−$1773	Not wrong at all	White	3
5	Fair	20	Protestant	$0	Always wrong	Other	2
6	Good	25	Catholic	$3549	Almost always wrong	White	1
7	Fair	27	Protestant	$1243	Sometimes wrong	Black	2

2. a. How many variables are in the table?
   b. What is the sample size in the table?
   c. What is the value of d_{32}?

3. a. What level of measurement is "Self-Rated Health"?
   b. What level of measurement is "Race"?
   c. What level of measurement is "Income Change in the Past Year"?

4. a. What level of measurement is "Age at Birth of First Child"?
   b. What level of measurement is "Religious Preference"?
   c. What level of measurement is "Attitude toward Premarital Sex"?

5. If every person reported the same "Age at Birth of First Child" for two years in a row, would that indicate that the quality of measurement is reliable, valid, or both?

6. If people underreport their total income in surveys because they tend to forget about their unearned incomes (stock dividends, interest on savings accounts, and so on), does that indicate that the measurement of income is unreliable or invalid?

7. Are the attributes of "Religious Preference" shown in the table exhaustive?

8. Are the attributes of "Attitude toward Premarital Sex" shown in the table mutually exclusive?

9. If an investigator reports that 50 percent of the respondents are Protestant, is that a descriptive statistic or an inferential statistic?

10. If an investigator reports that there is a 95-percent probability that the average person watches between 2 and 4 hours of TV per day, is that a descriptive statistic or an inferential statistic?

11. Assume the eight people in the table were interviewed after responding to a local newspaper ad that offered people $25 to take part in a survey. Would these people constitute a random sample of the local population?

12. Write six questions that are to be answered by a sample of five students in your class. Two of the questions should be at the nominal level of measurement, two should be at the ordinal level of measurement, and two should be at the interval/ratio level of measurement. Make five copies of your questionnaire. Your instructor will randomly assign students to groups, with six students per group. Give each of the other five students in your group a copy of your questionnaire to fill out anonymously. (You will be expected to fill out the questionnaires that they have prepared.) Collect your set of completed questionnaires and make a data matrix for your sample and questions. Turn in the completed questionnaires, together with your data matrix.

2

Distributions of
Variable Attributes

THE AGE PARADE

My professor invited the members of our class to attend the *Age Parade* in Washington, DC, on Saturday, October 16, 1999. She explained that the parade was being sponsored by the Population Association of America (PAA) to enhance awareness of generational differences in an aging population entering the next millennium. She said it would be very interesting because one out of every 1,000 people in the United States had been randomly selected to march in the parade, which would amount to a total of about 265,000 persons. Because she offered us extra credit for attending and writing a short report on our observations, a number of us decided to go. We agreed to meet at 7:50 AM at the corner of 14th and Constitution Avenue. The parade participants were to organize at the Washington Monument in ascending order of age and march down Constitution Avenue to the U.S. Capitol.

The parade stepped off precisely at 8 AM. As it approached me, I counted ten marchers per row. A white flag in the first row was emblazoned with a scarlet "0" to indicate these were the less-than-1-year-olds. First came rows of babies less than a month old, each being carried by its father or mother. The rows were about five feet apart, and by my count 35 rows passed by me in a minute's time. Gradually, as the passing infants became noticeably older, more and more of them were being pushed in strollers. When a flag with a red "1" on it went by, I glanced at my watch and calculated that it had taken almost eleven minutes for all of the less-than-1-year-olds to be carried and pushed in front of me. At 35 rows per minute, I had just seen about 3,800 infants. I decided to keep a record of how long it took each age cohort to march by.

Next came the toddlers and older preschoolers. Most were still being pushed in strollers to keep up the pace. Only among the 4-year-olds were many marching on their own. When the last of them passed by, I recorded that it had taken 55 minutes for the under-5-year-olds to go by. I calculated that I had now seen about 19,000 children. Because it had taken almost an hour to get to the 5-year-olds, I feared I had a long day ahead of me. I was right. It was another 55 minutes before the last of the 9-year-olds arrived. The good news was that the age cohorts were not getting larger; the bad news was their size wasn't getting smaller, so I settled in for the long haul.

Three and a half hours later I finally saw the end of the 29-year-olds. The last four 5-year cohorts had taken 54, 53, 50, and 54 minutes, respectively, to pass by my spot. Although I noted that these times were slightly shorter than those of the youngest two 5-year cohorts, I was not heartened by this statistic. It was now almost 1:30 P.M., and I was only beginning to see the folks in their 30s.

I gradually perceived that it was taking longer for the 30-to-34-year-olds to pass by. Finally, the flag for the 35-year-olds was abreast of me. I checked my watch. It had taken the 30-to-34-year-olds 61 minutes. That's six minutes more than the longest previous time for a 5-year cohort. The cohorts were getting larger! In a little over an hour (64 minutes to be precise), the 35-to-39 cohort finished passing. Another 5-year cohort record! Would the 40-to-44-year-old cohort be even larger? In another 59 minutes, the last of the 44-year-olds was abreast of me. Good—the cohort size had decreased. However, it was still the third largest that I had seen.

Although it was late afternoon and I was getting tired, the pace of the marchers didn't slacken. I still counted 35 rows per minute passing by. That was good news. More good news: The 45-to-49 cohort took *only* 52 minutes to go by. That was the second 5-year cohort decline in a row. Then a hypothesis occurred to me. Because the marchers were now in their 50s, there should be increasing declines in size as older and older cohorts pass due to increasing mortality rates. For my sake, I hoped it would be true—not that I wanted to see anyone drop dead right in front of me.

Sure enough, the 50-to-54s marched past in only 40 minutes! When the 55-to-59-year-olds passed by in a mere 32 minutes, I was really getting excited about the way my hypothesis test was going. Even though the 60-to-64-year-olds walked past in just 28 minutes, time seemed to be dragging. It was now 7 PM, and I was anxious to finish and head home. The next cohort was 65-to-69 years of age, the first post-retirement cohort. When I checked my watch as the last of them stepped sprightly by, their time was 28 minutes, the same as the previous cohort! That was disappointing, but it also made me think. There must be more behind differences in cohort sizes than differential mortality rates. I'd have to think about this some more next week.

Now I was observing those in their 70s, and they showed it. There were lots of wrinkles, quite a few limps and canes, and increasing numbers in self-propelled wheelchairs, compliments of the PAA. After recording that the 70-to-74s and the 75-to-79s took only 25 and 20 minutes, respectively, I knew that my hypothesis was back on track and the parade should be nearly over. The oldest-of-the-old, those 80 and above, were now gamely working their way past me, some with the assistance of wheelchairs. And I also noticed for the first time that there were many more women than men, perhaps twice as many. The 80-to-84-year-olds marched past in a mere 13 minutes, followed by the 85-to-89-year-olds, who passed me in only 7 minutes. All right!

Now the end was in sight. The 90-to-94-year-olds wheeled by (literally) in just 3 minutes, followed quickly by the 95-to-99-year-olds, who whizzed past in less than a minute (49 seconds). I was now watching the second hand on my watch very carefully in order to get a precise time for the last cohort, the centenarians (100 years old or older). I timed them at a seemingly blinding speed of 10 seconds. That was it! It was 8:35 PM, and the Age Parade was finally over.

As I mulled over my excruciatingly long day on the way home, the technical term for what I had seen popped into my mind. I had witnessed an *age distribution!* A real, live age distribution. Not some dry statistical table that we probably have some place in our text, but a living, breathing, moving age distribution. I didn't think I would like to see another one, but I was glad I saw just this one.

Although staging a parade to illustrate the distribution of age, or any other variable, would be a creative and effective way to communicate its characteristics, it would be highly impractical. In this chapter we will learn more conventional means of reporting and displaying the distribution of a variable's attributes; that is, we will learn how to make tables and graphs that are effective means of describing distributions. The structure of the tables and graphs depends somewhat on the type of variable that is being described—that is, whether it is a nominal, ordinal, or interval variable. We will start with the most elementary type of variable, the nominal variable.

NOMINAL ATTRIBUTES

Are you currently married, widowed, divorced, separated, or have you never been married? Your answer to this question is likely to indicate one of the most important roles in your social identity—your marital status. Social scientists call it the marital-status variable. There is little doubt that society places a high value on marriage as an ideal. Although there may be many unmarried persons in society, this status is usually seen as a temporary or unsatisfactory state of affairs. Most young, single adults are hoping to find a spouse, divorced people often remarry, and the loss of a spouse through death is a source of much unhappiness. Yet, there are a number of social trends that suggest marriage is on the decline. People are postponing marriage to pursue higher education; divorce is becoming more common and less stigmatizing; cohabitation is gaining in social acceptability; and an aging society is causing an increase in widowhood.

So, how common is being married in comparison with the other marital statuses? And is the incidence of marriage on the decline in American society? We will use data from the General Social Survey (GSS) to answer these questions. The GSS is a large sample survey of the adult population (18 years of age or older) of the United States, and it has been conducted by the National Opinion Research Center every year or two since 1972. The preceding question on marital status is asked of every respondent in the GSS. Because there were 2,903 respondents in the 1996 survey, which will be used in this chapter, we need to summarize their responses if we are to reach precise conclusions. The first step in conducting a statistical analysis is to make a frequency distribution for the variable or variables being investigated. We will cover tabular and graphical ways of displaying frequency distributions.

Tabular Summaries

Tabular displays use numbers to describe or summarize the attributes of a variable. You have to "read" the numerical summaries to arrive at a proper interpretation. Numerical reports are the most common way of presenting the results of statistical analyses. Visual or graphical displays, however, offer another way of showing statistics. The difference between numerical and graphical reports is analogous to the difference between a written description of a person and a photograph of a person. We will cover tabular displays of frequency distributions first and then show how to display them visually with graphs.

Frequency Distributions

A simple way to present data on the attributes of a variable such as marital status is to make a list of each person's marital status. A list of the marital statuses of the 2,903 individuals in the 1996 GSS would take an inordinate amount of printed space, and it would be extremely difficult to reach any accurate conclusions by visually examining the entire list. To illustrate the unwieldy nature of listings, Table 2.1 shows the marital statuses of the first 35 respondents in the GSS data file. Even with only 35 cases, it is difficult to make a good assessment of the marital statuses by simply "eyeballing" the list.

A **frequency distribution** is a count of the number of times each attribute of a variable occurs in a sample of observations. The construction of a frequency distribution is illustrated by showing how it is done manually for the cases in Table 2.1. Simply go down the list of attributes and make a running tally of the frequency (f) with which each attribute is encountered. Then, sum the frequencies to get the total number of cases (n). Table 2.2 shows the results.

Each frequency in the distribution indicates how many cases have the corresponding attribute. For example, 7 persons were married, 7 were divorced, and so on. The individual frequencies standing alone, however, do not tell us anything meaningful about the attributes. For example, if I conducted a study and reported that 234 persons were married, that statistic would not be informative. The person receiving the report would

Table 2.1 Marital Status of the First 35 Cases in the 1996 GSS

Married	Never married	Divorced	Married	Never married
Never married	Separated	Never married	Never married	Separated
Divorced	Separated	Married	Married	Separated
Married	Never married	Never married	Married	Never married
Separated	Divorced	Married	Separated	Divorced
Separated	Never married	Never married	Separated	Never married
Separated	Divorced	Divorced	Separated	Divorced

Table 2.2 Tallies of Marital Statuses for the First 35 Cases in the 1996 GSS

	Tallies	Frequency (f)
Married	IIII II	7
Widowed		0
Divorced	IIII II	7
Separated	IIII IIII	10
Never married	IIII IIII I	11
Total (n)		35

probably ask for more information, such as "How many persons were in the study?" or "How many persons were not married?" A frequency alone does not speak for itself. It must be compared with another appropriate number to make it meaningful. An easy way to interpret each frequency is to compare it with the total number of cases on which observations are made, as follows:

$$f \text{ out of } n \text{ cases has attribute } \mathbf{a}$$

$$f = \text{frequency}$$

$$n = \text{number of cases}$$

Thus, 7 out of 35 persons were married; 0 out of 35 persons were widowed; 7 out of 35 persons were divorced; 10 out of 35 persons were separated; and 11 out of 35 persons had never been married.

Percentage Distributions

While these interpretations are accurate, they are not as meaningful as we would like because we are probably not accustomed to thinking in batches of 35, or whatever number n happens to represent. A first step in making f more informative is to convert it into a **proportion** (p):

$$\text{Proportion} = p = \frac{f}{n}$$

For the first 35 cases in the 1996 GSS, the proportion for each marital status is shown in Table 2.3. Proportions always sum to 1, as illustrated in this table.

Proportions indicate the number out of a certain base that has a given attribute. If the proportion, which is always a fraction, is given to three decimal places, it equals the number of units out of 1,000 that have the attribute. Thus, the proportion of married people equals .200 = 200/1000, or 200 out of 1,000 persons were married. The maximum proportion equals 1.000 (every case has the attribute), and the minimum proportion equals .000 (none of the cases have the attribute).

Proportions are more meaningful than raw frequencies because they are expressed as fractions in the decimal system, the mathematical system that we are used to working with. Our reliance on the decimal system is almost certainly connected to the fact that we have ten fingers. The decimal system utilizes batches of ten; ten batches of ten equals one hundred, ten batches of one hundred equals one thousand, and so on.

A statistic that is even more meaningful than the proportion is the **percentage.** The percentage (%) indicates the number of cases out of every one hundred cases that have a given attribute. The percentage can be calculated from the proportion as

$$\text{Percentage} = \% = p \cdot 100 = \frac{f}{n} \cdot 100$$

For the first 35 cases in the 1996 GSS, the percentage for each marital status is shown in Table 2.4. Twenty percent, or 20 out of every 100 persons, were married; zero percent were widowed; 20 out of every 100 persons were divorced; 28.6 out of every 100 persons were separated; and 31.4 percent had never been married. Thus, among the first 35 cases in the 1996 GSS data file, there is a greater percentage of never-married persons than persons of any other status. Separated persons have the second highest percentage, and the percentage that are divorced is equal to the percentage married.

Common sense tells us that these percentages are not representative of the country as a whole. To see if our hunch is true, let's examine the frequency and **percentage distributions** of marital status for the entire sample of persons in the 1996 General Social Survey (Table 2.5).[1]

We see in Table 2.5 that 1,645 out of 2,903 GSS respondents, or 56.7 percent, were married at the time of the survey. Thus, a little more than half of the respondents were married, and this status was far more common than any other marital status. The separated status was the least common, whereas it was the second most frequent among the first 35 respondents in the data file. Although it was convenient to use the first 35 cases to illustrate the computation of frequency and percentage distributions, the discrepancies between the results for the entire data file and those for the first 35 cases highlight the dangers of drawing conclusions on the basis of convenience samples which, in this case, happened to be very unrepresentative of all the respondents in the survey.

---

1. Because just one person is randomly selected to be interviewed in each household, the GSS overrepresents persons in single-adult households and underrepresents persons living with one or more other adults. This bias has been adjusted for by weighting (w_i) each case by the number of adults in the household (n_i) divided by the mean number of adults in all households, which is $w_i = n_i/1.841$ in the 1996 GSS. Setting the denominator of the weight equal to the mean causes the sum of the weights to equal the number of respondents in the survey. Consequently, the sum of the fs in Table 2.5 equals the actual number of respondents in the survey.

Table 2.3 Marital Status Proportions for the First 35 Cases, 1996 GSS

	$f/n =$	p
Married	7/35 =	.200
Widowed	0/35 =	.000
Divorced	7/35 =	.200
Separated	10/35 =	.286
Never married	11/35 =	.314
Total		1.000

Table 2.4 Marital Status Percentages for the First 35 Cases, 1996 GSS

	$p \cdot 100 =$	%
Married	.200 · 100 =	20.0
Widowed	.000 · 100 =	.0
Divorced	.200 · 100 =	20.0
Separated	.286 · 100 =	28.6
Never married	.314 · 100 =	31.4
Total		100.0

Table 2.5 Marital Status, 18+ Years of Age, 1996 GSS

	Frequency (f)	Percent (%)
Married	1645	56.7
Widowed	203	7.0
Divorced	341	11.7
Separated	86	3.0
Never married	628	21.6
Total (n)	2903	100.0

   Although the percentage who were married is much higher than the combined percentage who were divorced or separated (11.7 + 3.0 = 14.7%), the marital status underestimates how common is the experience of divorce. The GSS asked the married and widowed respondents if they had ever been divorced or legally separated. Table 2.6 shows the frequencies and percentages who were married and widowed but had previously been divorced or separated, along with the figures for those who were married and widowed and had never been divorced or separated. The percentage who were married but had been previously divorced or separated (13.5%) is almost as high as the percentage who were divorced or separated (14.8%). This suggests that there is a rather high remarriage rate for persons who get divorced. When we sum the percentages of those who have ever been divorced or separated (13.5 + 1.4 + 14.8 = 29.7%), we find that about 3 out of every 10 persons have been divorced or separated at some time in their lives. This figure, although relatively high, is still substantially less than the percentage who have been married but never divorced or separated (43.2 + 5.4 = 48.6%).

Ratios

We have been comparing percentages of persons who have different statuses by noting whether one percentage is different from another and by making some vague judgment about whether the difference is large or small. We now take a more careful look at how to compare frequencies or percentages. Perhaps the most natural way to make a more precise comparison is to subtract the percentage in one category from the percentage

Table 2.6 Marital Status and Ever Divorced/Separated, 1996

	Frequency (*f*)	Percent (%)
Married, never divorced/separated	1248	43.2
Married, previously divorced/separated	389	13.5
Widowed, never divorced/separated	157	5.4
Widowed, previously divorced/separated	40	1.4
Divorced/separated	427	14.8
Never married	628	21.7
Total	2890	100.0

in another. In Table 2.5, for example, more persons are married than divorced, and the difference between the two percentages is 45.0 (56.7% − 11.7% = 45.0%). In other words, out of every 100 persons we would expect to find 45 more married persons than divorced persons. Is this a big difference? That depends somewhat on how large is the smaller of the two percentages. Consider a hypothetical society in which 72.5 percent are married, 27.5 percent are divorced, and 0.0 percent are separated, widowed, or never married. Now, the difference in percentages between the married and divorced persons is 45.0, the same as in Table 2.5. Although the percentage differences are identical, is the tendency to be married rather than divorced really the same in both societies?

Let's entertain this question in the context of a real-life issue. In which situation is there a greater opportunity for a divorced person to become romantically involved with a married person? The chances of such a relationship, all other things being equal, will depend more on the number of married persons to each divorced person than on the difference between the percentage who are married and the percentage who are divorced. That is, the ratio of the percentage of married persons to the percentage of divorced persons is the most appropriate measure. The ratios of married to divorced persons are

$$\text{Ratio} = \frac{\text{percent married}}{\text{percent divorced}} = \frac{56.7\%}{11.7\%} = 4.85 \qquad \text{(Table 2.5)}$$

$$\text{Ratio} = \frac{\text{percent married}}{\text{percent divorced}} = \frac{72.5\%}{27.5\%} = 2.64 \qquad \text{(Hypothetical)}$$

These ratios tell us that there were 4.85 married persons to each divorced person in the 1996 GSS, but there were only 2.64 married persons per divorced person in the hypothetical society. Thus, all other things being equal, a divorced person would be more likely to become romantically involved with a married person in the United States in 1996 than in the hypothetical society.

In general, *a* **ratio** *is defined as the percentage (or number) of units in category A divided by the percentage (or number) of units in category B.*

$$\text{Ratio} = \frac{\%_A}{\%_B} = \frac{(f_A/n)100}{(f_B/n)100} = \frac{f_A}{f_B}$$

We used percentages in the preceding examples, but dividing the corresponding frequencies would give the same value. For example, if we use the frequencies from Table 2.5, the ratio of married to divorced persons equals 1645/341 = 4.82, which differs slightly from the preceding ratio because the percentages were rounded to one decimal place. A ratio can be interpreted as *the number of units in A to each unit in B,* or it can be interpreted equivalently as *the number of times as many category A units as category B units.* For example, there were 4.82 times as many married persons as divorced persons in the 1996 GSS. Another name for the ratio of one category frequency to another is **odds.** The ratio of A to B gives the odds of belonging to category A versus category B. Thus, the odds of being married versus divorced are 4.82 to 1.

The values of ratios and odds may range from 0 to ∞ (infinity). A value of 1 indicates that both categories are of equal size. Values between 0 and 1 indicate that the category in the base (denominator) of the ratio is the larger group. If we compute the ratio of divorced persons to married persons, we get $341/1645 = .207$, which indicates that there is .207 divorced person to each married person.

Sometimes ratios are multiplied by 100 to help clarify their interpretation, which is especially helpful when the ratio is typically less than 1. The sex ratio is a case in point. Table 2.7 shows the frequency distribution for sex (gender) in the 1996 GSS. As Table 2.7 indicates, there were fewer males than females in the adult population. Because the sex ratio is typically computed with the number of females in the denominator, this practice produces a sex ratio of less than 1 if the ratio is not multiplied by 100. The sex ratio is

$$\text{Sex ratio} = \left(\frac{f_M}{f_F}\right)100 = \left(\frac{1360}{1544}\right)100 = (.881)100 = 88.1$$

Thus, there were 88.1 males for every 100 females in the 1996 GSS. In general, when the ratio of A to B is multiplied by 100, it indicates the number of members of category A per 100 members of category B.

We have been using ratios or odds to compare one category of persons to another within a single population (married versus divorced within the population of adult Americans). Sometimes it is desirable to compare a single category of persons across two different populations or across two different points in time within the same population. To illustrate the latter, let's compare the percentage of 1996 GSS respondents who were married to the percentage of married persons in the combined 1977 and 1978 surveys to see how much change has occurred in almost two decades. Table 2.8 shows the marital status distribution for 1977–78. Nearly 70 percent of the respondents in the 1977–78 GSS were married, in comparison with only 56.7 percent married in the 1996 GSS (Table 2.5). We can compute the ratio of the percent married in 1996 to that of 1977–78:

$$\text{Change ratio} = \frac{\%_{1996}}{\%_{1977-78}} = \frac{56.7\%}{69.9\%} = .811$$

Measures of change typically include the earlier point in time in the denominator. There was .811 of a percent married in 1996 for each percent married in 1977–78. A more common way of measuring change is to compute the **rate of change:**

$$\text{Rate of change} = \left(\frac{\%_{1996} - \%_{1977-78}}{\%_{1977-78}}\right)100 = \left(\frac{56.7\% - 69.9\%}{69.9\%}\right)100$$

$$= \left(\frac{-13.2}{69.9}\right)100 = -18.9\%$$

Table 2.7 Gender, 1996 GSS

	Frequency (f)	Percent (%)
Males	1360	46.8
Females	1544	53.2
Total (n)	2904	100.0

Table 2.8 Marital Status, 1977–78 GSS

	Frequency (f)	Percent (%)
Married	2132	69.9
Widowed	212	6.9
Divorced	151	4.9
Separated	72	2.4
Never married	485	15.9
Total (n)	3052	100.0

This statistic expresses the difference in percentages—the change in percentage married—as a percent of the earlier point in time. Because the percentage in a given category at the later point in time may be smaller than the percentage in the same category at the earlier point in time, the rate of change may be negative. This statistic indicates that there was an 18.9-percent decline in the percentage of married persons between 1977–78 and 1996.

Graphical Summaries

Tables of frequency and percentage distributions give precise and detailed summaries of variables. Consumers, however, must read each of the numbers and carefully compare them, perhaps with the use of ratios, to draw accurate conclusions about the relative sizes of the different attributes. This can be a demanding task, and it is often helpful to make graphical displays of the numerical information. When properly constructed, graphs may not only ease the burden of the person who is trying to digest the information (graphs are user friendly) but also help to ensure that he or she will draw the correct conclusions. Next we discuss some of the principles for constructing good graphs.

Bar Charts

Although there are many different kinds of graphs and charts, the **bar chart** is one of the most popular because it is both simple to construct and easy to read. It also does an excellent job of presenting an undistorted view of the most important features of a distribution. Figure 2.1 shows a bar chart that has been constructed for the percentage distribution of marital status that was given in tabular form in Table 2.5.

As illustrated in Figure 2.1, there is a bar for each category or attribute of the variable being graphed. Each bar should be of equal width. There is a gap of equal width between each pair of bars. The gaps illustrate the fact that the attributes are discrete or noncontinuous; that is, they do not blend into one another, as is the case with a continuous variable such as income. Nominal and ordinal variables always have discrete categories. Interval and ratio variables may be either discrete or continuous. Charts for continuous variables have the bars touching one another (no gaps) and are called *histograms*. We discuss histograms later in the chapter. The gaps between the bars should be wide enough to emphasize the distinctness of each category but not as wide as the bars themselves. The gaps in Figure 2.1 are one-fourth (.25) of the width of the bars.

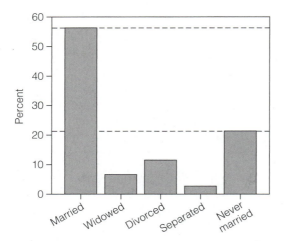

FIGURE 2.1 Marital Status, 18+ Years of Age, 1996 GSS

The most common construction of the bar chart involves marking and labeling the horizontal axis with the attributes of the variable and scaling the vertical axis in terms of either frequency counts or percentages. In Figure 2.1 the height of each bar indicates the percentage of cases in each category. In order to "read" the percentage associated with each bar, the viewer's eyes must move from the top of the bar on a line parallel to the horizontal axis to the vertical axis that has been scaled (typically the left vertical axis). Two dotted reference lines have been plotted in Figure 2.1 to illustrate the path that the eyes must follow. Usually the height of the bar does not coincide exactly with one of the labeled tick marks, and the reader must estimate the frequency or percentage. This emphasizes the fact that the main purpose of the bar chart is not to present the numerical information contained in a table. Instead, the chart is designed to provide a visual comparison of the heights of the bars for the different attributes.

On the basis of the different heights of the bars, the viewer should be able to perceive quickly and easily the most important differences among the attributes. In Figure 2.1 the viewer will quickly notice that married is by far the most frequent attribute and never married is the second most frequent attribute. To discern the overall ordering of the attributes, however, the eyes have to do a considerable amount of darting back and forth from bar to bar. We can make the viewer's job easier by placing the attributes in descending order on the horizontal axis in terms of their frequencies or percentages (see Figure 2.2). It is entirely valid to rearrange the placement of the attributes, because there is no inherent order for the attributes of a nominal variable such as marital status. When the attributes are arranged in descending order of frequency, the viewer can more quickly grasp the position of each attribute in terms of its incidence relative to each of the other attributes.

Notice that Figure 2.2 includes a bar chart for the frequency distribution as well as for the percentage distribution. It is somewhat arbitrary as to which distribution should be presented as a bar chart. In terms of the main objective of presenting the viewer with a visual

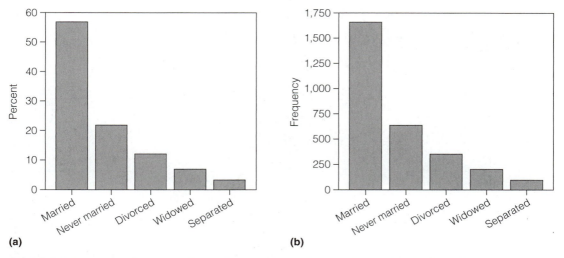

FIGURE 2.2 Percent and Frequency Distributions of Marital Status in Descending Order of Frequency

comparison of the relative sizes of the attributes, it makes no difference whether the percentages or the frequencies are used to scale the vertical axis. The ratio of one bar's height (married) to a second bar's height (never married) is the same in both charts in Figure 2.2.

What should be the minimum and maximum values displayed on the vertical axis? Figure 2.3 shows three different scales that may be used for the percentage distribution of sex. Figure 2.3a uses 0 as the minimum value and 60, the next higher multiple of 10 after the maximum observed percentage of 53.2 percent for males, as the maximum value. This is the most common way of scaling the axis. But sometimes you will see a scale like the one in Figure 2.3b, where 40, the next smaller multiple of 10 below the minimum value of 46.8 percent for males, is chosen as the minimum value. The percentage of females relative to males looks much larger in Figure 2.3b because the bar representing females is almost twice as high as the bar representing males. Which method of scaling is better? Based on the principle discussed earlier that ratios should be used when comparing attributes, the vertical axis needs to start at 0 in order for the eye to form a "visual ratio" of the height of the female bar to the height of the male bar. Thus, Figure 2.3b presents a biased picture of the relative sizes of the male and female categories.

Figure 2.3c uses 100 percent, the maximum possible value of a percentage, as the maximum value on the vertical scale. Consequently, the size of the female category relative to that of the male category appears more similar than in Figure 2.3a. It is as if we are standing farther away from Figure 2.3c than from Figure 2.3a, and that is why the difference in chart c looks smaller than the one in chart a. An advantage of Figure 2.3c is that we can visually compare the height of a bar to 100 percent to see just how large a category is relative to its maximum possible value. However, this is not usually the primary purpose of a bar chart. The primary purpose is to present visual comparisons of

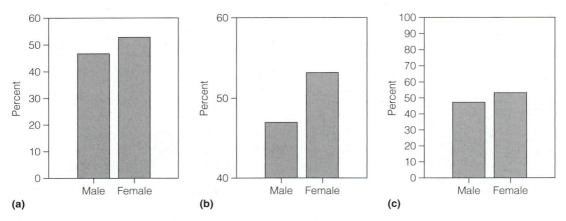

FIGURE 2.3 Percentage Distribution of Respondent's Sex, 18+ Years of Age, 1996 GSS

one attribute to another. Thus, we prefer chart a to chart c. This choice is not entirely clear-cut, however, and there is room for some legitimate debate on this issue. You should be aware of how the choice of minimum and maximum values can influence the reader's perception of a graph.

The person constructing a bar graph must also decide about the relative lengths of the horizontal and vertical axes. Should the length of the horizontal axis be greater than, equal to, or less than that of the vertical axis? The decision may have an important effect on how the viewer interprets the data. In Figure 2.4a the horizontal axis is shorter than the vertical axis, whereas in Figure 2.4b the horizontal axis is considerably longer than the vertical axis. In which chart do the differences in marital status appear greater? Stretching the horizontal axis and shrinking the vertical axis as in Figure 2.4b makes the differences in the heights of the bars appear smaller than in Figure 2.4a, where the vertical axis is stretched and the horizontal axis is shrunken. It is difficult to give rigorous recommendations regarding the relative lengths of the horizontal and vertical axes. More often than not in practice, the horizontal axis is longer than the vertical axis. Ratios of width to height ranging from 1.2 to 1.6 are common. The greater the number of bars that are to be plotted along the horizontal axis, the longer the horizontal axis tends to be. When constructing graphs or viewing graphs, remember to keep in mind how the ratio of width to height influences the viewer's perception of the differences being displayed.

Grouped Bar Charts When we want to examine the attributes of a certain variable for members of two or more groups or for two or more points in time, we may make a grouped bar chart. For example, we may want to compare the marital statuses of males and females, or we may want to compare marital status in 1996 with data from 1977–78. Figure 2.5 shows a grouped bar chart for the percentage distributions given in Tables 2.5 and Table 2.8. The bars for the same attribute measured at two different points in time are touching, but there are still gaps between the different attributes. The eye naturally compares the bars that are touching, which is the objective of the graph. The bars for the different points in time are shaded differently. The shading for one year

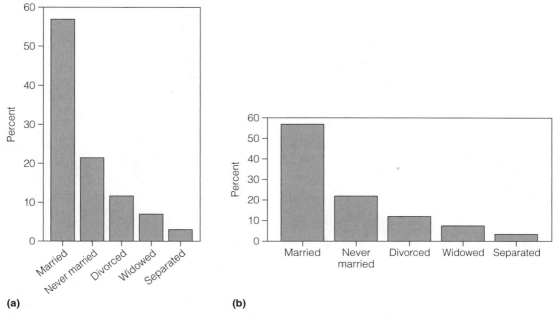

FIGURE 2.4 Percentage Distribution of Marital Status, 18+ Years of Age, 1996 GSS

should not be significantly stronger or more striking than the shading for the other year; otherwise, the group with the more vibrant shading will dominate the graph. The shadings in Figure 2.5a give a soft contrast resulting in a more balanced presentation of the two points in time. In Figure 2.5b, however, the black bar for 1996 overwhelms the white bar for 1977–78, causing the 1996 attributes to dominate.

Pie Charts

A type of graph that is popular in newspapers and magazines is the pie chart. In a **pie chart,** a circle is divided into a number of "slices" (one for each attribute) that are proportionate in size to the number of cases with each attribute. Figure 2.6 shows a pie chart for marital status in 1996.

The viewer is expected to form an impression of the relative frequencies of the attributes by comparing the sizes of their slices. There is no scale in a pie chart to use for estimating the percentage or frequency of each attribute. Often the percent belonging to each category is simply printed beside the label for the category. It is common to shade the different slices with different colors or, as in Figure 2.6, with different shades of gray. Again, care must be exercised not to make the shading of one category so strong or vibrant that it dominates the viewer's perception of the graph. When different colors or shadings are used for different groups, the color differences may rival the size differences for the viewer's attention. It can be legitimately argued that each slice should be of the same color or shade so that the only differences among them are the sizes of the slices. Remember that all of the bars are the same color in a simple bar chart.

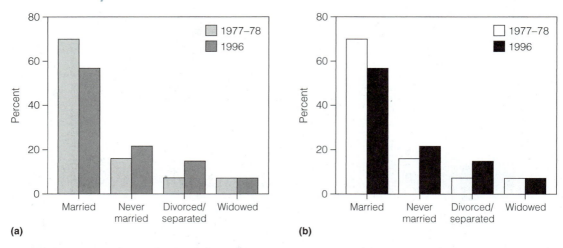

FIGURE 2.5 Percentage Distributions of Marital Status, 1977–78 and 1996 GSS

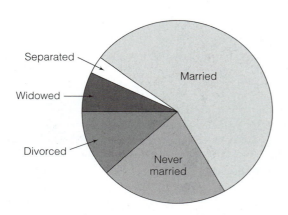

FIGURE 2.6 Percent Distribution of Marital Status,
18+ Years of Age, 1996 GSS

Figure 2.7 is a pie chart with the same shading for each marital status and with the percentages for each category listed. A disadvantage of uniform shading is that it may not capture the interest of the reader as readily as variable-shaded slices.

3-D Charts
Both the bar charts and the pie charts that we have examined have been constructed in two dimensions. With modern computer software it is now relatively easy to add a third dimension, depth, to both types of charts. To illustrate three-dimensional (3-D) charts, we will look at responses to a question asked in the GSS about political party identification: "Generally speaking, do you usually think of yourself as a Republican, Democrat, Independent, or what?" Figure 2.8 shows a 3-D pie chart of the

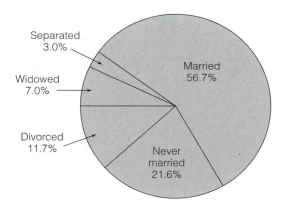

FIGURE 2.7 Percentage Distribution of Marital Status, 1996 GSS

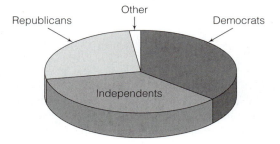

FIGURE 2.8 Political Party Identification, 1996 GSS

percentage distribution of political party preference in the United States. The two-dimensional pie chart (a circle) is now shown as a three-dimensional cylinder. The third dimension, height, adds no new information to the pie chart because the height is the same for each political party. Does the third dimension help us to accurately interpret the chart, or is it simply a device to enhance the attractiveness of the chart? Based on your viewing of Figure 2.8, which party do you think enjoyed the greatest support in 1996?

Because the Independents' slice of the pie is in the foreground, it appears larger than the Democrats' slice. The actual percentages for Democrats, Independents, Republicans, and Others are 36.2, 36.4, 25.8, and 1.6 percent, respectively. Thus, for all practical purposes, the Democrats and Independents are equal in numbers. In this case, the 3-D chart gives a false impression of the relative numbers of Independents and Democrats. The 3-D chart may look more interesting than the two-dimensional chart, but you must try not to mislead the viewer if you choose to use one.

Figure 2.9 shows a three-dimensional bar chart for the same percentage distribution of party identification. Each 3-D bar has the same depth and width; they vary only in height. The perspective is from above and to the right of the bars. Two problems are created by

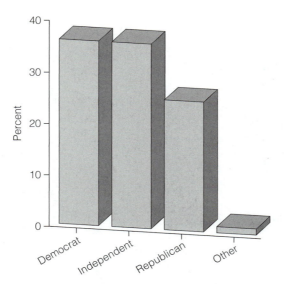

FIGURE 2.9 Political Party Identification, 1996 GSS

the three-dimensional rendering of the bars. First, the three-dimensional perspective makes it more difficult to accurately estimate the heights of the bars along the vertical scale of the graph. Second, comparisons of the heights of bars can be affected by which aspect of the bars' heights you compare. If you compare the top front or back edges of the bars, the Independents' bar appears slightly taller than the Democrats' bar, as it should. But if the viewer compares the edges of the two bars that are closest to one another (the top right edge of the Democrats' bar compared with the top left edge of the Independents' bar), the Democrats' bar looks taller than the Independents' bar.

Principles of Graphs

Remember that the primary purpose of a graph is to present a clear picture of the most important features of your data. The graph should quickly give the viewer a clear, unbiased visual impression of the information you wish to communicate. Therefore, the graph should be constructed as simply as possible.

Do not include more information than is necessary to get across the main point(s). For example, it is generally not necessary to label the bars with the values (frequencies or percentages) that they represent. The purpose of the bar chart is to provide a visual comparison of the heights of the bars in order to let the viewer "see" which attributes are most common. The viewers are more likely to remember the image of the different heights of the bars, or of the different-sized pieces of the pie, than they are to remember the frequencies or percentages that correspond to each attribute. If you want the readers to have access to the actual numbers, include them in a table. This is what is generally done in a scientific, government, or business report. If the information is being presented in the mass media for the general public, however, the table may be unnecessary. If the viewers are interested in the frequencies, they can be estimated from the

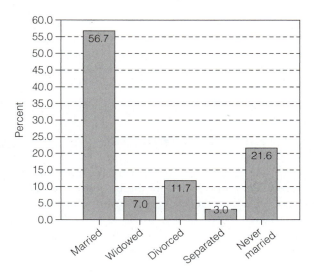

FIGURE 2.10 Overly Detailed Graph of Marital Status, 1996 GSS

scale that forms one of the axes of a bar chart. Consistent with the principle of simplicity and economy of information, it is unnecessary to list numerous scale values along the axis. A few scale values, perhaps supplemented with some unlabeled tick marks at regular intervals, will do. Additionally, although it may be helpful to include a few reference lines in a bar chart (see Figure 2.1), especially when there are numerous bars to compare, they should be used judiciously. Including reference lines at each tick mark is distracting. In sum, including value labels for each bar, many value labels for the scaled axis, and many reference lines results in extraneous information that distracts attention from the main focus of the bar chart (see Figure 2.10).

Another aspect of the simplicity principle is that the size of the graph should be proportionate to the amount of information presented. The more information you try to pack into a small space, the denser will be the graph and the harder it will be to read. The main factor that determines the amount of information in bar and pie charts is the number of attributes to be displayed. Thus, marital status, with five attributes, will need significantly more space than gender, with only two attributes.

The principle of simplicity also implies that two-dimensional charts are preferable to three-dimensional charts. The disadvantage of three dimensions is not that such charts present extraneous information (they don't) or that they look cluttered (they may be more pleasing or attractive). The liability of a 3-D chart, as discussed earlier, is that it may lead to inaccurate perceptions of the magnitudes of the attributes. If three dimensions are more attractive or interesting to the eye, however, then their careful use may motivate readers to examine them closely. This aspect may be particularly valuable in mass media communications.

The same ideas apply to the use of colors (and shades of gray). When multiple colors are used for bars or slices of pie, they may distort perceptions. If color is used judiciously, either a single color or softly contrasting colors, it may enhance the attractiveness of the chart and thus motivate the viewer to study it.

Finally, research on perception has shown that charts involving the comparison of lengths along a common scale, such as bar charts, lead to more accurate perceptions of the underlying numerical values than charts that involve judging angles or comparing areas, such as pie charts. Thus, generally speaking, bar charts are preferable to pie charts. We have discussed pie charts mainly because you are likely to see many of them in the mass media.

ORDINAL AND INTERVAL ATTRIBUTES

Did human beings evolve from animals? The teaching of evolutionary theory in public schools has produced considerable controversy in many communities over the years. Creationists who believe in some sort of divine intervention in the creation of the universe and the human species have often either opposed the teaching of evolution or called for creationism to receive equal treatment in the curriculum. Are the opponents of evolution a small but vocal minority, or is there a real split in society concerning the theory of evolution?

The 1993 and 1994 GSSs asked whether the statement "Humans evolved from animals" was definitely true, probably true, probably not true, or definitely not true. These response alternatives form an ordinal variable with those saying don't know in the middle or neutral position. Unlike the responses to the nominal marital-status variable, these belief attributes can be rank-ordered from those who are most favorable to evolution (definitely true) to those who are least favorable (definitely not true). The ordinal nature of this variable provides an additional type of information that a nominal variable does not possess.

Tabular Summaries

Cumulative Distributions

The frequency and percent distributions that are given in Table 2.9 for the evolution question are calculated in the same manner as the ones described for marital status. They indicate that there are only slightly more persons who do not think humans evolved from animals (32.4 + 15.1 = 47.5%) than believe we did evolve from animals (14.0 + 30.3 = 44.3%). Thus, these pooled percentages indicate that society is rather evenly divided on this controversial issue. This ordinal variable, however, measures differences in the certainty of the respondents' beliefs (definitely versus probably true or not true). We see that although believers and nonbelievers are about equal in numbers, the nonbelievers are much more certain in their beliefs. The ratio of the percentages of definitely not true to definitely true is 32.4/14.0 = 2.3; that is, there are more than twice as many nonbelievers who are certain in their belief as there are believers who are certain. Perhaps this helps explain the intensity of creationists' opposition to the teaching of evolution in public schools.

We have seen that ordinal variables allow us to discern gradations of attributes along the continuum that is being measured by the variable. In the case of the GSS measure of belief in evolution, the attributes are rank-ordered from high to low in terms of certainty of belief in the truth of evolution: definitely true is higher than probably true, probably true is higher than don't know, don't know is higher than probably not true,

Table 2.9 Did Humans Evolve from Animals? 1993–1994 GSS

	f	Pct	cf	cPct
Definitely not true	910	32.4	910	32.4
Probably not true	424	15.1	1334	47.5
Don't know	233	8.3	1567	55.7
Probably true	851	30.3	2418	86.0
Definitely true	393	14.0	2811	100.0
Total	2811	100.0		

and probably not true is higher than definitely not true. In the case of the evolution variable, as with most ordinal or interval variables, it is somewhat arbitrary as to which end of the continuum is defined as the high end. We could just as logically say that the variable is measuring certainty of the falseness of evolution, in which case definitely not true is the highest-ranked attribute and definitely true is the lowest-ranked attribute. No matter which end is defined as high, the arrangement of the attributes will always be the same: probably true will be between definitely true and don't know, don't know will be between probably true and probably not true, and probably not true will be between don't know and definitely not true.

We can capitalize on the order of the attributes to create a new kind of frequency distribution, a **cumulative frequency distribution** (*cf*). *The cumulative frequency of an attribute equals the number of cases that are equal to or less than the attribute.* Thus, the cumulative frequency equals the frequency of an attribute plus the sum of the frequencies of all lower attributes. In Table 2.9 the attributes are arranged in ascending order from low to high. The cumulative frequency for definitely not true equals 910, the *f* of the attribute, because there are no lower attributes. The calculation of the cumulative frequencies is illustrated in Table 2.10. Because the sum of *f* below a category is equal to the *cf* of the next lower category, the *cf* of a category is also equal to the *f* of that category plus the *cf* of the next lower category. The cumulative frequency of the highest attribute (definitely true) always equals the number of cases ($n = 2811$) and thus is not particularly meaningful.

The cumulative frequencies in Table 2.9 indicate that 1,334 persons believe that evolution is probably not true or definitely not true; 1,567 persons either don't know about the truth of evolution or believe that it is not true; and 2,418 believe that evolution is probably true, don't know if it is true, or believe it is not true. Just as percents are usually more meaningful than raw frequencies, **cumulative percents** (*cPct*) are often more meaningful than cumulative frequencies. *The cumulative percent of an attribute equals the percentage of cases that are equal to or less than that attribute.* Cumulative percents may be calculated by accumulating the percents in a percentage distribution in the same manner that the cumulative frequencies are calculated. As an alternative method, each cumulative frequency may be directly converted to a cumulative percent:

$$cPct = \frac{cf}{n} \times 100$$

Table 2.10 Calculation of Cumulative Frequencies for "Did Humans Evolve from Animals?"

	f	+	(Sum of f Below)	=	cf
Definitely not true	910	+	(0)	=	910
Probably not true	424	+	(910)	=	1334
Don't know	233	+	(910 + 424)	=	1567
Probably true	851	+	(910 + 424 + 233)	=	2418
Definitely true	393	+	(910 + 424 + 233 + 851)	=	2811

The cumulative percent for probably not true, for example, equals $(1334 \div 2811) \times 100 = 47.5\%$. The cumulative percents in Table 2.9 indicate that 47.5 percent believe that humans did not evolve from animals, 55.7 percent either don't know or believe humans did not evolve from animals, and 86.0 percent have at least some doubt about the truth of human evolution.

Grouped-Score Distributions

Ordinal variables rank the attributes but do not measure the intervals or distances between attributes. Interval and ratio variables are higher levels of measurement than ordinal variables because they provide information about how much the attributes differ from one another; they measure the interval or distance between attributes. The additional information provided by interval variables, however, does not affect the way cumulative frequencies and percents are calculated. Interval variables often create some practical problems with respect to the display of the distributions. We illustrate this with the fertility rate variable.

Our previous examples of variables all involved individuals as the units of analysis. The fertility rate, however, is a variable that is measured at the level of the nation-state; nations are the units of analysis. The *total fertility rate* of a nation is the average number of children per woman in that country.[2] Table 2.11 shows the frequency, percent, and cumulative percent distributions of the fertility rates (rounded to two decimal places) of 107 nations around the year 1995. The average number of children ranges from 1.30 (Italy) to 8.19 (Rwanda), a huge range. The fertility rates with the highest frequencies are 1.80 children per woman ($f = 5$), 2.00 children per woman ($f = 4$), and 2.80 children per woman ($f = 4$). The cumulative percents are interpreted in the same manner as earlier; they indicate the percentage of nations that have a certain fertility rate or lower. The middle of the distribution is about 3.05 because 50.5 percent of the countries have a fertility rate of 3.05 or less.

As you can see, the frequency distribution for the fertility rate is quite a long one. There are a large number of different values (attributes) of the fertility rate and a relatively small number of countries at any one value. Because of its length, this frequency distribution has been compressed into three columns in Table 2.11 to facilitate reading it. Still, it is not easy to comprehend in this form.

---

2. The total fertility rate is defined as the average number of births that women would have in their lifetime if all women lived to the end of their childbearing years, and at each year of age, they experienced the birthrates occurring in the specified year.

Table 2.11 Fertility Rate: Average Number of Children per Woman, 107 Nations, Circa 1995

Fertility	f	Pct	cPct	Fertility	f	Pct	cPct	Fertility	f	Pct	cPct
1.30	1	.9	.9	2.70	1	.9	43.9	4.70	1	.9	72.9
1.40	2	1.9	2.8	2.80	4	3.7	47.7	4.76	1	.9	73.8
1.47	1	.9	3.7	2.83	1	.9	48.6	4.90	1	.9	74.8
1.50	3	2.8	6.5	2.90	1	.9	49.5	5.10	1	.9	75.7
1.55	1	.9	7.5	3.05	1	.9	50.5	5.42	1	.9	76.6
1.58	1	.9	8.4	3.08	1	.9	51.4	5.64	1	.9	77.6
1.60	1	.9	9.3	3.10	1	.9	52.3	5.70	1	.9	78.5
1.65	2	1.9	11.2	3.11	1	.9	53.3	5.81	1	.9	79.4
1.70	2	1.9	13.1	3.19	1	.9	54.2	5.91	1	.9	80.4
1.78	1	.9	14.0	3.20	1	.9	55.1	5.94	1	.9	81.3
1.80	5	4.7	18.7	3.21	1	.9	56.1	6.10	1	.9	82.2
1.82	2	1.9	20.6	3.33	1	.9	57.0	6.20	1	.9	83.2
1.83	2	1.9	22.4	3.35	1	.9	57.9	6.29	1	.9	84.1
1.84	2	1.9	24.3	3.39	1	.9	58.9	6.33	1	.9	85.0
1.88	2	1.9	26.2	3.51	1	.9	59.8	6.40	2	1.9	86.9
1.90	2	1.9	28.0	3.73	1	.9	60.7	6.43	1	.9	87.9
1.94	1	.9	29.0	3.77	1	.9	61.7	6.53	1	.9	88.8
1.99	1	.9	29.9	3.78	1	.9	62.6	6.65	1	.9	89.7
2.00	4	3.7	33.6	3.83	1	.9	63.6	6.67	1	.9	90.7
2.03	1	.9	34.6	3.96	1	.9	64.5	6.68	1	.9	91.6
2.06	1	.9	35.5	3.97	1	.9	65.4	6.71	1	.9	92.5
2.10	2	1.9	37.4	4.00	1	.9	66.4	6.77	1	.9	93.5
2.11	1	.9	38.3	4.21	1	.9	67.3	6.80	2	1.9	95.3
2.18	1	.9	39.3	4.30	1	.9	68.2	6.81	1	.9	96.3
2.40	1	.9	40.2	4.33	1	.9	69.2	6.90	1	.9	97.2
2.44	1	.9	41.1	4.37	1	.9	70.1	6.94	1	.9	98.1
2.47	1	.9	42.1	4.48	1	.9	71.0	7.25	1	.9	99.1
2.50	1	.9	43.0	4.50	1	.9	72.0	8.19	1	.9	100.0

Lengthy distributions such as this one are typical of **continuous variables.** Continuous variables, as opposed to **discrete variables,** may have an infinite number of different values. A discrete variable may take on only a finite number of different values. If individual women were the units of analysis, the number of children each woman has is a discrete variable because there is only a fixed number of different values that it can take; decimal fractions, such as 2.5 or 3.7, are not possible and thus it is not possible to have values between 2 and 3 children or between 3 and 4 children. Fertility rate is a continuous variable because there are an infinite number of different decimal fractions that it can theoretically take (before rounding). For example, between any two observed values in Table 2.11 you can imagine that another value is possible. This is a characteristic of continuous variables. Between any two values that you might specify, it is theoretically possible for another value to occur.

Because continuous variables may take on so many different values, their frequency distributions are often long and sparse, especially when there is a relatively large number of cases. Thus, it is usually beneficial to group the values of a continuous variable into a relatively small number of categories or classes before constructing the frequency distribution.

The distributions for the fertility variable in Table 2.11 are actually **grouped-score distributions.** Because the reported fertility rate scores are rounded to two decimal places (hundredths), countries with slightly different fertility rates may be grouped together. For example, the five countries with reported fertility rates of 1.80 probably did not have exactly the same value before rounding. *The* **real limits** *of a reported number are one-half of a measurement unit below the reported number and one-half of a measurement unit above the reported number.* Because the fertility rates are reported in hundredths, the unit of measurement is one hundredth (.01). Thus, the real limits are .005 below and .005 above the reported rate. For the lowest fertility rate, which is 1.30, the *lower real limit* is $1.30 - .005 = 1.295$ and the *upper real limit* is $1.30 + .005 = 1.305$. The real limits of the lowest rate are 1.295–1.305. Although grouping of scores has already occurred to a small extent through the process of rounding, further grouping is necessary to make the distribution more comprehensible.

Three factors have to be determined when constructing a grouped-score frequency distribution: (1) the number of groups to use; (2) the interval size for each group; and (3) the lowest score of the first interval. There are no hard and fast rules for determining how the groups are to be constructed. We will discuss several guidelines, but you may have to experiment with different values to determine the best results.

The basic principle that should guide the choice of the number of groups is that the number should provide a good representation of the most important features of the distribution. Various sources recommend different numbers of groups, with recommendations ranging from five to fifteen groups. If you use too few groups (e.g., less than five groups), some of the important features of the shape of the distribution may not be represented. Two groups, for example, will never be sufficient. If we divide the fertility rates in Table 2.11 into two groups of equal intervals, we get the distribution with the intervals expressed in real limits as shown in Table 2.12. This distribution tells us that there are many more countries with low fertility rates than with high fertility rates. Although this grouping provides some information, it does not specify very well where in the lower range the decline in fertility is sharpest and whether there are any significant variations in frequencies among those 78 nations with low fertility. Also, the cumulative percentages do not provide any additional information when there are only two groups. If we choose too many groups, such as those shown in Table 2.11, the distribution will be cluttered with numerous minor variations in frequencies.

The recommendation of five to fifteen groups is quite a large range. One factor that may help narrow the range is the average number of cases per group. The number of cases divided by the number of groups should be *at least three.* There are 107 nations and 84 groups in Table 2.11, so the average number of cases per group is $107 \div 84 = 1.27$, which is much too small. To have an average of three nations per group, we need $107 \div 3 = 35.7$ groups, which is far too many groups.

Let us consider setting the number of groups equal to the middle of the recommended range—that is, to ten groups. Once we have selected a number, the next step

Table 2.12 Average Number of Children per Woman, 107 Nations

Fertility Rate	f	Pct	cPct
1.295–4.745	78	72.9	72.9
4.745–8.195	29	27.1	100.0
Total	107	100.0	

is to determine the interval size, which should be the same for all groups. The **interval size** *equals the upper real limit of the distribution minus the lower real limit divided by the number of groups.* In our fertility rate data the interval would be $(8.195 - 1.295) \div 10 = 6.9 \div 10 = .69$. As is often the case, the computed interval is an unusual or "unnatural" number. It is typical to choose natural intervals that build on the numbers 1, 2, or 5, or alternatively 10, 20, 25, or 50 for larger measurement units. For scales involving decimal units, such as the fertility rate, we can use .1, .2, .25, or .5. In our case, if we round .69 down to the next lower natural unit, we have .50 as one possible group interval, and if we round up, we have 1.00 as a second possible interval. We now examine both of these intervals.

We have to select a lower limit for the first group. We could start with the lower limit of the entire distribution—that is, 1.30. It is typical, however, to begin with a number that is an even multiple of the selected group interval. For an interval of 1.00, the first even multiple of 1.00 that is less than or equal to the lower limit of the distribution is 1.00. If we use 1.00 as the lower limit of the first group, the first interval will be 1.00–1.99. These are the **stated limits** of the interval because the upper and lower limits are expressed in the units of measurement (i.e., hundreths). The real limits of the interval are .995–1.995. The remainder of the groups (expressed in the stated limits of each interval) and their associated frequency and percentage distributions are shown in Table 2.13. Notice that when stated limits are used, there is a slight gap between each interval that is equal to the unit of measurement.

Because the selected interval of 1.00 is greater than the interval would be for ten groups (.69), we have fewer than ten groups in our grouped frequency distribution. The group with the greatest number of countries is that with fertility rates of 1.00–1.99, the first group. In these countries, which comprise about 30 percent of all nations, women are averaging less than two children per woman. This is less than the *replacement fertility rate* of 2.11. If a country averages less than 2.11 children per woman for a sufficient period of time, its population will decline. The replacement fertility rate is slightly greater than 2.00 children per woman because not all women live to the end of their childbearing years. The United States, with a fertility rate of 2.06, is slightly below the replacement rate. Looking back at Table 2.11, we can see that 37.4 percent of the nations are below the replacement rate. Thus, more than one-third of the nations are on track for declining populations.

Another important feature of the fertility rate distribution in Table 2.13 is the decline in the number of countries in each successively higher group until we reach the group

Table 2.13 Average Number of Children per Woman, 107 Nations

Fertility	f	Pct	cf	cPct
1.00–1.99	32	29.9	32	29.9
2.00–2.99	21	19.6	53	49.5
3.00–3.99	17	15.9	70	65.4
4.00–4.99	10	9.3	80	74.8
5.00–5.99	7	6.5	87	81.3
6.00–6.99	18	16.8	105	98.1
7.00–7.99	1	.9	106	99.1
8.00–8.99	1	.9	107	100.0
Total	107	100.0		

with rates of 6.00–6.99, at which point there is a significant increase in the frequency. That is, as the fertility rate increases, the number of nations decreases until we reach rates of greater than six children per woman. Thus, there are a large number of countries with very low fertility rates and there are also a substantial number with extremely high fertility rates. The world tends to be polarized between the high- and low-fertility nations.

Now, let's construct a grouped frequency distribution using the narrower interval .50. What should the lower limit of the first group equal? The first even multiple of .50 that is less than or equal to the lower limit of the distribution (1.30) is 1.00. If we use 1.00 as the lower limit of the first group, the first interval will be 1.00–1.50. This interval, however, is not a good choice because there are no nations with fertility rates that fall in the bottom half of this group; 1.3 is the smallest fertility rate in this group. Thus, the frequency for this group is likely to be relatively small simply because the lower limit is so much less than the smallest value in the distribution. A better choice for the lower limit of the first group is 1.25. The interval for this group will thus be 1.25–1.75, and the lower limit of each group will be an even multiple of .25.

The groups (expressed in terms of the real limits of the intervals) and the frequency distribution for these intervals are given in Table 2.14. There are fourteen groups with interval widths of .50. The midpoints of the intervals are even multiples of the interval widths. Sometimes it is just as useful to make the midpoints of the intervals natural numbers as it is to make the lower limits of the intervals natural numbers. The midpoints may provide a single meaningful number with which to characterize the entire group. Thus, in Table 2.14, we may think of the groups as representing countries where women average 1.5 children, 2.0 children, 2.5 children, and so on.

Can we learn anything important about the world distribution of fertility from the fourteen-group distribution in Table 2.14 that we didn't learn from the eight-group distribution in Table 2.13? You must recognize that when the number of groups is significantly increased that some relatively unimportant variations in frequencies from group to group may appear. In broad strokes, there are still more countries at the lower end of the distribution than in the middle. And, there are still a relatively large number of countries with high fertility rates in the range of 6.00–7.00. Now, however, the lowest fertility category

Table 2.14 Average Number of Children per Woman, 107 Nations

Midpoint	Real Interval	f	Pct	cPct
1.50	1.245–1.745	14	13.1	13.1
2.00	1.745–2.245	28	26.2	39.3
2.50	2.245–2.745	5	4.7	43.9
3.00	2.745–3.245	13	12.1	56.1
3.50	3.245–3.745	5	4.7	60.7
4.00	3.745–4.245	7	6.5	67.3
4.50	4.245–4.745	6	5.6	72.9
5.00	4.745–5.245	3	2.8	75.7
5.50	5.245–5.745	3	2.8	78.5
6.00	5.745–6.245	5	4.7	83.2
6.50	6.245–6.745	10	9.3	92.5
7.00	6.745–7.245	6	5.6	98.1
7.50	7.245–7.745	1	.9	99.1
8.00	7.745–8.245	1	.9	100.0
Total		107	100.0	

(midpoint of 1.50) is no longer the group with the most nations. That honor goes to the second lowest group, the group with fertility rates centered on 2.00. Thus, we can now see that extremely low fertility—that which is far below the population replacement rate—is relatively rare. The greatest concentration of countries is clustered roughly around the replacement fertility rate. An additional piece of information revealed by the fourteen-group distribution is the sizable group of countries in the category centered on three children per woman. Whether there is something meaningful about this concentration or whether it is simply a chance anomaly produced by the selected cutting points is not easy to discern. Because the fourteen-group distribution reveals that the next to the lowest fertility group, and not the lowest group, is by far the most common in the distribution, the distribution shown in Table 2.14 is preferred over the eight-group distribution in Table 2.13.

Unequal-Interval Groups

We have stressed constructing equal-sized intervals for all groups when variables are measured at the interval or ratio level of measurement. The advantage of equal intervals is that the relative frequencies of groups may be compared without differences in widths of group intervals biasing the comparison. All other things being equal, the wider the interval of a group, the greater will be the frequency associated with the group. There may be situations, however, where the use of **unequal intervals** may be justified. If there are one or a few cases that have much greater or much smaller values than the next higher or next lower cases, called **outliers,** problems may arise from creating equal-sized intervals. For one thing, the intervals may turn out to be so wide that the grouped distribution makes too few distinctions between the great majority of cases. Or, if we try to compensate for this by keeping the intervals small, there may be a number of empty groups between the interval containing the outlier(s) and the remainder of the frequency distribution. One solution may be to create an open-ended category at the upper end of the distribution labeled "*X* or more" or at the lower end of the distribution labeled "*X* or less." *X* would equal the real upper limit of the highest category with a non-zero frequency or the lower real limit of the lowest category with a non-zero frequency. Alternatively, we might create a single wide interval at the upper or lower end of the distribution that is a multiple of the normal interval width large enough to include the outlier(s). The case of outliers is one example of many possible situations where we can legitimately consider using some unequally sized intervals.

Graphical Summaries

Just as with nominal variables, graphical presentations of ordinal- and interval-variable distributions may help readers to see the most important aspects of these distributions. Although graphs for ordinal variables are the same as those for nominal variables, there are special types of graphs for continuous interval variables, called histograms and frequency polygons, that reflect the continuous nature of their attributes.

Bar Charts

We previously considered the construction of bar charts for nominal variables. Bar charts are also used to graph ordinal variables, and they are constructed in the same manner as for nominal variables. The attributes of ordinal variables are discrete, or non-continuous. Therefore, there should be gaps between the bars for the ordered attributes to indicate that they do not form a continuum. Unlike nominal variables, whose attributes are not ordered, the attributes should be presented in ascending order on the horizontal axis. Figure 2.11 shows the bar chart for the percentage distribution of the human evolution question. The chart shows at a glance that there are far more definitely not true responses than definitely true responses.

The distributions of discrete interval-level variables, such as the number of children a woman has, should logically be graphed in the same fashion as ordinal variables. That is, there should be gaps between the bars to indicate that the variable is not continuous. In practice, however, it is often the case that a different type of bar chart, called a *histogram,* is used for discrete interval variables, especially when the variable takes on a large number of different values.

Histograms

A **histogram** is a type of graph that is used to display the distribution of a continuous variable. It is constructed like a bar chart except the bars for adjacent groups of attributes touch each other. The touching bars indicate that the attributes in each group merge with the attributes above and below it. The cutting points that are used to divide the distribution of continuous scores into groups for the purpose of displaying the distribution are rather arbitrary choices by the investigator. Histograms for two alternative ways of grouping the cross-national fertility data are shown in Figure 2.12. The investigator has some freedom in how the horizontal axis is labeled. In Figure 2.12a, the values at which the distribution has been cut to construct the groups of scores are marked. These values form the upper and lower limits of each group. In Figure 2.12b, the midpoint of each group is listed on the horizontal axis.

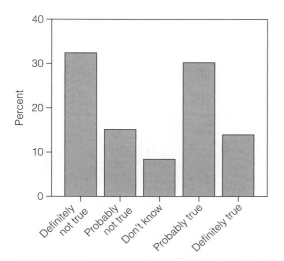

FIGURE 2.11 Did Humans Evolve from Animals? 1993–1994 GSS

Figure 2.12a uses a smaller number of groups, each of which is twice as wide as those in Figure 2.12b. As a result, Figure 2.12a presents a simpler picture of the fertility distribution because it smoothes out some of the bumps in Figure 2.12b. The picture presented by Figure 2.12a is oversimplified, however, because it doesn't reveal that the most common fertility rate is centered approximately around two children per woman and, it gives no indication of the spike in the distribution at three children per woman.

Frequency Polygons

A frequency polygon is another way to graph the distribution of a continuous variable. To construct a **frequency polygon,** first scale the horizontal axis in the same way you would for a histogram. Then make a dot above the midpoint of each group of scores at the same height that the group's histogram bar would be. Last, connect each pair of adjacent dots with a straight line. The frequency polygon for the total fertility rate is given in Figure 2.13.

The picture of a distribution provided by the frequency polygon is basically the same as that provided by a histogram. The frequency polygon, however, provides a smoother view than the histogram. In a histogram, the transitions from one bar to another are all vertical (90-degree angles). In a frequency polygon, the transitions are less steep (less than 90-degree angles), although there can still be some rather steep ascents or descents, as in the drop from 2.00 children to 2.5 children in Figure 2.13. Because the frequency polygon is smoother, it is usually more faithful to the transitions in the data.

Shapes of Distributions

When the sample size is small, a histogram can have only a small number of groups and still have a sufficiently large number of cases per group. As a consequence, the histogram will look jagged when only a small number of groups have been formed, as in Figure 2.14a where 100 observations on a hypothetical variable have been divided into seven

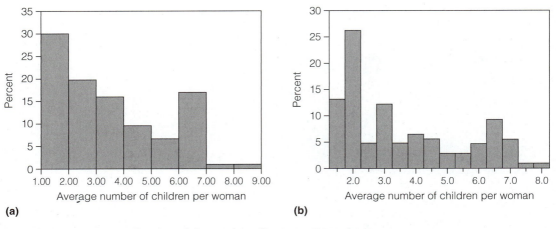

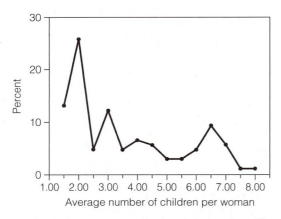

FIGURE 2.12 Percent Distributions of the Total Fertility Rate, 107 Nations

FIGURE 2.13 Frequency Polygon of the Total Fertility Rate

categories. As the number of sample cases increases, the continuous variable can be divided into smaller and smaller intervals. When this is done, the histogram will look smoother and smoother. Figure 2.14b illustrates this for 10,000 cases distributed across 35 categories (a few of the extreme categories have so few cases that the bars are not high enough to be noticed). You can imagine that as the number of observations approaches infinity ($n \to \infty$), the intervals can be made so small that the curve will look perfectly smooth, as in Figure 2.14c. When the interval widths become infinitely small, the histogram will resemble a smooth frequency polygon.

In reality, you will never see a smooth curve like Figure 2.14c because social researchers never have samples of observations that large. But statisticians usually draw smooth curves to illustrate the shapes of different distributions, partly because it is eas-

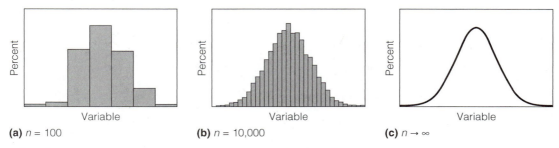

(a) n = 100   **(b)** n = 10,000   **(c)** $n \to \infty$

FIGURE 2.14 Smoothness of Histograms as a Function of Sample Size

ier than drawing the jagged histograms that we actually see in our research. Be prepared to see lots of smooth histograms or frequency polygons.

Let's examine the distribution shown in Figure 2.14c more closely in the form of Figure 2.15a. This distribution is a **bell-shaped symmetrical distribution.** It is symmetrical because the shape of the distribution above the middle of the distribution (marked by the dashed line) is exactly the reverse of the shape below the middle. The tails of a distribution refer to the upper and lower portions of the distribution where the curve approaches the horizontal axis (a frequency of 0). The right and left tails of a symmetrical distribution are the same length. In a bell-shaped distribution, this means that extremely high and extremely low scores are equally uncommon. Scores in the middle of the distribution are the most common or normal attributes.

Figure 2.15b shows a second kind of symmetrical distribution, the **uniform distribution.** In a uniform distribution, all scores, whether high, middle, or low, are equally common. Because the shape of the distribution is the same throughout—that is, flat—it is also a member of the family of symmetrical distributions. The uniform distribution, however, does not have tails because the curve does not approach zero at either the upper or lower end of the distribution.

Both bell-shaped and uniform distributions play major roles in statistical theory. You will read much more about the bell-shaped curve later in this text. Surprisingly, given the amount of attention they receive in statistics, neither type of symmetrical distribution is common empirically—that is, among observed distributions of social variables. Figure 2.16a shows the distribution of the length of telephone interviews in a social survey. Surveys, either administered by interviewers or self-administered, are the most frequently used method of collecting data in the social sciences. A popular type of survey is the telephone survey. Many people are surprised to learn how long telephone interviews often last. As can be seen in Figure 2.16a, the most frequent interview length is in the range of 45 to 50 minutes. It can also be seen that the distribution of interview length is approximately symmetrical. As times move away from the center of the distribution, either becoming longer or shorter, there is a similar decrease in their frequency. The left-hand tail and the right-hand tail are about the same length.

In addition to examining the distribution of interview length in minutes, we can also look at the distribution of the ranks of the interview lengths. If we rank order the 560 interviews from the shortest (rank = 1) to the longest (rank = 560), we can make

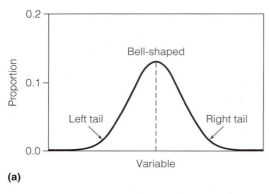

FIGURE 2.15 Illustrative Symmetrical Distributions

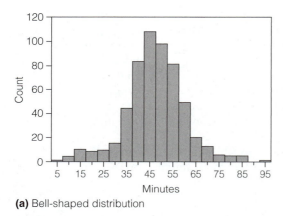

(a) Bell-shaped distribution

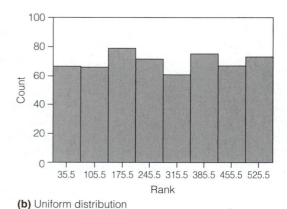

(b) Uniform distribution

FIGURE 2.16 Length of Telephone Interviews of 560 Respondents to the 1985 Akron Area Survey

a grouped-score histogram of these ranks. Figure 2.16b shows the histogram for eight groups of ranks. The ranks, which range from 1 to 560, are divided into eight equal intervals. The interval for the first group is 1–70 and the midpoint of the interval is 35.5. The interval for the highest group is 491–560, and the midpoint of this interval equals 525.5. The bars for the eight groups are all about the same height, indicating that the ranks have a nearly uniform distribution. If there were no ties, the count for each group would be exactly 70, but many different interview lengths had frequencies greater than 1. For example, the most frequent interview time was 46 minutes, which occurred 34 times. Because all interviews of the same length are given the same rank score, some groups have slightly more cases than other groups. You need not be concerned with how rank scores are created now. That topic will be taken up later in the book. The important point for now is that all rank-scored variables have uniform distributions.

Next we consider two important types of nonsymmetrical distributions (Figure 2.17). Distributions that are not symmetrical are called **skewed distributions.** If the left tail is longer than the right tail, it is called a *left-skewed distribution* (Figure 2.17a), and

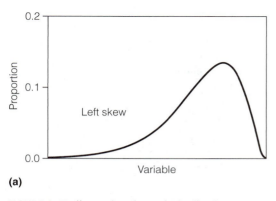
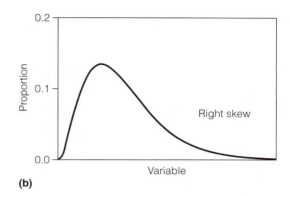

(a)   **(b)**

FIGURE 2.17 Illustrative Skewed Distributions

if the right tail is longer than the left, it is a *right-skewed distribution* (Figure 2.17b). In a skewed distribution, the most frequently occurring attributes are near one end of the distribution, whereas in a bell-shaped distribution, the statistical norm is in the center of the distribution.

Skewed distributions are easy to find in the social sciences. Figure 2.18a shows the distribution of literacy scores (percent of the population that can read) for 107 nations. The literacy distribution is an example of an extremely left-skewed distribution. The statistical norm is at the literate end of the continuum, and nations with low percentages of literacy are very uncommon in the world today. The distribution is so skewed that the histogram does not even have a right-hand tail.

The age distribution in the United States is an example of a right-skewed distribution (Figure 2.18b). The statistical norm (the bump in the distribution) is people in their 30s and 40s. This distribution applies only to persons 18 years of age and older. Because the right tail is substantially longer than the left, the elderly are further from the norm than are young adults. Even if children were included in the distribution, the right tail would still be longer than the left tail. As the population grows older, the bump in Figure 2.18b will move further to the right. Assuming life expectancy doesn't expand proportionately, the right-hand tail will grow shorter and the left-hand tail will grow longer. That is, in 20 years or so, the age distribution may be nearly bell-shaped, with the statistical norm occurring at a significantly older age.

SUMMARY

In this chapter we saw how to make tabular and graphical summaries of the entire distribution of a variable's attributes. The type of summary depends somewhat on the level of measurement of a variable. Frequency and percentage distributions may be used to describe all levels of measurement—nominal, ordinal, and interval. When a variable has many attributes, the attributes are grouped into a smaller number of categories to produce a grouped-score frequency distribution. Frequency distribution tables give the number and percentage of cases for each attribute or group of attributes. Attributes may

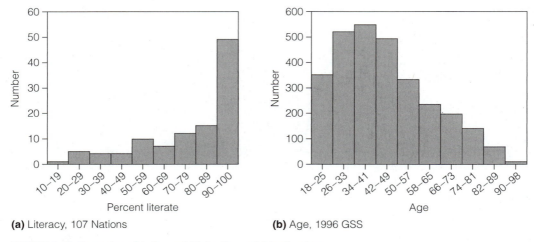

(a) Literacy, 107 Nations　　　　　　　　　　**(b)** Age, 1996 GSS

FIGURE 2.18 Examples of Left- and Right-Skewed Distributions

be compared by taking the ratio of their frequencies or percents. If a variable is measured at the ordinal or interval level, cumulative distributions that report the number of cases at or below each attribute are also used.

Bar charts are frequently used to describe nominal, ordinal, and discrete interval variables. A bar chart consists of nontouching vertical bars whose heights are proportional to the frequency or percentage of the attributes. If the variable is continuous (interval or ratio), the bars of adjacent attributes or groups of attributes touch one another and the graph is called a histogram. Frequency polygons replace the bars of a histogram with a line connecting points plotted at the height of each attribute. Pie charts display each attribute as a slice of the pie proportional to the attribute's percentage of the total cases. Bar and pie charts may also be drawn with a three-dimensional perspective, but care must be exercised not to distort the relative sizes of the attributes. Finally, smooth frequency polygons were used to illustrate symmetrical distributions, either bell-shaped or uniform, and skewed distributions in which either the right tail or the left tail is longer than the other.

KEY TERMS

frequency distribution	cumulative frequency distribution	histogram
proportion		frequency polygon
percentage	continuous variables	bell-shaped symmetrical distribution
percentage distribution	discrete variables	
ratio	grouped-score distribution	uniform distribution
rate of change	real limits	skewed distribution
bar chart	unequal intervals	
pie chart	outliers	

SPSS PROCEDURES

SPSS is a popular statistical program that your instructor may ask you to use in this course. All of the statistics and graphs shown in this text can be computed with SPSS. At the end of each chapter there is an SPSS section that shows you how to use SPSS to compute the statistics covered in that chapter. These instructions can be used as examples for conducting other statistical analyses that may be assigned by your instructor, or you can follow them to replicate the statistical results reported in the text.

When you start SPSS, an empty *Data Editor* window will appear. To open your data file, click on *File* in the main menu at the top of the window and click on *Open* and *Data* in the submenus. When the *Open File* window appears, find your data folder in the *Look In* box, click on your file name (e.g., *gss96.sav*) in the main box, and click on Open. A portion of your data file will appear in the *Data Editor* window. You can use the horizontal and vertical scroll bars to see other parts of the data set.

Gss96 - SPSS Data Editor					
File Edit View Data Transform Analyze Graphs Utilities Window Help					
1 : year		1996			
	year	id	wrkstat	hrs1	hrs2
1	1996	1	5	-1	-1
2	1996	2	1	40	-1
3	1996	3	4	-1	-1
4	1996	4	2	20	-1
5	1996	5	7	-1	-1
Data View / Variable View /					
		SPSS Processor is ready			

Before conducting a statistical analysis, you should check the variable to see if there are any codes that should be defined as missing values. Examples of codes that should be defined as missing are $8 = $ DK (don't know) and $9 = $ NA (no answer). Some SPSS data files that you use, such as the General Social Survey, may already have defined the appropriate missing values. To check the variable, click on *Variable View* at the bottom of the *Data Editor* window and find the row that contains your variable. Look in that row under *Values* to see if there are any codes that should be defined as missing, and look in the adjacent column under *Missing* to see if they have been specified as missing. If not, click on the *Missing* cell to open a *Missing Values* window and enter the codes you want to define as missing.

To make a frequency distribution for the marital-status variable, named *marital,* click on *Analyze* in the main menu and click on *Descriptive Statistics* and *Frequencies* in the submenus.

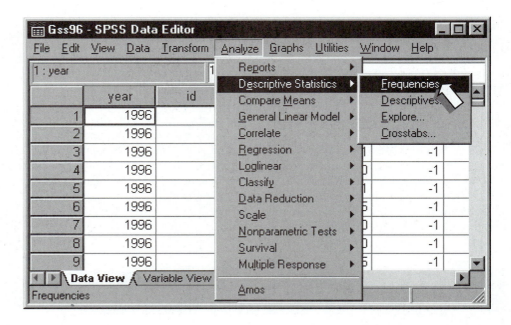

A *Frequencies* window will open. Click the name of your variable (*marital*) in the left-hand box and click on the pointer in the middle to move *marital* to the right-hand *Variable(s)* box. The result is shown below. You may move another variable, such as *marjew*, to the *Variable(s)* box by clicking on it in the left-hand list and clicking on the pointer.

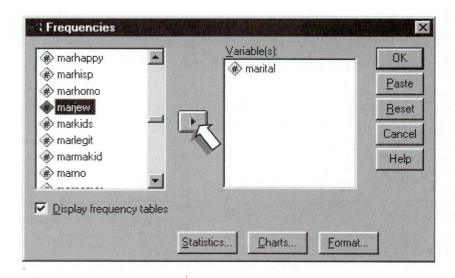

After you have selected the variable(s) you desire, click on the *OK* button and a frequency distribution table will appear in the *Output* window. To get the frequencies shown in the first output table and in Table 2.5, the cases must be weighted as described in footnote 1 of Chapter 2. The second output table shows the frequency distribution for the unweighted cases. It shows considerably fewer married persons and more divorced, widowed, separated, and never-married persons. Because most variables are not as closely related to household size as is marital status, case weighting is not absolutely necessary for most variables and thus the details of implementing weighting are not described.

MARITAL

		Frequency	Percent	Valid Percent	Cumulative Percent
Valid	MARRIED	1645	56.7	56.7	56.7
	WIDOWED	203	7.0	7.0	63.6
	DIVORCED	341	11.7	11.7	75.4
	SEPARATED	86	3.0	3.0	78.4
	NEVER MARRIED	628	21.6	21.6	100.0
	Total	2903	100.0	100.0	
Missing	NA	1	.0		
Total		2904	100.0		

MARITAL (unweighted)

		Frequency	Percent	Valid Percent	Cumulative Percent
Valid	MARRIED	1390	47.9	47.9	47.9
	WIDOWED	282	9.7	9.7	57.6
	DIVORCED	455	15.7	15.7	73.3
	SEPARATED	118	4.1	4.1	77.3
	NEVER MARRIED	658	22.7	22.7	100.0
	Total	2903	100.0	100.0	
Missing	NA	1	.0		
Total		2904	100.0		

In addition to the frequencies for each marital status, the tables indicate that there is one case with a *Missing* value called *NA* (i.e., no answer). Because of the possibility of one or more missing values, SPSS prints two percentage distributions. The first one, *Percent,* uses the number of valid cases plus the number of missing cases as the denominator for the percent. The second percentage distribution is based on only valid cases. The valid percents are presented in Table 2.5. The cumulative percents are not valid for nominal variables such as marital status.

To make a **bar chart** for marital status like the one in Figure 2.1, click on *Graph* in the main menu and then click on *Bar* in the submenu. A *Bar Chart* window will open, offering choices of several types of bar charts. Because the default options that are already selected (*Simple* and *Summaries for groups of cases*) are what we want, just click on *Define* to open the following window:

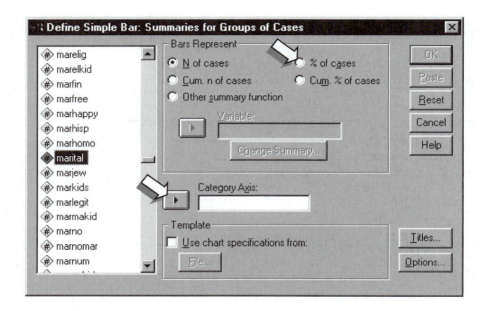

Click on % of *cases* under *Bars Represent,* click on *marital* in the left-hand variable list, and click on the pointer button under *Category Axis* to move the marital variable into that box. You can also click on *Options* to get a window where you can click on *Display groups defined by missing values* to deselect the default option that would make a bar for cases with missing values. Click the *Continue* button to close the *Options* window, and then click on *OK* to get the following graph.

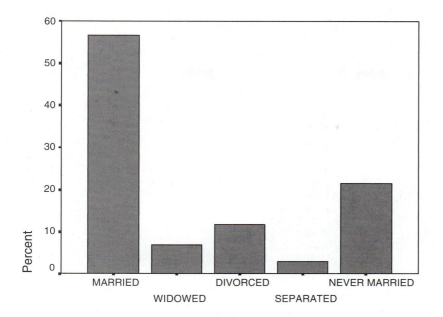

MARITAL STATUS

You may **edit the graph** to look like Figure 2.1 by double-clicking on the bar chart to place it in the chart editor shown next. The main menu of the chart editor has two submenus that may be used to edit your graph, *Chart* and *Format*. Click on the *Chart* option to get the submenu of editing options shown below. Click on *Axis* to get an *Axis Selection* window in which you can choose to edit the vertical axis (*Scale*) or the horizontal axis (*Category*). Alternatively, if you double-click directly on either axis in the graph, you will also get a window for editing that axis (see the *Scale Axis* and *Category Axis* windows that follow). In the *Scale Axis* window, choose *Center* in the *Title Justification* box to center the title as in Figure 2.1. If you want to change the scaling of the vertical axis (see Figure 2.3), use the *Scale Axis* window and type the minimum and maximum values that you want for the axis in the boxes for *Displayed Range*. In the *Category Axis* window, click on the *Labels* button to get a window in which you can change the labels for the bars to lowercase so they will all fit on one horizontal line as in Figure 2.1, instead of being staggered. To add the horizontal dashed reference lines shown in Figure 2.1, click on *Reference Line* in the *Chart* menu to get a window in which you may specify where you want the lines drawn.

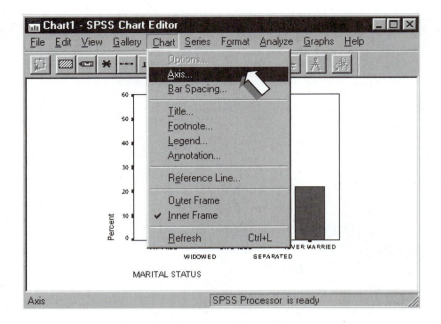

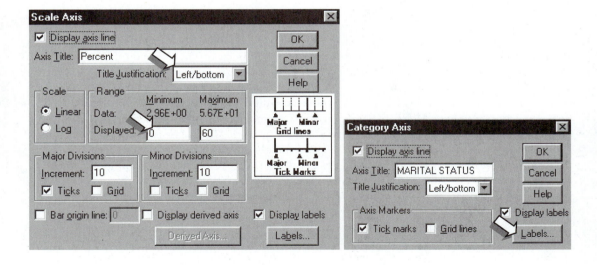

To change the **ratio of width to height** of a chart, as in Figure 2.4, click on *Edit* in the main menu and choose *Options* in the submenu. In the *Options* window, click on the *Charts* tab and type the ratio you desire in the *Chart Aspect Ratio* box.

If you want to make a **grouped bar chart** such as the one shown in Figure 2.5, click on *Graphs* in the main menu, *Bar* in the submenu, and *Clustered* and *Define* in the next window. In the *Define Clustered Bar* window, move the variable name to be graphed (e.g., *marital*) into the *Category Axis* box and the grouping variable (e.g., the year variable) into the *Define Clusters by* box. Then click on *OK*.

Pie charts like those in Figures 2.6 and 2.7 are made by clicking on *Graphs* in the main menu and *Pie* in the submenu. In the *Define Pie* window, enter the variable name in the *Define Slices by* box and click *OK*.

3-D pie charts and bar charts are made by clicking on *Graphs* in the main menu, *Interactive* in the submenu, and on *Bar* or *Pie* in the next submenu. In the *Create Bar Chart* or *Create Pie Chart* window, click on the *2-D Coordinate* button and select *3-D Effect*. Making *Interactive* graphs gives you a lot more flexibility in the way the graph is designed, but the process is more complex and will not be described here.

Cumulative distributions for ordinal and interval variables (Table 2.9) are standard output from the *Frequencies* procedure. **Histograms** (Figure 2.12) for interval variables are made by clicking on *Histogram* in the submenu of *Graphs*.

EXERCISES

1. A random sample of 30 persons was asked how often they attended church each year: never (0), several times a year (1), monthly (2), or weekly (3). The region of the United States in which they lived was recorded as Northeast (NE), Midwest (MW), South (S), or West (W). The survey found the following attributes for each person:

Region	Church	Region	Church	Region	Church	Region	Church
NE	0	W	0	MW	0	NE	1
W	2	MW	1	NE	3	MW	3
S	3	MW	3	S	0	S	3
S	1	S	2	S	1	MW	1
MW	2	S	3	W	3	S	0
W	0	NE	1	NE	0	W	0
NE	2	W	1	S	0	S	2
S	3	MW	0				

Note: The data are based on the GSS 2000.

   a. Construct a frequency distribution and a percentage distribution for Region.
   b. Interpret the frequency for the attribute South.
   c. Interpret the percentage for the attribute West.
   d. Is it valid to compute a cumulative frequency distribution for Region, and why?

2. Use the data in Exercise 1 to:

   a. Construct a frequency distribution and a percentage distribution for Church.
   b. Interpret the frequency for the attribute coded 0.
   c. Interpret the percentage for the attribute coded 3.

3. Using Table 2.6, compute the ratio of married persons who have never been divorced/separated to remarried persons (i.e., married persons who have been divorced/separated). Interpret the ratio.

4. Using Table 2.9, compute the ratio of the number of persons who believe evolution is definitely not true to the number who believe it is definitely true. Interpret the ratio.

5. The following table shows opinions about the morality of premarital sex from the GSS 2000. "If a man and a woman have sexual relations before marriage, is it . . . ?"

Response	f	Pct
Always wrong	502	28.0
Almost always wrong	157	8.8
Sometimes wrong	384	21.4
Not wrong at all	749	41.8
Total	1792	100.0

   a. Compute the cumulative percentage distribution, treating not wrong at all as the highest response.
   b. Interpret the cumulative percentage for sometimes wrong.
   c. Compute the ratio of those who believe premarital sex is not wrong at all to those who believe it is always wrong. Interpret the ratio.

6. a. Construct a cumulative percentage distribution for Church in Exercise 1.
   b. Interpret the cumulative percentage for the attribute coded 2.

7. Construct a bar chart of the percentages in Exercise 5.

8. Construct a bar chart of the percentages for Region in Exercise 1.

9. A class of 36 students has the following grade point averages (GPA).

GPA	GPA	GPA	GPA
3.99	3.20	2.87	2.67
3.88	3.06	2.86	2.44
3.82	3.03	2.84	2.43
3.68	3.01	2.84	2.40
3.57	3.01	2.84	2.31
3.54	3.00	2.83	2.12
3.52	2.99	2.82	2.07
3.48	2.98	2.78	1.94
3.35	2.89	2.77	1.81

   The guidelines for constructing grouped frequency distributions suggest that nine groups with intervals of .25 would be appropriate. The real lower limit of the first group should be 1.745. Construct the grouped frequency distribution and percentage distribution for these groups.

10. Construct a grouped frequency distribution for GPA in Exercise 9 with six categories and intervals of .4.

11. For the GPA percentage distribution in Exercise 9:

   a. Construct a histogram.
   b. Construct a frequency polygon.

12. The following grouped frequency distribution gives the percentage of the population below the poverty level for the 50 states and the District of Columbia in 1996 (*Statistical Abstract of the United States: 1998*).

Percent in Poverty	f
6.0–8.9	7
9.0–11.9	20
12.0–14.9	10
15.0–17.9	8
18.0–20.9	4
21.0–23.9	0
24.0–26.9	2
Total	51

a. Construct a histogram for the distribution.
b. Construct a frequency polygon for the distribution.

13. Respondents in the GSS 2000 (*gss2000b*) were asked which of the following statements "comes closest to expressing what you believe about God":

I don't believe in God.
I don't know whether there is a God, and I don't believe there is any way to find out.
I don't believe in a personal God, but I do believe in a Higher Power of some kind.
I find myself believing in God some of the time, but not at others.
While I have doubts, I feel that I do believe in God.
I know God really exists, and I have no doubts about it.

Use SPSS to make a frequency distribution and a bar chart for the variable *god*. What percentage of those who made a choice among the statements have no doubts that God exists? Interpret this percentage. Compute this percentage from the frequencies in the table. What percentage of those who were asked the question (NAP means not applicable—that is, not asked), gave no answer (NA), or said don't know (DK)?

14. Use SPSS to make a frequency distribution and a histogram for the variable *childs* (number of children a person has ever had) in the GSS 2000 (*gss2000a* or *gss2000b*). What percent of those who responded have had two children? What is the cumulative percent for two children and how do you interpret it? Compute the cumulative percent for two children by summing two numbers in the frequency-distribution table. Is the distribution left-skewed, right-skewed, or more or less symmetrical? Why?

3

Central Tendency

THE AVERAGE PERSON

They were average. Quite an average young woman and man, no more and no less tall than the average person. Their weight was also average. And when it came to intelligence and grades in school, they, of course, were about average. They also had average musical voices. And in athletics, they were of average speed afoot, and yes, possessed average leaping ability as well. Neither attractive nor ugly, they were quite average in appearance. The length of their hair was average, and they both parted it down the middle. Neither loners nor joiners, they had average numbers of friends. And, of course, they were an average couple. Their home had an average mortgage value and they drove average-priced cars. Their tastes in food and drink were of an average sort. And they attended church as frequently as the average person. They had two children who were. . .*

By the time you finished reading the preceding description of a hypothetical couple (a social vignette), you probably felt you had a good impression of what they were like. Average. Although it was mentioned that they parted their hair down the middle and had two children, no other specific details were given. They were simply described as being average on each of a number of attributes that are often used to describe people. You developed a good notion of what they are like because, as a functioning member

---

*Vignette continues in Chapter 4.

of society, you are familiar with these traits and what is typical or average for each of them. That is, you were capable of making relatively accurate attributions simply by being told they were average.

One of the points of presenting this vignette is to demonstrate that you are already in possession of a fundamental statistical concept, the **average.** As members of society, we learn through social experience, through the media, and perhaps in sociology classes what is extraordinary or above average, what is ordinary or average, and what is less than average. The English language has many synonyms for the word *average:* ordinary, usual, typical, medium, mediocre, and common. You can probably think of more. Although these words mean more or less the same thing as average, there are subtle distinctions among them that become important when they are used to describe people. For example, would you rather be described as a mediocre student or an average student? An average performance implies both sufficiency and lack of distinction, whereas a mediocre performance stresses lack of distinction. Statistics also has several different synonyms for average, and they carry with them important differences of meaning. That is, there are several different statistical measures of what is average, each pointing out something different about averageness.

Before turning to these statistical concepts of the average, let's consider another aspect of the above social vignette. You may have developed a feeling about this couple as a result of reading that they were average in every way. If people are consistently average across the board, you probably have a somewhat negative image of them. They are too plain, too much the same. In our society, perhaps especially in our society, value is placed on individuality. Nothing in the vignette distinguishes this couple, either positively or negatively, from others. They have no unique or individualistic characteristics. There are two points of relevance for statistics in this regard. In statistics, we learn different ways to measure what is average, but statistics makes no value judgments about whether the average attribute is good or bad. But, more importantly, statistics measures what is average one variable at a time. In these averages there are no implications that people who have average incomes, for example, will also be the same people who have average educational achievement or average physical attractiveness. In fact, it is very unusual for people to have average attributes on such a large number of variables as those mentioned in the vignette. Perhaps that is why there is a strangeness about the couple. The vignette paints a caricature of a couple who would seldom, if ever, be encountered in society.

MEASURES OF CENTRAL TENDENCY

Chapter 2 covered ways to construct frequency distributions, both tabular and graphical, to summarize the distribution of attributes on a variable. The objective was to choose a relatively small set of numbers (frequencies or percentages) that would accurately summarize a large set of observations. In this chapter we simplify our description of an observed variable even further by considering ways of choosing a single number, an average, to summarize the variable attributes. How can we possibly choose a single number that will accurately represent hundreds or thousands of observed attributes? In a sense, we can't, so let's rephrase the question slightly. What number would *best* represent all the observations of a variable? Such a number will have to be around the middle of the distribution if it is to fairly represent all scores—including high, middle, and low values. Statistical summaries that are measures of the middle of the distribution are called measures of **central tendency.** There are actually three measures of the middle, and it may be no accident that their names all begin with the letter "m." They are the **mode,** the **median,** and the **mean.** We begin with the mode.

The Mode

The **mode** *is the attribute that occurs most frequently.* That is, the mode is the attribute that has the highest frequency or percent in a frequency or percentage distribution. To compute the mode we must therefore first construct a frequency or percentage distribution. Let's start with the example of a nominal variable, religious preference. The GSS asks respondents: "What is your religious preference? Is it Protestant, Catholic, Jewish, some other religion, or no religion?" Table 3.1 shows the frequency and percentage distributions for the respondents in the 1998 GSS. The figures confirm the fact that the United States is a predominantly Christian country, with four out of five persons (79.7 percent, to be exact) claiming either a Protestant denomination or the Catholic faith as their religious preference. There are, however, more than twice as many Protestants as Catholics. People of Jewish faith are a very small minority of the population (1.8 percent). Of those who profess some other religious faith, the most frequent preferences are Islamic, Buddhist, and Hindu (not shown). Perhaps surprisingly, 14 percent say they have no religious preference. Because Protestant has the greatest frequency of the major religious categories shown in Table 3.1, Protestant is the mode or the modal religious preference. Note that for a nominal variable such as religious preference, the distribution has no middle because the attributes are not ranked from high to low. Thus, central tendency is a misnomer for the mode of a nominal variable.

Let's turn to an example of an ordinal variable, self-rated health. Respondents to the 1998 GSS were asked: "Would you say your own health, in general, is excellent, good, fair, or poor?" As indicated in Table 3.2, good has the highest frequency, and therefore good is the mode. When the response excellent is added to that of good, we see that more than three-fourths (78.7 percent) of the adult population consider their health to be at least good.

The modal health attribute is nearer the highest attribute than the lowest attribute (Figure 3.1). That is, there are more attributes to the left of the mode than to the right. Thus, the health distribution is skewed to the left. When there are an even number of attributes, such as the four attributes of self-rated health, there is no middle attribute and the distribution must be skewed.

The amount of time spent viewing television provides an example of an interval/ratio variable. In the 1998 GSS, respondents were asked: "On an average day, about how many hours do you personally watch television?" The modal number of hours that people reported was 2 (Table 3.3). The number of hours ranged all the way up to 21. Thus, the mode is much closer to the minimum value of 0 than to the maximum value of 21. The distribution of hours watching television has a long tail toward the upper end (Figure 3.2). That is, it is skewed to the right.

Figure 3.3 summarizes the position of the mode in differently skewed and symmetrical distributions. When the mode is below the middle of the distribution (i.e., to-

Table 3.1 Religious Preference, 1998 GSS

Religion	f	Pct
Protestant	1524	54.5
Catholic	705	25.2
Jewish	50	1.8
Other	122	4.4
None	396	14.2
Total	2797	100.0

Table 3.2 Health Self-Reports, 1998 GSS

Health	f	cf	Pct	cPct
Poor	136	136	4.8	4.8
Fair	466	602	16.5	21.3
Good	1345	1947	47.7	69.0
Excellent	874	2821	31.0	100.0
Total	2821		100.0	

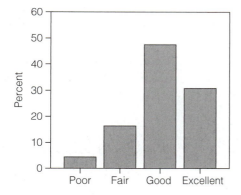

FIGURE 3.1 Health Self-Reports, 1998 GSS

ward the left end), the distribution is skewed to the right; when the mode is in the center of the distribution, the distribution is symmetrical; and when the mode is above the middle of the distribution, the distribution is skewed to the left.

It is possible for two or more attributes in a distribution to be tied or nearly tied for the highest frequency. If two attributes are tied for the highest frequency, the distribution is **bimodal.** For the political-party identification variable, Independents and Democrats are virtually tied for the highest frequency (Figure 2.9). Thus, political-party identification has a bimodal distribution. For the human-evolution variable, definitely not true has only a slightly higher frequency than probably true (Figure 2.11). Thus, belief in human evolution is also a bimodal variable. The term *bimodal* can also be applied to continuous distributions with two peaks separated by a "valley," even when one peak is clearly higher than the other. The fertility rate variable, for example, actually has three peaks (multimodal), one at 2 children per woman, another at 3, and a third at 6.5 (Figure 2.12b). In such a case, the attribute with the highest frequency is a *major mode* and the other peaks are *minor modes* (Figure 3.4).

It is important to note that the three variables that we have observed in this chapter are examples of nominal, ordinal, and interval levels of measurement. Because we may validly

Table 3.3 Hours per Day Watching Television, 1998 GSS

Hours	f	Pct	cPct
0	119	5.1	5.1
1	499	21.4	26.4
2	642	27.5	53.9
3	399	17.1	71.0
4	318	13.6	84.6
5	157	6.7	91.3
6	89	3.8	95.1
7	18	.8	95.9
8	47	2.0	97.9
9	7	.3	98.2
10	17	.7	98.9
11	3	.1	99.1
12	6	.3	99.3
13	1	.0	99.4
14	1	.0	99.4
15	7	.3	99.7
18	2	.1	99.8
20	4	.2	100.0
21	1	.0	100.0
Total	2337	100.0	

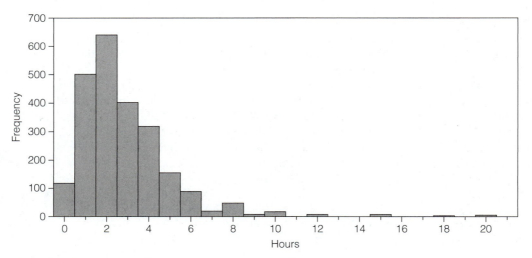

FIGURE 3.2 Hours per Day Watching Television, 1998 GSS

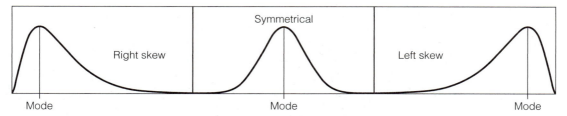

FIGURE 3.3 Position of Mode by Shape of Distribution

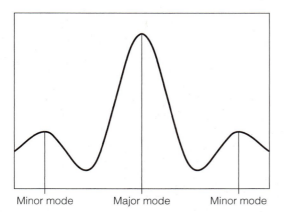

FIGURE 3.4 Multimodal Continuous Distribution

inquire about the most frequent attribute for each of these levels of measurement, *the* **mode** *is a measure of central tendency that can be used for variables at any level of measurement.*

The Median

Defining the Median

Whereas the mode or most frequent attribute will not necessarily be in the middle of the distribution, a second measure of central tendency, the **median,** focuses squarely on the center of the distribution. In general, the word *median* pertains to the middle of something. You are familiar with the median strip, a strip of land between lanes of traffic going in opposite directions. In statistics, *the* **median** *is the middle attribute in a distribution.* To find the median of a distribution, the cases are first arranged in ascending order according to the values of the variable. The median rank is then computed to determine the middle case in the distribution:

$$\text{Median rank} = \frac{n + 1}{2}$$

The median is the value of the variable for the case with the **median rank.**

As an example, a sample of nine women 45 years of age or older from the 1998 GSS who had ever had children are listed in Table 3.4 in ascending order according to the age at which they gave birth to their first child. The variable is the mother's age at birth of her first child. The median rank for the nine women equals $(9 + 1)/2 = 10/2 = 5$. That is, the median or middle case is the fifth woman. The median itself is the value of the age-at-first-birth variable for the fifth woman, which is 25 years of age. There are equal numbers of cases (i.e., four) above and below the median.

If another woman is drawn from the GSS and if she was 27 years of age at the birth of her first child, we will have the sample of ten women shown in Table 3.5. When there is an even number of cases, there is not a single middle case. Instead, we can think of the distribution as having two middle cases. In general, the median ranks of the two middle cases are $n/2$ and $n/2 + 1$. In this example, cases $10/2 = 5$ and $10/2 + 1 = 6$ are the two middle cases. We can also use the formula for the median rank to compute a single median rank of $(10 + 1)/2 = 11/2 = 5.5$, which is the average of the two middle ranks. With an even number of cases, the median itself is computed as the average of the values of the cases above and below the median rank—that is, the average of the two middle cases. In Table 3.5, the values of the cases just above and just below the median rank are 25 and 27 years of age, respectively, and thus the median equals $(25 + 27)/2 = 52/2 = 26$ years of age. Although the median is a value that is not observed for any of our cases, it divides the distribution into two equal parts, with five cases above and five cases below the median.

If the number of cases is large, it will be much more convenient to determine the median from a frequency distribution than from a case-by-case listing of values. The same rules apply, however, for determining the median. If there is an odd number of cases, the median is the value of X for the case with the median rank [i.e., $(n + 1)/2$]. If there is an even number of cases, the median equals the average of the two middle cases, the cases just above and below the median rank. You can use the cumulative frequency distribution of X to find the value of the middle case or of the two middle cases. Table 3.6 shows the frequency distribution for age at birth of first child for the 631 women 45 years of age and older in the 1998 GSS. The cfs indicate the ranks of the cases. For example, the cf for the youngest age (i.e., 14) is 2, which means the two cases at age 14 have ranks of 1 and 2. Because the cf for next older age (i.e., 15) equals 5, the three cases at age 15 have ranks 3, 4, and 5. The cf for age 16 equals 27, and thus the cases at age 16 have ranks 6–27. The median rank equals $(631 + 1)/2 = 316$. Which year of age contains the 316th case? The cf for age 20 is 249, which is less than the median rank, but the cf for age 21 is 323, which is greater than the median rank of 316. Thus, age 21 contains the case with the median rank, and the median is thus 21 years of age.

What if there had been an even number of cases in Table 3.6? For example, imagine that there was only one woman, not two, who had her first child at age 42. Then, there would have been 630 cases and the median rank would have been $(630 + 1)/2 = 315.5$. The median would have been the average value of the cases with ranks 315 and 316. Because the cfs for ages 20 and 21 would still have been 249 and 323, respectively, age 21 would have contained the 315th and 316th cases. Thus, the median would equal the average value of these two cases, which is $(21 + 21)/2 = 21$. This illustrates the fact that if the median rank falls within the ranks associated with a certain value of X, the median will equal that value of X. It is possible, however, that when there is an even number of cases, the median rank may fall between two different values of X. If so, the median will

Table 3.4 Women's Age at First Birth

	Rank	Age	
	1	15	
	2	16	
	3	18	
	4	21	
Median rank →	5	25	← Median
	6	28	
	7	29	
	8	31	
	9	42	

Table 3.5 Women's Age at First Birth

	Rank	Age	
	1	15	
	2	16	
	3	18	
	4	21	
	5	25	
Median rank →	(5.5)	(26)	← Median
	6	27	
	7	28	
	8	29	
	9	31	
	10	42	

equal the average of the two different values. For example, assume that the frequencies and cumulative frequencies for ages 20, 21, and 22 are as shown in Table 3.7. The median rank of 315.5 falls halfway between ranks 315 and 316, and the ages corresponding to these ranks are 21 and 22, respectively. Thus, the median equals $(21 + 22)/2 = 21.5$ years old.

★The Median by Interpolation[1]

In Table 3.6 there are 249 women whose age at first birth was less than 21 and $631 - 323 = 308$ women whose age was greater than 21. Therefore, 21 does not divide the distribution into two equal parts and thus does not lie exactly in the middle of the distribution. This is an unfortunate result that often occurs when there are multiple cases with the same value in the middle of the distribution. If the median rank lies within a group of cases that all have the same value, but it is not exactly in the middle of those cases, this value will not divide the distribution into groups of equal size. In Table 3.6, the ranks of the women who were age 21 at first birth range from 250 to 323. (Although technically they all have the same rank because they all have the same reported age at first birth, theoretically they could be ranked from 250 to 323 because age is a continuous variable.) The median rank of 316 is near the upper end of this range, and thus more women had their first child after age 21 than before age 21.

There is a solution to this problem that may be used for continuous variables. It is customary to assume that the observed values have been rounded off. If X is rounded to the nearest whole number, the real limits of each observed value are $X \pm .5$, or $(X - .5)$ to $(X + .5)$. If X is rounded to the nearest tenth, the real limits are $X \pm .05$. Because the observed values of the age-at-first-birth variable are whole years, we assume that the real limits are age $\pm .5$. The real limits for those women whose reported age at first birth was 21 are $21 \pm .5$, or 20.5–21.5. Thus, the median age lies somewhere between 20.5 and 21.5. To estimate this median, it is conventional to assume that the

---

1. An asterisk (★) is used throughout the book to indicate optional material.

Table 3.6 Age at Birth of First Child, Women 45+ Years Old, 1998 GSS

Age	f	Pct	cf	cPct
14	2	.3	2	.3
15	3	.5	5	.8
16	22	3.5	27	4.3
17	37	5.9	64	10.1
18	53	8.4	117	18.5
19	70	11.1	187	29.6
20	62	9.8	249	39.5
21	74	11.7	323	51.2
22	51	8.1	374	59.3
23	45	7.1	419	66.4
24	26	4.1	445	70.5
25	41	6.5	486	77.0
26	25	4.0	511	81.0
27	19	3.0	530	84.0
28	21	3.3	551	87.3
29	14	2.2	565	89.5
30	18	2.9	583	92.4
31	10	1.6	593	94.0
32	10	1.6	603	95.6
33	4	.6	607	96.2
34	6	1.0	613	97.1
35	4	.6	617	97.8
36	5	.8	622	98.6
37	3	.5	625	99.0
38	2	.3	627	99.4
41	2	.3	629	99.7
42	2	.3	631	100.0
Total	631	100.0		

Table 3.7 Hypothetical Frequencies for Age at First Birth

Age	f	cf
⋮	⋮	⋮
20	62	249
21	66	315
22	59	374
⋮	⋮	⋮
Total	630	

true ages of the 74 women that are reported to be 21 are uniformly distributed across this interval. This assumption is illustrated in Figure 3.5 by showing the ranks of the 74 women evenly distributed across the one-year interval.

The problem is to find the value between 20.5 and 21.5 that is proportional to the position of the median rank (316) in the assumed uniform distribution of these 74 women. This is a problem of **interpolation.** If the women are uniformly distributed across the one-year interval, the subinterval between each woman will equal $1/74 = .0135$ years. As you move across the axis in Figure 3.5, each vertical bar indicates an age that is .0135 of a year higher than the previous bar. The woman with the median rank would be the $316 - 249 = 67$th woman in the interval. The value of age proportionate to the 67th position in the interval would be $20.5 + (67 \times .0135) = 21.4054$. That is, the age of the median

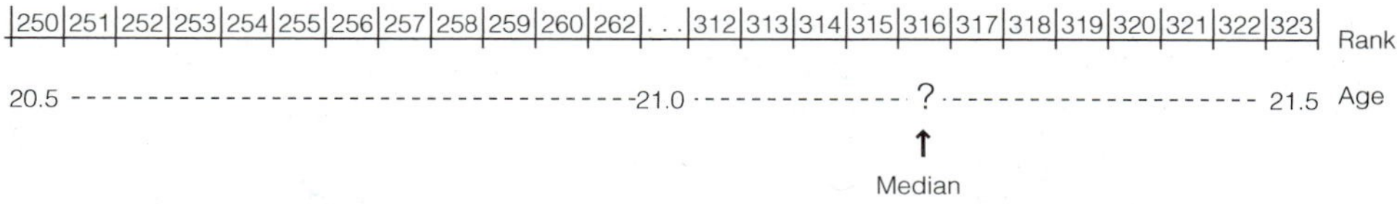

|250|251|252|253|254|255|256|257|258|259|260|262|. . .|312|313|314|315|316|317|318|319|320|321|322|323| Rank

20.5 -21.0 - - - - - - - - - - - - - ? - - - - - - - - - - - - - - - - - - - 21.5 Age

↑
Median

FIGURE 3.5 Locating the Median Age at First Birth

woman would be 21 plus 67 of the .0135 subinterval. Notice, however, that for each woman this method results in an age that corresponds to the position of the bar to the right of her rank. That is, it assigns to each woman an age equal to the upper limit of each subinterval. If 323 happened to be the median rank (i.e., if the last woman in the interval was at the median rank), her age would be 20.5 + (74 × .0135) = 21.5, which is not a valid value because it is the lower limit of the next higher age at first birth. This right-shifting of the ages can be corrected by adding 66.5 subintervals instead of 67 subintervals to 21.0 to get the median age. Thus, the median = 20.5 + (66.5 × .0135) = 21.3986. This puts the median age at first birth in the middle of the subinterval for the median rank. The following equation describes the computations that we have carried out:

$$\text{Median} = \text{lower limit} + \left(\frac{n}{2} - cf_B\right) \times \frac{w}{f} \tag{3.1}$$

$$= 20.5 + \left(\frac{631}{2} - 249\right) \times \frac{1}{74} = 20.5 + (315.5 - 249) \times .0135$$

$$= 20.5 + (66.5 \times .0135) = 20.5 + .8986 = 21.3986$$

where

Lower limit = real lower limit of the interval containing the median

cf_B = cumulative frequency below the interval containing the median

w = width of the interval containing the median

f = frequency of the interval containing the median

In Equation 3.1, $w/f = 1/74 = .0135$ is the subinterval between each case in the interval containing the median, and $(n/2) - cf_B = (631/2) - 249 = 315.5 - 249 = 66.5$ is the number of subintervals from the lower limit of the interval to the median. Study Equation 3.1 and the calculation of the median to make sure you understand the logic underlying this formula. Equation 3.2 is an alternative but equivalent equation that is frequently given for the median:

$$\text{Median} = \text{lower limit} + \frac{(n \div 2) - cf_B}{f} \times w \tag{3.2}$$

One fruitful way of studying a formula is to examine the consequences of changing a distribution by adding, removing, or changing a single observation. For example, what if we randomly removed one woman from the distribution in Table 3.6? What effect would that have on the median and why? The answer depends somewhat on the age at first birth for the woman who was removed. But no matter what that age, the median rank = $(630 + 1) \div 2 = 315.5$ and $n \div 2 = 315$. Thus, the median will now be located between the 315th and 316th women. The effect of the age at first birth of the case removed depends on whether that age is greater than 21 (the median category), less than 21, or equal to 21. If it is greater than 21, the only term that changes in the formula is $n \div 2$. Thus, the median will be located 66 subintervals above the lower limit, instead of the previous value of 66.5 subintervals, and therefore the median will be smaller:

$$\text{Median} = 20.5 + (315 - 249) \times (1/74) = 20.5 + (66 \times .0135) = 21.3919$$

If the age at first birth of the case removed is less than 21, $cf_B = 248$ and the median will be located 67 subintervals above the lower limit, causing the median to be larger than in the original data:

$$\text{Median} = 20.5 + (315 - 248) \times (1/74) = 20.5 + (67 \times .0135) = 21.4054$$

And if the age at first birth of the case removed equals $21, f = 73$, the subinterval equals .0137, and thus the median also will be larger than in the original data:

$$\text{Median} = 20.5 + (315 - 249) \times (1/73) = 20.5 + (66 \times .0137) = 21.4041$$

Notice that the effect of dropping one case is very slight, because one case is only a very small proportion of all the cases. But more importantly, notice that it is only the ordinal position of the omitted case relative to the category containing the median that is important for how the median is affected. Once we have determined whether the omitted case is above or below the median interval, it doesn't matter what the age at first birth of that case was. The effect on the median is the same whether the age was 22 or 42, for example. And the effect on the median is also the same whether the omitted age was 14 or 20. This means that *the median is relatively insensitive to changes in the distribution of scores.*

Like age at first birth, hours per day spent watching television is also a continuous variable. Thus, we can use Equation 3.1 to calculate the median from Table 3.3 on p. 66. Under the assumption that respondents round fractions of hours to the nearest hour when reporting their viewing time, the real limits of each interval are the reported number of hours $\pm$ .5. Table 3.3 contains a *cPct* distribution but not a *cf* distribution. However, we can use the *cPct* distribution in conjunction with Equation 3.1 by assuming that $n = 100$. This assumption is legitimate because the sum of the percents = 100%. Thus, $n \div 2 = 100 \div 2 = 50$. In place of *f*, we use the percent in the category containing the median (*Pct*). The median category is the smallest value with a cumulative percent greater than or equal to 50. And instead of cf_B, we use the cumulative percent below the median category ($cPct_B$). In this case, 2 is the median category because $cPct = 26.4$ for one hour of television, and $cPct = 53.9$ for two hours. Therefore, $2 \pm .5 = 1.5 - 2.49$ is the true interval for the median category. Using Equation 3.1, median hours per day watching television equals:

$$\text{Median} = 1.5 + (50 - 26.4)\,\frac{1}{27.5} = 1.5 + (23.6 \times .0364)$$
$$= 1.5 + .8582 = 2.3582$$

So, the median hours per day watching television equal 2.3582, a value closer to the true upper limit than to the middle of the interval.

In the examples so far, the formulas for the median of continuous variables have been applied to ungrouped frequency distributions. The same formulas can also be used when the raw data have been grouped into a smaller number of categories. Table 2.14 on p. 45 shows the grouped frequency and percentage distribution for the total fertility rate. The smallest interval with a cumulative percent greater than or equal to 50 is the interval 2.745–3.245. The median for the grouped total fertility rate is

$$\text{Median} = 2.745 + (50 - 43.9)\,\frac{3.245 - 2.745}{12.1} = 2.745 + \left(6.1 \times \frac{.5}{12.1}\right)$$

$$= 2.745 + .252 = 2.997$$

The median lies almost exactly in the middle of the grouped interval.

*Real Limits Revisited

When we used Equations 3.1 and 3.2 for the median, we assumed that the values of X had been rounded off. In the case of age at first birth, for example, this assumption means that the real limits are $X \pm .5$, and the lower limit used in Equations 3.1 and 3.2 is $X - .5$. Although this is the conventional method used for computing medians for continuous variables, it is not always valid to assume the given values of X have been rounded off. When women in the GSS were asked how old they were when their first child was born, they undoubtedly reported their ages as of their last birthday prior to the birth of their first child. This is the usual way we give our age. We do not round it off to the nearest year. For example, a woman who was 21 years and 9 months old would not round off and say she was 22 years old; instead, she would say she was 21 years old. This is called *truncating* the true value to a whole number. Thus, the real limits for 21 years of age are 21.0 and 21.99, and the lower limit to use in Equations 3.1 and 3.2 is 21.0 instead of 20.5. When we used 20.5 as the lower limit, the median for the distribution in Table 3.6 was

$$\text{Median} = 20.5 + \left(\frac{631}{2} - 249\right) \times \frac{1}{74} = 21.3986$$

If we use 21.0 as the lower limit, the median is

$$\text{Median} = 21.0 + \left(\frac{631}{2} - 249\right) \times \frac{1}{74} = 21.8986$$

Thus, when we do not assume that age at first birth is rounded off, which is an invalid assumption, the estimated median years of age at first birth is .5 year older than it would be with the usual assumption. We used the conventional assumption in our previous computations because that is the way textbooks show how to compute the median (but see Blalock, 1979, p. 46). The point of this discussion is that we need to ask ourselves whether reported values have been rounded, truncated, or otherwise altered before we blindly apply the assumption that they have been rounded.

Median of Ordinal Variables

When a variable is not continuous, Equations 3.1 and 3.2 should not be used to interpolate the median. Ordinal variables are a case in point. The median is a valid measure for ordinal variables because it is defined as the attribute of the middle case. To find the middle case, it is only necessary to rank order the cases, an operation that is valid for ordinal variables. To find the median of an ordinal variable in a frequency distribution, use the *cfs* or *cPcts* to locate the attribute that contains the median rank. The self-reported health status in Table 3.2 on p. 65 provides an example. "Good" is the median attribute because the median rank of $(2821 + 1)/2 = 1411$ lies within the ranks associated with good. While it is true that good health does not divide the distribution into two groups of equal size (the number who report excellent health is greater than the sum of the numbers who report poor or fair health), good is the most precise estimate we can make of the median.

If the median rank happens to fall between two ordinal attributes when there is an even number of cases, the situation is ambiguous. We cannot determine a median by the usual rules because we cannot take the average of ordinal attributes. (No numbers associated with the attributes of an ordinal variable can be summed to take an average.) The best we can do is to say that the median equals the two adjacent attributes combined—for example, fair to good health.

The median cannot be calculated for nominal variables. The attributes of a nominal variable are not ordered, which is a necessary criterion for determining the middle attribute.

Comparing the Median and Mode

For hours watching television and age at first birth, the median was greater than the mode when the median was determined by interpolation. Both of these variables were skewed to the right (a long tail above the mode). Because the mode is toward the lower end of the distribution in a right-skewed distribution while the median is in the middle, the median will be greater than the mode in such distributions. The opposite pattern exists for left-skewed distributions. In *unimodal* symmetrical distributions, the mode and the median will be equal. These patterns are illustrated in Figure 3.6.

The fact that the median and the mode are often different indicates that they should not be thought of as interchangeable measures. True, they are both indicators of central tendency or averageness and thus have much in common. But the median and mode distinguish between different aspects of averageness, and these distinctions are important. For example, because the mode is the most frequently occurring attribute, if we blindly drew one name from a hat containing the names of everyone in a sample, the attribute of the selected person would be more likely to equal the mode than any other value. The mode is the most likely value. Although the meanings of words in everyday English are not very precise, the mode appears to be equivalent to words such as *typical* or *common*. Also, the mode may be closest to the meaning that most people have in their heads when they think of average. Thus, when people talk about the common man and the ordinary or typical person, they may be thinking of the modal person.

Just because the mode is the most likely value and the measure of central tendency that is closest to the meaning of average that people have in their heads, it is not necessarily better than the median. On the side of the median is the fact that as the middle attribute, it is closest to the attributes of all the other cases. Thus, if you wanted to represent a group by picking the person with an attribute that was most similar to those of all the other members as a whole, you would pick a person with the median attribute. For example, a person with median income would be closer to all the other people's incomes than a person with modal income because the income distribution is

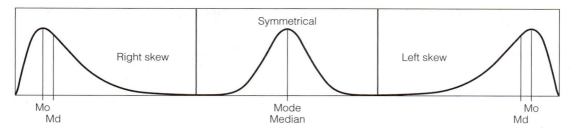

FIGURE 3.6 Relative Positions of Mode (Mo) and Median (Md) by Shape of Distribution

right-skewed, and the median person is right in the center whereas the modal person is more toward the lower end of the distribution. We will return to this characteristic of the median in the next chapter. For now, suffice it to say that the median and mode both single out important aspects of a distribution and thus should both be reported as measures of central tendency.

The Mean

Defining the Mean

Whereas the mode and the median each equal a single attribute that meets a certain criterion—the most frequent attribute and the middle attribute, respectively—as the measure of central tendency, the third measure of central tendency, the **mean,** equals *the arithmetic average of the attributes of all the cases in a distribution.* More precisely, *the* **mean** *is the sum of values across all cases in a distribution divided by the number of cases.* You are probably already familiar with this way of taking the average of a group of numbers. For the nine women in Table 3.4, the mean age at first birth is

$$\text{Mean} = \frac{15 + 16 + 18 + 21 + 25 + 28 + 29 + 31 + 42}{9} = \frac{225}{9} = 25$$

Although the mean of 25 equals the median in Table 3.4, this is just a coincidence. The mean does not always equal the median (or the mode, for that matter).

We will now introduce some important mathematical notation in the formula for the mean:

$$\text{Mean} = \overline{X} = \frac{\Sigma X}{n} \tag{3.3}$$

The mathematical symbol for the mean is $\overline{X}$, which is pronounced "X bar." Get used to seeing $\overline{X}$ and hearing "X-bar"; you will see and hear it often. Σ is capital sigma, the Greek letter S, or just sigma. It is the summation operator. X is a generic symbol for any variable. It refers to all of the attributes of variable X in a distribution. ΣX is read "sum of X." It means sum the values of X across all the cases in the distribution. The last symbol in Equation (3.3) is n, which again refers to the number of cases in the sample distribution. Let's use Equation (3.3) to compute the mean of X for the ten women in Table 3.5, where X is age at first birth:

$$\overline{X} = \frac{\Sigma X}{n} = \frac{15 + 16 + 18 + 21 + 25 + 27 + 28 + 29 + 31 + 42}{10}$$

$$= \frac{252}{10} = 25.2$$

The mean age-at-first-birth is 25.2 years, a little less than the median value of 26 years.

*Finding the Mean from Frequency Distributions

When data are in the form of a frequency distribution, we can use a different formula to calculate the mean:

$$\overline{X} = \frac{\Sigma f X}{n} \tag{3.4}$$

In Equation 3.4, X represents each distinct value in the frequency distribution and f indicates the frequency of each X, as before. The formula instructs us to sum the products of each distinct value of X and its frequency. To illustrate, Table 3.8 gives again the frequency distribution for the age-at-first-birth variable along with the product of each age times its frequency ($f X$).

The mean equals the sum of the products divided by n:

$$\overline{X} = \frac{\Sigma f X}{n} = \frac{14274}{631} = 22.6212$$

It is important to recognize that $f X$ for some value of X is simply the sum of X for all of the cases that have that value of X. For example, for the three women who were 15 years of age when they had their first child,

$$f X = 3 \cdot 15 = 15 + 15 + 15 = 45$$

Thus, $f X$ is just a quick way to take the sum for all the cases that have the same value of X. This means that Equation 3.4 is equivalent to Equation 3.3. That is, the mean, whether computed directly from the raw data with Equation 3.3 or computed from a frequency distribution with Equation 3.4, is the sum of all the values of X divided by the number of cases—that is, the arithmetic average of X. Equation 3.4 is just a computational formula for the mean, a method for more efficiently computing the mean. It is not at odds with the fundamental definition of the mean.

When the data are in the form of a grouped frequency distribution, a slightly different formula for the mean is needed:

$$\overline{X} = \frac{\Sigma f X_{\text{mid}}}{n} \tag{3.5}$$

When values of X have been grouped together to form a more compact frequency distribution, the exact values of X for each case are not shown. If we want to use the data in the grouped frequency distribution to compute the mean, it will be necessary to make an estimate of X for each interval or grouping in the distribution. The usual esti-

Table 3.8 Age at Birth of First Child (X), Women 45+ Years of Age, 1998 GSS

X	f	fX	X	f	fX
14	2	28	28	21	588
15	3	45	29	14	406
16	22	352	30	18	540
17	37	629	31	10	310
18	53	954	32	10	320
19	70	1330	33	4	132
20	62	1240	34	6	204
21	74	1554	35	4	140
22	51	1122	36	5	180
23	45	1035	37	3	111
24	26	624	38	2	76
25	41	1025	41	2	82
26	25	650	42	2	84
27	19	513	Total	631	14274

mate is based on the same logic used for the median. It assumes that within each group, the cases are uniformly distributed across the interval. Based on this assumption, the midpoint of the interval (X_{mid}) is the best estimate of X. The midpoint of an interval is

$$X_{mid} = \frac{\text{upper limit} + \text{lower limit}}{2}$$

Equation 3.5 instructs us to take the sum of the products of the midpoint of each interval and the frequency for the interval.

Table 3.9 shows a grouped frequency distribution for age at first birth. The classes were constructed on the basis of the guidelines discussed in Chapter 2. The width of each interval is five years. For example, the true age at first birth for the first interval is $9.5 \le$ age < 14.5. Thus, the midpoint of the first interval is $(14.5 + 9.5)/2 = 24/2 = 12$ years. The last column of Table 3.9 takes the product of f times the midpoint, providing an estimate of the sum of ages in each interval.

Using the sum of products in the last column of Table 3.9, the estimated mean of the distribution equals

$$\overline{X} = \frac{14307}{631} = 22.6735$$

Notice that this grouped formula mean is slightly different from the mean of 22.6212 calculated from Table 3.8 using Equation 3.4. Remember that the grouped data formula provides an *estimate* of the sample mean, and thus it will not necessarily equal the true sample mean.

Table 3.9 Calculation of Grouped Data, Mean Age at First Birth (X), 1998 GSS

X	X_{mid}	f	fX_{mid}
10–14	12	2	24
15–19	17	185	3145
20–24	22	258	5676
25–29	27	120	3240
30–34	32	48	1536
35–39	37	14	518
40–44	42	4	168
Totals		631	14307

The grouped data formula for the mean, Equation 3.5, is to be used when the available data are in the form of a grouped frequency distribution. You may, for example, have seen the grouped frequency distribution in a research report or a government document that did not report the mean. If you have the raw data or an ungrouped frequency distribution, you should use it to calculate the mean. For example, you may choose to include a grouped frequency distribution in a research report or article, such as Table 2.14 for the fertility rate. But the mean fertility rate should be calculated from the ungrouped frequency distribution (Table 2.11) or from the raw data file itself. You may feel intimidated by the task of calculating the mean from Table 2.11 and thus may prefer to use the shorter grouped distribution in Table 2.14. You should remember, however, that most of the time you will be using a computer to do your calculations, and thus it is not difficult to calculate the mean from the raw data. Grouped frequency distributions are used to provide a less complicated description of a distribution, not to ease the calculation of the mean and other descriptive statistics.

Interpreting the Mean

The mean may equal a value that no case in the distribution is observed to have, or even a value that could not even possibly be observed. For example, the mean usually contains a fraction, even if fractions cannot possibly exist on the true values. A discrete variable such as number of siblings can have only values that are whole numbers (0, 1, 2, . . . siblings). Yet the mean more often than not will equal a whole number plus a fraction. The mode, on the other hand, will by definition always be equal to an observed attribute in the distribution. The median is theoretically equal to an observed value (the middle attribute) although Equation 3.1 may give an unobserved value (but one that could be observed if the variable had been measured with more precision). The fact that there is not a mathematical or conceptual requirement for the mean to equal an observed or even a possible value may be viewed as a defect of the measure—one that makes it an artificial descriptor. This is not a deficit, however, because the mean has several useful properties that more than offset this artificiality.

One useful property of the mean is that it is like the center of gravity or, perhaps more meaningfully, the center of mass of the distribution. Imagine that each bar in a histogram has a weight (or more accurately, a mass) that is proportional to its height (Figure 3.7). The greater the height of the bar, the greater its weight. Also, imagine that each bar rests on a beam that is scaled in the units of the variable being described (e.g., hours of TV watching). The position of each bar is directly above the value whose frequency the bar represents. Now, if a fulcrum is placed under the beam at the point of the mean, the beam will rest perfectly balanced on the fulcrum (Figure 3.7). It will tip neither to the left nor to the right. If, however, the fulcrum were moved to the left, the beam would tip to the right, and if the fulcrum were moved to the right, the beam would tip to the left. The mean is the only value at which the histogram will balance perfectly on the fulcrum. Neither the median nor the mode will balance the histogram, unless the median or mode happens to equal the mean. When the histogram is balanced on its mean, it is as if the total mass of the distribution (the sum of the values, ΣX) is concentrated at one point, the mean. Thus, we have the concept of the **mean as the center of mass.**

Influence on the Mean

The fact that the mean is the center of mass of a distribution entails the notion that every case in the distribution influences the value of the mean. This aspect can be seen readily in the definition of the mean as the arithmetic average of *all* the values in the distribution. Because the mean involves summing all the values of X (i.e., ΣX), each case necessarily influences the mean. It may not be as obvious, however, that each value of X does not have an equal influence on X. The relative influence of each value of X is directly proportionate to its distance from the mean. The greater the difference between a value of X and the mean, the more influence the value has on the mean. The difference between X and the mean is called a **deviation score:**

$$\text{Deviation score} = X - \overline{X}$$

Table 3.10 shows again the frequency distribution for mother's age at birth of her first child. The third column in the table gives the deviation scores, $X - \overline{X}$. These scores were computed using the mean of the reported scores that we calculated with Equation 3.5—that is, $\overline{X} = 22.6212$ years of age. (This mean, remember, is not based on the true midpoints of the reported ages.) The deviation score for each of the two women who were 14 years of age at the birth of their first child equals $14 - 22.6212 = -8.6212$. The negative sign means that the women are below the mean, in this case 8.6212 years below the mean. At the other end of the scale, the deviation score for the two women who were 42 years of age when they had their first child equals $42 - 22.6212 = 19.3788$. The positive value indicates that the women were 19.3788 years above the mean. Because the deviation scores of the two women who were 42 at first birth are $19.3788 \div 8.6212 = 2.25$ times as large as the deviation scores of the women who were 14 at first birth, the former values of the variable will have 2.25 times more influence on the mean than the latter.

A second factor to be considered in assessing influence is the number of cases in the distribution. The greater the number of cases (n), the less will be the influence of any single case. To specify the exact functional form for the influence of $X - \overline{X}$ and n, we must give an operational definition of the concept of **influence on the mean.** *The*

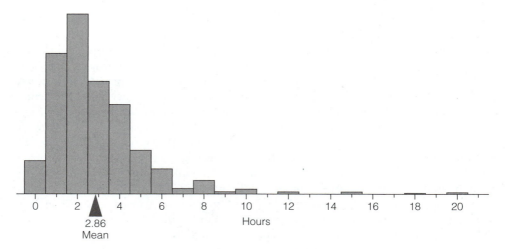

FIGURE 3.7 Histogram (Hours Watching TV) Balanced on a Fulcrum at the Mean

Table 3.10 Influence of Each Value of Age at First Birth (X) on the Mean of X

X	f	$X - \bar{X}$	$-\dfrac{X - \bar{X}}{n - 1}$	X	f	$X - \bar{X}$	$-\dfrac{X - \bar{X}}{n - 1}$
14	2	−8.6212	.0137	28	21	5.3788	−.0085
15	3	−7.6212	.0121	29	14	6.3788	−.0101
16	22	−6.6212	.0105	30	18	7.3788	−.0117
17	37	−5.6212	.0089	31	10	8.3788	−.0133
18	53	−4.6212	.0073	32	10	9.3788	−.0149
19	70	−3.6212	.0057	33	4	10.3788	−.0164
20	62	−2.6212	.0042	34	6	11.3788	−.0180
21	74	−1.6212	.0026	35	4	12.3788	−.0196
22	51	−.6212	.0010	36	5	13.3788	−.0212
23	45	.3788	−.0006	37	3	14.3788	−.0228
24	26	1.3788	−.0022	38	2	15.3788	−.0244
25	41	2.3788	−.0038	41	2	18.3788	−.0291
26	25	3.3788	−.0054	42	2	19.3788	−.0307
27	19	4.3788	−.0069				

influence *of a single value of X on the mean equals the change in the mean that will occur if a case with that value is removed from the distribution.* The formula for the influence of X on the mean is

$$\text{Influence of } X \text{ on mean} = -\frac{X - \bar{X}}{n - 1} \qquad (3.6)$$

The influence of each value of age at first birth is given in the fourth column of Table 3.10. Because $X - \overline{X}$ is in the numerator, the greater the deviation score, the greater is its influence on the mean. Because $n - 1$ is in the denominator (we are dividing by $n - 1$), the greater the n, the smaller will be the influence of X on the mean.

Equation 3.6 gives both the magnitude and the direction of the influence. Because $n - 1$ is always positive, the negative of the deviation score determines the sign or the direction of influence. For the two women who were 14 at first birth, their influence equals

$$-\frac{14 - 22.6212}{631 - 1} = -\frac{-8.6212}{630} = -(-.0137) = .0137$$

The deviation score is negative (-8.6212), the deviation score divided by $n - 1$ is also negative ($-.0137$), and the negative of a negative is positive ($.0137$). Thus, if we remove one of the women who was 14 at first birth, the mean will increase by .0137. This makes sense because if we remove a low score, one that is below the mean, the mean will increase. With respect to the oldest women at first birth, the influence score is as follows:

$$-\frac{42 - 22.6212}{631 - 1} = -\frac{19.3788}{630} = -.03076$$

If one of these women were removed from the distribution, the mean would decrease by .0308—that is, from 22.6212 to 22.5904. These two cases have the greatest influence in the distribution, but their influence on the mean is not great because no single case can be very influential when there are hundreds of cases in a distribution. If $n = 64$, however, and the mean is still 22.6212, the influence of $X = 42$ will be $-19.3788/63 = .3076$. The influence is ten times as great because the sample size is only $64/631 = .101$ (about one-tenth) as large. The most important point, however, is that the greater the deviation score, the greater will be the influence. The influence of $X = 42$ is $.0308/.0137 = 2.25$ times as great as the influence of $X = 14$. If a case equals the mean, its deviation score will be zero, and thus, its influence will be zero.

Comparing the Mean, Median, and Mode

Cases with large deviation scores, either positive or negative, are often referred to as **outliers.** (Cases lying far from the median or mode are also called outliers.) In a **skewed distribution,** there will be more extreme outliers at one end of the distribution than the other, depending on the direction of the skew. As discussed in the section on the median, age at first birth is skewed to the right. That is, its distribution has a longer tail above the mode than below the mode. With respect to deviation scores from the mean, the cases with the largest positive deviation scores in a right-skewed distribution will be further from the mean than the cases with the largest negative deviation scores. For age at first birth, for example, women who were 32 years of age and older all have larger deviation scores, in absolute value, than the largest deviation score among those women who were below the mean. There are more extreme outliers above the mean than below the mean in a right-skewed distribution. The opposite pattern holds for left-skewed distributions. This characteristic is typical of skewed distributions.

As we have seen, cases with the largest deviation scores have the greatest influence on the mean. In a right-skewed distribution, the most influential cases lie above the mean. They pull the mean upward, as evidenced by the fact that if they are removed

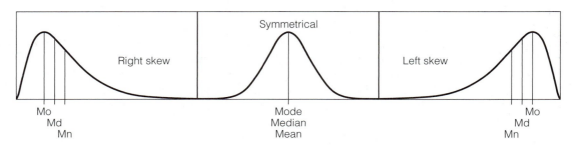

FIGURE 3.8 Relative Positions of Mode (Mo), Median (Md), and Mean (Mn) by Shape of Distribution

from the distribution, the mean will fall. The consequence is that the outliers in the upper tail of a right-skewed distribution pull the mean above both the mode and the median. In a left-skewed distribution, there are more extreme outliers in the lower tail, and they exercise the greatest influence on the mean. Consequently, the mean is pulled below the mode and the median. In a symmetrical distribution, the outliers in the upper and lower tails are equally extreme, and thus the mean is not pulled to either side of the mode or median. The **relative positions of the mode, median, and mean** in skewed and symmetrical distributions are shown in Figure 3.8.

In the two skewed distributions in Figure 3.8, the mean is located closest to the outliers in the longest tail. This is indicative of the fact that the mean is unusually sensitive to or influenced by these scores. The greater the length of the long tail relative to the length of the short tail, the more the mean is pulled toward the long tail of the distribution.

In the discussion of the median, we saw that if we remove a case that is located above the median, the median will decrease. But all cases above the median have the same influence, no matter how far they are from the median. Likewise, all cases below the median have the same influence; if one is removed, the median will increase. Consequently, the median is not at all sensitive to outliers.

Of the three measures of central tendency, the mode is probably least sensitive to outliers. If a case is removed from either tail, the mode will not change unless there is a difference of only 1 between the modal frequency and the frequency of an attribute in the tail (a very unusual distribution).

The differential sensitivity of measures of central tendency to very high scores and very low scores (i.e., outliers) should not be seen as indicating the superiority of one measure or another. The best measure to use will depend on the problem at hand and the purpose for which a measure of central tendency is being used. The relative merits of the measures will become clearer as we take up various topics in subsequent chapters.

PERCENTILES

Defining Percentiles

Measures of central tendency describe the center or middle of a distribution of scores. Percentiles are another family of measures that are used to describe other important positions in a distribution. A percentile is the value of a variable below which a specified percentage of cases falls. In general, *the Pth* **percentile,** *or X_P, equals the value of X below* *which P percent of the cases lie.* For example, several important percentiles for describing a distribution are X_{90} (90th percentile), X_{75} (75th percentile), X_{50} (50th percentile), X_{25} (25th percentile), and X_{10} (10th percentile).

The 50th percentile, X_{50}, is equal to the median. Thus, we have already used one percentile as a measure of central tendency. Because of the identity of the median and X_{50}, we can adapt the procedures for determining the median to determine any percentile. First, we must find the rank of the percentile, which is

$$X_P \text{ rank} = \frac{P}{100}(n + 1)$$

If $n = 10$, some illustrative percentile ranks are

$$X_{10} \text{ rank} = \frac{10}{100}(10 + 1) = .10(11) = 1.1$$

$$X_{25} \text{ rank} = \frac{25}{100}(10 + 1) = .25(11) = 2.75$$

$$X_{50} \text{ rank} = \frac{50}{100}(10 + 1) = .50(11) = 5.5$$

$$X_{75} \text{ rank} = \frac{75}{100}(10 + 1) = .75(11) = 8.25$$

$$X_{90} \text{ rank} = \frac{90}{100}(10 + 1) = .90(11) = 9.9$$

In general, the percentile will be the value X for the case with the calculated percentile rank. Often, however, the percentile rank will be a fractional value, and thus there will not be a case with that exact percentile rank. When this occurs, we will select an appropriate value of X that lies between the case with the next lower rank and the case with the next higher rank. To illustrate, we return to the example of ten women and their ages at first birth that was given in Table 3.5. Table 3.11 shows the five percentile ranks just computed and their associated percentiles for women's age at first birth.

The X_{10} rank (i.e., 1.1) is one-tenth of the way between ranks 1 and 2. Therefore, we select a value of X that is one-tenth of the way between the corresponding ages of 15 and 16, which is

$$X_{10} = 15 + .1(16 - 15) = 15 + .1(1) = 15.1$$

Thus, the 10th percentile equals 15.1 years. To find the 90th percentile, we note that the X_{90} rank (i.e., 9.9) is nine-tenths of the way between ranks 9 and 10:

$$X_{90} = 31 + .9(42 - 31) = 31 + .9(11) = 40.9$$

As with the median, instead of using a case-by-case listing of the data, it is usually more convenient to determine a percentile from a frequency distribution. Use the cumulative frequency distribution to find the value of X that includes the rank of the desired percentile. The percentile will equal that value of X. If the rank of the percentile

Table 3.11 Women's Age at First Birth

	Rank	Age	
	1	15	
X_{10} rank →	(1.1)	(15.1)	← X_{10}
	2	16	
X_{25} rank →	(2.75)	(17.5)	← X_{25}
	3	18	
	4	21	
	5	25	
X_{50} rank →	(5.5)	(26.0)	← X_{50}
	6	27	
	7	28	
	8	29	
X_{75} rank →	(8.25)	(29.5)	← X_{75}
	9	31	
X_{90} rank →	(9.9)	(40.9)	← X_{90}
	10	42	

falls between the ranks associated with two different values of X, the percentile will equal an appropriate value between those two values of X, as illustrated in Table 3.11.

X_{25}, X_{50}, and X_{75} are also known as the 1st, 2nd, and 3rd **quartiles,** respectively, because they divide the distribution into four equal parts, or quarters. The notation for quartiles is

$$Q_1 = X_{25}$$
$$Q_2 = X_{50}$$
$$Q_3 = X_{75}$$

*Percentiles by Interpolation

It is often the case that the Pth percentile as determined by the preceding procedures will not divide the distribution into two groups with P percent of the cases below X_P and $(100 - P)$ percent above X_P. If so, and if X is a continuous variable, we may adapt Equation 3.1 for the median to give the following formula for computing the percentile for any value of P:

$$P\text{th percentile} = X_P = LL + (P - cPct_B)\frac{w}{Pct} \tag{3.7}$$

where

$$LL = \text{true lower limit of the interval containing } X_P$$

$$cPct_B = \text{cumulative percent below the interval containing } X_P$$

$$w = \text{width of the interval containing } X_P$$

$$Pct = \text{percent in interval containing } X_P$$

As with the median, we first locate the category of X that contains X_P by finding the smallest value of X that has a cumulative percent greater than or equal to P. We can then determine values of LL, $cPct_B$, w, and Pct to enter into Equation 3.7. For example, to find the 10th percentile (X_{10}) for the age-at-first-birth variable in Table 3.6, we note that the smallest value of X with a cumulative percent greater than or equal to $P = 10$ is 17 years of age. The real limits of 17 years of age are $17.0 \leq$ age < 18.0, so LL $=$ 17.0 (see the discussion of real limits in the section titled "Real Limits Revisited"). The cumulative percent below 17.0 years of age is 4.3 percent, so $cPct_B = 4.3$. The width of the interval containing X_{10} is $w = 1$, and the percent in the X_{10} category is $Pct = 5.9$ percent. Thus, the 10th percentile equals

$$X_{10} = 17.0 + (10 - 4.3)\frac{1}{5.9} = 17.0 + 5.7(.17) = 17.0 + .97 = 17.97$$

This result indicates that 10 percent of the women were younger than 17.97 years of age at the birth of their first child. The 25th percentile lies in the category 19 years of age, and thus the 25th percentile equals

$$X_{25} = 19.0 + (25 - 18.5)\frac{1}{11.1} = 19.0 + 6.5(.09) = 19.0 + .59 = 19.59$$

Twenty-five percent of the women were younger than 19.59 years of age at the birth of their first child. Other important percentiles, including the median (X_{50}), are as follows:

$$X_{50} = 21.0 + (50 - 39.5)\frac{1}{11.7} = 21.0 + 10.5(.085) = 21.0 + .90 = 21.90$$

$$X_{75} = 25.0 + (75 - 70.5)\frac{1}{6.5} = 25.0 + 4.5(.154) = 25.0 + .69 = 25.69$$

$$X_{90} = 30.0 + (90 - 89.5)\frac{1}{2.9} = 30.0 + .5(.345) = 30.0 + .17 = 30.17$$

★Percentile Ranks

The opposite side of the coin of a percentile is a percentile rank. *The* **percentile rank** (PR_X) *equals the percentage of cases that are less than a specified value of* X. For each percentile (X_P), there is a corresponding percentile rank (PR_X). For each percentile X_P, the corresponding percentile rank is $PR_X = P$. Table 3.12 gives the percentiles that were computed earlier along with their corresponding percentile ranks.

Keep in mind that *percentile ranks are percentages, and percentiles are values of* X.

You can determine the percentile rank of any value of X, not just those percentiles that have already been determined. For example, if you want to know what percentage of women had their first child before they were out of their teens, you would ask about the percentile rank of $X = 20.0$; that is, what is the percentage of women whose age at first birth was less than 20.0? We can derive a formula for the percentile rank by solving Equation 3.7 for P (getting P to the left of the equal sign and all other quantities to the right). After solving for P and replacing it with PR_X, we have

$$PR_X = cPct_B + \frac{X - LL}{w} \times Pct \tag{3.8}$$

Table 3.12 Selected Percentile Ranks and Percentiles for Age at First Birth

PR_X	X_P
10	17.97
25	19.59
50	21.90
75	25.69
90	30.17

The terms in Equation 3.8 have the same meaning as those in Equation 3.7. The percentile rank for $X = 20.0$ is

$$PR_{20} = 29.6 + \frac{20 - 20.0}{1} \times 9.8 = 29.6 + (.0 \times 9.8) = 29.6 + .0 = 29.6$$

Thus, 29.6 percent of the women had their first child before they were 20 years old.

The percentile rank is closely related to the cumulative percents in a percentage distribution. But whereas the percentile rank equals the percentage of cases less than some X, the cumulative percent equals the percentage of cases less than or equal to X. In Table 3.6, for example, the cumulative percent for $X = 20$ is listed as 39.5 percent; 39.5 percent were less than or equal to 20 years old when their first child was born. To be more precise, however, the cumulative percent is the percentage less than or equal to the upper real limit of some value of X. Since 20.99 is the upper real limit expressed to two decimal places, the cumulative percent associated with $X = 20$ means that 39.5 percent were less than or equal to 20.99 years of age at the birth of their first child. An equivalent and even more precise way of defining the cumulative percent of some X is that it is the percentage of cases less than the lower real limit of the next higher category of X. Thus, the cumulative percent for $X = 20$ means that 39.5 percent were less than 21.0 years of age at first birth. This is the percentile rank for 21.0 years of age ($PR_{21.0}$). Therefore, the cumulative percent for some X equals the percentile rank of the lower real limit of the next higher value of X.

SUMMARY

In this chapter we considered ways of choosing a single number that *best* represents all the observations on a variable. Because such a number would have to be at or near the middle of the distribution to represent all scores fairly, these summaries are called measures of central tendency. There are three measures of the middle: the mode, the median, and the mean. The mode is the attribute that occurs the most frequently. Although the mode may fall approximately in the middle of the distribution, this does not always happen. When the mode is below the middle of the distribution (i.e., toward the left end), the distribution is skewed to the right; when the mode is in the center of the distribution, the distribution is symmetrical; and when the mode is above the middle of the distribution, the distribution is skewed to the left. The mode is a valid measure for all levels of measurement—nominal, ordinal, and interval/ratio.

The median is the middle attribute in a distribution. To find the median of a distribution, the cases are first arranged in ascending order according to the values of the variable. The median rank, which is $(n + 1)/2$, is then computed to determine the middle case in the distribution. The median is the value of the variable for the median case. (When there are multiple cases with the same value in the middle of the distribution, the reported value of the median case may not divide the distribution into two equal parts. If the variable is continuous, we can correct this by first determining the real limits of the score for the median case and then using a special formula to find the median value within these limits by interpolation. This procedure can also be used when the data are in a grouped frequency distribution.) In a right-skewed distribution, the median will be greater than the mode, whereas in a left-skewed distribution, the mode will be greater than the median. The median is a valid statistic for ordinal and interval/ratio variables, but not for nominal variables.

The mean equals the arithmetic average of all the scores in a distribution. In other words, the mean is the sum of all values in a distribution divided by the number of cases. Thus, the mean can be envisioned as the center of mass of a distribution. Although all cases influence the mean, the greater the deviation of the score from the mean, the greater is the influence of the score. Cases with large deviation scores are often referred to as outliers. In a right-skewed distribution there will be larger outliers above the middle of the distribution than below it, and thus the mean will be larger than the median (and the mode). In a left-skewed distribution, the mean will be smaller than the median. The mean is a valid measure for interval/ratio variables but not for nominal and ordinal variables.

Measures of central tendency describe the center of a distribution. Percentiles are a family of measures that are used to describe other important positions in a distribution. A percentile is the value of a variable below which a specified percentage of cases falls. For example, the 25th percentile is the value of X below which 25 percent of the cases fall. The 25th, 50th, and 75th percentiles are also called the first, second, and third quartiles, respectively. The 50th percentile equals the median. The opposite side of the coin of a percentile is a percentile rank, which equals the percentage of cases that are less than a specified value of X.

KEY TERMS

central tendency	mean	relative positions of the mode, median, and mean
mode	mean as the center of mass	
bimodal	deviation score	percentile
median	influence on the mean	quartiles
median rank	skewed distribution	★percentile rank

SPSS PROCEDURES

In addition to making frequency distributions, we can also use *Frequencies* to compute measures of central tendency. To get these statistics for the distribution of the age-at-first-birth variable (*agekdbrn*) shown in Table 3.6, first open the 1998 GSS file and select respondents who are female and 45 years of age or older (click on *Data, Select Cases . . . , If condition is satisfied,* and *If . . .* to open the *Select Cases: If* window, enter sex = 2 and age >= 45 in the box, and click on *Continue* and *OK* to execute the selection). Then click on *Analyze* in the main menu and click on *Descriptive Statistics* and *Frequencies* in the submenus. In the *Frequencies* window, move *agekdbrn* to the *Variable(s)* box and click off *Display frequency tables* if you don't want to see the frequency distribution. Then click on *Statistics* to open the *Frequencies: Statistics* window. Under *Central Tendency,* click on *Mean, Median, and Mode* and then click *Continue.* Back in the *Frequencies* window, click *OK* and a *Statistics* table will appear in the output window containing the mode, median, and mean. The mean is the same value that was computed from Table 3.6. The median of 21 is the value of the category in which the median rank is located. This median is not calculated by interpolating into the real limits of the category (Equations 3.1 and 3.2). To interpolate, click on *Values are group midpoints* in the *Frequencies: Statistics* window. Because SPSS uses a formula that differs from the standard textbook formula, the value of the median that is produced will not quite equal the value calculated in Chapter 3 with Equation 3.1.

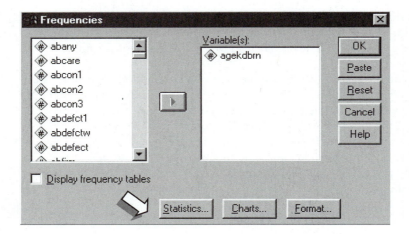

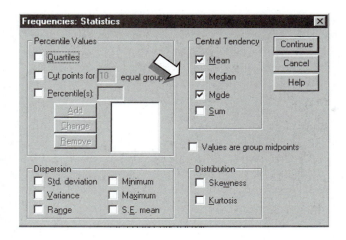

Statistics

AGEKDBRN

N	Valid	631
	Missing	114
Mean		22.62
Median		21.00
Mode		21

EXERCISES

1. A group of patients with Alzheimer's disease survived the following number of years from onset of the disease until death:

   9   6   12   8   9   11   10   15

   a. Find the modal survival time.
   b. Find the median survival time.
   c. Find the mean survival time.

2. After an additional patient with Alzheimer's disease died after 19 years, the survival times (years) are

   9   6   12   8   9   11   10   15   19

   a. Find the modal survival time.
   b. Find the median survival time.
   c. Find the mean survival time.

3. If the mean number of home runs hit by professional baseball players is 22.4, if the mode is 19, and if the median is 20.7, is the distribution of home runs left-skewed, symmetrical, or right-skewed?

4. The median IQ of the software engineers in a firm is 111.2, the mean is 108.6, and the mode is 114. Is the distribution of IQ scores left-skewed, symmetrical, or right-skewed?

5. The table below shows suicide rates (per 100,000) in 24 Western nations, circa 1993. (Example rates: U.S. = 11.9, Italy = 7.1, Russia = 41.7.)

Suicide Rate	f	cf
5.0–9.9	5	5
10.0–14.9	9	14
15.0–19.9	5	19
20.0–24.9	2	21
25.0–29.9	1	22
30.0–34.9	1	23
35.0–39.9	0	23
40.0–44.9	1	24
Total	24	

SOURCE: Statistical Abstract of the United States: 1998.

a. Find the modal suicide rate.
b. Find the median suicide rate.
c. Find the mean suicide rate.

6. The table below shows the percentages of the voting-age population who voted in the 50 states and the District of Columbia in 1994. (Example percentages: West Virginia = 33.9%, Ohio = 46.6%, South Dakota = 63.9%.)

Percent Who Voted	f	cf
30.0–34.9	3	3
35.0–39.9	4	7
40.0–44.9	13	20
45.0–49.9	13	33
50.0–54.9	9	42
55.0–59.9	4	46
60.0–64.9	5	51
Total	51	

SOURCE: Statistical Abstract of the United States: 1998.

a. Find the modal percent who voted.
b. Find the median percent who voted.
c. Find the mean percent who voted.

7. For the suicide rates in Exercise 5,

a. What is the 25th percentile?
b. What is the percentile rank of a suicide rate of 16.0?

8. For the voting percentages in Exercise 6,

a. What is the 80th percentile?
b. What is the percentile rank at which 50.0 percent voted?

9. For the survival times of persons with Alzheimer's disease in Exercise 1,

   a. Which survival time has the greatest influence on the mean in absolute value?
   b. What is the value of the influence statistic with the maximum influence for the survival time?
   c. Interpret in words the meaning of the maximum influence.

10. For the suicide rates in Exercise 5,

   a. How much influence does Russia have on the mean suicide rate?
   b. Interpret in words the meaning of the influence of Russia.

11. After the instructor returned the graded exams to a class, the student with the highest score in the class discovered a grading error that increased the score after the grading error was corrected. As a result of the correction of the grading error in the student's score,

   a. Would the mode increase, decrease, or stay the same?
   b. Would the median increase, decrease, or stay the same?
   c. Would the mean increase, decrease, or stay the same?

12. Several days after the instructor had graded the exams in a class and computed the mode, median, and mean, a student took a makeup exam and got the lowest score in the class. As a result of adding the new score to the distribution,

   a. Would the mode increase, decrease, or stay the same?
   b. Would the median increase, decrease, or stay the same?
   c. Would the mean increase, decrease, or stay the same?

13. The GSS 2000 (*gss2000b*) asked persons who use a computer at home, at work, or at some other location to estimate how many hours per week they spend sending and answering electronic mail or email. Use SPSS to find the mode, median, mean, and 75th percentile of the variable *emailhr*. Interpret the values of these statistics. (*Note:* 0 hours means less than 1 hour per week, including not at all. You could use the variable *emailmin* to find the number of minutes devoted to email each week by those who use email less than 1 hour per week.)

14. The GSS 2000 (*gss2000a*) asked respondents to evaluate the contributions that have been made to this country by different races, nationalities, and religions. For each group, respondents were asked to pick one of the following responses: (1) one of the most important positive contributions to this country; (2) an important contribution; (3) some contribution; or (4) little positive contribution to this country. Use SPSS to compute the appropriate measures of central tendency for the contributions made by the Irish (*contirsh*). Interpret the values of these statistics.

4

Dispersion

DIFFERENT AS NIGHT AND DAY

(Continued from Chapter 3) . . . as different as night and day. He did poorly in high school, and his sister excelled in her studies. But whereas she was tall and gangly, a shy bookworm with only a few friends, he was athletic and extroverted, the most popular guy in his class. After graduation, however, their fortunes took different turns. She excelled in science at Dartmouth and blossomed socially in the Ivy League environment; he lost his basketball scholarship at Emporia State due to poor grades and eventually dropped out of college. Today, she is a successful bio-medical researcher and community leader in Boston while he ekes out a living as an insurance salesman with a tendency to drink too much.

As this vignette illustrates, there are often great differences among members of groups, including siblings reared in the same family. On the other hand, groups such as country clubs, the armed services, and informal cliques may be more striking for their uniformity than for their diversity of traits. In general, groups of all kinds experience a tension between pressures toward conformity to attitudinal and behavioral norms and a recognition of the need to tolerate and even encourage diversity and uniqueness; that is, there is a struggle within all groups between being average and being different, between central tendency and **dispersion** or deviation.

In this chapter we will cover several different measures of the amount of dispersion within a group or distribution. In the preceding vignette, we saw a number of differences between a pair of siblings. Most groups are larger than two, and thus many pairs of individuals have to be taken into account when describing dispersion within a group. All of the measures of dispersion, however, can be viewed as being based on how much one of a pair of individuals differs from the other. The difference can be either between a certain pair or a composite of the differences across all pairs of individuals. After covering measures of the amount of dispersion for interval, ordinal, and nominal variables, we will present measures of the shape of dispersion—in particular, how skewed and how peaked the distribution of scores is.

INTERVAL VARIABLES

People differ significantly in their educational achievements and incomes. These variables represent highly valued attributes in our society. Education is valued in its own right and is also an extremely important means by which people obtain better jobs and thus higher incomes. Income is highly valued for what it can buy and for the prestige it may bestow on the earner. The educational and income differences among people are thus very important aspects of social inequality in our society. In these cases, education and income differences, or dispersion, are synonymous with educational and income inequality. How unequal are people in terms of their years of education and incomes, and which attribute is more unequally distributed? Both of these variables are typically measured as interval/ratio variables. Thus, we will begin with measures of dispersion for interval/ratio variables.

Education and Income Distributions

The GSS asks each respondent how many years of elementary school, high school, and college they completed for credit. Although this variable does not indicate the quality of the schools, what the person studied, or what degrees, if any, they earned, this simple ratio variable is strongly predictive of many other attributes of people, including their incomes. The distribution of years of schooling is shown in Table 4.1 and Figure 4.1.[1] Notice the large number of people with 12 years of schooling (the mode); this number represents high school graduates who did not complete any years of college. The second largest group is at 16 years of schooling; these people are primarily college graduates who did not complete any graduate school. At the lower end of the scale are a few persons with no formal schooling at all, or only a couple of years of education. Two percent have completed no more than an elementary school education (6 years or

---

1. The education variable in the GSS data file contains an open-ended upper category, which is 20 or more years of schooling. To estimate the educational distribution of the 69 persons in the upper category, a random number between 0 and 1 was assigned to each person in this category. Those with random numbers less than .5 were coded as having 20 years of schooling, those with random numbers from .5 to .833 were coded as 21 years of schooling, and those with random numbers greater than .833 were coded as 22 years of schooling. This resulted in 26, 32, and 11 persons being assigned 20, 21, and 22 years of schooling, respectively.

Table 4.1 Years of Schooling, 1998 GSS[a]

Years	f	Pct	cf	cPct
0	2	.1	2	.1
2	5	.2	7	.2
3	10	.4	17	.6
4	9	.3	26	.9
5	8	.3	34	1.2
6	23	.8	57	2.0
7	21	.7	78	2.8
8	82	2.9	160	5.7
9	75	2.7	235	8.3
10	113	4.0	348	12.3
11	138	4.9	486	17.2
12	851	30.2	1337	47.4
13	270	9.6	1607	57.0
14	350	12.4	1957	69.4
15	146	5.2	2103	74.6
16	412	14.6	2515	89.2
17	86	3.0	2601	92.2
18	109	3.9	2710	96.1
19	41	1.5	2751	97.6
20	26	.9	2777	98.5
21	32	1.1	2809	99.6
22	11	.4	2820	100.0
Total	2820	100.0		

[a]Median = 13; mean = 13.27.

less). At the opposite end of the scale, 2.4 percent have at least 4 years of post–graduate education (20 years or more of schooling).

Notice that the median educational attainment (12.77) is larger than the mode (12), and the mean (13.27) is larger than the median. This is typically indicative of a right–skewed distribution. The tail below the mode, however, is somewhat longer than the upper tail, which is typical of a left skew. Clearly, the education distribution is not typical. We will return to the topic of skewness later in the chapter.

Respondents to the GSS were also asked to indicate their personal earnings in 1997. Only 5.8 percent (not shown) refused to give their income, a percentage that is probably lower than most people would expect. This variable indicates gross income (before taxes and deductions) from wages, salary, or self–employment of the respondent alone. It does not include any unearned income from stocks, bonds, rents, or other types of capital. Table 4.2 and Figure 4.2 show the earnings distribution in intervals of $10,000

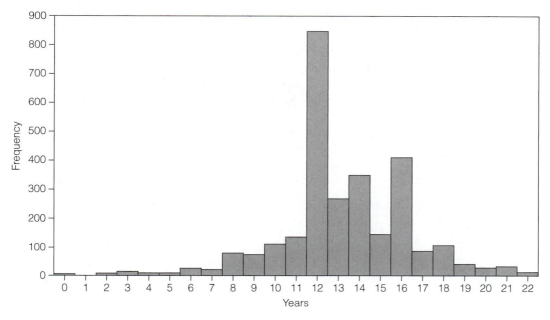

FIGURE 4.1 Years of Schooling Completed, 1998 GSS

for full-time workers, defined as working 35 or more hours per week.[2] The income distribution is clearly right-skewed. The long tail is to the right, the median (29,817) is above the middle of the modal category (20,000–29,999), and the mean (34,817) is above the median.

Notice that quite a few of these full-time workers (7.9 percent) had incomes from their occupation that were less than $10,000 in 1997. The poverty threshold for a person living alone in 1996 was $7,995. Almost three-fourths of these respondents earning less than $10,000 had incomes that fell below the poverty threshold. Some probably did not work at their occupation throughout all of 1997. Perhaps other special circumstances explain their low incomes. This figure, however, is consistent with the fact that many people work full time at year-round jobs that do not pay enough to keep them out of poverty.

2. Respondents to the GSS were asked in which of 23 different income groups, with intervals of widely varying widths, their earnings fell in 1997 (the intervals are not shown here). The groups included an open-ended upper category of $110,000 or more. A growth curve regression model was used to predict the number of persons in the open-ended category who would have incomes of $120,000, $140,000, $160,000, $180,000, and $200,000 (the predictions were virtually 0 beyond $200,000). These incomes were then randomly assigned to the 34 persons in the open-ended category with probabilities proportionate to the predicted number having each income. This resulted in 20, 9, 2, and 3 persons being assigned incomes of $120,000, $140,000, $160,000, and $180,000, respectively. (No one was randomly selected to receive an income of $200,000). The persons in each of the 26 income groups were then randomly assigned incomes within their group from a uniform distribution equal to the interval of their income group. With these constructed incomes, it was then possible to create the equal-interval income groups shown in Table 4.2.

Table 4.2 Personal Earnings in 1997 of Full-Time Workers, 1998 GSS[a]

Interval	f	Pct	cPct
$0–9,999	105	7.9	7.9
$10,000–19,999	262	19.7	27.7
$20,000–29,999	331	24.9	52.6
$30,000–39,999	232	17.5	70.1
$40,000–49,999	156	11.8	81.8
$50,000–59,999	98	7.4	89.2
$60,000–69,999	40	3.0	92.2
$70,000–79,999	26	2.0	94.2
$80,000–89,999	26	2.0	96.2
$90,000–99,999	9	.7	96.8
$100,000–109,999	8	.6	97.4
$110,000–119,999	9	.7	98.1
$120,000–129,999	11	.8	98.9
$130,000–139,999	5	.4	99.3
$140,000–149,999	4	.3	99.6
$150,000–159,999	1	.1	99.7
$160,000–169,999	1	.1	99.8
$170,000–179,999	1	.1	99.8
$180,000–189,999	2	.2	100.0
Total	1327	100.0	

[a]Median = 29,110.25; mean = 34,817.16.

It is also worth noting that the maximum income in Table 4.2 ($184,000) is much lower than the multimillion dollar incomes of the athletes, entertainers, and entrepreneurs that we hear so much about in the media. First, many high incomes are based largely on unearned income from property or capital, which is not included in our distribution. And second, persons with super-high occupational earnings represent only a small fraction of a percent of all workers and thus are unlikely to be included in a sample survey of only 1,327 workers such as the GSS.

Ranges

A natural way of measuring how much a distribution is dispersed is to calculate the difference between the maximum and minimum values in the distribution. This simple, straightforward statistic is called the **range.**

$$\text{Range} = X_{\max} - X_{\min} \tag{4.1}$$

The range is the difference between the pair of persons with the highest and lowest scores, the most extreme cases in the distribution. Thus, it is the maximum amount by which any pair of persons differs.

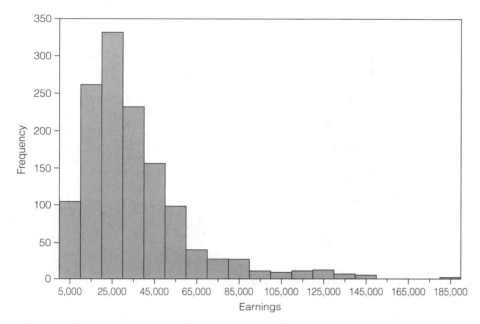

FIGURE 4.2 Personal Earnings in 1997, Full-Time Workers, 1998 GSS

As we can see in Table 4.1, the minimum and maximum values of education are 0 and 22 years, respectively. Thus, the range equals

$$\text{Education range} = 22 - 0 = 22$$

The minimum and maximum values of earnings for the distribution in Table 4.2 are $349 and $183,951 (not shown). The income range is as follows:

$$\text{Earnings range} = 183,951 - 349 = 183,602$$

As intuitive as the range appears to be as a measure of dispersion, and it is certainly one statistic that we should use, the fact that it takes into account only the extreme endpoints of the distribution is a major weakness of the measure. Two groups could have the same range of values, but in one group there might be only a few cases at the extremes and many clustered around the middle of the distribution, whereas in the other group there might be many persons at both ends of the distribution and only a few in the middle. The second group would clearly be characterized by greater dispersion or inequality than the first, even though the range is the same in both.

The **interquartile range** (IQR) is a different type of range that focuses on dispersion in the middle of the distribution instead of between the extremes of the distribution. The IQR measures the range of values for the middle 50 percent of the distribution. More specifically, the interquartile range equals the third quartile minus the first quartile:

$$\text{IQR} = Q_3 - Q_1 \tag{4.2}$$

To determine the IQR for education, we first find the ranks of the first and third quartiles of education:

$$\text{Education} \begin{cases} Q_1 \text{ rank} = .25(n + 1) = .25(2820 + 1) = 705.25 \\ Q_3 \text{ rank} = .75(n + 1) = .75(2820 + 1) = 2115.75 \end{cases}$$

We can now use the cumulative frequency distribution in Table 4.1 to find the value of education associated with each of these ranks. The first quartile equals the value of education that is one-fourth of the way between the values associated with the 705th and 706th ranks. Because education is 12 for both of these ranks, $Q_1 = 12$ years of education. Similarly, because education is 16 for both the 2115th and 2116th ranks, $Q_3 = 16$ years of education. Thus, the interquartile range is

$$\text{Education IQR} = Q_3 - Q_1 = 16 - 12 = 4$$

The interquartile range is 4 years, which represents the range of the middle 50 percent of the education distribution. It is equivalent to the difference between a person who finished high school and someone who finished college. This difference is substantial in modern societies in which education greatly affects occupational choices, incomes, and lifestyles.

The income groups in Table 4.2 ($w = 10,000$) could be used to *estimate* the 1st and 3rd income quartiles. However, because the income variable is a continuous variable with no ties (each case is a unique value), we can obtain more accurate figures for the income quartiles by using the ungrouped data. First, we would have a computer program sort the cases in ascending order of earnings. Second, we would determine the ranks of the 25th and 75th percentiles, which equal $.25(n + 1) = .25(1328) = 332$ and $.75(n + 1) = .75(1328) = 996$, respectively. The incomes of the 332nd and 996th persons (sorted in ascending order by income) are the 1st and 3rd quartiles. These are

$$Q_1 = \$18,891.74$$

$$Q_3 = \$44,459.92$$

Using these figures, the income interquartile range equals

$$\text{Income IQR} = Q_3 - Q_1 = 44,459.92 - 18,891.74 = 25,568.18$$

Thus, the difference or range separating the person whose income is higher than 75 percent of the group and the one whose income is higher than only 25 percent is $25,568. Whereas this measure of dispersion is much smaller than the entire range of incomes, the range across the middle of the distribution (i.e., the IQR) is more than the amount earned by the person at the 25th percentile, and it is almost as much as the median income itself ($29,110).

The interquartile range is an improvement over the simple range as a measure of dispersion. However, it still has the limitation that it does not take into account the spread of all the scores to measure dispersion. In this sense, the IQR is analogous to the median. The median uses only one score (Q_2, the 50th percentile), the one that is exactly in the middle of the distribution, to measure central tendency, whereas the IQR uses only two scores (Q_3 and Q_1), the two that frame the middle 50 percent of the distribution, to measure the spread of the distribution. Thus, just as we supplemented the median with the mean, which uses all the scores to measure central tendency, we now supplement the IQR with measures of dispersion that use all of the scores in the distribution.

*Standard Difference

Instead of using the difference between two selected values of X to measure dispersion, we will create a composite of the differences between all pairs of scores as a measure of the spread of the distribution. It will simplify the presentation if we begin with only a small sample of cases. Table 4.3 contains an illustrative sample of five cases for the education variable we have been considering.

We want to calculate the difference in number of years of schooling for each pair of persons. First, how many pairs of persons are there in a distribution? In general, a distribution of n cases has the following number of pairs:

$$\text{Number of pairs of cases} = \frac{n(n-1)}{2}$$

For the five cases in Table 4.3, there are $5(5-1)/2 = (5 \cdot 4)/2 = 20/2 = 10$ pairs of cases. The number of pairs grows exponentially as sample size increases. For the 1,327 persons in the income distribution of Table 4.2, there are $(1327 \cdot 1326) / 2 = 879{,}801$ pairs, and for the 2,820 persons in the education distribution of Table 4.1 there are 3,974,790 pairs!

Table 4.4 shows the differences in education (D_{ij}) for each of the ten pairs of cases, where i and j equal the case numbers in each pair. For example, the difference between the first and second persons is $D_{12} = 12 - 9 = 3$; case 1 has 3 more years of schooling than case 2. The order of the cases in Table 4.3 is arbitrary, and thus the order of the differences in Table 4.4 is also arbitrary. Notice that some of the differences are positive and some negative. If we add up all the differences to get a summary, the positive and negative differences will partially cancel each other. In this instance, the sum is negative (-14), but if the cases had been sorted differently the sum might have been positive. We could insure that all Ds are positive by sorting the cases into descending order before computing the Ds. Or, we could simply set the difference to the absolute value (drop all negative signs) for each pair no matter in which order the cases happen to be sorted. However, the solution adopted in statistics is to square the differences (D_{ij}^2) to eliminate the negative signs. An advantage of squared differences is that squared terms can be mathematically manipulated in equations, whereas terms consisting of absolute differences ($|D_{ij}|$) cannot. There is, however, a more important statistical reason for using squared differences. They minimize sampling error. This advantage will become clearer when we cover sampling issues in the next chapters.

As sample size increases, the sum of squared differences (ΣD^2) will increase because the number of pairs increases. Thus, the sum of squared differences is not a good measure of dispersion because it can't be used to compare the dispersion of a variable among groups that have different numbers of cases. The solution is to divide the sum of squared differences by the number of pairs to get the **mean squared difference** (MSD):

$$\text{MSD} = \frac{\Sigma D^2}{.5n(n-1)} \tag{4.3}$$

$$\text{Education MSD} = \frac{136}{.5 \cdot 5(5-1)} = \frac{136}{10} = 13.6$$

Thus, the mean or average of the squared differences between all pairs of persons is 13.6 years. This may seem like a large difference because the maximum years of schooling is

Table 4.3 Example Scores for Education[a]

Case Number	Education
1	12
2	9
3	14
4	16
5	12

[a]Mean education = 12.6.

Table 4.4 Educational Differences (D_{ij}) for All Pairs, $n = 5$

Pairs i, j	X_i	−	X_j	=	D_{ij}	$|D_{ij}|$	D_{ij}^2
1, 2	12	−	9	=	3	3	9
1, 3	12	−	14	=	−2	2	4
1, 4	12	−	16	=	−4	4	16
1, 5	12	−	12	=	0	0	0
2, 3	9	−	14	=	−5	5	25
2, 4	9	−	16	=	−7	7	49
2, 5	9	−	12	=	−3	3	9
3, 4	14	−	16	=	−2	2	4
3, 5	14	−	12	=	2	2	4
4, 5	16	−	12	=	4	4	16
Total					−14	32	136

22 years. But remember, this figure refers to *squared* differences between persons. The MSD can be converted back to the scale of measurement of the variable by taking its square root. The square root of the mean squared difference is named the **standard difference:**

$$\text{Standard difference} = \sqrt{\text{MSD}} = \sqrt{\frac{\Sigma D^2}{.5n(n-1)}} \qquad (4.4)$$

$$\text{Education standard difference} = \sqrt{13.6} = 3.69$$

The standard difference in education between persons is 3.69 years of schooling, which is nearly the difference between having a high school education or a college education.

The standard difference is not equivalent to the mean or average difference. The mean difference equals

$$\text{Mean difference} = \frac{\Sigma |D_{ij}|}{.5n(n-1)} = \frac{32}{10} = 3.2$$

The educational mean difference between persons is 3.2 years, which is somewhat smaller than the standard difference. Although the standard difference is generally larger than the mean difference, it usually does little injustice to the data to think of the standard difference as an "average" difference, with quotation marks around average.

Turning to earnings, Table 4.5 gives five illustrative cases of earnings. The differences in earnings for the ten pairs of these cases are shown in Table 4.6. The following statistics summarize the differences:

$$\text{Earnings mean difference} = \frac{\Sigma |D_{ij}|}{.5n(n-1)} = \frac{216,000}{10} = 21,600$$

$$\text{Earnings MSD} = \frac{\Sigma D^2}{.5n(n-1)} = \frac{6,359,660,000}{10} = 635,966,000$$

$$\text{Earnings standard difference} = \sqrt{635,966,000} = 25,218.37$$

Table 4.5 Example Scores for Earnings[a]

Case Number	Earnings
1	35,200
2	17,500
3	64,800
4	40,100
5	26,700

[a]Mean income = 38,860.

Table 4.6 Earnings Differences (D_{ij}) for All Pairs, $n = 5$

Pairs i, j	X_i	−	X_j	=	D_{ij}	$|D_{ij}|$	D_{ij}^2
1, 2	35,200	−	17,500	=	17,700	17,700	313,290,000
1, 3	35,200	−	64,800	=	−29,600	29,600	876,160,000
1, 4	35,200	−	40,100	=	−4,900	4,900	24,010,000
1, 5	35,200	−	26,700	=	8,500	8,500	72,250,000
2, 3	17,500	−	64,800	=	−47,300	47,300	2,237,290,000
2, 4	17,500	−	40,100	=	−22,600	22,600	510,750,000
2, 5	17,500	−	26,700	=	−9,200	9,200	84,640,000
3, 4	64,800	−	40,100	=	24,700	24,700	610,090,000
3, 5	64,800	−	26,700	=	38,100	38,100	1,451,610,000
4, 5	40,100	−	26,700	=	13,400	13,400	179,560,000
Total					−11,200	216,000	6,359,660,000

The mean squared difference in earnings is a whopping $635,966,000. The square root of the MSD results in a standard difference of earnings equal to $25,218.37. That is, the standard or "average" amount by which workers differ in their occupational earnings is $25,218.37. The actual mean difference, which is $21,600, is again smaller than the standard difference.

Five-example cases have been used to illustrate the concepts of the mean squared difference and the standard difference. Table 4.7 shows the values of these statistics for the actual respondents in the GSS. The true standard differences are much greater than the standard differences for the example data (35,712.66 versus 25,218.37 and 4.21 versus 3.69 for income and education, respectively), especially for income. The standard earnings difference between people is slightly greater than the mean earnings, and the standard difference in people's education is greater than a four-year college education. Also, the standard differences for these variables are greater than their interquartile ranges (25,568.18 and 3.77). That is, the standard differences are greater than the differences between the persons at the top and bottom of the middle 50 percent of the distribution (the 75th and 25th percentiles). There is considerable dispersion or

**Table 4.7 Mean and Difference Statistics for
Earnings and Education, 1998 GSS**

	Earnings	Education
Mean	34,817.16	13.27
Mean squared difference	1,275,393,741.10	17.71
Standard difference	35,712.66	4.21
n	1,327	2,280

inequality in these variables. We will take up later the question of how to determine which variable—education or income—shows the greatest inequality.

We have used difference scores to measure the dispersion of a distribution: the difference between the maximum and minimum values of a variable (the range); the difference between the 75th percentile and the 25th percentile (the interquartile range); and an aggregate measure of the differences between all pairs of persons (the standard difference). The standard difference was introduced because it is a straightforward extension of the range and interquartile range as well as an intuitively appealing way to measure dispersion. The standard difference, however, is seldom used in statistical analyses. Instead, a closely related statistic called the standard deviation is the most commonly used measure of dispersion.

Standard Deviation

The standard deviation is based on a perspective on dispersion that is somewhat different from the range and the interquartile range (and the standard difference). Instead of using the difference between the observed scores of a pair of cases as the basic unit of dispersion, such as the difference between the maximum and minimum or the difference between the third quartile and the first quartile (or the difference between each pair of cases), the standard deviation uses the difference between a case's score and the mean of all the scores as the unit of dispersion. The difference between a score and the mean $(X - \overline{X})$ is a **deviation score.** Deviation scores were introduced in Chapter 3 in the discussion of the influence of each value of X on the mean; the greater the absolute value of a case's deviation score, the greater its influence. The standard deviation statistic is based on the difference between each case's score and the mean. Thus, the mean is a reference point to which each score is compared. The greater the deviations of the scores from the middle of the distribution, as measured by the mean, the greater is the dispersion of the distribution.

The basic units of this statistic, the deviation scores, are given in Table 4.8 for five illustrative values of education and earnings (Table 4.3 and Table 4.5). The deviation score indicates how much a case is above the mean (a positive deviation) or below the mean (a negative deviation). For example, the first education deviation score (-3.6) indicates that the person with 9 years of schooling is 3.6 years below the mean of 12.6; the fourth earnings deviation score (27,940) means that the worker with $64,800 in earnings is $27,940 above the mean of $36,860.

Table 4.8 Example Deviation Scores for Education and Earnings, *n* = 5

	Education					Earnings			
X	−	*X̄*	=	*X* − *X̄*	(*X* − *X̄*)²	*X*	−	*X̄*	=	*X* − *X̄*	(*X* − *X̄*)²

X	−	*X̄*	=	*X* − *X̄*	(*X* − *X̄*)²	*X*	−	*X̄*	=	*X* − *X̄*	(*X* − *X̄*)²
9	−	12.6	=	−3.6	12.96	17500	−	36,860	=	−19,360	374,809,600
12	−	12.6	=	−.6	.36	35200	−	36,860	=	−1,660	2,755,600
12	−	12.6	=	−.6	.36	26700	−	36,860	=	−10,160	103,225,600
14	−	12.6	=	1.4	1.96	64800	−	36,860	=	27,940	780,643,600
16	−	12.6	=	3.4	11.56	40100	−	36,860	=	3,240	10,497,600
		Total		0.0	27.20			Total		0	1,271,932,000

Notice that the **sum of deviations** from the mean equals 0 for both education and earnings. This is true for every variable:

$$\Sigma(X - \bar{X}) = 0$$

This is the mathematical equivalent of the idea of the mean as the center of mass of a distribution that was illustrated in Figure 3.7. The consequence of this fact is that we cannot use the sum of the raw deviation scores as a measure of dispersion because the sum is always the same—that is, zero. The solution to this problem is to use squared deviation scores. The reasons for preferring squared deviations over absolute values ($|X - \bar{X}|$) are that squared deviations are more mathematically tractable than absolute values and, more importantly, squared deviations minimize sampling error (more will be said on this topic later). After computing the squared deviation for each case, we sum them over all cases. This sum is called the **sum-of-squares** (SS):

$$\text{Sum-of-squares} = \text{SS} = \Sigma(X - \bar{X})^2 \qquad (4.5)$$

$$\text{Education SS} = 27.20$$

$$\text{Earnings SS} = 1,271,932,000$$

We divide the sum of squares by *n* to get the mean squared deviation, which is named the **variance:**

$$\text{Variance} = \frac{\text{SS}}{n} = \frac{\Sigma(X - \bar{X})^2}{n} \qquad (4.6)$$

$$\text{Education variance} = \frac{27.20}{5} = 5.44$$

$$\text{Earnings variance} = \frac{1,271,932,000}{5} = 254,386,400$$

Thus, the variance of education indicates that the mean of the squared deviations from the education mean is 5.44 years. The variance of earnings shows that the mean of all the squared earnings deviations from the mean of earnings is $254,386,400. This is a huge variance, but it indicates what can happen when we square deviations. The statistic is named *variance* because it measures how much cases *vary* from the mean, on the average, in squared units.

As we will see, the variance is used frequently in other statistics. As a measure of dispersion, however, it is not very helpful because it is expressed in squared deviations rather than the actual measurement scale of the variable in question. We can convert back to the original scale of measurement by taking the square root of the variance. The square root of the variance is the **standard deviation,** which is symbolized by *s*:

$$\text{Standard deviation} = s = \sqrt{\frac{\text{SS}}{n}} = \sqrt{\frac{\Sigma(X - \overline{X})^2}{n}} \tag{4.7}$$

$$\text{Education } s = \sqrt{5.44} = 2.33$$

$$\text{Earnings } s = \sqrt{254{,}386{,}400} = 15{,}949.50$$

The standard deviations of education and earnings indicate that the standard amount that persons deviate from the mean of education is 2.33 years and the standard amount that persons deviate from the earnings mean is $15,949.50. When interpreting the standard deviation, remember that it is always positive even though individual deviation scores can be either positive or negative. Thus, the standard deviation of education does not indicate that persons are 2.33 years *above* the mean, on the "average." It means that the standard amount that cases deviate from the mean is 2.33 years above or below the mean. Also, the greater the value of *s*, the greater the dispersion of the distribution. The larger the *s*, the wider the dispersion around the mean, while the smaller the *s*, the closer cases are clustered around the mean of the distribution.

Because the standard deviation is the square root of the variance, if we square the standard deviation, we will get the variance. Thus, the variance is symbolized by s^2. To summarize, the symbols for the measures of dispersion based on deviations from the mean are

$$\text{SS} = \text{Sum-of-squares}$$

$$s^2 = \text{Variance}$$

$$s = \text{Standard deviation}$$

The variance and standard deviation of earnings and education for the actual cases in the GSS are given in Table 4.9. The real values of these statistics are substantially greater for both variables than the values for the five-case example calculated earlier.

★Frequency Distribution Formula

When the data are in a frequency distribution, the standard deviation may be computed by a formula analogous to Equation 3.5 for the mean of a frequency distribution:

$$s = \sqrt{\frac{\Sigma f (X - \overline{X})^2}{n}} \tag{4.8}$$

For each value of *X* in the frequency distribution, the squared deviation from the mean is multiplied by its frequency. Summing these products across all values of *X* in the table gives the sum of squared deviations (SS). The variance is then computed by dividing by *n*, and the standard deviation equals the square root of the variance.

Table 4.9 Mean and Deviation Statistics for Earnings and Education, 1998 GSS

	Earnings	Education
Mean	34,817.16	13.27
Variance	637,216,315.26	8.85
Standard deviation	25,243.14	2.98
n	1,327	2,280

*Relationship between Standard Difference and Standard Deviation

The standard deviation (and variance) is based on the difference between each case and the mean, whereas the standard difference (and mean squared difference) uses the difference between each pair of cases. Although the deviation and difference statistics are based on different conceptualizations of dispersion, there is a mathematical relationship between the two types of measures. Notice that the variances and standard deviations in Table 4.9 are each smaller than the comparable values of the mean squared difference and the standard difference in Table 4.7. The variances of earnings and education are each about one-half the values of the mean squared difference, and the standard deviations are approximately 70 percent as large as the standard differences. The mathematical relationships between the deviation and difference statistics are

$$s^2 = \frac{n-1}{2n} \cdot \text{MSD} \tag{4.9}$$

$$s = \sqrt{\frac{n-1}{2n}} \cdot \text{Standard difference} \tag{4.10}$$

Substituting the sample size, MSD, and standard difference of earnings and education from Table 4.7 into Equations 4.9 and 4.10 gives

$$\text{Earnings } s^2 = \frac{1327-1}{2 \cdot 1327} \cdot \text{MSD} = \frac{1326}{2654} \cdot \text{MSD} = .499623 \cdot 1,275,393,741.10$$
$$= 637,216,315.26$$

$$\text{Earnings } s = \sqrt{.499623} \cdot \text{standard difference} = .70684 \cdot 35,712.66 = 25,243.14$$

$$\text{Education } s^2 = \frac{2820-1}{2 \cdot 2820} \cdot \text{MSD} = \frac{2819}{5640} \cdot \text{MSD} = .499823 \cdot 17.71 = 8.85$$

$$\text{Education } s = \sqrt{.499823} \cdot \text{standard difference} = .706981 \cdot 4.21 = 2.98$$

These values are equal to those in Table 4.9. Notice in the preceding equations the term $(n-1)/2n$ is only slightly less than .5 in these large sample sizes. When $n \geq 50$, $.49 \leq (n-1)/2n < .50$; that is, when n is greater than or equal to 50, $(n-1)/2n$ will be greater than or equal to .49 but less than .50. Thus, for reasonably large sample sizes, the variance will always be just slightly less than half the MSD. In computing the standard deviation, $\sqrt{(n-1)/2n} > (n-1)/2n$ because the square root of a fraction equals a larger fraction. Thus, as we have just seen, the standard deviation will be approximately .7 as large as the standard difference.

Equations 4.9 and 4.10 indicate that if you know either of the difference statistics, you can compute both of the deviation statistics. Also, if you know one of the deviation statistics, you can compute the difference statistics. For example,

$$\text{MSD} = \frac{2n}{n-1} \cdot s^2$$

The implication of this relationship is that the difference and deviation measures of dispersion are two sides of the same coin. They are interchangeable. Both have been introduced so you can have two somewhat different ways of thinking about measures of dispersion, either as differences between pairs of cases or as deviations of cases from the mean.

Goodness-of-Fit

In Chapter 3 we discussed three different measures of central tendency. Now we return to the question of which is better and what criterion is appropriate for drawing such a conclusion. Let's start with the question: "What single number M provides the best description of an entire distribution?" That number might be the mean, the median, the mode, or any other possible measure of the middle of the distribution—that is, its central tendency.

To answer the question, it is necessary to specify some criterion for measuring how well the value describes the distribution. This is referred to as measuring the goodness-of-fit of the statistic. A logical principle for a measure of goodness-of-fit is that the statistic should fit the observations as closely as possible *overall*—that is, taking into account all the scores in the distribution. We have already covered the basic unit for measuring fit or, more precisely, lack of fit: the **deviation score.** Whereas we considered only deviations from the mean, however, we now want to consider deviations from all possible measures of the "middle" (M) of the distribution—that is, $X - M$.

We will use two ways of summarizing the deviations from M:

$$\text{Sum of absolute deviations} = \Sigma |X - M|$$

$$\text{Sum of squared deviations} = \Sigma (X - M)^2$$

For both of these criteria of fit, we want to know what value of M minimizes the sum. The value of M that minimizes each sum is the value that provides the best fit or the best description of the distribution of X, according to each criterion. Minimizing the sum of absolute deviations is called the **least absolute deviations criterion,** and minimizing the sum of squared deviations is the **least–squares criterion.**

To illustrate, each measure of fit is calculated for both the mean and the median of the five-case earnings example (Table 4.10). The sum of absolute deviations (drop all minus signs before summing the deviations) from the mean equals $62,360, and the sum of absolute deviations from the median (the middle score in the five-case distribution is $35,200) equals $60,700. Thus, the sum of absolute deviations from the median is less than the sum of absolute deviations from the mean. Therefore, the median fits the earn-

**Table 4.10 Absolute and Squared Deviations
of Earnings from Mean and Median,[a] $n = 5$**

X	$|X - \bar{X}|$	$|X - Md|$	$(X - \bar{X})^2$	$(X - Md)^2$
17,500	19,360	17,700	374,809,600	313,290,000
35,200	1,660	0	2,755,600	0
26,700	10,160	8,500	103,225,600	72,250,000
64,800	27,940	29,600	780,643,600	876,160,000
40,100	3,240	4,900	10,497,600	24,010,000
Sum	62,360	60,700	1,271,932,000	1,285,710,000

[a]Mean = $36,860; median = $35,200.

ings distribution better than the mean, according to the absolute-deviations measure of goodness-of-fit.

Not only does the median fit better than the mean by this criterion, the median fits better than any other possible value of M. In general, $\Sigma|X - M|$ is minimized when M equals the median:

$$\Sigma|X - \text{median}| \text{ is minimal}$$

The least-absolute-deviations criterion specifies that the median provides the best description of the distribution of any interval/ratio variable.

The conclusion, however, is different for the least-squares criterion. To illustrate again with the earnings data in Table 4.10, the sum of squared deviations from the mean equals $1,271,932,000 compared with a sum of squared deviations from the median of $1,285,710,000. Therefore, when the goodness-of-fit is measured by the sum of squared deviations, the mean fits the data better than the median. In general, the mean provides a smaller sum of squared deviations than any other possible value of M. In other words, the $\Sigma(X - M)^2$ is minimized when $M = \bar{X}$:

$$\Sigma(X - \bar{X})^2 \text{ is minimal}$$

According to the least-squares criterion, the mean provides the best description of any interval/ratio distribution.

We have seen that the measure of central tendency that is best depends on the criterion selected to measure the goodness-of-fit, so the question of fit comes down to which criterion is best. On the face of it, the least-absolute-deviations criterion may appear to be best. Why square the deviations? Large deviations have even more influence on the measure of fit as a consequence of squaring (if one deviation is twice as large as a second deviation, it will be four times as large after squaring). However, in much research, the least-squares criterion is preferable to the least-absolute-deviations criterion. The reason has to do with sampling. The statistic that minimizes the squared deviations (the mean) will have less sampling error than the statistic that minimizes the absolute deviations (the median). Again, this justification will become clearer in the following chapters. If you are conducting research that doesn't involve sampling, however, the median may be the preferred measure of central tendency. The U.S. Census Bureau, for example, uses the median to describe distributions of variables such as age and income because it is describing the entire population of the United States instead of a sample from that population.

Coefficient of Variation

Which variable—education or income—has greater dispersion or is distributed more unequally? Because the earnings standard deviation equals $25,243 and the education standard deviation equals 2.98 years (Table 4.9), would you conclude that there is greater inequality of earnings than of education? No. The two variables are measured on different scales (dollars and years), and thus comparing their standard deviations, or their means for that matter, would be like comparing apples and oranges.

The problem, however, goes even deeper than whether the measurement scale is the same for the variables to be compared. Suppose you wanted to know whether there is more or less income inequality for females than for males. Earnings are measured on the same scale for each gender. Table 4.11 gives the 1997 occupational earnings of males and females who were working 35 or more hours a week in 1998. The male workers have a much higher mean (no surprise) and a much higher standard deviation (surprised?). Would you conclude that income dispersion or inequality is greater for males than females because males have a higher standard deviation?

A male who is 1 standard deviation above the mean would have an income of $40,022.90 + 28,349.49 = \$68,372.39$. Is his income greater relative to the mean than a woman's earnings that are 1 standard deviation above the female mean—that is, $28,987.73 + 19,707.09 = \$48,694.82$? The ratios of each person's earnings to the mean earnings are $68,372.39/40,022.90 = 1.71$ (male) and $48,694.82/28,987.73 = 1.68$ (female). The male who is 1 standard deviation above the mean has earnings that are 1.71 times as much as the mean earnings, while a female in the same position has earnings 1.68 times the mean female earnings. Thus, when you take into account the higher male mean, the male standard deviation is not greater relative to the male mean than the female standard deviation is relative to the female mean.

The point of this exercise is that you must control for any difference in means before comparing the standard deviations of two variables. This is done by dividing the standard deviation by the mean to get the **coefficient of variation** (CV):

$$CV = \frac{s}{\overline{X}} \tag{4.11}$$

$$\text{Female CV} = \frac{19,707.09}{28,987.73} = .68$$

$$\text{Male CV} = \frac{28,349.49}{40,022.90} = .71$$

The CVs indicate that the male and female standard deviations are virtually identical relative to their respective means. Thus, there is no difference in income inequality between males and females.

Now, let's return to the initial question of whether income or education inequality is greater. One consequence of measuring income in dollars and education in years is that the income mean is greater than the education mean. If education were measured in total

Table 4.11 Earnings by Sex, 1998 GSS

Sex	n	Mean	Standard Deviation
Female	626	28,987.73	19,707.09
Male	701	40,022.90	28,349.49
Total	1327	34,817.16	25,252.66

lifetime *minutes* of schooling, however, the education mean would be greater than the earnings mean. If we divide the income and education standard deviations by their respective means, we will correct for the effects of the difference in scales of measurement, giving

$$\text{Education CV} = \frac{2.98}{13.27} = .225$$

$$\text{Income CV} = \frac{25243.14}{34817.16} = .725$$

The coefficient of variation for income is more than three times as great as that for education, which indicates that income is distributed much more unequally than education. Although educational differences are undoubtedly one important cause of why people have different incomes, the dispersion of income is much greater than we would expect on the basis of educational dispersion.

Assumption We must make one important qualification about the use of the coefficient of variation. *The coefficient should be used only for ratio variables.* Remember, ratio variables have a true zero point. The true zero value is a feature of a number scale that is necessary for conducting division of the numbers. When you divide one number by another number from the same numeric scale, the ratio of the two indicates how many times greater the numerator is than the denominator. This makes sense only if there is an absolute zero value. Thus, to divide the standard deviation of a distribution of scores by the mean of the distribution to get the CV, the scores must be based on a ratio level of measurement. Education, as measured by years of schooling completed, is a ratio variable. Earnings, strictly speaking, is not, because self-employed workers may lose money in their occupations; that is, they may have earnings less than zero. Only a small percentage of workers, however, are self-employed, and only a small minority of these probably have a negative net income during any given year. Therefore, practically speaking, comparing earnings inequality with education inequality is a valid operation.

ORDINAL VARIABLES

The computation of deviation scores used in the standard deviation and related statistics requires interval/ratio variables. With ordinal variables we can know only the ranks of the attributes, and thus we cannot calculate the difference between attributes or the deviation of an attribute from a measure of central tendency, such as the median.

There are no measures of ordinal or rank dispersion. That is because the ranks of every ordinal variable are uniformly distributed. Thus, all ordinal variables have an equal amount of ordinal dispersion. The measure of dispersion described in the next section on nominal variables will have to suffice for measuring the dispersion of ordinal variables.

NOMINAL VARIABLES

Qualitative Dispersion

To illustrate the measurement of dispersion of interval variables, we looked at the earnings of full-time workers. In this section we will examine how common it is to work full time and what is the labor-force status of those who are not working full-time. The GSS determines the labor-force status of each person by their response to the question: "Last week, were you working full time, part time, going to school, keeping house, or what?" The labor-force status of males in the 1998 GSS is shown in Table 4.12. The first three categories in the table are considered to be members of the labor force. Some of those in the third category have jobs but are temporarily not working due to illness, vacation, strike, or some other reason. Others in the third category do not have jobs but are considered to be unemployed members of the labor force because they want to work and have recently taken some action to find a job. Although some of the categories may be ranked (working full time is higher than part time, and part time is higher than all of the other categories), there is no rank ordering among the last four statuses, all of whose members are currently not working for income. Thus, labor-force status is a nominal variable.

Although there are six categories of labor-force status, slightly more than two-thirds of the men are concentrated in a single status, working full time. Thus, the differences between men in their work statuses are not extensive because a large majority of them are doing the same thing: working full time. We can readily perceive what the distribution of a nominal variable (or any variable) would look like if there were no variation at all; all cases would be in one category. What would a distribution look like if it had the maximum possible variation? Figure 4.3 shows the distribution for no variation (100% working full time, or in any category), the observed variation (Table 4.12), and the maximum variation (16.17% in each category). The most dispersed distribution of a *nominal* variable would have an equal percentage of cases in each category (16.67% for six categories). Is the observed distribution closer to no variation or to maximum variation? It appears to be closer to no variation. Now, let's develop a quantitative measure of nominal dispersion.

Index of Qualitative Variation

When we introduced the measurement of dispersion for interval variables, we used pairs of individuals to measure dispersion—that is, the range and the interquartile range (and the standard difference). For a particular pair of individual scores, for example, the maximum and minimum or the third quartile and the first quartile, we computed the difference in their values. The measure of dispersion for nominal variables also uses pairs

Table 4.12 Labor-Force Status of Males, 1998 GSS

	f	Pct
Working full time	826	68.4
Working part time	88	7.3
Temporarily not working or unemployed	65	5.4
Retired	179	14.8
In school	33	2.7
Keeping house	16	1.3
Total	1207	100.0

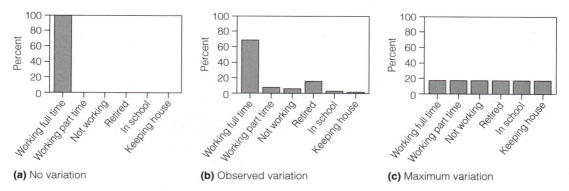

(a) No variation   (b) Observed variation   (c) Maximum variation

FIGURE 4.3 Percentage Distributions of Labor Force Status by Amount of Variation

of individuals as the unit of dispersion. Unlike interval variables, however, with nominal variables we cannot compute differences in values of X because there is no quantified interval between nominal attributes. With a nominal variable we are able to determine only whether the persons in a pair both have the same attribute or have different attributes. One pair of persons, for example, may both be full-time workers (i.e., same attribute), whereas another pair may consist of a part-time worker and a retired person (i.e., different attributes). Thus, in our measure of nominal-variable dispersion, we simply count the total number of pairs that have different attributes. The total number of pairs that are different will provide a measure of the variation or differences in the observed distribution (e.g., Figure 4.3b). The observed number of differences are then divided by the maximum possible number of differences, which is the number of differences there would be in Figure 4.3c. Thus, the **index of qualitative variation** (IQV) can be determined as follows:

$$\text{Index of qualitative variation} = \frac{\text{number of pairs of cases that differ}}{\text{maximum number of pairs that can differ}}$$

Table 4.13 shows how to determine the number of pairs of persons that have different attributes. For each pair of *attributes,* multiply the number of cases with one attribute (f_i) times the number with the second attribute (f_j). For example, there are $f_1 = 826$ males who are full-time workers and $f_2 = 88$ males who are part-time workers. The product, $f_1 f_2 = 72{,}688$, indicates the number of pairs of persons in which one male is a full-time worker and the other is a part-time worker. Thus, there are 72,688 pairs that exhibit this specific difference in work status. The second pair of attributes shows that there are $f_1 f_3 = 826 \cdot 65 = 53{,}690$ pairs in which one person is a full-time worker and the other is not working. This represents 53,690 more pairs that differ in work status. We continue through all additional pairs of attributes, taking the product of the frequencies of each attribute for each pair. The last product indicates that there are $f_5 f_6 = 33 \cdot 16 = 528$ pairs in which one person is in school and the other is keeping house. In total there are fifteen pairs of work statuses and thus fifteen pairs of products that must be computed:

$$\text{Number of pairs of attributes} = \frac{g(g-1)}{2} = \frac{6 \cdot 5}{2} = 15$$

where g equals the number of attributes or groups in a nominal variable.

After computing all products of pairs of attribute frequencies, the products are summed to give the total number of pairs of cases having different attributes. The formula is thus

$$\text{Number of pairs with different attributes} = \Sigma f_i f_j$$

This sum equals 358,889 pairs for work status (Table 4.13).

The distribution that would have the maximum possible number of differences is one in which there are an equal number of cases that have each attribute (Figure 4.3c). For example, a work-status distribution in which the frequencies for working full time, working part time, not working, retired, in school, and keeping house are all equal (a very, very unlikely distribution) would have the maximum number of different pairs. Such a distribution is a *uniform distribution.* How many persons would there be in each category of a uniform distribution if each category had an equal frequency (f_i)? In general, the mean number of cases per category equals n/g. Thus, if each category had an equal number of cases, there would be $n/g = 1207/6 = 201.17$ cases in each category of the work-status distribution. Even though it is not possible to observe a fractional number of cases (f_i), the digits to the right of the decimal point should be retained in the computations.

If there are $f_i = n/g$ cases per category, then the product $f_i f_j = (n/g)(n/g) = (n/g)^2$ for each pair of categories. Because there are $g(g-1)/2$ pairs of categories, the sum of products for all pairs of categories in a uniform distribution is

$$\Sigma f_i f_j = \frac{g(g-1)}{2} \cdot \left(\frac{n}{g}\right)^2 \text{ in a uniform distribution}$$

And thus,

$$\text{Maximum number of pairs that can differ} = \frac{g(g-1)}{2} \cdot \left(\frac{n}{g}\right)^2$$

$$= \frac{6(6-1)}{2} \cdot \left(\frac{1207}{6}\right)^2 = 15 \cdot (201.17)^2$$

$$= 15(40{,}468.03) = 607{,}020.42$$

Table 4.13 Number of Pairs that Differ in Work Status (Males)

i, j	X_i	X_j	f_i	×	f_j	=	$f_i f_j$
	Pairs of Attributes						
1, 2	Full time	Part time	826	×	88	=	72,688
1, 3	Full time	Not working	826	×	65	=	53,690
1, 4	Full time	Retired	826	×	179	=	147,854
1, 5	Full time	In school	826	×	33	=	27,258
1, 6	Full time	Keep house	826	×	16	=	13,216
2, 3	Part time	Not working	88	×	65	=	5,720
2, 4	Part time	Retired	88	×	179	=	15,572
2, 5	Part time	In school	88	×	33	=	2,904
2, 6	Part time	Keep house	88	×	16	=	1,408
3, 4	Not working	Retired	65	×	179	=	11,635
3, 5	Not working	In school	65	×	33	=	2,145
3, 6	Not working	Keep house	65	×	16	=	1,040
4, 5	Retired	In school	179	×	33	=	5,907
4, 6	Retired	Keep house	179	×	16	=	2,864
5, 6	In school	Keep house	33	×	16	=	528
Total							358,889

Taking the ratio of the observed number of pairs that differ to the maximum number that can differ gives the formula for the Index of Qualitative Variation (IQV):

$$IQV = \frac{\Sigma f_i f_j}{\frac{g(g-1)}{2} \cdot \left(\frac{n}{g}\right)^2} = \frac{358,889}{607,020.42} = .591 \tag{4.12}$$

The index of qualitative variation for the work status distribution of males equals .59, which indicates that their work statuses are dispersed only .59 as much as the maximum possible dispersion.

The maximum and minimum values of IQV are 1 and 0, respectively. If $\Sigma f_i f_j$ equals maximum dispersion, then the numerator of IQV will equal its denominator and IQV will equal 1. If there is no dispersion—that is, if all the cases are in one category—then for every pair of attributes $f_i f_j = 0$ because either f_i, f_j or both equal 0. Thus, $\Sigma f_i f_j = 0$ and IQV = 0. When we interpret IQV, the closer its value is to 0, the less the dispersion; the closer it is to 1, the greater the dispersion.

Now, let's compare the labor-force status of males to that of females. Compare the percentages of females in the various work statuses (Table 4.14) with those of males (Table 4.12), and compare the bar chart for females (Figure 4.4) with that of males (Figure 4.3b). Who do you think has the more varied or diverse work statuses, females or males?

A smaller percentage of females are working full time, the status in which two-thirds of males are concentrated. The smaller percent of full-time workers for females is compensated for primarily by higher percents working part time and keeping house—two

**Table 4.14 Labor-Force Status
of Females, 1998 GSS**

	f	Pct
Working full time	742	47.3
Working part time	216	13.8
Temporarily not working or unemployed	39	2.5
Retired	215	13.7
In school	46	2.9
Keeping house	312	19.9
Total	1570	100.0

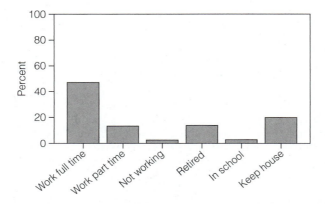

FIGURE 4.4 Labor Force Status of Females, 1998 GSS

categories that are quite small for males. Thus, females are more equally dispersed across the work statuses than males, which means they should have a higher index of qualitative variation than males. This is in fact the case. The female IQV = .838 (calculation not shown), indicating that the female variation in work status is .838 of the maximum possible dispersion. And the female variation is .838/.591 = 1.42 times greater than the male variation.

STANDARDIZED SCORES

One of the uses of a measure of variation is to compare the dispersion of one variable with that of another. We saw, however, that it was inappropriate to directly use a measure of dispersion, such as the standard deviation, to compare variables measured on different scales (such as education and earnings) or with significantly different means. Instead, the standard deviation was divided by its mean to control for differences in scales or means, and the resulting statistic, the coefficient of variation, was used to compare the dispersion of different variables. We also used the index of qualitative variation to compare the dispersion of different groups (i.e., males and females) on the same nominal variable (work status), which is appropriate because the IQV expresses the observed differences as a proportion of the maximum possible differences.

We will now examine how to make another type of comparison across variables. In this case, however, it is of interest to compare the score of an individual case on two or more variables. For example, how can we compare a person's education with the person's earnings? Does a woman with 16 years of education and an annual income of $50,000 stand higher in the education distribution or in the earnings distribution? How is it possible to compare a man's years of schooling with his earnings?

To compare scores on different variables, we must take into account any difference in dispersion between the variables. The basic principle is that the relative position of any specific value of X is inversely related to the dispersion of X; the greater the dispersion of X, the closer any X will be to the mean relative to the other cases in the distribution. For example, Figure 4.5 shows two hypothetical earnings distributions (they are not realistic because earnings is right-skewed), each with a mean of $35,000. One distribution, however, has a standard deviation of $15,000, which is 50 percent greater than the standard deviation of $10,000 for the other distribution. Consider a person with earnings of $65,000. In both distributions that person is $30,000 above the mean in absolute dollars. In the distribution with the smaller variation, however, $65,000 is virtually at the top of the distribution, whereas in the other distribution it would take an income of $80,000 to be in the same relative position. Thus, a person with $65,000 will be better off financially in the distribution with a standard deviation of $10,000 because there will be fewer persons with an income that high.

Now, consider a person at the other end of the distribution who earned only $5,000. That person would be at the bottom of the distribution with a standard deviation of $10,000, whereas a person would have to lose $10,000 to be at the bottom of the distribution with a standard deviation of $15,000. The person with $5,000 is worse off income-wise in the distribution with a standard deviation of $10,000. Thus, Figure 4.5 illustrates the principle that the greater the standard deviation from the mean, the closer to the mean will be a case's position relative to others.

We will now consider how to adjust raw values of X mathematically to take this principle into account. First, we calculate deviation scores by subtracting the mean from each value of X. The deviation scores tell us how far above or below the mean is each value of X *in units of its scale of measurement.* Then we adjust for the amount of dispersion in the distribution by dividing the deviation score by the standard deviation. The result is a **standardized score,** which is symbolized by **z:**

$$z = \frac{X - \overline{X}}{s} \tag{4.13}$$

As a consequence of dividing the deviation score by s, the larger the dispersion of the distribution, the smaller will be the z score, in absolute value, for any value of X.

The calculation of the standardized scores for the five-case example involving education and earnings is shown in Table 4.15. For education, the deviation scores range from 3.6 years below the mean of 12.6 for the case with 9 years of schooling to 3.4

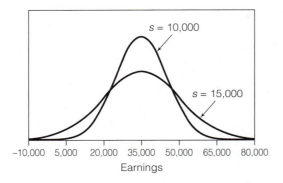

FIGURE 4.5 Hypothetical Earnings Distributions with Different Standard Deviations and Equal Means

Table 4.15 Standardized Scores (z) for Example Scores of Education and Earnings, $n = 5$

Education					Earnings				
$\dfrac{X - \bar{X}}{s}$	=	...	= z	z^2	$\dfrac{X - \bar{X}}{s}$	=	...	= z	z^2
$\dfrac{9 - 12.6}{2.33}$	=	$\dfrac{-3.6}{2.33}$	= -1.543	2.382	$\dfrac{17,500 - 36,860}{15,949.50}$	=	$\dfrac{-19,360}{15,949.50}$	= -1.214	1.473
$\dfrac{12 - 12.6}{2.33}$	=	$\dfrac{-.6}{2.33}$	= $-.257$	.066	$\dfrac{32,200 - 36,860}{15,949.50}$	=	$\dfrac{-1,660}{15,949.50}$	= $-.104$	.011
$\dfrac{12 - 12.6}{2.33}$	=	$\dfrac{-.6}{2.33}$	= $-.257$	.066	$\dfrac{26,700 - 36,860}{15,949.50}$	=	$\dfrac{-10,160}{15,949.50}$	= $-.637$	.406
$\dfrac{14 - 12.6}{2.33}$	=	$\dfrac{1.4}{2.33}$	= .600	.360	$\dfrac{64,800 - 36,860}{15,949.50}$	=	$\dfrac{27,940}{15,949.50}$	= 1.752	3.069
$\dfrac{16 - 12.6}{2.33}$	=	$\dfrac{3.4}{2.33}$	= 1.458	2.125	$\dfrac{40,100 - 36,860}{15,949.50}$	=	$\dfrac{3,240}{15,949.50}$	= .203	.041
Totals			.001	4.999				.000	5.000

years above the mean for the case with 16 years of schooling. Each deviation score is divided by the standard deviation of 2.33 years. The first person deviates from the mean by -3.6 years, which is greater than the standard deviation from the mean (2.33). Dividing -3.6 by 2.33 gives a z of -1.543, which indicates that this person is 1.543 times further from the mean than the standard deviation from the mean. The negative sign for $z = -1.543$ indicates that the person is below the mean. This value can be interpreted thus: *a person with 9 years of schooling is 1.543 standard deviations below the mean.* The z of 1.458 for the top person in the distribution indicates that this person is 1.458 standard deviations above the mean.

With respect to the standardized earnings in Table 4.15, the person with the lowest earnings is 19,360 below the mean, which is greater than the standard deviation of 15,949.50. To be specific, the person's z score indicates that he or she is 1.214 times the standard deviation below the mean. The highest standardized earnings score is for $64,800, which is 1.75 standard deviations above the mean earnings. In general, *a stan-dardized score* *indicates how many standard deviations a case is above or below the mean.*

The objective of transforming raw scores into standardized scores is to make comparisons across distributions. The person with the fewest years of education (9) also happens to have the lowest income (17,500). Although the z scores for this case are similar for education and earnings, 9 years of schooling is a little further below the mean (-1.543 standard deviations) than $17,500$ is below the mean (-1.214 standard deviations). When we compare the z scores of the person with the highest earnings, we see that whereas the earnings of $64,800 is 1.752 standard deviations above the mean, that person's education (14 years) is only .6 of a standard deviation above the mean. There is no necessity for a case's standardized scores on two variables to be identical or even similar. They may even have opposite signs, indicating the case is above the mean on one variable and below the mean on the other.

We will now consider some characteristics of the *distribution of standardized scores.* Notice in Table 4.15 that the sum of the z scores equals 0 for both education and earnings ($\Sigma z = .001$ for education due to rounding) because the sum of the deviation scores is always 0 [$\Sigma(X - \overline{X}) = 0$] as shown in Table 4.8. It is the reason the mean is described as the *center of mass.* Dividing each deviation score by a constant, the standard deviation, does not alter this zero–sum trait. Because the z scores sum to zero, their mean will equal zero:

$$\overline{z} = \frac{\Sigma z}{n} = \frac{0}{n} = 0 \qquad (4.14)$$

The mean of the standardized scores equals zero for every variable. This is one reason the z scores are called *standardized* scores and can be compared across variables. It is also the reason that the sign of the standardized score indicates whether the case is above or below the mean.

What will the **standard deviation of z** scores equal? Notice in Table 4.15 that the sum of the squared z's is 5, which is the value of n. It is always the case that $\Sigma z^2 = n$. We will now substitute z for X in Equation 4.7 to determine the standard deviation of z:

$$s_z = \sqrt{\frac{\Sigma(z - \overline{z})^2}{n}} = \sqrt{\frac{\Sigma z^2}{n}} = \sqrt{\frac{n}{n}} = \sqrt{1} = 1 \qquad (4.15)$$

Because $\overline{z} = 0$ and $\Sigma z^2 = n$, the formula for the standard deviation reduces to $s_z = 1$. That is, the standard deviation of a standardized-score distribution is always equal to unity. This is the second reason z scores are called *standardized* scores and the reason we may legitimately compare a standardized score from the distribution of one variable with the standardized value of a score from an entirely different variable.

Now, let's use the means and standard deviations of education and earnings from the 1998 GSS given in Table 4.9 to standardize some values of these variables. Would 16 years of education (a college graduate) be comparable to annual earnings of $100,000 in 1998, on the average?

$z = (16 - 13.27)/2.98 = 2.73/2.98 = .92$ (education = 16)

$z = (100,000 - 34,817.16)/25,243.14 = 65,182.84/25,243.14 = 2.58$
(earnings = $100,000)

The college graduate would be almost a standard deviation above the mean ($.92s$, to be precise), but \$100,000 of earnings would be about 2.5 standard deviations ($2.58s$) above the mean of the earnings distribution. Thus, \$100,000 is much higher in the earnings distribution than a college graduate is in the education distribution. The college graduate would have to be doing a great deal better in the labor market than he or she did in school to earn \$100,000. This doesn't mean that it can't be done, but it does suggest that it is unlikely for the average graduate.

So, what would the earnings be of a college graduate who is doing just as well, but no better, in the earnings distribution than in the education distribution? That is, because a college graduate is .92 standard deviation above the education mean, what would the graduate earn if he or she were also .92 standard deviation above the earnings mean? We can re-express the formula for z scores to show X as a function of z to help us find the answer to this question.

$$X = z \cdot s + \overline{X}$$

We should substitute the earnings mean and standard deviation and the z score for 16 years of education into this equation to find the earnings that would be comparable to 16 years of education.

$$\text{Earnings} = (.92)\,(25{,}243.14) + 34{,}817.16 = \$58{,}040.85$$

This result indicates that *if* college graduates do as well in earnings as in education, they would earn \$58,040.85 (although not necessarily right after leaving school). Later in the book we will consider statistical techniques to find out if it approximates reality.

MEASURES OF THE SHAPE
OF A DISTRIBUTION

Thus, far, we have been concerned with measuring the amount or magnitude of dispersion in a distribution. Distributions with equal amounts of dispersion, however, may differ dramatically in the shape or pattern of that dispersion. We have considered graphically one aspect of the shape of a distribution, namely, its *skewness* or lack of symmetry. In the next section we will present a quantitative measure for the degree of skewness. The following section will take up a measure of a new characteristic of a distribution's shape, its *peakedness*.

Skewness

As we saw in Chapter 2, a histogram that has a longer tail to the right or upper end of its distribution is *right-skewed*. When the left tail is longer than the right, the distribution is *left-skewed*. When the tails are identical on each side of the center, the distribution is *symmetrical*. The mathematical counterpart of a longer tail to one side or the other is greater *deviation scores* on the side with the longer tail. There are more extreme deviations from the *center* of the distribution on one side than there are on the other. Thus, deviation scores may be used to develop a quantitative measure of skewness. Because we would like to use our measure to compare the skewness of different variables, however, standardized deviation scores, or z scores (i.e., deviation scores divided by the standard deviation) will be used instead.

Table 4.16 gives the standardized scores and their second, third, and fourth powers for the five-example cases of education and income. Note, as review, that the sum of the z's equals zero, and thus the means of the z scores equal zero for both education and earnings. Now, compare the most extreme z scores on each side of the mean for both variables. The highest and lowest z scores for education are about equal in absolute value; 9 years of education and 16 years are each about one-and-a-half standard deviations from the mean, which would suggest that education is a fairly symmetrical variable. The highest z score for earnings, however, is substantially greater than the lowest; the highest earnings are about 1.7 standard deviations above the mean, whereas the lowest earnings are only 1.2 standard deviations below the mean. Thus, earnings appear to be right-skewed.

We cannot, however, use the z scores themselves to measure skewness because, again, they sum to zero; the greater extreme positive z for earnings is offset exactly by a greater number of negative z scores. We can, however, use the squared standardized scores to construct a measure of skewness. Squaring increases z scores that have large absolute values (including large negative values) proportionately more than it increases small z scores if $|z|$ is greater than 1. For example, for earnings of $17,500, z^2 is $1.473/1.214 = 1.213$ times greater than z in absolute value. But, for earnings of $64,800, z^2 is $3.069/1.752 = 1.752$ times greater than z. Squaring stretches the largest positive z proportionately more than it stretches the smallest negative z. If $|z| < 1$, squaring shrinks larger values *less* than it shrinks smaller values. The consequence is that the sum of the squared negative z's does not equal the sum of the squared positive z's. If the more extreme z's are above the mean, the sum of the squared positive z's will be greater than the sum of the squared negative z's. For example, the sum of z^2 for $z < 0$ is $1.473 + .011 + .406 = 1.890$, whereas the sum of z^2 for $z > 0$ is $3.069 + .041 = 3.110$. We can, therefore, construct a measure of skewness by subtracting the sum of squared values for the negative z's from the sum of the squared values for the positive z's and dividing the difference by the number of cases. For example, earnings skewness would equal $(3.110 - 1.890)/5 = 1.220/5 = .244$. The positive result indicates that the earnings variable is right-skewed.

The conventional measure of skewness, however, uses a mathematically more straightforward approach. **Skewness** *is defined as the mean of the cubed standardized scores:*

$$\text{Skewness} = \frac{\Sigma z^3}{n} = \overline{z^3} \tag{4.16}$$

In Table 4.16 notice what happens when the scores are cubed or raised to the third power. The cubed values have the same sign as the original z scores. For education,

$$(-1.543)^3 = (-1.543)(-1.543)(-1.543) = (2.382)(-1.543) = -3.677$$

And, in general,

$$z^3 = z \cdot z \cdot z = z^2 \cdot z$$

Because z^2 is always positive, the sign of z^3 is determined by the sign of z itself.

Table 4.16 Second, Third, and Fourth Powers of Standardized (z) Education and Earnings, n = 5

		Education					Earnings		
X	z	z^2	z^3	z^4	X	z	z^2	z^3	z^4
9	−1.543	2.382	−3.677	5.676	17,500	−1.214	1.473	−1.788	2.171
12	−.257	.066	−.017	.004	35,200	−.104	.011	−.001	.000
12	−.257	.066	−.017	.004	26,700	−.637	.406	−.258	.165
14	.600	.360	.216	.130	64,800	1.752	3.069	5.376	9.417
16	1.458	2.125	3.098	4.516	40,100	.203	.041	.008	.002
63	.001	4.999	−.397	10.330	184,300	.000	5.000	3.336	11.755

Now, if we sum the cubed scores, the sign of the sum will depend upon which is greater, the sum of the negative cubed z's or the sum of the positive cubed z's. Because raising z to the third power will stretch extreme values of z even more than it will stretch less extreme z's (in absolute value), the sign of the sum will indicate whether the distribution is right-skewed or left-skewed. If the sum is positive, the distribution is right-skewed. If the sum is negative, the distribution is left-skewed. And if the sum equals zero, we have a symmetrical distribution. The values of the skewness statistic for education and earnings in Table 4.16 are

$$\text{Education skewness} = \frac{-.397}{5} = -.079$$

$$\text{Earnings skewness} = \frac{3.336}{5} = .667$$

The small negative value for education indicates that it is slightly left-skewed but nearly symmetrical. The substantially larger and positive value for earnings indicates that earnings is skewed appreciably more than education and in the opposite direction.

Because the skewness statistic is positive for right-skewed distributions, this skewness is often called a *positive skew,* and because the statistic is negative for left-skewed distributions, such distributions are said to have a *negative skew.* So, right skew and positive skew refer to the same type of skewness, and left skew and negative skew indicate the same skewness.

Table 4.17 gives the descriptive statistics for the real earnings and education distributions in the 1998 GSS. The skewness for education is close to zero or symmetry, but slightly negative, just as in the preceding example data. The skewness for earnings is positive, just as in the example data, but it is much stronger, which indicates that ultra-high earnings, which stretch the right tail far away from the mean, relative to the length of the left tail, is a striking feature of the earnings distribution in the United States.

Peakedness

The second measure of the shape of a distribution refers to its peakedness or **kurtosis.** Peakedness deals with the steepness of a histogram's slopes on either side of the center of the distribution. The steeper the slopes, the more peaked the distribution. In Figure 4.5 on p. 116, the distribution with a standard deviation of 10,000 has steeper slopes,

**Table 4.17 Descriptive Statistics
for Earnings and Education, 1998 GSS**

	Earnings	Education
Mean	34,817.16	13.27
Standard deviation	25,243.14	2.98
Skewness	2.05	−.14
Kurtosis	5.96	1.19
n	1,327	2,280

and so it is tempting to conclude that it has greater kurtosis. That is an incorrect conclusion, however, because there is a second aspect to kurtosis that must be considered. The steeper slopes must also be accompanied by longer tails. So, kurtosis refers to the degree to which cases are packed around the center creating steep slopes *and* also distributed far away from the center as evidenced by long tails. In Figure 4.5, the distribution that is more peaked does *not* have longer tails; it has shorter tails. Its greater peakedness results from its smaller standard deviation, which, in this case, also pulls in its tails. The concept of peakedness or kurtosis must not be confused with dispersion.

Table 4.18 shows two distributions that help to illustrate the concept of kurtosis. The first distribution has five values of X with intervals of 1 between each successive value (a uniform distribution). The sum of squared deviations around the mean of 3 equals 10, and thus the variance equals $10/5 = 2$. In the second distribution, the two scores on either side of the mean are half a point closer to the mean. Thus, the distribution is more closely clustered around the center (i.e., it is more peaked). The highest and lowest values, however, are a little further from the mean than in the first distribution (i.e., it has longer tails). The sum of squared deviations is the same as for the first distribution. Thus, both distributions have the same amount of variance and the same standard deviation. Notice, however, that the fourth power of the highest and lowest deviation scores is greater in the second distribution. As a result, the sum of the fourth power of the deviation scores is greater in the second distribution. Thus, we have two distributions with equal variances. The second variable fits our concept of a distribution with greater kurtosis. And most importantly, the sum of the fourth power of the deviations reflects the greater kurtosis of the second distribution.

The measure of kurtosis equals the mean of the fourth power of the standardized scores (z) minus 3:

$$\text{Kurtosis} = \frac{\Sigma z^4}{n} - 3 = \overline{z^4} - 3 \qquad (4.17)$$

Standardized scores are used in place of the raw deviation scores so that the measure of kurtosis can be used to compare the kurtosis of different variables that may have different standard deviations. Because the fourth power of any variable is always positive (just as squared scores are positive), $\Sigma z^4 \geq 0$ and thus the minimum value of kurtosis equals -3. The greater the value of the kurtosis statistic, the more peaked the distribution. The values of kurtosis for the education and earnings variables in Table 4.16 are

Table 4.18 Example Distributions for Illustrating Kurtosis

Less Peaked Distribution				More Peaked Distribution			
X	$X - \bar{X}$	$(X - \bar{X})^2$	$(X - \bar{X})^4$	X	$X - \bar{X}$	$(X - \bar{X})^2$	$(X - \bar{X})^4$
1	-2	4	16	.82	-2.18	4.75	22.5625
2	-1	1	1	2.50	$-.50$	.25	.0625
3	0	0	0	3.00	.00	.00	.0000
4	1	1	1	3.50	.50	.25	.0625
5	2	4	16	5.18	2.18	4.75	22.5625
Totals	0	10	34		.00	10.00	45.2500

$$\text{Education kurtosis} = \frac{10.330}{5} - 3 = 2.066 - 3 = -.934$$

$$\text{Earnings kurtosis} = \frac{11.754}{5} - 3 = 2.351 - 3 = -.649$$

These results show that although earnings is slightly more peaked than education, these two hypothetical distributions have relatively similar amounts of kurtosis. For the real education and earnings variables summarized in Table 4.17, the kurtosis for earnings (5.96) is considerably greater than that for education (1.19).

In Chapter 5 we will describe the characteristics of a distribution called the **normal distribution.** The normal distribution plays an important role in *inferential statistics.* The shape of the normal distribution also serves as a model to which other distributions may be compared. *The kurtosis statistic equals 0 in a normal distribution.* Thus, positive values of kurtosis indicate greater peakedness than a normal distribution, and negative values of kurtosis indicate less peakedness than a normal distribution. Because the values of kurtosis for earnings and education in Table 4.17 are both greater than 0, earnings and education are more peaked than a normal distribution.

SUMMARY

In this chapter we learned ways to measure the amount of dispersion in a distribution. Three measures of dispersion were covered for interval/ratio variables: the range, the interquartile range, and the standard deviation (or the standard difference). The range equals the difference between the highest and lowest scores in the distribution. The interquartile range equals the difference between the third quartile (75th percentile) and the first quartile (25th percentile). The square root of the mean squared difference between all pairs of cases is called the standard difference. And finally, the square root of the mean of the squared deviations between each score and the mean is called the standard deviation. The standard difference and the standard deviation are interchangeable measures of dispersion because each statistic can be easily computed from the other; the standard deviation, for example, is approximately equal to .7 times the standard difference. Although the standard deviation is chosen much more often by investigators, the standard difference shows that all these measures of dispersion are based on differences of scores. The discussion of goodness-of-fit indicated that the sum of squared deviations around the mean, which is the basis for the standard deviation, is smaller than the sum of squared deviations around any other value of the variable, including the median. The coefficient of variation, which equals the standard deviation divided by the mean, can be used to compare the amount of dispersion on one variable with that of another.

The preceding measures of dispersion for interval/ratio variables cannot be used for ordinal or nominal variables because subtraction is not a valid arithmetic operation for the latter two levels of measurement. The index of qualitative variation is a measure of dispersion that can be used for ordinal and nominal variables. The index is based on comparing all pairs of cases, just as the standard difference does. Instead of computing the difference between each pair, however, the index simply counts the number of pairs that have different attributes. The index of qualitative variation equals the number of pairs that are different divided by the maximum number of pairs that could be different. Because it is a ratio, it ranges from 0 to 1.

Standardized scores are not used to measure dispersion but instead are used to measure each case's position in a distribution. The standardized score of a case equals the case's observed score minus the mean (a deviation score) divided by the standard deviation. As such, it indicates how many standard deviations the case is above or below the mean, depending on the sign of the standardized score. Standardized scores can also be used to compare a case's positions on two or more variables. Two measures of the shape of a distribution are based on standardized scores. The amount of skewness in a distribution equals the mean of the cubed standardized scores. Values greater than zero indicate a right-skewed distribution, and negative values indicate a left-skewed distribution. The measure of peakedness of the distribution, called kurtosis, equals the mean of the fourth power of the standardized scores minus 3. Values greater than 0 are more peaked than a normal distribution, and values less than 0 are less peaked.

KEY TERMS

dispersion	sum of deviations	index of qualitative variation
range	sum-of-squares	
interquartile range	variance	standardized score (z)
★mean squared difference	standard deviation	standard deviation of z
★standard difference	least-squares criterion	skewness
	coefficient of variation	kurtosis

SPSS PROCEDURES

We will compute measures of dispersion and shape for the years of education variable, named *educ22,* that was created by recoding the open-ended upper category as described in footnote 1. We can use *Frequencies* to compute these statistics. Click on *Analyze, Descriptive Statistics,* and *Frequencies.* In the *Frequencies* window, move *educ22* to the

Variable(s) box, click off *Display frequency tables,* click on *Statistics* to open the *Frequencies: Statistics* window, and click on *Quartiles, Std. deviation, Variance, Range, Skewness,* and *Kurtosis.* The first and third quartiles can be used to compute the interquartile range.

We also can use *Descriptives* to get measures of dispersion and shape, plus standardized scores. Click on *Analyze, Descriptive Statistics,* and *Descriptives.* In the *Descriptives* window, move *educ22* to the *Variable(s)* box, click on *Save standardized values as variables,* and click on *Options.* In the *Descriptives: Options* window, click on *Std. deviation, Variance, Range, Skewness,* and *Kurtosis* and then click *Continue.* The selected statistics will appear in the *Output* window, and the standardized scores will appear as a new variable in your *SPSS data editor* named *zeduc22.*

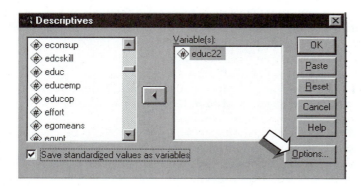

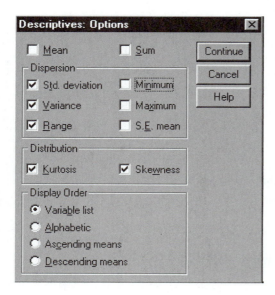

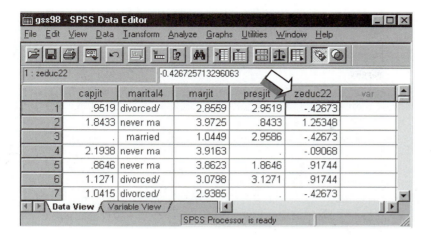

Descriptive Statistics

	N	Range	Std.	Variance	Skewness	Kurtosis
	Statistic	Statistic	Statistic	Statistic	Statistic	Statistic
EDUC22	2820	22.0000	2.9758182	8.855	-.142	1.225
Valid N (listwise)	2820					

Even if you recoded the education variable as described in footnote 1, you would not get exactly the same results shown here for *educ22* because your random numbers would not be the same as those used to create *educ22*. The descriptive statistics for the unrecoded education variable (*educ*) are shown below for those who want to replicate them.

Descriptive Statistics

	N	Range	Std.	Variance	Skewnes	Kurtosis
	Statistic	Statistic	Statistic	Statistic	Statistic	Statistic
EDUC	2820	20	2.928	8.570	-.256	1.069
Valid N (listwise)	2820					

SPSS divides the sum of squared deviations by $n - 1$ when computing the standard deviation and variance. The rationale for doing this is discussed later in the textbook. The result is that these statistics will be slightly larger than if the division was by n, as is done in Chapter 4. SPSS also uses different formulas for skewness and kurtosis than those in Chapter 4, which results in slightly different values for these statistics.

EXERCISES

1. For five women who had at least one female sex partner, the number of female sex partners for each one since age 18 is

   1   3   18   1   2

   (These numbers are representative of those for the variable *numwomen* in GSS 1998 for women aged 30–50.)

   a. What is the sum of squares for number of female sex partners?
   b. What is the variance?
   c. What is the standard deviation?
   d. What is the coefficient of variation?

2. For five men who had at least one male sex partner since age 18, the number of male sex partners for each one since age 18 is

   3   19   2   125   1

   (These numbers are representative of those for the variable *nummen* in GSS 1998 for men aged 30–50.)

   a. What is the sum of squares for number of male sex partners?
   b. What is the variance?
   c. What is the standard deviation?
   d. What is the coefficient of variation?

3. Interpret the value of the standard deviation for number of female partners in Exercise 1.

4. Which variable—number of female sex partners of women, or number of male sex partners of men—has the greatest dispersion? Why?

5. a. Compute the z scores for the female sex partner distribution in Exercise 1.
   b. Compute the mean of the z scores.
   c. Compute the standard deviation of the z scores.
   d. Do your results for the mean and standard deviation of the z scores indicate that you have correctly calculated the z scores for number of female sex partners? Why?

6. a. Compute the z scores for the distribution of male sex partners in Exercise 2.
   b. Compute the mean of the z scores.
   c. Compute the standard deviation of the z scores.
   d. Do your results for the mean and standard deviation of the z scores indicate that you have correctly calculated the z scores of number of male sex partners? Why?

7. Who is a greater number of standard deviations from the mean, the women who had 1 female sex partner or the woman who had 18 female partners? Why?

8. Who is more deviant in number of same-sex partners in terms of standard deviation units, the woman with 18 partners or the man with 125 partners? Why?

9. a. What is the skewness of the number of female partners in Exercise 1?
   b. Is the distribution right-skewed or left-skewed?
   c. What is the kurtosis of the number of female partners in Exercise 1?
   d. Is the distribution more peaked or less peaked than the normal distribution?

10. a. What is the skewness of the number of male partners in Exercise 2?
    b. Is the distribution right-skewed or left-skewed?

    c. Is the distribution of male sex partners more skewed, less skewed, or about equally skewed in comparison with that of female partners?

    d. What is the kurtosis of the number of male partners in Exercise 2?

    e. Is the distribution more peaked or less peaked than the normal distribution?

    f. Is the distribution of male sex partners more peaked, less peaked, or about equally peaked in comparison to that of female partners?

11. The following table shows the distribution of votes for the three major presidential candidates in the 1996 election among GSS 1998 respondents.

PRES96

		Frequency	Percent	Valid Percent	Cumulative Percent
Valid	Clinton	923	53.9	53.9	53.9
	Dole	589	34.4	34.4	88.4
	Perot	199	11.6	11.6	100.0
	Total	1711	100.0	100.0	

    a. Calculate the index of qualitative variation for the voting distribution.

    b. How dispersed are the votes in comparison to the maximum possible dispersion?

12. The following table shows the distribution of votes for the three major presidential candidates in the 1992 election among GSS 1998 respondents.

PRES92

		Frequency	Percent	Valid Percent	Cumulative Percent
Valid	CLINTON	817	48.0	48.0	48.0
	BUSH	672	39.5	39.5	87.5
	PEROT	212	12.5	12.5	100.0
	Total	1701	100.0	100.0	

    a. Calculate the index of qualitative variation for the voting distribution.

    b. Is the voting distribution in 1992 more or less dispersed than the distribution in 1996 from Exercise 11?

13. In the GSS 2000 (*gss2000a* or *gss2000b*), the variable *sexprtnr* is the number of sex partners a respondent has had since the age of 18. (*Sexprtner* is the sum of *numwomen* and *nummen*.) Use SPSS to calculate measures of central tendency, dispersion, skewness, and peakedness for *sexprtnr*. Interpret the results.

14. In the GSS 2000 (*gss2000a* or *gss2000b*), respondents evaluated how wrong or not wrong they think it is to have homosexual sex relations (*homosex*) and to have sex with a person other than one's spouse (*xmarsex*). Use SPSS to make frequency distributions for these two variables. Compute appropriate measures of dispersion for these two variables. Interpret and compare the dispersion of *homosex* and *xmarsex*.

5

Sampling Error

EVERYONE SPEAKS ENGLISH IN FRANCE

A colleague of mine recently returned from a short vacation in Paris. He had taken two years of French as an undergraduate years ago and was looking forward to the opportunity to practice his French during the visit. He even took a review course in conversational French in preparation for the trip. During his stay in Paris, he thoroughly enjoyed seeing the standard tourist attractions—Notre Dame, the Louvre, the Eiffel Tower, the Arc de Triomphe, the Champs Élysées, the Seine, a few galleries—and sampling the cuisine and wine in charming little restaurants. When I asked if his French had improved during his stay, however, he expressed his single disappointment about the trip: "Everyone speaks English in France! Whenever I went into a place—the hotel, a gallery, a shop, a café, a restaurant—I always tried to engage the clerks or waiters in conversation with my best French. But, almost invariably, they immediately recognized my accent and began speaking to me in English. I guess they wanted to practice their English as much as I wanted to speak French. It almost feels as if I wasted my time studying French. Most French people speak English today! It takes some of the romance out of going to France."

My colleague was making a generalization, and to generalize is a human trait. We know that every person is unique, but we couldn't get through a single day without making generalizations about others. Generalizations tell us what to expect from particular types of persons in particular situations, even when they are total strangers. Thus, generalizations are useful because they enable us to efficiently navigate the myriad social interactions we encounter from day to day.

Where do our generalizations come from? One source is authorities. We learn generalizations from our parents, teachers, and leaders. Another important source is our own social experiences or observations of others' behavior. Based on a number of empirical observations, we make inferences about the characteristics of whole groups of people, most of whom we have never encountered. For better, and sometimes for worse, we are continuously constructing our own generalizations and inferences. It appears to be such a universal human trait that it can be called the *inferential imperative.*

Is this drive to generalize a wise thing? Although it is true that social life would be very inefficient without generalizations, that does not mean that we always make good inferences. We are prone to several types of inferential errors. *Overgeneralizations* are one of these. We may generalize to broader populations than we have actually had contact with. My colleague seemed to be making inferences about the French in general—not just the Parisian population that he had observed. A second type of inferential error is a *biased generalization* based on a nonprobability sample. In a nonprobability, or nonrandom, sample, not all of the members of the population about which inferences are made have a chance of being observed. In this case, the people who are observed may not be representative of the population. My colleague based his generalization mainly on contacts with Parisians who were employed in the service industry (clerks, guides, and waiters) and were unlikely to be representative of Parisians in general in terms of their English proficiency. Samples of this type are referred to as *convenience samples.* The only good thing about a convenience sample is the fact that it is convenient! The final type of inferential error that we are prone to make is an *unreliable generalization.* Even if we select a random sample from a specified population, we may have too few observations to draw a reliable inference. My colleague's generalization was based on a small sample of only fifteen to twenty persons whom he encountered during his short stay in Paris. A small sample is unreliable because quite a different inference might have been drawn if a different sample of the same number of persons had been surveyed.

Although we are prone to make inferential errors, in everyday life we have an opportunity to correct these errors based on encounters with additional people. If our social encounters are unsatisfactory because we have applied erroneous generalizations, we can normally modify or diversify our generalizations based on additional information that we receive. Some erroneous inferences, however, are more pathological. Rigid stereotypes and prejudices about racial and religious groups are examples of this category. Because these biases or overgeneralizations promote intergroup conflict or fulfill unconscious psychological needs, they are not amenable to correction by contradictory observations. Such generalizations are supported by observations that have been filtered through selective lenses that screen out noncorroborating facts.

Making generalizations is as important in the social sciences as it is in everyday life. In social science and statistics, however, the process of making inferences is far more self-conscious and systematic, even more cautious, than in everyday life. It involves, ideally, a careful specification of the population about which generalizations are to be drawn, the use of random sampling procedures to insure that representative observations are made, and a precise specification of the reliability of and the confidence in the inferences. As you develop an understanding of these principles, this knowledge will not only make you a better social scientist but will help you make better generalizations in your personal life.

POPULATIONS AND SAMPLES

Populations

In Chapter 1 we made a distinction between the theoretical **population** that we would like to make inferences about and the study population. *A study population is the aggregate of all units of analysis that have a chance of being included in the sample to be observed.* In Chapters 2 through 4, we used data from the General Social Survey to illustrate frequency distributions, measures of central tendency, and measures of dispersion. The GSS consists of a probability sample of individuals; it is a sample of people who were randomly selected to respond to the survey. But what are these respondents a sample of? In other words, from what *study population* were they selected? This question is important because researchers who analyze the GSS are interested in more than the characteristics of the survey respondents. They are analyzing these data to learn about the population from which these persons were drawn. For example, investigators may want to make inferences about the means or the standard deviations of variables in the population. The values of these statistics in the population, however, are unknown because we do not have observations or data on the entire population. We have observed only the sample of the population that responded to the GSS. Therefore, we must generalize from the sample to the population. To understand the limits or boundaries of these generalizations, we need a careful definition of the population. Preparing this definition will help us to avoid *overgeneralizations.*

The study population from which the GSS respondents were selected consists of English-speaking adults residing in households in the continental United States. Spanish is the language of a large majority of the non-English speakers who are not included in this population. Adults are operationally defined as persons who are eighteen years of age or older. The specification that this population is limited to persons residing in households excludes college students living on campus, members of the armed forces residing in military facilities, nursing home residents, and prisoners in jails and penitentiaries. The continental United States restriction excludes residents of Alaska and Hawaii. The restriction of the population to English-speaking residents of households in the continental United States is undoubtedly done for practical and economic reasons. In other words, the GSS is drawn from a *convenience population.* We must be careful not to overgeneralize its data to encompass all adults in the United States.

Not all samples are probability or random samples. The sample of 107 nations used in Table 2.11 is not a random sample of all nations. There were approximately 208 nations in 1995. The 107 nations in the sample comprise those with relatively complete and accurate data on an array of socioeconomic variables such as those shown in Table 1.3. Thus, they undoubtedly overrepresent the more modern, industrialized nations of the world because these nations are most likely to collect extensive and accurate data on their populations. As a result, descriptive statistics for this sample of nations may not provide accurate generalizations about the nations of the world as a whole.

Probability Sampling

How do we go about selecting a random sample of cases from a population? The most elementary type of random sampling is **simple random sampling** (SRS). The first step is to prepare a list of all the cases in the population. The cases can be listed in any order. Next, we number the cases in the list from 1 to N, where N equals the population size. Thus, each case in the population is given a unique identification number. Third, a set of n unique random numbers between 1 and N is taken or generated, where n is the desired sample size. (The topic of how to determine n is not addressed in this text.) Finally, the n cases in the population list whose identification numbers match the n random numbers compose the selected random sample.

Before we look at an example of simple random sampling, let's consider what randomness and **random numbers** mean. We are all familiar with examples of randomness from the realm of gambling, such as flipping a coin, rolling dice, or spinning a roulette wheel. If the gaming device is fair, the outcome of each flip, roll, or spin is *unpredictable*. That is, each of the possible outcomes is equally likely to occur (a head or a tail is equally likely to be up; each of the six sides of a die is equally likely to be on top). Another characteristic of a random event is that its outcome is independent of any previous outcomes. For example, if four straight heads are flipped (which is possible but not very likely), the outcome of a fifth flip of the coin is still equally likely to be a head or a tail. This aspect of randomness is one that is difficult for the average person to accept. If you called tails four times in a row and each time a head came up, you would probably have a strong inclination to think that your luck was bound to change on the fifth flip. This is the *gambler's fallacy*. If the coin is fair, you still have no better than a 50-50 chance of getting a tail on the fifth outcome. A run of bad luck or good luck in the *short run* is not evidence of a lack of randomness in the game. In the *long run,* however, the number of times that each outcome occurs will even out. In ten flips of a fair coin you may get seven heads and three tails, but in 10,000 flips, there should be approximately 5,000 heads and 5,000 tails (or more accurately, the percentages of heads and tails should each be about 50 percent).

Now let's look at a table of random numbers that can be used to select a probability sample. Table 5.1 contains 1,000 single-digit random numbers generated from a uniform distribution. Moving across any row or down any column, each digit is randomly generated. That is, the numbers between 0 and 9 are each equally likely to occur at any point in the table. You should not see any patterns to the numbers as you look through the table. This does not mean that if we were to examine some relatively small group of adjacent numbers, each digit would appear approximately an equal number of times. In the first block of 10 digits in the last row, there are three 6s and no 2s, 4s, or 8s. And in the third block of the fourteenth row, there are seven 0s! These unequal outcomes do not indicate nonrandomness. Anything can occur in the short run of a random process. In the relatively long run, however—as in Table 5.1 as a whole—there should be approximately equal percentages of each digit. Table 5.2 shows that this is the case.

A small population of physicians specializing in hematology (diseases of the blood) is used to illustrate the method of taking a simple random sample. The names of the ten physicians listed in Table 5.3 were taken from the Yellow Pages of a local telephone directory. As such, they constitute the population of hematologists in that local area. Imagine that we wanted to conduct in-depth interviews with a random sample of these physicians to explore their motivations for taking up this specialty. The physicians are listed in alphabetical order, although this is not necessary when taking a random sample. They are numbered from 1 to 10 (ID), where $N = 10$. Assume we had the time and resources to conduct five interviews ($n = 5$). Thus, we need to get five random numbers. If we allow the number 0 to represent physician number 10, we can draw five single-digit random numbers from Table 5.1. Because the numbers in Table 5.1 are

Table 5.1 1,000 Random Numbers

6395895072	3979676405	9253237185	5425966995	9172822930
2678493328	6389181120	5528495121	5539971832	4173187927
0861253378	3261448940	3191504213	0649383494	8774745221
3567120191	9194727310	0504255777	4941246455	5605744647
5761247330	1444903454	4223667250	8356964582	8559853143
6456648392	6791726364	5838193981	7067538864	9603804418
0882138283	8321602415	3377983263	5333220856	7646179822
9747560305	0052438125	5711429041	1362754205	7165068289
7949535989	1383063510	0106642955	8133265665	6907068137
8985838119	8150018163	6118405643	5739146050	2949367801
1462163401	1061795270	1722692307	1010668347	2705693571
5935523940	3667214340	7012678703	2781269151	9526680458
1219266698	9587340159	6723608567	1379475053	4967007851
6444205707	1497366345	0500086000	0219655029	8863928201
3229541904	9487741217	3819846761	0769744475	6917103348
0711607221	2580786428	6496043438	1718420082	8378613743
6717835152	0289801119	8238952974	0304100694	0458366513
0924490257	5352924690	3786889741	6849871029	0440270917
2124687190	7520405586	9635083163	0191447859	6817676506
5666093517	9882546950	6199078527	8174235895	5667684902

random, we can start anywhere in the table, move up or down, or to the right or the left, and take the first five unique numbers that we encounter. Let's start in the upper left-hand corner of the table and move across the first row. The first five numbers are 6, 3, 9, 5, and 8. Each of these numbers uniquely identifies one of the ten physicians. An **X** has been placed to the right of the names in Table 5.3 that have ID numbers matching these five random digits. These indicate that Drs. Hazra, Marquinez, Mubashir, Rehmus, and Ross are the five physicians who were randomly selected for our interview sample. Because the Yellow Pages give their addresses and phone numbers, we can call them, send them letters, or visit their offices to attempt to gain their participation in our study.

What would have happened if we had decided to select a sample of six physicians instead of five? The sixth random digit in the first block of numbers in Table 5.1 is a 9, the ID number of Dr. Ross. We want a sample of six unique cases, because we do not want to interview the same doctor twice. Thus, we ignore any random numbers that point to a case that has already been selected and go on to the next random number. Because that number is 5, which points to Dr. Marquinez who also has been previously selected, we ignore it and move to the next number, which is 0. Because 0 stands for 10, which points to Dr. Trochelman, who has not yet been selected, we now have our sixth member of the sample. The procedure of skipping random numbers that have already been used so as not to select the same case more than once is called **sampling without replacement.** Sampling *with replacement* means that selected cases are put

Table 5.2 Frequency Distribution of 1,000 Random Numbers

Number	f	Pct
0	103	10.3
1	107	10.7
2	94	9.4
3	95	9.5
4	99	9.9
5	103	10.3
6	108	10.8
7	98	9.8
8	97	9.7
9	96	9.6
Total	1000	100.0

Table 5.3 Physicians (MDs) Specializing in Hematology, Listed in Yellow Pages

ID	Name	Random Selection
1	Croft, Herbert E.	
2	Haas, Andrew Jr.	
3	Hazra, Sandra V.	X
4	Koenig, Joseph M.	
5	Marquinez, Frederick P.	X
6	Mubashir, Bashar A.	X
7	Petrus, John J.	
8	Rehmus, Esther	X
9	Ross, Charles E.	X
10	Trochelman, R. Douglas	

back in the population and are eligible to be selected again. Sampling *without replacement* is the usual practice in studies of human populations.

What if the population to be sampled has more than ten cases? For example, what if $N = 75$? Because the population size is a two-digit number, we have to use two-digit random numbers to select the sample. Because each single digit in Table 5.1 is random, each pair of adjacent numbers is a two-digit random number. Thus, we could use, for example, the first two columns in Table 5.1 to get our random numbers. If we wanted a sample of $n = 9$, we would use the first nine unique numbers between 1 and 75, the range of ID numbers assigned to our population. The first six numbers in columns one and two are unique numbers in the desired range (63, 26, 08, 35, 57, 64) and thus they would determine the first six members of the sample. The next number (08) has already been used, so it would be skipped. The next three numbers (97, 79, and 89) are outside the range, so they would be skipped as well. Finally, the following three numbers (14, 59, and 12) are within the range and have not been used before, so they point to the final three cases to be included in the sample.

The simple random sampling method that we have described is conceptually the simplest type of probability sampling. This type of sampling is the foundation for all the inferential statistics to be covered in this text. Another way of saying the same thing is to note that these inferential statistics are based on the assumption that simple random sampling has been used to select the cases that have been observed and on which statistics such as the mean and standard deviation have been computed.

Although simple random sampling is relatively easy to understand, it is not a practical or efficient method in some studies. Let us return to the case of the GSS, a survey of persons residing throughout the continental United States. One problem in selecting a simple random sample from this population is that there is no available list of all persons residing in the United States. The U.S. Bureau of the Census attempts to enumerate all persons residing in the United States every ten years. The results of their efforts, however, are confidential and not generally available to researchers who need a national sample. Also, many people are not located by the Census Bureau (an undercount) and its list of the population becomes outdated relatively rapidly between the decennial censuses.

Even if a good list of the national population were available, it would not be efficient to survey a simple random sample of the nation. A sample of approximately 3,000 persons, such as the GSS now surveys every two years, would result in people being selected from hundreds of communities and rural areas scattered across the entire continental United States. Such a sample would require interviewers to spend an enormous amount of travel time and money to locate and interview the selected respondents in their households. Consequently, simple random sampling is never used in surveys of populations that are geographically widely dispersed, such as the GSS. Instead, much more complex methods of sampling are used to generate a random sample of households that is more efficient to study. These methods are called **multistage cluster samples.** Roughly speaking, such samples involve the random selection of a relatively small number of communities and other areas that are representative of all cities and places. Within these communities, a small number of blocks are randomly selected, and then a small sample of households is drawn from each selected block. Finally, a random sample of one or more residents is selected within each selected household. Because interviewers can visit several households within the same block and can go to several blocks within the same community, it is much more practical and economically feasible to collect data from such a probability sample. Although the details of how multistage samples are collected is beyond the scope of this book, suffice it to say that the samples generated by these methods *are* probability samples that are "representative" of the population to be studied. Thus, such samples can be used to make valid inferences from the sample to the population.

Sampling Error

How do we make inferences from our sample of observations to the population we want to study? On the surface, this appears to be a very simple procedure. If we want to know about the central tendency of a population variable, we simply use a measure of central tendency computed on the sample as the best estimate of the population central tendency. For example, we use the sample mean as an estimate of the population mean. Or, if we want to make an inference about the dispersion of a variable in the population, we use the value of the standard deviation in the sample as the prediction of the population standard deviation.[1]

To illustrate this inference, we return to the mean and standard deviation of years of education as computed from the GSS (Table 5.4). The mean and standard deviation of education in the sample are 13.3 years and 3.0 years, respectively. The symbols that we have used for these statistics are $\overline{X}$ and s. We now stipulate that these symbols refer to the values of the mean and standard deviation *in the sample.* Thus, they are called **sample statistics.** The values of sample statistics are known because we can compute them from our sample observations or data.

---

1. In this case, we have to make a slight change in the formula for the standard deviation to get an unbiased estimate of the population standard deviation. This formula will be discussed later.

Table 5.4 Mean and Standard Deviation of Years of Education, 1998 GSS

Sample Statistics	Population Parameters
$\overline{X} = 13.3$	$\mu = \text{mu} = ?$
$s = 3.0$	$\sigma = \text{sigma} = ?$

The values of these statistics in the population from which the sample is selected are called **population parameters.** Instead of the Roman letters that are used for the symbols for sample statistics ($\overline{X}$ and s), Greek letters are used to symbolize population parameters. μ, the lowercase Greek letter m, is the symbol for the population mean; μ is named *mu* and pronounced "mew." σ (sigma), the lowercase Greek s, is the symbol for the population standard deviation. In general, Roman letters are used as the symbols for sample statistics, and Greek letters are used as the symbols for population parameters. Most statistics have pairs of symbols: one for the value of the sample statistic and one for the value of the population parameter.

The values of population parameters are unknown unless a census of the entire population has been conducted. When we work with *sample* data, we do so because it is impossible or too expensive to collect data from the entire population. If we do not have observations on the entire population, we cannot know the value of the population parameters with certainty. We must make an inference from the known sample statistic to the unknown population parameters. Perhaps that is why the well-known Roman letters are used for known sample statistics and the less well-known Greek letters are used for unknown population parameters.

If we select a random sample from the population, then we can generally say that the sample statistic is the best estimate of the population's parameter. Thus, we can say that 13.3 years is our estimate of the mean years of education among the population of the United States that is 18 years of age or older (but remember the more technical definition of this population) and 3.0 years is the estimated standard deviation of the population's education. This is true, as far as it goes, but a good inference is more complex than this. No matter how good a sample we have, it only provides an *estimate* of the unknown parameter. The fact that it is an estimate means that it is not perfect. The fact that it is not perfect means that there is likely to be an error to a greater or lesser extent.

The error we are concerned about is called sampling error. **Sampling error** *is the difference between a sample statistic and a population parameter:*

$$\text{Sampling error} = \text{sample statistic} - \text{population parameter}$$

Each sample statistic has its own sampling error. For example, there is the sampling error of the mean and the sampling error of the standard deviation:

$$\text{Sampling error of mean} = \overline{X} - \mu$$

$$\text{Sampling error of standard deviation} = s - \sigma$$

Sampling error can be either positive or negative. If the sample statistic is greater than the population parameter, then sampling error is positive and we have an overestimate of the parameter. If sampling error is negative, we have an underestimate of the parameter. However, because the population parameter is unknown, the sampling error is unknown. The sampling errors for the mean and standard deviation of education in the GSS can be expressed as follows:

$$\text{Sampling error of education mean} = 13.3 - \mu = ?$$

$$\text{Sampling error of education standard deviation} = 3.0 - \sigma = ?$$

The fact that the sampling error is unknown for any specific sample does not mean that we have no theoretical knowledge about sampling error. Arguably, the greatest achievement of the science of statistics is the development of a strong theory of sampling error that can be used to accurately *estimate* the sampling errors involved in specific inferences. This is why statistics, or more precisely inferential statistics, can be called the *measurement of uncertainty*. To lay the foundation for the measurement of sampling error, we must look more closely at how sampling error occurs and what influences the magnitude of this error.

Sampling Distributions

Many people are surprised to learn that monks make beer. There are five Trappist monasteries in Belgium, however, that brew some of the finest ales in the world, both for their own consumption and for sale to the public. Only a small number of monks in each monastery enjoy the privilege of working in the brewery. Because these closed communities are rather secretive about their affairs, the public does not know the exact numbers. *Imagine,* however, that the true number of brewer monks are as given in Table 5.5.

The five abbeys listed in Table 5.5 are the *population* of all beer brewing abbeys in Belgium. The *unit of analysis* is the abbey, not each individual monk. The variable X to be studied is the number of brewer monks per abbey, and the *parameter* to be investigated is μ, the mean number of monks involved in brewing. Imagine that we are in the unusual position of knowing the value of the population parameter, which is

$$\mu = \frac{\Sigma X}{n} = \frac{35}{5} = 7$$

Now, suppose a researcher who does not know the population values given in Table 5.5 plans to do a study to estimate μ. Because of the difficulty of acquiring the information about the number of brewers and the fact that the abbeys are dispersed throughout Belgium, the researcher decides to select a simple random sample of two abbeys to investigate. The researcher will number the five abbeys from 1 to 5 and take two different random numbers between 1 and 5 to select the sample of abbeys. Suppose that whatever pair of abbeys was randomly selected, the investigator was eventually able to ascertain the true number of brewers in each of the two abbeys. The sample mean would then be computed and used as the estimate of the population mean, which the researcher does not know. What would be the researcher's estimate, and how great, if any, would be the sampling error?

Table 5.5 Number of Monks Working as Brewers in Belgian Trappist Abbeys

Abbey	Number of Brewer Monks
Westvleteren	4
Rochefort	6
Orval	7
Westmalle	8
Chimay	10
Total	35

The answer depends on which pair of abbeys falls into the random sample.

$$\frac{N(N-1)}{2} = \frac{5 \cdot 4}{2} = 10 \text{ unique pairs}$$

Ten unique pairs (see Chapter 4) can be selected. Because each abbey has an equal probability of being selected, each pair of abbeys also has an equal probability. The ten pairs of abbeys that can be selected, along with the sample mean for each pair, is given in Table 5.6. As you can see, the value of the sample mean depends to a great extent on which pair of abbeys is selected for the sample. The means range from 5 to 9. *The distribution of the values of the sample mean for all unique samples that can be selected is the* **sampling distribution** *of the mean.* A graph of the sampling distribution of the mean number of brewer monks (for $n = 2$) is shown in Figure 5.1.

You can see at once that the distribution of means is *symmetrically distributed* around its center, which equals 7. The value 7 occurs twice, which is more than any other sample mean. That is, 7 is the mode. You may remember that in a symmetrical distribution the median and mean should both equal the mode. The median equals 7 because there are an equal number of samples above and below the value 7. The mean of the sampling distribution of ten means is

$$\mu_{\overline{X}} = \frac{\Sigma \overline{X}}{m} = \frac{70}{10} = 7$$

where m equals the number of distinct samples that can be selected. $\mu_{\overline{X}}$ is the symbol for the mean of the sampling distribution of the mean. The lowercase Greek letter mu is used rather than $\overline{X}$ because the mean of the sampling distribution is the mean of a *population*—the population of means of *all* distinct samples of size $n = 2$ that can be selected. This population is different from the population of all abbeys: It is the population of all unique samples that can be drawn from the population of abbeys. The subscript $\overline{X}$ is used to indicate that $\mu_{\overline{X}}$ is the mean of the sampling distribution of $\overline{X}$, not the mean of X.

As expected, the mean of this symmetrical distribution equals 7, the value of the mode and the median. What other important quantity equals 7? The mean of the five abbeys in the population (Table 5.5) is also 7, which illustrates that the *mean of the sampling distribution of the mean equals the population mean* (μ).

**Table 5.6 Means of Number of Brewer Monks for All Unique
Samples of $n = 2$**

Sample	Abbeys	$\Sigma X / n = \overline{X}$	$\overline{X} - \mu$ = Sampling Error
1	Westvleteren, Rochefort	(4 + 6) / 2 = 5.0	5.0 − 7.0 = −2.0
2	Westvleteren, Orval	(4 + 7) / 2 = 5.5	5.5 − 7.0 = −1.5
3	Westvleteren, Westmalle	(4 + 8) / 2 = 6.0	6.0 − 7.0 = −1.0
4	Westvleteren, Chimay	(4 + 10) / 2 = 7.0	7.0 − 7.0 = 0.0
5	Rochefort, Orval	(6 + 7) / 2 = 6.5	6.5 − 7.0 = −0.5
6	Rochefort, Westmalle	(6 + 8) / 2 = 7.0	7.0 − 7.0 = 0.0
7	Rochefort, Chimay	(6 + 10) / 2 = 8.0	8.0 − 7.0 = 1.0
8	Orval, Westmalle	(7 + 8) / 2 = 7.5	7.5 − 7.0 = 0.5
9	Orval, Chimay	(7 + 10) / 2 = 8.5	8.5 − 7.0 = 1.5
10	Westmalle, Chimay	(8 + 10) / 2 = 9.0	9.0 − 7.0 = 2.0
Total		70.0	0.0

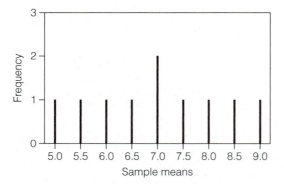

FIGURE 5.1 Sampling Distribution of Mean
Number of Brewer Monks, $n = 2$

This fact indicates that on the average, our sample means for $n = 2$ are right on target. However, in real research we do not work with the mean of the sampling distribution (i.e., the mean of the means of all possible samples). In a real research project, we have
only one sample and thus only one sample mean to use as an estimate of the population
mean. Table 5.6 shows the sampling error that results for each sample when its mean is
used as an estimate of the population mean μ. There are four samples where
$\overline{X} - \mu < 0$, that is, the sampling error is negative. These samples result in underestimates
of the mean number of brewer monks. For example, if the random sample selected by the
researcher consisted of the abbeys Westvleteren and Rochefort, there would be a serious
underestimate; that is, the sampling error would equal negative 2. There are also four samples with positive sampling error. These samples would result in overestimates of the mean.
A sample consisting of Westmalle and Chimay would produce the greatest overestimate.

Finally, two samples hit the target right on the nose! Samples 4 and 6 have no sampling error. Because there are two samples with a mean of 7 (sampling error equals 0) and each of the other values of the sample mean occurs only once, the sample mean is more likely to equal the population mean (i.e., $\mu = 7$) than it is to equal any other single value. In other words, a sample mean equal to the population mean has the maximum likelihood of occurring. For this reason, the sample mean is called a *maximum likelihood estimator of the population mean.*

Maximum likelihood, however, does not mean that the sample mean is more likely to equal the population mean than not. The population mean is simply more likely than any other value. But in our example, 8 of 10 samples do not equal the population mean. Thus, 80 percent of the samples that can be selected will have at least some sampling error. But again, the amount of this error varies considerably from one of these samples to another. So, although we are more likely to make an error than not, how large is our sampling error likely to be?

To answer this question, compute the standard deviation of the sampling distribution, which is defined as

$$\sigma_{\overline{X}} = \sqrt{\frac{\Sigma(\overline{X} - \mu_{\overline{X}})^2}{m}} \tag{5.1}$$

This formula for the standard deviation of the sampling distribution of the mean simply adapts the basic defining formula for the standard deviation (Equation 4.7) to the notation for the sampling distribution of the mean. It consists of summing the squared deviations of each sample mean $\overline{X}$ from the mean of the means ($\mu_{\overline{X}}$), dividing by the number of sample means (m), and taking the square root. The lowercase Greek sigma is used as the symbol instead of the Roman s because we are dealing with a population parameter—in this case, the population of all unique samples that can be selected. The subscript $\overline{X}$ of σ tells us that we are dealing with the standard deviation of the sampling distribution of $\overline{X}$ instead of the standard deviation of X.

Let's use the sampling distribution of the mean in Table 5.6 to illustrate the standard deviation of the mean. As we saw, the mean of the means equals the population mean— that is, $\mu_{\overline{X}} = \mu = 7$. Thus, the Sampling Error column in Table 5.6 is equal to $\overline{X} - \mu_{\overline{X}}$ in Equation 5.1. We need to square this quantity, getting the results in Table 5.7.

The sum of squared deviations (SS) for the mean equals 15. Entering this number into Equation 5.1 gives the standard deviation of the means:

$$\sigma_{\overline{X}} = \sqrt{\frac{15}{10}} = \sqrt{1.5} = 1.225$$

This result indicates that the standard amount that the sample means deviate from the mean of the sampling distribution is 1.225. Because each deviation from the mean of the sampling distribution, or μ, is a sampling error, *the standard deviation of the sampling distribution of the mean is named the* **standard error** *of the mean.*

$$\text{Standard error of mean} = \sigma_{\overline{X}} = \sqrt{\frac{\Sigma(\overline{X} - \mu_{\overline{X}})^2}{m}} \tag{5.2}$$

Thus, $\sigma_{\overline{X}} = 1.225$ is the standard sampling error that we are likely to make in using the mean of a random sample of $n = 2$ as an estimate of the mean number of brewer

Table 5.7 Squared Deviations of Sample Means

$\overline{X} - \mu_{\overline{X}}$	$(\overline{X} - \mu_{\overline{X}})^2$
−2.0	4.00
−1.5	2.25
−1.0	1.00
0.0	.00
−0.5	.25
0.0	.00
1.0	1.00
0.5	.25
1.5	2.25
2.0	4.00
0.0	15.00

monks in a population of $N = 5$. That is, our sample mean is expected to miss the population mean by 1.225—either too high or too low.

What if we selected a larger sample, such as $n = 3$? How would that affect our estimate of the mean number of brewer monks? As Table 5.8 shows, there are also ten unique samples of $n = 3$ abbeys from the population of $N = 5$ abbeys.

The mean of the sample means is again

$$\mu_{\overline{X}} = \frac{\Sigma \overline{X}}{m} = \frac{70}{10} = 7$$

The sample means now range only from 5.67 to 8.33, and the standard error of the mean is

$$\sigma_{\overline{X}} = \sqrt{\frac{6.66}{10}} = \sqrt{.666} = .816$$

Thus, the standard error is now smaller, being only about two-thirds as large as it was for samples of $n = 2$. A quick visual comparison of the dispersion of the sampling distribution for $n = 3$ in Figure 5.2 with that in Figure 5.1 for $n = 2$ readily confirms this. It illustrates the principle that larger sample sizes will have smaller standard errors. The fact that you can reduce sampling error by selecting a larger sample should be intuitively logical.

In the preceding illustrations, we used a small population of $N = 5$ to produce a small sampling distribution to simplify the examples. As the population size increases, the number of distinct samples m of size n grows very rapidly. For example, if $N = 10$ and $n = 3$,

$$m = \frac{N!}{n!(N-n)!} = \frac{10!}{3! \, 7!} = \frac{10 \cdot 9 \cdot 8 \cdot 7 \cdot 6 \cdot 5 \cdot 4 \cdot 3 \cdot 2 \cdot 1}{(3 \cdot 2 \cdot 1)(7 \cdot 6 \cdot 5 \cdot 4 \cdot 3 \cdot 2 \cdot 1)} = 120$$

Table 5.8 Means of Number of Brewer Monks for All Unique Samples of n = 3

Sample	Abbeys	$\Sigma X/n = \bar{X}$	$\bar{X} - \mu$ Sampling = Error	$(\bar{X} - \mu)^2$
1	Westvleteren, Rochefort, Orval	(4 + 6 + 7) / 3 = 5.67	5.67 − 7.0 = −1.33	1.78
2	Westvleteren, Rochefort, Westmalle	(4 + 6 + 8) / 3 = 6.00	6.00 − 7.0 = −1.00	1.00
3	Westvleteren, Rochefort, Chimay	(4 + 6 + 10) / 3 = 6.67	6.67 − 7.0 = −0.33	.11
4	Westvleteren, Orval, Westmalle	(4 + 7 + 8) / 3 = 6.33	6.33 − 7.0 = −0.67	.44
5	Westvleteren, Orval, Chimay	(4 + 7 + 10) / 3 = 7.00	7.00 − 7.0 = 0.00	.00
6	Westvleteren, Westmalle, Chimay	(4 + 8 + 10) / 3 = 7.33	7.33 − 7.0 = 0.33	.11
7	Rochefort, Orval, Westmalle	(6 + 7 + 8) / 3 = 7.00	7.00 − 7.0 = 0.00	.00
8	Rochefort, Orval, Chimay	(6 + 7 + 10) / 3 = 7.67	7.67 − 7.0 = 0.67	.44
9	Rochefort, Westmalle, Chimay	(6 + 8 + 10) / 3 = 8.00	8.00 − 7.0 = 1.00	1.00
10	Orval, Westmalle, Chimay	(7 + 8 + 10) / 3 = 8.33	8.33 − 7.0 = 1.33	1.78
Total		70.00	0.00	6.66

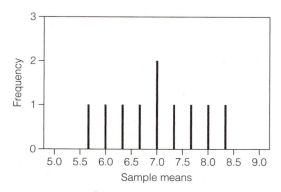

FIGURE 5.2 Sampling Distribution of Mean
Number of Brewer Monks, n = 3

That is, there are 120 unique samples in comparison with the 10 samples used in the illustrations. If $N = 100$ and $n = 3$,

$$m = \frac{100!}{3!(97)!} = \frac{100 \cdot 99 \cdot 98}{3 \cdot 2 \cdot 1} = 161,700$$

and if $N = 1,000$ and $n = 3$,

$$m = \frac{1,000!}{3!(997)!} = \frac{1000 \cdot 999 \cdot 998}{3 \cdot 2 \cdot 1} = 166,167,000$$

there are more than 166 million distinct samples of size 3 from a population of 1,000!

Now, consider the GSS. In 1997 the U.S. Census Bureau estimated that there were approximately 198,108,000 people in the United States who were 18 years of age or older. If the GSS were to take a sample of $n = 3,000$ from this population, the number of distinct samples that could be selected is

$$m = \frac{198,108,000!}{3,000! \cdot 198,105,000!} = ?$$

The number is so incredibly huge (more than 307 digits long) that it cannot begin to be computed with a pocket calculator or a standard software package like Excel.

The point of this example is that the preceding illustrations of sampling distributions are not realistic. You can't view a realistic sampling distribution because it is too large; there are too many samples. And furthermore, we don't know the values of X in the population, so even if we could display all the distinct samples, we couldn't know the values of X that are needed to calculate the sample means in each sample.

Although we can't observe a realistic sampling distribution of the mean, statistical theory can tell us what the characteristics of the sampling distribution will be *under certain specified conditions.* In the next section we will review certain characteristics of the theoretical sampling distribution of the mean and the conditions under which they hold.

THEORETICAL SAMPLING DISTRIBUTION
OF THE MEAN

Characteristics of the Sampling Distribution

Just as we can describe any variable or distribution in terms of its central tendency, dispersion, and shape (i.e., skewness and peakedness), so also can we describe the sampling distribution of the mean. Statistical theory tells us very precisely what the values of the mean, standard deviation, skewness, and peakedness will be under certain general conditions. We will first describe very carefully the characteristics of the distribution and then discuss the conditions under which these characteristics hold.

We will use the age distribution of the population of the United States to provide examples of the characteristics of the sampling distribution of the mean. The mean (μ_X) and standard deviation (σ_X) of age, along with the population size (N), are given in Table 5.9. Normally, it is sufficient to use simply μ and σ without the subscript X when we are referring to the population mean and standard deviation of X. In this case, however, the subscript X is used to clearly identify the mean and standard deviation of the sampling distribution of $\overline{X}$. Mu (μ) and sigma (σ) are being used because the mean and standard deviation of age are *population parameters.* Remember that it is unusual to know the values of population parameters; they are almost always *unknowns.* We have chosen a variable (i.e., age) with known parameters so that we can use numerical values to illustrate the characteristics of the sampling distribution of the mean.

The mean, standard deviation, and shape of the sampling distribution of the mean are

$$\mu_{\overline{X}} = \mu_X \tag{5.3}$$

$$\sigma_{\overline{X}} = \frac{\sigma_X}{\sqrt{n}} \tag{5.4}$$

$$\overline{X} \sim \mathrm{N} \tag{5.5}$$

Table 5.9 Age of U.S. Population, 1997

N	267,235,000
μ_X	35.6
σ_X	22.4

Note: Computed from Table 16, *Statistical Abstract of the United States,* 1997, based on the 1990 Decennial Census and annual updates from the Current Population Surveys.

Equation 5.3 indicates that *the* **mean of the sampling distribution of the mean** *(i.e., a mean of means) equals the mean of X in the population.* We saw examples of this relationship earlier in the case of the sampling distributions for the mean number of brewer monks in samples from a population of five Trappist monasteries. Equation 5.3 says that this relationship will always be the case for the sampling distribution of the mean, no matter what the variable X is and no matter what the sample size n equals. Thus, if X is age, the mean of the sampling distribution of the mean age equals 35.6 years:

$$\mu_{\overline{AGE}} = \mu_{AGE} = 35.6$$

The fact that the mean of the sampling distribution equals the population mean does not mean that we know its value. In most cases (unlike age), the population mean is *unknown.* It is because the population mean is unknown that we selected a sample. The sample mean is used as an estimate of this unknown parameter.

If the mean of a sampling distribution of a statistic equals the population parameter, the statistic is unbiased. Or, more precisely, the sample statistic is an **unbiased estimator** of the population parameter. Thus, Equation 5.3 says that the sample mean is an unbiased estimator of the population mean. Not all sample statistics are unbiased estimators like the mean (see Box 5.1). The fact that the sample mean is unbiased does not indicate that the mean of X in any single sample that happens to be drawn will equal the population mean. The means of some samples that can be selected will be too high, others will be too low, and some will be right on target, or approximately so. That is, there is a distribution of the sample means, and it is only *on the average* that the mean is unbiased.

Equation 5.4 says that the standard deviation of the sampling distribution of the mean equals the population standard deviation of X divided by the square root of n. Earlier, we defined the standard deviation of the mean, called the **standard error of the mean,** as the standard deviation of the means from all unique samples of size n that can be selected. It is seldom, if ever, possible to calculate the standard error of the mean by taking all possible samples of size n, but if we could do this, the value we would get would equal that provided by Equation 5.4. Let's take the age of the U.S. population as an example. The population standard deviation of age is 22.4 years (Table 5.9). If, for example, we selected a random sample of 125 people, Equation 5.4 tells us the standard error of the mean would be

$$\sigma_{\overline{AGE}} = \frac{\sigma_{AGE}}{\sqrt{n}} = \frac{22.4}{\sqrt{125}} = \frac{22.4}{11.2} = 2.0$$

BOX 5.1 Sampling Distribution of *s*

Equation 4.7 defined the standard deviation as

$$s = \sqrt{\frac{\Sigma(X - \bar{X})^2}{n}}$$

The standard deviation also has a sampling distribution. The mean of the sampling distribution of *s*, however, does not equal the population standard deviation. The mean of the distribution of *s* is less than the population standard deviation.

$$\mu_s < \sigma$$

Thus, *s* is a biased estimator of σ and s^2 is a biased estimator of σ^2; they provide underestimates of σ and σ^2.

There is a simple alternate version of the formula for the sample variance, $\hat{s}^2$, that is an unbiased estimator of the population variance σ^2:

$$\hat{s}^2 = \frac{\Sigma(X - \bar{X})^2}{n - 1}$$

That is, if the sum of squares is divided by $n - 1$ rather than n, $\hat{s}^2 > s^2$ and the mean of the sampling distribution of $\hat{s}^2$ will equal the population variance:

$$\mu_{\hat{s}^2} = \sigma^2$$

The square root of $\hat{s}^2$ is

$$\hat{s} = \sqrt{\frac{\Sigma(X - \bar{X})^2}{n - 1}} \qquad\qquad (5.6)$$

Despite the fact that dividing the sum of squares by $n - 1$ gives an unbiased estimate of the population variance, $\hat{s}$ is not an unbiased estimate of the population standard deviation. Specifically, the mean of the sampling distribution of $\hat{s}$ is less than the population standard deviation. Although $\hat{s}$ provides an underestimate of σ, it is not as biased as *s*. Thus,

$$\mu_s < \mu_{\hat{s}} < \sigma$$

Further corrections may be made to make $\hat{s}$ an unbiased estimate of *s* (Loether and McTavish, 1976, pp. 420–421).

In sum, although Equation 4.7 provides the definition of a standard deviation, if the sample observations are to be used to estimate the population standard deviation, as is often the case, $\hat{s}$ provides a better estimate than *s* and additional corrections may be made to $\hat{s}$ to get an unbiased estimate. As you can see, estimating the population standard deviation is more complex than estimating the population mean. Researchers, however, are more interested in making inferences about the mean than in making inferences about the standard deviation.

The standard error of the mean of age is 2 years. Or, in other words, the standard amount that sample means would deviate from the population mean is 2 years. Some sample means would be right on target and others would be too high or too low by varying amounts. But the *average* sample would be off target by 2 years. This might sound like a contradiction of Equation 5.3, which says the mean of all sample means equals the population mean (the target) because the sum of the sample means that are too high is offset exactly by the sum of all those that are too low. With the standard deviation of the means, however, the sample means that have negative deviations are squared, making them positive, before they are added to the squared deviations of the means that are too high; that is, the negative deviations do not offset the positive deviations because the standard deviation is based on squared deviations.

Now, let's consider what would happen to the standard error of the mean if we took a sample of $n = 500$ from the U.S. population:

$$\sigma_{\overline{AGE}} = \frac{\sigma_{AGE}}{\sqrt{n}} = \frac{22.4}{\sqrt{500}} = \frac{22.4}{22.4} = 1.0$$

The standard error of the mean would be 1 year, which is half of the value we just found. By taking a sample that is four times as large ($n_2/n_1 = 500/125 = 4$), we reduce the standard error by 50 percent. The reason can be seen in Equation 5.4, in which n is located in the denominator. Thus, because we are dividing by the square root of n, the larger the value of n, the smaller the standard error. This should seem intuitively reasonable. The larger the sample we draw, the smaller our sampling error will tend to be. It is probably not intuitively obvious that the standard error is inversely proportionate to the *square root of n*. Thus, to cut the standard error by one-half, we have to quadruple the sample size. The relationship between sample size and the standard error of the mean is illustrated in Figure 5.3 for the variable age. As you can clearly see, there are diminishing returns to increases in sample size in terms of reductions in the standard error. There are large reductions in the standard error as n increases from 2 to 100 or 200 but only small reductions in the standard error as n increases from 200 or 300 to 1,000.

The other term in Equation 5.4 that affects the standard error of the mean is the population standard deviation of X. Because σ_X is in the numerator of the equation, the larger σ_X, the larger the standard error $\sigma_{\overline{X}}$. If population A has a standard deviation of X that is twice as large as population B, then A will have a standard error of the mean that is twice as large as that of B for any given n. The more dispersed the population distribution, the more likely it is that a sample mean will deviate from the population mean due to random sampling error. Figure 5.4 shows two hypothetical populations, A and B, with $\sigma_X = 2$ and $\sigma_X = 1$, respectively.

Assume that samples of $n = 3$ were selected from each population and that each sample resulted in a large positive sampling error in which the three selected cases came from the upper right-hand tail of the population distribution. The sample from population A would result in a larger sampling error ($\overline{X}_A - \mu = 4 - 0 = 4$) than the sample from population B ($\overline{X} - \mu = 2 - 0 = 2$) because there are longer tails in A from which cases with extremely high values can be selected by chance. This example is intended to illustrate the principle that populations with larger standard deviations will have more samples with large sampling errors of the mean. Consequently, the greater the population standard deviation, the greater the sampling error of the mean.

The final characteristic of the sampling distribution of the mean is specified by Equation 5.5. The symbol $\sim$ means distributed, and N means normal distribution.[2] Thus, Equation 5.5 states that *the sample mean is distributed normally,* or equivalently, *the sampling distribution of the mean has a* **normal distribution.** A normal distribution is one with a specific shape. This distribution is very important to the theory of inferential statistics, and thus we will spend a good deal of time studying the characteristics of the normal distribution. It is important because it is very useful in helping us to specify the accuracy of our sample estimates. *Normal* does not mean that most variables in

---

2. In this context, N does not refer to the population size, as it did before. Even in statistics, you often have to pay attention to the context to understand the meaning of symbols

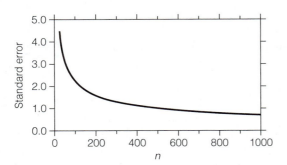

FIGURE 5.3 Standard Error of Mean Age as a Function of Sample Size *n*

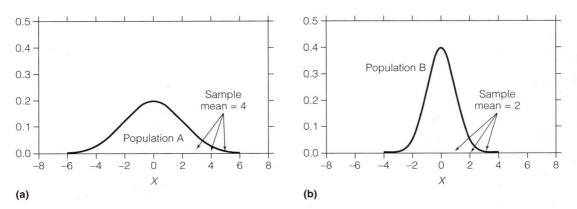

(a) **(b)**

FIGURE 5.4 Samples of *n* = 3 from Populations with Different Standard Deviations

the social sciences have this form of distribution, although it was once mistakenly thought this might be the case. Without overgeneralizing too much, we might say that this shape is normal for sampling distributions. More accurately, though, many important statistics, but not all, have sampling distributions that have a normal distribution.

To summarize the characteristics of the sampling distribution of the mean discussed thus far, Equation 5.3 specifies the central tendency of the distribution ($\mu_{\overline{X}} = \mu_X$), Equation 5.4 specifies the dispersion of the distribution ($\sigma_{\overline{X}} = \sigma_X/\sqrt{n}$), and Equation 5.5 specifies the shape of the distribution ($\overline{X} \sim N$). We now turn to a careful description of the shape of a normal distribution.

The Normal Distribution

Standard Normal Distribution

Figure 5.5 illustrates the normal distribution of a variable *X* and its standardized score *Z*. You can see immediately that it is a *symmetrical* distribution; the shape of the curve above the mode is exactly the reverse of the shape below the mode. The fact that it is symmetrical means that the mode, median, and mean will all be equal. Not all sym-

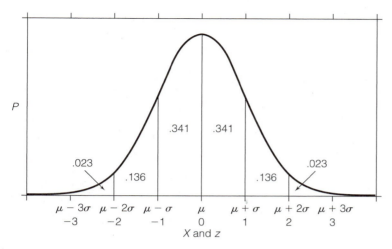

FIGURE 5.5 Normal Distribution

metrical distributions, however, have a normal distribution. A normal distribution is a symmetrical distribution with a certain amount of peakedness or kurtosis (Equation 4.17). Specifically, kurtosis equals 0 in a normal distribution. Distributions with kurtosis greater than 0 are more peaked than a normal distribution, and distributions with kurtosis less than 0 are less peaked than a normal distribution.

To make the shape of a normal distribution more concrete and useful, it is necessary to look at the proportion of cases that lie between the mean and a given number of standard deviations above or below the mean. Or, alternatively, we can specify the proportion of cases that lie more than a given number of standard deviations above or below the mean of a normal distribution. We will examine a normal distribution both ways.

The total area A under the curve describing a distribution may be defined as equal to 1.0 (see Figure 5.6). That is, if you created a group-score frequency distribution for a continuous variable (see Chapter 2) and summed the proportions of cases lying in each interval, the sum of proportions would equal 1.0. If we made each interval in the grouped-score distribution 1 standard deviation wide, or a width of 1 for standardized (Z) scores, the proportion of cases in each interval would equal that shown in Table 5.10 (the lower and upper intervals are $Z < -3$ and $Z > 3$, respectively) and Figure 5.5 (the lower and upper intervals are $Z < -2$ and $Z > 2$, respectively). Notice that slightly more than one-third of the area ($p = .341$) lies in the interval μ to $\mu + \sigma$ (between the mean and 1 standard deviation above the mean). Because the normal distribution is symmetrical, the same proportion falls in the interval $\mu - \sigma$ to μ (between 1 standard deviation below the mean and the mean). Thus, about two-thirds ($p = .682$) of the area or cases in a normal distribution lie in the interval $\mu \pm \sigma$ (within 1 standard deviation of the mean).

The cumulative distribution (cp) in Table 5.10 indicates that $p = .0228$ of the cases are more than 2 standard deviations below the mean, and the same proportion ($p = 1 - .9772 = .0228$) are more than 2 standard deviations above the mean. In other words, in a normal distribution, less than 5 percent of the cases ($p = .0228 + .0228 = .0456$) are more than 2 standard deviations away from the mean. Thus, in a normal distribu-

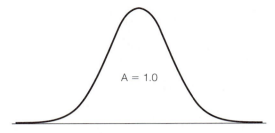

FIGURE 5.6 Area under a Distribution

Table 5.10 Standard (Z) Normal Distribution

Z	p	cp
Below −3	.0013	.0013
−3 to −2	.0214	.0228
−2 to −1	.1359	.1587
−1 to 0	.3413	.5000
0 to 1	.3413	.8413
1 to 2	.1359	.9772
2 to 3	.0214	.9987
Above 3	.0013	1.0000
Total	1.0000	

tion it is very unusual for a case to be more than 2 standard deviations from the mean. For example, if you randomly selected one case from a normally distributed variable X, it is very unlikely that the sample case would be more than 2 standard deviations from the mean of X.

Equation 5.7 describes a normally distributed X:

$$P = \frac{1}{\sigma\sqrt{2\pi}}\, e^{-(1/2)[(X-\mu)/\sigma]^2} \tag{5.7}$$

Equation 5.7 describes the height of the curve (P) shown in Figure 5.5 for any value of X. The formula for the normal distribution looks very imposing, but there are a few important things worth noticing in the formula. First, there are three constants in the equation (2, $\pi = 3.14159$, and $e = 2.71828$), so it is not quite as formidable as it looks at first glance. There are also three variables in the equation (X, μ, and σ). Although for any given X, μ and σ are constants, they vary from variable to variable. Thus, the height of the normal curve depends on the mean and the standard deviation of X.

Now, observe that the exponent of e in Equation 5.7 contains $[(X - \mu)/\sigma]^2$. Because X minus the mean divided by the standard deviation is equivalent to a z score, which indicates how many standard deviations X is above or below the mean, the height of the curve is a function of the squared number of standard deviations that X is from the mean. Although it may not be obvious, the fact that $[(X - \mu)/\sigma]^2$ is always positive means that the exponent of e (i.e., $-(1/2)[(X - \mu)/\sigma]^2$) is always negative. Consequently, the greater the number of standard deviations that X is from the mean, the lower will be the height of the curve (P). Second, notice the location of σ in the term $1/(\sigma\sqrt{2\pi})$, which means that the greater the standard deviation of X, the lower will be the curve for any value of X. That is, the more X is spread out or dispersed, the smaller the proportion of cases in any specific interval of X.

The effects of X, μ, and σ in Equation 5.7 are illustrated in Figure 5.7. The highest curve is for Z, a standardized variable with a mean of 0 and a standard deviation of 1. The second highest curve is for X_1, which has a standard deviation of 2. X_2 has the lowest curve because it has the highest standard deviation ($\sigma = 4$). The higher the mean of X, the more its normal curve is moved to the right. And the higher the standard deviation

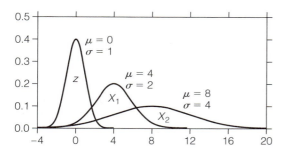

FIGURE 5.7 Normal Distributions with Different
Means and Standard Deviations

of X, the lower is the curve. Although the curves have different locations and shapes, they
are all normal distributions. Although they have different standard deviations, they all
have the same proportion of cases within any given number of standard deviations from
the mean. That is, $p = .682$ of the cases will be within 0 ± 1, 4 ± 2, and 8 ± 4 for Z,
X_1, and X_2, respectively. And if X_1 and X_2 were standardized to have means of 0 and standard deviations of 1, each would have the same curve as Z in Figure 5.7.

A standardized variable that has a normal distribution is named the **standard
normal distribution.** The lowercase z is used as the symbol for a standardized variable. If the standardized variable is normally distributed, however, an uppercase Z is
used instead for this special variable, as in Figure 5.7. When $\mu = 0$ and $\sigma = 1$ are
entered into Equation 5.7, it simplifies to the following formula for the standard normal distribution:

$$P = \frac{1}{\sqrt{2\pi}}\, e^{-Z^2/2} \tag{5.8}$$

A detailed description of the area under the standard normal distribution is given in
Table A.1 in the appendix. Because the normal distribution is symmetrical, only the upper half of the distribution is shown in Table A.1. For any positive $Z (Z \geq 0)$, this table
gives the area under the normal curve lying to the right of Z—that is, the proportion
of all cases with values greater than Z. Technically, this area equals $1 - cp$ (1 minus the
cumulative proportion) for Z. For example, if you wanted to know the proportion of
all cases with $Z > 1.5$ in a standard normal distribution, you would look for the area
A shown in Figure 5.8. The fact that $A = .06681$ for $Z = 1.5$ indicates that only $p =
.06681$ of the cases are more than 1.5 standard deviations *above* the mean in a normal
distribution. To find this area in Table A.1, go down the left-hand column labeled Z to
the row containing 1.5 and go across to the column labeled .00. The cell at this location represents $Z = 1.50$ and contains the number .06681, the desired value of A.

Because the normal distribution is symmetrical, the area above Z equals the area below negative Z; that is, $1 - cp_Z = cp_{-Z}$. For $Z = 1.5$, $1 - cp_{1.5} = cp_{-1.5} = .06681$ (Figure 5.9). If you want to know the total area under the curve that is more than Z
standard deviations above or below the mean, simply sum A for Z and $-Z$, or double
A for Z. The proportion of all cases that are more than $Z = 1.5$ from the mean equals
$.06681 + .06681 = 2 \times .06681 = .13362$. Thus, .13362 or 13.362 percent of the cases
are more than 1.5 standard deviations from the mean of a normal distribution.

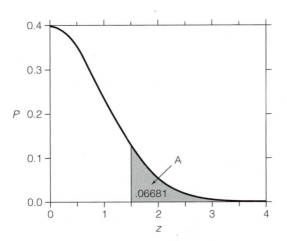

FIGURE 5.8 Area (A) under Normal Distribution above $Z = 1.5$

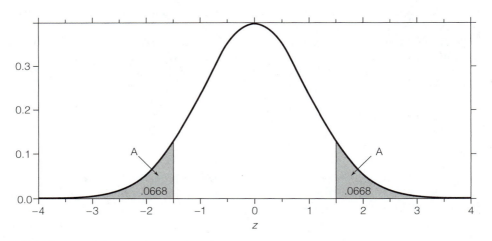

FIGURE 5.9 Areas (A) under Normal Distribution
above $Z = 1.5$ and Below $Z = -1.5$

The opposite of the last question is: What proportion of cases are within Z standard deviations from the mean of a normal distribution? For $Z = 1.5$, this question asks what proportion lies in the interval $-1.5 < Z < 1.5$. The proportion equals $1 - 2A = 1 - 2 \times .06681 = 1 - .13362 = .86638$. In general, to find the area under the normal curve in the interval $0 \pm Z$, look up the value of A corresponding to Z in Table A.1, double A, and subtract the product from 1.

Let's determine the areas or proportions for another value of Z, $Z = 1.96$. This may look like a strange value of Z to be interested in, but, as we shall see, it has great statistical significance. In Table A.1, go down the first column to $Z = 1.9$ and go across to the column headed .06, where you will find .02500, the desired value of A. Thus, $Z >$

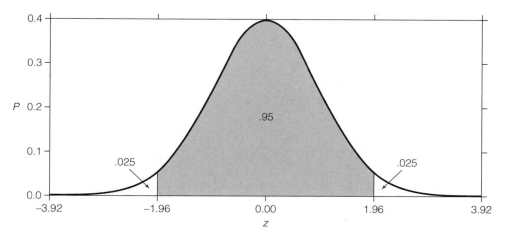

FIGURE 5.10 The Middle .95 of the Standard Normal Distribution

1.96 for .025, or 2.5 percent of the cases in a standard normal distribution. Next, $2A = 2 \times .025 = .05$, or 5 percent of the cases are more than 1.96 standard deviations above or below the mean of a normal distribution. Finally, $1 - 2A = 1 - .05 = .95$, or 95 percent of the area under the normal curve is less than (i.e., within) 1.96 standard deviations of the mean. These areas are shown in Figure 5.10.

Exactly 5 percent of the cases are located more than 1.96 standard deviations away from the mean in a normal distribution. This is a small minority of the cases. In terms of sampling, this means that if we selected a random sample of $n = 100$ cases from a population with a normal distribution, only 5 out of 100 would be expected to be located in either the upper or lower .025 of the distribution—that is, more than 1.96 standard deviations from the mean. We say, therefore, that the probability of randomly selecting a case from these regions of the distribution is $P = .05$. Statisticians have come to use *a probability of .05 as a useful definition of an unlikely event.* As a consequence, hypotheses that are found to have a probability of only .05 of being true are usually rejected because of an insufficient likelihood of being true. But we are getting a bit ahead of the story. Suffice it to say at this point that .05 is a *significant* probability for hypothesis testing, and the associated $Z = 1.96$ is also an important number.

Now, of course, 1.96 is only slightly less than 2.00. The area A under the normal curve lying beyond $Z = 2$ is .02275 (Figure 5.5 and Table A.1) and $2A = .0455$, which equals .05 when rounded off to 2 significant digits. Thus, $Z > 2.00$ or $Z < -2.00$ have about the same probabilities of occurring as $Z > 1.96$ or $Z < -1.96$. Consequently, you can think of the areas in the tails of a normal distribution that are more than 2 standard deviations from the mean as defining outcomes that have a probability of approximately .05.

Normal Sampling Distribution

We now return to the discussion of the sampling distribution of the mean. Remember that Equation 5.5 indicates that the sampling distribution of the mean will have a normal distribution under certain conditions. The standard deviation of this distribution, the standard error of the mean, is given by Equation 5.4. We saw, for example, that the

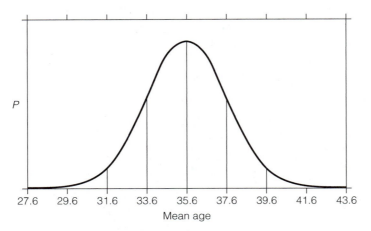

27.6   29.6   31.6   33.6   35.6   37.6   39.6   41.6   43.6

Mean age

FIGURE 5.11 Normal Sampling Distribution of Mean Age, *n* = 125

standard error of the mean of age equals 2.0 for samples of $n = 125$. Given Equations 5.3 through 5.5, we expect the sampling distribution of mean age to have a normal distribution like the one shown in Figure 5.11. The distribution is centered around the population mean, $\mu = 35.6$. The intervals marked on the age axis are 2 years wide, which equals the standard error of age. Samples with means greater than 39.6 have standardized scores that are greater than

$$Z_{\overline{X}} = \frac{\overline{X} - \mu_X}{\sigma_{\overline{X}}} = \frac{39.6 - 35.6}{2} = \frac{4}{2} = 2$$

So, the area A under the normal sampling distribution to the right of 39.6 years is .02275, and the area below 31.6 years [$Z = (31.6 - 35.6)/2 = -2$] equals .02275. Thus, the probability of getting a sample mean greater than 39.6 or less than 31.6 (i.e., more than 2 standard errors from the mean) is .02275 + .02275 = .0455 ≈ .05. This means that we *could* randomly select a sample with a mean that is more than 4 years above or below the population mean, but the low probability of being this far off target indicates this outcome is not very likely ($P < .05$). We can say that we have a very high probability ($P \approx .95$) of getting a sample mean that is in the range of 31.6 − 39.6—that is, $31.6 < \overline{X} < 39.6$. In terms of a probability of *exactly* .95, we are 95 percent confident that the mean of a randomly selected sample will lie in the interval $\mu \pm 1.96\sigma_{\overline{X}}$. This is the *95-percent confidence interval for the sample mean.*

The concept of a *confidence interval* is a very important part of inferential statistics. It will be treated in the next chapter. We should, however, note two important qualifications about the 95-percent confidence interval just described. First, this interval is usually *unknown* because in most research situations, unlike the case of age, we do not know the population mean and standard deviation of X (μ and σ). In the next chapter, however, we will discuss a method for *estimating* a confidence interval. Second, we want to estimate a confidence interval for the population mean rather than the sample mean because it is the population mean about which we want to make inferences.

Assumptions

Certain conditions must be satisfied before Equations 5.4 and 5.5 can be treated as valid. These conditions are **assumptions** that researchers make about the method of sampling, the sampling fraction, and the shape of the distribution of X in the population. Actually, it is somewhat of a misnomer to call them assumptions because that suggests that we cannot check the truth of these conditions, which is not the case. We will specify three conditions or assumptions for the validity of Equations 5.4 and 5.5 and discuss the consequences for our knowledge about the sampling distribution of the mean if these conditions are not true. The three assumptions are

$$\text{SRS} \tag{A5.1}$$

$$\frac{n}{N} \approx 0 \tag{A5.2}$$

$$X \sim N \tag{A5.3}$$

The meaning and significance of each assumption is discussed, as well as the consequences if the assumption is not true.

Simple Random Sampling

The first assumption is that simple random sampling (SRS) is the method used to select the sample of cases that are observed. This assumption is important for two reasons. First, if random sampling is not used, Equation 5.3 will not be true. That is, the mean of the sampling distribution will not equal the population mean. In this case, we cannot claim that the sample mean is *unbiased*. Actually, it is not necessary that *simple* random sampling be used in order for the mean to be unbiased. It is only necessary that some form of random sampling with *equal probability of selection* be used. Equal probability of selection means that each case in the population has an equal probability of being selected for the sample. Simple random sampling guarantees this, but so do other types of random sampling with equal probability of selection, such as the method of *cluster sampling* used in the GSS.

A second consequence of assumption A5.1 does depend on *simple* random sampling. Equation 5.4 ($\sigma_{\bar{X}} = \sigma_X / \sqrt{n}$) for the standard error of the mean (the standard deviation of the sampling distribution) will not be true unless simple random sampling is used. If cluster sampling is used, the observations are not *independent* of each other as they are in simple random sampling. For example, if a sample of twenty blocks (twenty clusters) is first selected randomly in a community, if a sample of five households is randomly selected in each block, and if one person is selected from each household, we have a cluster sample of $n = 100$. These are not equivalent to 100 independent observations, however, because the people living on the same block tend to have many similar characteristics (similar Xs). If, for example, $X = $ income and you find that one household on a block has a total income of \$20,000, you may expect the other four households to have similar incomes. The five people randomly selected from the same block are more likely to have similar incomes than would a simple random sample of five people from the entire community. As a consequence, a cluster sample of 100 cases will have a *larger* standard error than a simple random sample of 100 cases. Thus, Equation 5.4, being based on the assumption of simple random sampling, will *underestimate* the standard error of the mean for a cluster sample. On the other hand, if the sample was selected by *stratified sampling,* a method used to reduce sampling error, Equation 5.4 would *overestimate* the standard error of the mean. Specialized methods of estimating standard errors are available for cluster and stratified sampling, but they are beyond the scope of this book.

Sampling Fraction Approximately Equals 0

The second assumption (A5.2) states that the **sampling fraction** n/N (sample size divided by population size) should be approximately equal to 0. The sampling fraction can never equal 0 because $n > 0$. And the sampling fraction can never equal 1 because $n < N$. Therefore, $0 < n/N < 1$. In the two examples of sampling from a population of $N = 5$ Trappist monasteries, $n_1/N = 2/5 = .4$ and $n_2/N = 3/5 = .6$. Thus, the two sampling fractions were .4 and .6 (or 40 and 60 percent) of the population size. These are actually very large sampling fractions for social research. Consider, instead, the GSS, which has a sample size of approximately $n = 3{,}000$ from a population of approximately $N = 198{,}108{,}000$ people 18 years of age or older. Therefore, the sampling fraction for the GSS is about $3{,}000/198{,}108{,}000 = .000015$, which is very close to 0. In the GSS, only $.000015 \times 100 = 0.0015$ percent of the population is selected into the sample.

The importance of assumption A5.2 is that the larger the sampling fraction, the more Equation 5.4 *overestimates* the standard error of the mean. When n/N is approximately equal to 0, however, Equation 5.4 has no appreciable bias. There is a correction factor that can be used to eliminate the bias provided by Equation 5.4, no matter how large or small the bias. This factor is called the **finite population correction** (FPC):

$$\text{FPC} = \sqrt{\frac{N - n}{N - 1}} \tag{5.9}$$

Notice that the numerator $(N - n)$ will always be positive and less than the denominator $(N - 1)$, except for the extremely unusual case of $n = 1$. Therefore, $0 < \text{FPC} < 1$; the FPC is always a fraction. In the example involving $N = 5$ Trappist monasteries, the finite population corrections for the sample sizes of 2 and 3 are

$$\text{FPC} = \sqrt{\frac{N - n_1}{N - 1}} = \sqrt{\frac{5 - 2}{5 - 1}} = \sqrt{\frac{3}{4}} = \sqrt{.75} = .866$$

and

$$\text{FPC} = \sqrt{\frac{N - n_1}{N - 1}} = \sqrt{\frac{5 - 3}{5 - 1}} = \sqrt{\frac{2}{4}} = \sqrt{.5} = .707$$

For the GSS, the FPC equals

$$\text{FPC} = \sqrt{\frac{198{,}108{,}000 - 3{,}000}{198{,}108{,}000 - 1}} = \sqrt{\frac{198{,}105{,}000}{198{,}107{,}999}} \sqrt{.999985} = .999992$$

Thus, for the GSS, the FPC is almost exactly equal to 1.

These examples illustrate the fact that as the sampling fraction approaches 0 $(n/N \rightarrow 0)$, the finite population correction approaches 1 (FPC $\rightarrow$ 1). This can be seen in Equation 5.9 if we note that for large values of N, $N - 1$ is only slightly less than N. Thus, if the denominator of Equation 5.9 is changed to N, the finite population correction is approximately

$$\text{FPC} \approx \sqrt{\frac{N - n}{N}} = \sqrt{\frac{N}{N} - \frac{n}{N}} = \sqrt{1 - \frac{n}{N}} \tag{5.10}$$

Equation 5.10 indicates that the FPC equals the square root of 1 minus the sampling fraction. So, as $n/N \to 0$, the FPC $\to 1$. This means that when Assumption 2 is true, FPC ≈ 1.

The way that the FPC is used to calculate the corrected standard error of the mean ($\tilde{\sigma}_{\overline{X}}$) is to multiply the FPC times the usual formula (Equation 5.4) to get

$$\tilde{\sigma}_{\overline{X}} = \sqrt{\frac{N - n}{N - 1}} \frac{\sigma_X}{\sqrt{n}} \tag{5.11}$$

Because Equation 5.11 equals $\sigma_X/\sqrt{n}$ multiplied by the FPC, as the FPC approaches 1 due to the sampling fraction approaching 0, the corrected value of the standard error is virtually equal to the uncorrected standard error. Thus, as long as the sampling fraction is very small (assumption 5.2), we can ignore the FPC and use Equation 5.4 for the standard error. Practically speaking, as long as $n/N < .1$, the finite population correction may be ignored. This is the case in much social research where the sample size is only a small fraction of the population size.

You should remember, however, that there may be times when using the FPC will work to your advantage because it may significantly reduce the calculated standard error, which is a good thing. Let's take the example of samples of $n = 3$ from the population of $N = 5$ monasteries. The population standard deviation of X, which can be calculated from the five population values of X in Table 5.5 (calculations not shown), is $\sigma_X = 2$. According to Equation 5.4, the standard error would equal

$$\sigma_{\overline{X}} = \frac{\sigma_X}{\sqrt{n}} = \frac{2}{\sqrt{3}} = \frac{2}{1.7321} = 1.1547$$

But, applying Equation 5.11 to get the corrected standard error, we have

$$\tilde{\sigma}_{\overline{X}} = \sqrt{\frac{N - n}{N - 1}} \frac{\sigma_X}{\sqrt{n}} = \sqrt{\frac{5 - 3}{5 - 1}} \frac{2}{1.7321} = .707 \times 1.1547 = .8165$$

In this case, ignoring the FPC would result in a substantial overestimate of the standard error. It is also important to note that the value that was just obtained by applying Equation 5.11 is the same value that was obtained earlier in the chapter when we calculated the mean for each of the ten unique samples of size $n = 3$ and then calculated the standard deviation of the ten sample means. That is, Equation 5.11 gives the same value of the standard error as would be obtained by calculating the standard error directly from the sampling distribution.

X Is Distributed Normally

The last assumption (A5.3) is that X has a normal distribution in the population. If X is not normally distributed, we cannot conclude, strictly speaking, that the sampling distribution of the mean will have a normal distribution. This limitation is potentially very damaging because the most important techniques in inferential statistics are derived

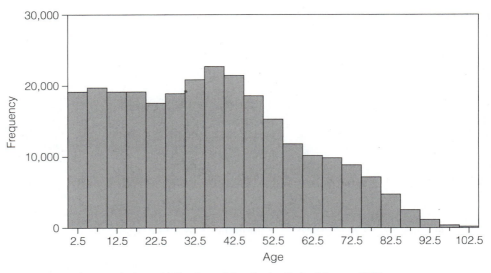

FIGURE 5.12 The Population Distribution of Age in the United States, 1997

from the idea that the sampling distribution of the mean, and of many other statistics as well, is normally distributed. Because many, if not most, Xs do *not* have normal distributions, it would appear that we are on very tenuous grounds in treating the sampling distribution as being normally distributed. We gave, for example, the sampling distribution of *mean age* as an example of a normal sampling distribution. Figure 5.12 shows very clearly that age itself, however, does not have a normal distribution. It is not symmetrical and thus not normally distributed. The skewness value of age equals .357, verifying that it is right-skewed. The kurtosis value equals −.682, which means that age is less peaked than a normal distribution (in which kurtosis equals 0).

Central Limit Theorem Fortunately, the central role that the normal distribution plays in inferential statistics is validated by a remarkable statistical principle, the **central limit theorem.** This theorem states that as sample size n increases, the sampling distribution of the mean approaches a normal distribution even though X itself does not have a normal distribution. The normal distribution is the limiting shape of the sampling distribution of the mean as n becomes very large. Thus, even though age itself is not normally distributed, the sampling distribution of its mean will have approximately a normal distribution when n is large. This means that if we invest our resources in selecting a sufficiently large sample, then we do not have to worry much about assumption A5.3. Many social science samples are sufficiently large to nullify concern about this assumption. Researchers conducting studies with sample sizes in the thousands (e.g., $n = 3000$ in the GSS) will almost certainly have little or no reason to worry about this assumption. Even if the sample size is only several hundred, the sampling distribution of the mean should be approximately normal for most variables. However, if you have a small sample of only 20 or 30 cases, *and* if you have a good reason to question the assumption that X is normally distributed in the population, then you should probably avoid the inferential statistics that are stressed in much of the remainder of the book. This warning should alert you to the importance of planning to obtain a relatively large sample when designing a study. There are additional good reasons for planning to obtain a large sample, such as the fact that the sampling error decreases as n increases.

SUMMARY

The objective of this chapter is to introduce some important concepts used in making inferences about population parameters from sample statistics. A population is a group from which a sample is selected. Sampling error is the difference between a sample statistic, such as the mean of the sample, and a population parameter, such as the mean of the population. Because the population parameter is unknown, the sampling error is unknown. However, we can estimate the sampling error if the sample is a probability sample—that is, a random sample. A random sample is one in which the elements to be selected cannot be predicted in advance. A sampling method that guarantees that every element in the population has an equal chance of being selected is necessary for this unpredictability to exist. Thus, tables of random numbers can be used to select such a sample. In a simple random sample, which is the basis for the inferential statistics covered in this textbook, each member of the population is given a unique identification number, and then n (the desired sample size) distinct random numbers between 1 and N (the population size) are used to select the members of the sample.

The sampling distribution of the mean is the distribution of the sample means from all distinct samples of size n that can be selected from a population of size N. That is, the sampling distribution of the mean gives the frequencies of all possible sample means that can result from a random sample of size n. Because the sampling distribution cannot be observed, we use statistical theory to define its central tendency, standard deviation, and shape. The mean of the sampling distribution of the mean equals the population mean, which indicates that the sampling method is unbiased, or representative. The standard deviation of the sampling distribution is the standard error, or the standard amount of sampling error that can be expected. The standard error of the mean equals the population standard deviation of X divided by the square root of n. The shape of the sampling distribution is a normal distribution. A normal distribution is a bell-shaped symmetrical distribution with a kurtosis value of 0. In a normal distribution, about two-thirds of the values are within plus or minus 1 standard deviation of the mean, and 95 percent of the values are within plus or minus 1.96 standard deviations of the mean.

There are three assumptions on which the characteristics of the sampling distribution of the mean are based. The first is random sampling. Without a random sample, the sampling distribution will not be unbiased. Furthermore, the formula for the standard error is based on a specific type of random sampling: namely, simple random sampling. The second assumption is that the sampling fraction (sample size divided by population size) is approximately equal to 0. When the sampling fraction is appreciably greater than 0, the formula for the standard error must be multiplied by the finite population correction. The third assumption is that the variable being studied is normally distributed in the population. This assumption is necessary for the sampling distribution to have a normal distribution. However, the central limit theorem tells us that even if the variable is not normally distributed, the sampling distribution of the mean approaches a normal distribution as sample size becomes larger and larger. Consequently, the assumption that the variable is normally distributed is a concern mainly for small samples.

The concepts and principles discussed in this chapter will be applied in the following chapter to make inferences, or estimates, about population parameters from sample statistics. In other words, this chapter has laid the theoretical foundation for the applied statistics to follow.

KEY TERMS

population	sampling error	normal distribution
simple random sampling	sampling distribution	standard normal distribution
random numbers	standard error	
sampling without replacement	mean of sampling distribution of mean	assumptions
sample statistic	unbiased estimator	sampling fraction
population parameter	standard error of mean	finite population correction
		central limit theorem

SPSS PROCEDURES

SPSS can be used to select a simple random sample. In the *Data editor*, enter the names or ids of the population from which you wish to take a sample. For example, enter the names of the five Trappist abbeys listed in Table 5.5. Click on *Data* in the main menu and *Select Cases* in the submenu, and the *Select Cases* window will open and you will click on *Random sample of cases*. In the *Select Cases: Random Sample* window, click on *Exactly*, enter 2 in the box following *Exactly*, and enter 5 in the next box. This will randomly select two abbeys from the population of five. After clicking *Continue* and *OK*, you will see a new variable in the *Data editor* window named *filter_$* on which selected cases have a score of 1 and nonselected cases have a 0. Also, there will be a slash drawn through the case number of the nonselected cases. You can see that Rochefort and Westmalle, which make two excellent ales, have been selected for the sample.

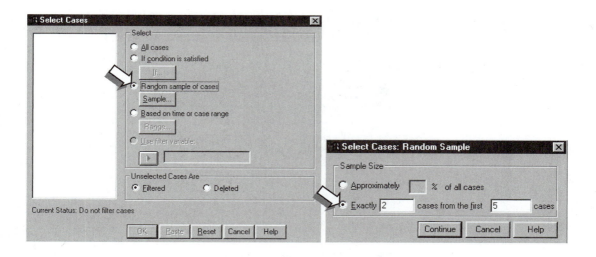

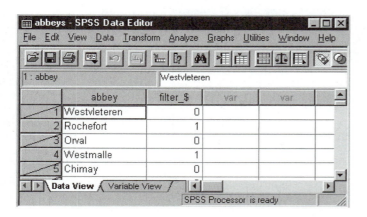

SPSS can also be used to look up the areas of the standard normal distribution that are given in Table A.1. Enter the standardized scores (z) in the *Data editor* for which you want to know the area under the normal distribution (e.g., 1.0, 1.5, 2.0, 2.5, 3.0). Click on *Transform* in the main menu and *Compute* in the submenu. In the *Compute Variable* window, find *CDFNORM(zvalue),* which stands for normal cumulative density function, in the *Function* box and click on it. Then move it into the *Numeric Expression* box by clicking on the pointer. Replace *zvalue* inside the parentheses of *CDFNORM(zvalue)* with z, the name of the variable containing the standardized scores you want to look up. Finally, enter the name of a new variable in the *Target Variable* box, which we named *cp* for cumulative proportion. After clicking on *OK,* you will find the cumulative proportions in the *cp* variable in the data editor. For example, the z score of 2.5 has a *cp* of .9938. If you multiply this by 100, you have the percentile rank of a z score of 2.5 in a normal distribution. This indicates that 99.38 percent of the distribution is below $z = 2.5$—that is, less than 2.5 standard deviations above the mean. Compute $1 - .9938 = .0062$, and you have the area above $z = 2.5$ that is listed in Table A.1.

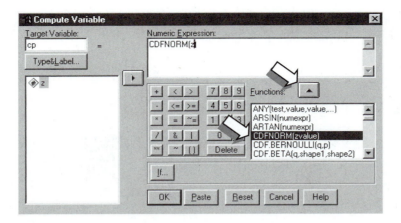

EXERCISES

1. Questionnaires were mailed to 500 households that were randomly selected from the telephone directory for Boise, Idaho, in order to measure household characteristics such as the number of adults and the number of children residing in each household. Describe the study population for this study.

2. All of the 120 students enrolled in Psychology 101 at Heinz College were offered extra credit for participating in a psychology experiment to be conducted outside of class. Fifty students accepted the offer and participated.

   a. Describe the study population for the experiment.
   b. Are the fifty students who participated in the experiment a probability sample? Why or why not?

3. Select a random sample of 8 hospitals to be surveyed about their costs per in-patient out of 37 hospitals in the ABC Standard Metropolitan Area. Starting at the top and going down the rightmost columns in Table 5.1, give the random numbers that will be used to identify the 8 hospitals to be included in the survey.

4. Select a random sample of 10 out of 437 employees in XYZ Services, Inc. to participate in a confidential focus group to discuss company morale issues. Starting at the top

and going down the leftmost columns in Table 5.1, give the random numbers that will be used to identify the 10 employees to be included in the focus group.

5. The mean age at first birth of the 631 women in Table 3.8 from the 1998 GSS is 22.6 years. Assume the mean age at first birth is 21.5 years for all English-speaking women aged 45 or older who reside in households in the continental United States.

   a. What is the value of the population parameter?
   b. What is the value of the sample statistic?
   c. What is the sampling error?

6. A student reporter wanted to know the average salary of professors at a private university. A simple random sample of twenty professors responded to an anonymous survey conducted by the reporter, and the mean of their salaries was $65,534. When the reporter asked a high-ranking administrator if this figure was accurate, the administrator said that there was a sampling error of −$3,241. What is the value of the population parameter?

7. There are 35 unique samples of size 3 that can be selected from a population of 7 persons. Assume that the mean age is calculated for each of the 35 samples. The mean of the 35 means is 47.6 years of age, and the standard deviation of the 35 means is 3.3 years.

   a. What is the standard error of the mean?
   b. What is the value of the population parameter?

8. There are four sororities at a commuter college. The number of women in the sororities are 26, 42, 35, and 49. For sample sizes of 3, construct the sampling distribution of the mean number of sorority members. Use the sampling distribution to compute the population mean and the standard error of the mean. Demonstrate that the sample means are unbiased.

9. When the midterm exams in a large chemistry class were returned, the instructor said that the scores were approximately normally distributed with a mean of 76.5 and a standard deviation of 11.3. LaWanda's score was 85.

   a. Approximately what percent of the students had higher scores than LaWanda?
   b. What percent were below LaWanda?
   c. If the instructor gave a grade of A to scores of 90 and above, what percent of the class received an A?
   d. What percent of the class was more than 2 standard deviations from the mean?

10. An IQ test is normally distributed, with a mean of 100 and a standard deviation of 15.

   a. What percent of persons have IQ scores greater than 145?
   b. What percent have IQs greater than 70?
   c. What percent have IQs between 80 and 120?

11. In planning a study of alcohol consumption among urban males, a sociologist estimated from previous research that the population standard deviation of pints of alcohol consumed per year is 24.6 pints. To help decide what sample size to collect, the sociologist wants to know the standard error of mean alcohol consumption for samples of 200, 400, 800, and 1,200 males from a population of approximately 120,000 adult males in an urban area.

   a. What is the finite population correction (FPC) for the sample of 1,200? Can it be ignored in calculating the standard error of the mean?
   b. What are the standard errors of the mean for the four sample sizes?

   c. When the sample size is quadrupled from 200 to 800, what should happen to the standard error? Is that what you found earlier?

   d. If the study were being conducted in an urban area with only 10,000 adult males, what are the standard errors with and without using the FPC for a sample of 1,200?

12. At a small liberal arts college of 1,500 students, a professor was planning to study how many hours per week students typically studied. He made an educated guess that the standard deviation is 10 hours per week.

   a. What would be the standard errors for samples of 100, 400, and 900, with and without the use of the FPC?

   b. What would be the standard error if all 1,500 students were included in the study, with and without the use of the FPC?

13. Assume that the respondents in the GSS 2000 (*gss2000a* or *gss2000b*) are a population and not a sample.

   a. Using SPSS, what are the mean age (μ) and the standard deviation of age (σ) in the population?

   b. Select a random sample of 30 out of N (N is the number of cases in *gss2000, gss2000a,* or *gss2000b*). What is the sample mean? What is the sampling error? Ignoring the FPC, what is the standard error of the mean? (Use only valid cases.) How many standard errors is the sample mean above or below the population mean? What percent of all possible samples of this size are expected to deviate at least this far from the population mean, assuming the sampling distribution of the mean has a normal distribution?

   c. Select a sample of 300 out of N. Answer the questions in Exercise 13b for this sample, using the FPC. Is the sampling error greater or less than for the sample of 30?

   d. Select a sample of 600 out of N. Answer the questions in Exercise 13b for this sample, using the FPC. Is the sampling error greater or less than for the sample of 300?

14. Again, assume that the respondents in the GSS 2000 (*gss2000a* or *gss2000b*) are a population and not a sample. Estimate the true sampling distribution of mean *age* by selecting ten random samples of size 300 and computing the mean age in each sample. How much does the mean of the ten sample means differ from the population mean? Is the sampling distribution approximately unbiased? Calculate the standard error of the mean from the sampling distribution. How much does it differ from the standard error calculated in Exercise 13c? (*Hint:* Ten samples may not give an accurate description of the sampling distribution.)

6

Inferences about
the Mean

TOMORROW'S WEATHER WILL BE . . .

Report from the Weather Channel, February 7, 2010: The weather will be highly variable across the U.S. tomorrow. You folks down in Mobile, Alabama, better take your raincoats and umbrellas to work 'cause the high will be only in the mid-to-upper 40s with a 90 percent chance of rain. That temperature, however, probably sounds pretty good to you nice people up in Fargo, North Dakota, because you will be lucky to see the temperature get above 20 and with 4 to 8 inches of fresh snow expected to fall overnight, you are certainly going to need your snowblowers. Meanwhile out west, my friends in LA will be laughing all the way to the beach because they don't need a weatherman to know there'll be no snow to blow. Is this a great country, or what?

What would we do without weather forecasts? Although we depend on them for making plans, we know the forecasts aren't rocket science. Sometimes they state the obvious. More often, they are full of hedges and qualifications. And, despite their generous margin of error, they are sometimes just plain wrong. But we learn to live with them, even depend on them, because we know that more often than not the predictions will be in the ballpark. Yes, weather science may not be rocket science, but it is better than no science at all.

   The forms of the forecasts will sound familiar to all of us. We are accustomed to hearing weather predictions expressed in terms of ranges of temperatures or precipitation, or in the form of a probability that there will be *some* rain or *some* snow. In these respects weather forecasts are very similar to what are called *confidence intervals* and *hypothesis tests* in inferential statistics. That is, both weather forecasters and social statisticians rely on probabilistic predictions about the ranges in which phenomena are believed to lie. Although weather forecasters usually predict what the *future* will be like, whereas social scientists are predicting what a *population* is like, you should feel comfortable with the kinds of inferential conclusions that social scientists typically draw.

CONFIDENCE INTERVALS

Theoretical Confidence Intervals for the Mean

Because of sampling error, we know that our sample statistic (e.g., the mean) will not necessarily equal the population parameter. We can, however, use the sample statistic to construct a confidence interval for the population parameter. *A* **confidence interval** *specifies a range of values that has a certain probability of including the target population parameter.* The notion of a confidence interval was alluded to in Chapter 5 when we examined the fact that 95 percent of the sample means would lie within 1.96 standard errors of the population mean ($\mu \pm 1.96\sigma_{\overline{X}}$). Because we usually compute a sample mean to estimate an unknown population mean, however, we would like to construct a confidence interval around the sample mean that has a certain likelihood (e.g., $P = .95$) of including the population mean. If the sample mean that we have calculated has a 95-percent probability of being within 1.96 standard errors of the population mean, then the reverse is also true. There is a 95-percent probability that the population mean is within 1.96 standard errors of our observed sample mean. Therefore, the 95-percent confidence interval (CI) for the population mean equals the sample mean plus or minus 1.96 standard errors:

$$95\% \text{ CI}_{\mu} = \overline{X} \pm 1.96\sigma_{\overline{X}} \tag{6.1}$$

or

$$P[(\overline{X} - 1.96\sigma_{\overline{X}}) \leq \mu \leq (\overline{X} + 1.96\sigma_{\overline{X}})] = .95 \tag{6.2}$$

Equation 6.2 is read: *The probability that the sample mean minus 1.96 standard errors is less than or equal to the population mean, which is less than or equal to the sample mean plus 1.96 standard errors, equals .95.*

Equation 6.2 is illustrated in Figure 6.1. Although we know that there is a 95-percent probability that the sample mean is within 1.96 standard errors of the population mean, we do not know whether the sample mean is above, equal to, or below the population mean. Figure 6.1 shows two sampling distributions, A and B, whose means form the lower and upper extremities of the 95-percent confidence interval. The mean of sampling distribution A is 1.96 standard errors below the sample mean, and the mean of sampling distribution B is 1.96 standard errors above the sample mean. We know that there is a 95-percent probability that our sample mean did not come from a sampling distribution that is further away than sampling distributions A and B. There is a probability of .025 that the sample mean came from a sampling distribution to the left of A and a probability of .025 that it came from a distribution to the right of B, but these are

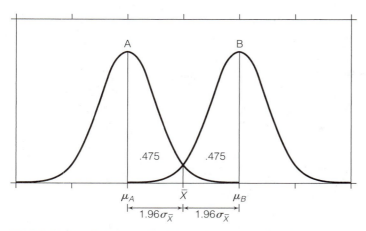

FIGURE 6.1 The 95-Percent Confidence Interval for the Mean

small probabilities. Therefore, we can be confident that the population mean μ is in this interval.

Although the 95-percent confidence interval is used frequently, investigators may choose other confidence intervals, such as 99-, 90-, or 75-percent. In general, the **theoretical confidence interval of the mean** equals

$$100P\% \text{ CI}_\mu = \overline{X} \pm Z_{\alpha/2}\sigma_{\overline{X}} \qquad (6.3)$$

where $\alpha = 1 - P$. *P is the* **confidence level,** *the probability that the population mean lies within the confidence interval, and thus,* α *is the probability that* μ *is outside of the interval.* That is, α is the probability that the interval is incorrect. If you want a probability of P that the mean is in a certain interval, which is the $100P$ percent confidence interval, then you would calculate $\alpha/2$, locate A $= \alpha/2$ in Table A.1, find the associated $Z_A = Z_{\alpha/2}$ in the table, and enter $Z_{\alpha/2}$ in Equation 6.3. Examples of the Z scores needed for four confidence intervals are shown in Table 6.1. Notice that the greater the desired confidence (i.e., P) that the interval contains μ, the larger will be $Z_{\alpha/2}$, and thus the wider will be the confidence interval. The 99-, 95-, and 90-percent intervals are used frequently, but the 75-percent interval is seldom used.

Remember that we locate these key values of $Z_{\alpha/2}$ in the standard normal table because the sample mean has a normally distributed sampling distribution, and thus

$$z_{\overline{X}} = \frac{\overline{X} - \mu}{\sigma_{\overline{X}}} \sim \text{N} \qquad (6.4)$$

This equation indicates that the standardized sample mean, which has a mean of 0 and a standard deviation of 1, is normally distributed. It is normally distributed because subtracting the population mean and dividing by the standard error of the mean (both constants) is a linear transformation of the sample mean. Although changing the mean and standard deviation, a linear transformation does not change the *shape* of the distribution. Therefore, it is appropriate to use a standard normal (Z) table to find the Z score associated with $\alpha/2$.

Table 6.1 Example Confidence Intervals

100P% CI	$\alpha/2$	$Z_{\alpha/2}$	$\overline{X} \pm Z_{\alpha/2}\ \sigma_{\overline{X}}$
99	.005	2.575	$\overline{X} \pm 2.575\ \sigma_{\overline{x}}$
95	.025	1.960	$\overline{X} \pm 1.960\ \sigma_{\overline{x}}$
90	.050	1.645	$\overline{X} \pm 1.645\ \sigma_{\overline{x}}$
75	.125	1.150	$\overline{X} \pm 1.150\ \sigma_{\overline{x}}$

Calculated Confidence Intervals

In a research project we want to calculate the 95-percent confidence interval described by Equation 6.2. Although we would always know the sample mean $\overline{X}$, we would seldom if ever know the standard error of the mean ($\sigma_{\overline{X}}$). Therefore, although Equation 6.2 is true, we do not have enough information to use this equation to calculate the 95-percent confidence interval. Despite the fact that we do not *know* the true value of $\sigma_{\overline{X}}$, we can *estimate* it with our sample observations and, as we shall see, this will allow us to use a formula that is closely related to Equation 6.2 to calculate the desired confidence interval.

Estimated Standard Error

According to Equation 5.4, the standard error of the mean equals

$$\sigma_{\overline{X}} = \frac{\sigma}{\sqrt{n}}$$

We know n, but we need to estimate σ, the population standard deviation of X, in order to estimate the standard error of the mean. Because the definitional formula for the sample standard deviation (Equation 4.7) gives a biased estimate of σ_X, we will use Equation 5.6 to reduce this bias:

$$\hat{s} = \sqrt{\frac{\Sigma(X - \overline{X})^2}{n - 1}}$$

The improved estimate of σ ($\hat{s}$) involves dividing the sum of squares by $n - 1$ rather than n.

When we substitute $\hat{s}$ for σ in the formula for $\sigma_{\overline{X}}$, we find that the **estimated standard error of the mean,** symbolized by $s_{\overline{X}}$, is

$$s_{\overline{X}} = \frac{\hat{s}}{\sqrt{n}} = \frac{\sqrt{\dfrac{\Sigma(X - \overline{X})^2}{n - 1}}}{\sqrt{n}} = \frac{\sqrt{\dfrac{\Sigma(X - \overline{X})^2}{n}}}{\sqrt{n - 1}} = \frac{s}{\sqrt{n - 1}} \qquad (6.5)$$

Equation 6.5 indicates that the standard error of the mean can be estimated from our sample data by simply dividing the sample standard deviation by the square root of $n - 1$. We use the symbol $s_{\overline{X}}$ because the Roman italic s indicates that it is computed from our sample observations and thus is not a population parameter. Consequently, we aren't talking theoretically anymore, because Equation 6.5 is a statistic to which we can actually pin a number.

As an example, we will consider the level of depression of caregivers of persons with Alzheimer's Disease (AD). We will use the data from a sample of 362 caregivers from the Cleveland metropolitan area who participated in a research project at the Alzheimer's Center of Case Western Reserve University. The participants were predominantly spouses or children of the person with AD who was being cared for at home. Caring for a person with AD is very stressful, especially when the person undergoing the inevitable loss of memory and mental capacities is a spouse or parent. It is, therefore, not unusual for AD caregivers to become significantly depressed. The measure of depression used was the CES-D scale described in Box 6.1. Each caregiver was administered the CES-D scale upon entry into the research project. The range of possible depression scores is 0-60. A score of 16 or greater indicates significant depression.

Figure 6.2 shows the distribution of caregiver depression and several descriptive statistics. You can see that the distribution is skewed to the right, which is confirmed by the positive measure of skewness (1.1). The kurtosis statistic (1.6) indicates that depression is more peaked than a normal distribution. The mean depression score (12.5) is less than 16, which indicates that the average caregiver in the sample is not significantly depressed. Remember, however, that the mean level of depression may be less than 16 in the sample because of sampling error even though the population mean for all AD caregivers is 16 or greater. We will need to compute a confidence interval for the population mean to determine whether 16 is included in the interval.

The first step in constructing the confidence interval is to estimate the standard error of the mean by using Equation 6.5:

$$s_{\overline{X}} = \frac{s}{\sqrt{n-1}} = \frac{9.5}{\sqrt{362-1}} = \frac{9.5}{19} = .5$$

Thus, the standard error of the mean is estimated to be .5, which indicates that .5 is the standard amount that a sample mean deviates from the population mean. This standard sampling error could be either above or below the population mean.

The next step is to select a confidence level—for example, $(100 \cdot P)\% = 95\%$—and then use the standard normal distribution (Table A.1) to find $Z_{\alpha/2}$ associated with $\alpha = 1 - P$. Unfortunately, it turns out that we cannot use the standard normal distribution when we don't know the true value of the standard error of the mean ($\sigma_{\overline{X}}$). Because we are using $s_{\overline{X}}$ to estimate $\sigma_{\overline{X}}$, we have to use another distribution, the t distribution, which is closely related to the normal distribution, to calculate our confidence interval for the population mean.

t Distributions

The t statistic stands for "Student's" *test* statistic (see Box 6.2), but it has come to be known simply as t. The ***t* statistic** is an estimate of the standardized score for a sample statistic. Remember that in general, a standardized score equals a score from a distribution minus the mean of the distribution with the difference divided by the standard deviation of the distribution. As an estimated standard score for the sample mean, t is calculated as follows:

$$t = \frac{\overline{X} - \mu}{s_{\overline{X}}} \qquad (6.6)$$

BOX 6.1 Centers for Epidemiological Studies Depression (CES-D) Scale

1. I was bothered by things that usually don't bother me.
2. I did not feel like eating: my appetite was poor.
3. I felt that I could not shake off the blues, even with help from my friends.
4. I felt that I was just as good as other people.
5. I had trouble keeping my mind on what I was doing.
6. I felt depressed.
7. I felt that everything I did was an effort.
8. I felt hopeful about the future.
9. I thought my life had been a failure.
10. I felt fearful.
11. My sleep was restless.
12. I was happy.
13. I talked less than usual.
14. I felt lonely.
15. People were unfriendly.
16. I enjoyed life.
17. I had crying spells.
18. I felt sad.
19. I felt that people dislike me.
20. I could not get going.

The response alternatives (0–3 points) for each of the above are as follows:

- *rarely or none of the time* {less than one day per week} (0)
- *some or a little of the time* {1–2 days per week} (1)
- *occasionally or a moderate amount of time* {3–4 days per week} (2)
- *most or all of the time* {5–7 days per week} (3)

For the positive statements (4, 8, 12, 16), the points for the response alternatives were reversed. The total CES-D score equals the sum of the points across all twenty statements. The possible range is 0–60. (Radloff 1977)

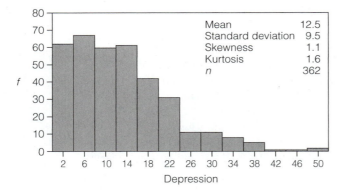

Mean	12.5
Standard deviation	9.5
Skewness	1.1
Kurtosis	1.6
n	362

FIGURE 6.2 Depression in Caregivers of Persons with Alzheimer's Disease

Equation 6.6 is very similar to Equation 6.4 for the standardized mean. The difference is that Equation 6.6 contains the estimate of the standard error of the mean ($s_{\overline{X}}$) in place of the true standard error of the mean ($\sigma_{\overline{X}}$). The consequence is that *although the **t distribution** is symmetrical and has a mean of 0, as does the standard normal distribution, it has a greater standard deviation ($\sigma_t > 1$) and is more peaked than the Z distribution.*[1] Because $s_{\overline{X}}$

---

1. The standard deviation of t equals $\sqrt{(n-1)/(n-3)}$ for samples of n greater than 3. Because the numerator $(n-1)$ is greater than the denominator $(n-3)$, the standard deviation of t is greater than 1. Thus, t is an *underestimate* of the number of standard deviations that the sample mean is away from μ. You can also see by examining the formula that as n increases, the ratio $(n-1)/(n-3)$ gets smaller and approaches 1, the standard deviation of the standard normal distribution, for large values of n.

BOX 6.2   Student's *t*

The statistician and chemist W. S. Gosset discovered the *t* distribution in 1908. At the time, Gosset was employed in the experimental unit of Guinness Breweries in Dublin, Ireland. He had only small samples available for several of his analyses for determining the best varieties of barley and hops for the brewing process. Due to company policy forbidding the publishing of trade secrets, Gosset used the pseudonym "Student" in articles he wrote about this result. His discovery of the *t* distribution is a good example of how the effort to solve an important practical problem like how to brew a better beer can result in the development of a scientific technique with widespread applications.

---

is an estimate of $\sigma_{\overline{X}}$ that varies somewhat from sample to sample (i.e., it has a sampling distribution), the use of $s_{\overline{X}}$ makes the distribution of *t* more dispersed as well as more peaked in the center and thicker in the tails than a normal distribution (i.e., it has greater kurtosis than *Z*).

There is actually a family of *t* distributions. Two illustrative *t* distributions are shown in Figure 6.3, where they can be compared with one another and with the *Z* distribution. A somewhat differently shaped *t* distribution occurs for each sample size *n*, or more precisely for each degree of freedom (*df*):

$$df = n - 1 \tag{6.7}$$

The **degrees of freedom** *of a statistic refers to the number of independent observations on which it is based.* This subject is complicated, but suffice it to say that *df* is always less than the number of sample observations *n*, in this case $n - 1$. You can understand this concept by thinking about the implications of the fact that the sample standard deviation used in the *t* statistic is based on squared deviation scores $(X - \overline{X})^2$. Because $\Sigma(X - \overline{X}) = 0$, once we know any $n - 1$ values of $X - \overline{X}$, the *n*th value will be determined. Thus, because of this statistical constraint, the *t* statistic has only $n - 1$ degrees of freedom. A related way of thinking about degrees of freedom of *t* is that 1 degree of freedom is used for each parameter that must be estimated to calculate *t*. Because a single parameter (μ) must be estimated to calculate the estimated standard error of the mean used in the *t* statistic, the degrees of freedom of *t* is $n - 1$. As we will see, more complicated statistics have fewer than $n - 1$ degrees of freedom.

In Figure 6.3, you can see that the *t* distribution with $df = 1$ has the greatest spread and the thickest tails. This is a rather unrealistic distribution, based on a sample size of only $n = 2$, and it is included for illustrative purposes to show the most extreme instance of the *t* distribution. This is the *t* distribution with the greatest standard deviation and kurtosis.[2] $t_{df=4}$ is also more dispersed than the normal distribution, but it is

---

2. You may ask why the peaks of the two *t* distributions are below the normal distribution if the *t* distribution has greater kurtosis than the normal distribution. The answer is that the larger standard deviations of *t* distributions spread them out more and pull their peaks down below the normal curve. If the *t* distributions had the same standard deviation as the normal distribution ($\sigma = 1$), their peaks would be higher than the normal peak but their tails would still be thicker than normal.

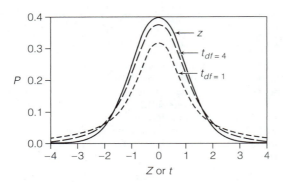

FIGURE 6.3 Selected *t* Distributions and the
Standard Normal Distribution (*Z*)

significantly closer to *Z*. This illustrates the point that as *df* increase, $\sigma_t \to 1$ and kurtosis $\to 0$. That is, *t* approaches the standard normal distribution (*Z*) as degrees of freedom get larger and larger. Because $n = df + 1$, as sample size gets larger and larger, the *t* distribution gets closer and closer to the normal distribution. *The Z distribution is the* **limiting case of the *t* distribution** as sample size becomes infinitely large ($n \to \infty$).

Table A.2 shows selected values from the *t* distributions that may be used to construct confidence intervals. Because there is a different *t* distribution for each *df*, we cannot provide a complete description of each *t* distribution as we did for the standard normal distribution in Table A.1. Table A.1 contains probabilities (A) in the body of the table associated with different *Z* scores, as indicated by the row and column labels. There were 400 different probabilities in Table A.1, one for each of 400 distinct values of *Z*. Table A.2, on the other hand, contains *t* scores in the body of the table. There are 40 rows in Table A.2, and each row represents a *different* *t* distribution, one for each of the listed degrees of freedom Each row contains ten values of *t*, one for each of the ten selected probabilities (A) shown in the column labels. Each probability indicates the proportion of all cases in a *t* distribution lying above or to the right of the 40 positive values of *t* in the column below A (Figure 6.4). $1 - A$ equals the proportion of all cases that are less than *t*. Thus, each *t* score in Table A.2 represents a *percentile*, the $100(1 - A)$th percentile. For example, the cell formed by row 10 (*df* = 10) and column 4 (A = .050) contains the 95th percentile, that is, $t_A = t_{.05} = 1.812$, which means that 5 percent of the cases in a *t* distribution with *df* = 10 (*n* = 11) are greater than 1.812. Because the *t* distribution is symmetrical, the table also indicates that 100A percent are less than $-t$. Therefore, $-t$ equals the 100Ath percentile, and the value in row 10 and column 4 represents $-t_A = -t_{.05} = -1.812$.

In Table A.2 we can also observe how the *t* distribution approaches the *Z* distribution as *df* increase. In the last row of Table A.2 (*df* = ∞), the values of *t* for A equal the values of *Z* for the same A in Table A.1. For example, if A = .025, $t_{.025} = 1.960$. If we look up A for *Z* = 1.96 in Table A.1 (row 20, column 7), we find A = .025. In Table A.2, all the values of *t* in each column are greater than the value in the last row. This

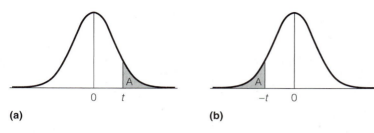

(a) (b)

FIGURE 6.4 t_A and $-t_A$ of the t Distributions

means that for any finite sample $(n < \infty)$, $t_A > Z_A$. For example, $t_{.10}$ is greater than $Z_{.10}$ for all sample sizes (df). This reflects the fact that the t distribution is more spread out or dispersed than the Z distribution.

In each row of Table A.2, however, the values of t decrease as df increases. How great does df have to be before the t distribution is virtually the same as the Z distribution? Certainly for samples of size $n = 501$ $(df = 500)$ there is no appreciable difference between the t_A in that row and the t_A (i.e., Z_A) in the bottom row of the table. Samples of $n > 100$ also have ts very close to the Zs. Some argue that $n > 30$ is large enough to treat the t and Z distributions as equal. As Figure 6.5 shows, you can barely *see* the differences between these two distributions. However, you can read some differences between the two from Table A.2, where $t_{.005}$ is 6.75 percent larger than $Z_{.005}$. This book does not specify a specific value of df at which the normal and t distributions are to be considered identical. When sample size is several hundred or more, however, we can safely assume that the t and Z distributions are interchangeable; that is,

$$\frac{\bar{X} - \mu}{s_{\bar{X}}} \sim N$$

t-Score Confidence Intervals

At this point, we are prepared to calculate a confidence interval for the mean. If we modify Equation 6.3, the theoretical equation for a confidence interval, by substituting $t_{a/2}$ for $Z_{a/2}$ and $s_{\bar{X}}$ for $\sigma_{\bar{X}}$ we get the **t-score confidence interval**:

$$100P\% \text{ CI}_\mu = \bar{X} \pm t_{\alpha/2} s_{\bar{X}} \tag{6.8}$$

where again, $\alpha = 1 - P$ is the probability that μ is outside of the interval. If you want a probability of P that the mean is in a certain interval, which is the $100P$ percent confidence interval, you would first calculate $\alpha/2$, go to the column for A $= \alpha/2$ in Table A.2, go down to the row for $df = n - 1$ to find $t_A = t_{\alpha/2}$, and enter $t_{\alpha/2}$ into Equation 6.8. After computing $s_{\bar{X}}$ and entering it into Equation 6.8, you are ready to calculate the confidence interval.

To illustrate, we continue with the example of depression among caregivers for persons with Alzheimer's Disease (Box 6.1) and calculate the 95-percent confidence interval for their mean depression score. Because $P = .95$, $\alpha = 1 - .95 = .05$ and $a/2 = .025$. The degrees of freedom are $df = n - 1 = 362 - 1 = 361$. We now go to the cell

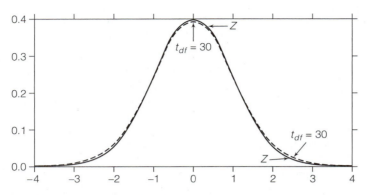

FIGURE 6.5 Normal and *t* Distributions

in Table A.2 that corresponds to the column A = .025 and the row *df* = 361 to find $t_{.025, 361}$. However, there is no row for *df* = 361, so we estimate $t_{.025}$ by taking a value that is about half way between those listed for *df* = 300 and *df* = 400, which are 1.968 and 1.966, respectively. Thus, we will enter $t_{.025}$ = 1.967 into Equation 6.8. We also enter the estimated standard error of the mean $s_{\bar{X}}$ = .5 and the sample mean $\bar{X}$ = 12.5 (Box 6.1) and calculate the 95-percent confidence interval as follows:

$$95\% \text{ CI}_\mu = \bar{X} \pm t_{.025}s_{\bar{X}} = 12.5 \pm 1.967 \cdot .5 = 12.5 \pm .9835$$
$$= 11.5165\text{--}13.4835 \approx 11.5\text{--}13.5$$

As indicated, we first calculate $t_{.025}s_{\bar{X}}$ = 1.967 · .5 = .9835. This product is often called the *margin of error*. In this case, the margin of error is approximately 1 point on the CES-D depression scale. The margin of error is both added to the mean and subtracted from the mean to get the upper limit and lower limit, respectively, of the confidence interval. The upper limit equals 12.5 + .9835 = 13.4835, and the lower limit equals 12.5 − .9835 = 11.5165. The upper and lower limits are expressed as an interval, in this case, 95% CI$_\mu$ = 11.5165–13.4835. The expression 11.5165–13.4835 does not mean 11.5165 *minus* 13.4835 because there are no spaces on either side of "–." It means 11.5165 *to* 13.4835, just as pp. 100–150 means pages 100 to 150. Because the limits of the interval round to 11.5 and 13.5, we can say *there is a 95-percent probability that the mean level of depression among AD caregivers in the Cleveland metropolitan population lies in the interval 11.5 to 13.5.*

In Chapter 4 we considered the range and the interquartile range as measures of dispersion. Similarly, we may also describe *confidence range.* The 95-percent confidence range (CR) equals the upper limit of the 95% CI minus the lower limit of the 95% CI:

$$95\% \text{ CR}_\mu = 13.4835 - 11.5165 = 1.967 \approx 2.0$$

In other words, we are 95 percent sure that the mean population depression is within a 2-point range. Because the total range of depression in the sample is $X_{max} - X_{min}$ = 52 − 0 = 52, our 95-percent confidence range for the mean looks quite small.

Although the 95-percent confidence interval is extremely popular, we can choose to compute a different one. The 75-, 90-, and 99-percent intervals are calculated below as examples:

$$75\% \text{ CI}_\mu = \overline{X} \pm t_{.125}s_{\overline{X}} = 12.5 \pm 1.152 \cdot .5 = 12.5 \pm .576$$
$$= 11.924\text{–}13.076 \approx 11.9\text{–}13.1$$

$$90\% \text{ CI}_\mu = \overline{X} \pm t_{.05}s_{\overline{X}} = 12.5 \pm 1.649 \cdot .5 = 12.5 \pm .8245$$
$$= 11.6755\text{–}13.3245 \approx 11.7\text{–}13.3$$

$$99\% \text{ CI}_\mu = \overline{X} \pm t_{.005}s_{\overline{X}} = 12.5 \pm 2.590 \cdot .5 = 12.5 \pm 1.295$$
$$= 11.205\text{–}13.795 \approx 11.2\text{–}13.8$$

These formulas differ from one another and from 95-percent CI only in the value of $t_{\alpha/2}$; $\overline{X}$ and $s_{\overline{X}}$ are the same in each equation. Notice that as the confidence level (P) increases, $t_{\alpha/2}$ becomes larger and thus the confidence interval is wider. To achieve greater confidence in your inference about the mean, we must be less precise. Or, to put it another way, the more precise we want to be, the greater will be the probability ($\alpha = 1 - P$) that the population mean lies outside the interval. We must, however, choose a single interval to report, and there is little more than convention to guide us in our choice. Social scientists tend to be very cautious in reporting their results and thus choose an interval with a very high confidence level. Most of the time the confidence level will be 95 percent or greater, and it will virtually never be less than 90 percent.

Remember that a score of 16 or greater on the CES-D scale indicates significant depression. The mean of 12.5 is 3.5 points below this level. Can we infer that the average AD caregiver is not depressed? Can we be sure that the mean in the sample is not below 16 due to sampling error? If the sample mean were below a population mean of 16 by chance, then the population mean should lie within the confidence interval we choose to construct. Looking at the conventional 95-percent CI, the upper limit of 13.5 points is still 2.5 points less than 16. Even if we had used a 99-percent interval, the upper limit is 2.2 points less than 16. So, no matter which interval we would have chosen to use in this case, a CES-D of 16 would still be well outside the interval. Thus, we can be very confident that the mean CES-D score is below the level of significant depression. There is still a small probability that the population mean is outside the confidence interval and, thus, that it might be 16 or higher, but the probability is so small that we can more or less discount it.[3]

HYPOTHESIS TESTS

A second type of inference about the mean that is very closely related to confidence intervals is hypothesis testing. In hypothesis testing, however, specific hypotheses about the value or values of the population mean are specified before carrying out the test and then the hypothesis is either rejected or not rejected. In general, *a* **hypothesis test**

---

3. Although the sample mean for caregiver depression (12.5) was less than 16, almost a third (30.9%) of the caregivers were significantly depressed (CES-D $\geq$ 16).

is used to determine whether a hypothesis about the value of a population parameter can be rejected in favor of a logical alternative with no more than some specified probability of error. As we have seen, confidence intervals specify a range of values in which the population mean is believed to lie, without focusing on any particular value of that mean. The same statistical concepts that were used to construct confidence intervals for the mean also will be used to test hypotheses about the mean. Consequently, we will not begin with a section on theoretical hypothesis tests. Unlike confidence intervals in which the same margin of error was added to and subtracted from the sample mean, two types of hypothesis tests will be used—namely, *one-tailed* and *two-tailed* tests.

Two-Tailed Hypothesis Tests

In a nutshell, to conduct a **two-tailed test** of the mean, the investigator hypothesizes that the population mean equals some value of interest, calculates the difference between the sample mean and the hypothesized mean, and rejects the hypothesis if there is only a small probability that the difference is due to chance. In other words, a hypothesis about the population mean is formed, the hypothesis is compared to the sample observations, and the hypothesis is rejected if there is only a small probability that it is true. It is necessary, however, to elaborate this simple three-step description by discussing the eight-step test outlined in Box 6.3.

Steps 2 through 6 in Box 6.3 each consist of quantities that were calculated to construct a confidence interval for the mean (although not necessarily in this order). Steps 1, 7, and 8 are important features of the hypothesis test that were not necessary for a confidence interval. The order of the steps is important. The values of α and *df* (steps 2 and 3) are necessary before step 4 can be accomplished. Steps 5 and 6 (the sample mean and the standard error) must be accomplished before the test statistic *t* can be calculated (step 7). And, steps 1 through 4 must be completed before steps 5 through 8 because the hypotheses and the criterion for reaching a decision must be specified before one looks at the data.[4] If, for instance, we examine the sample mean first, the observed value may influence the formation of the hypotheses. We will describe and discuss steps 1 through 8 using the example of depression among caregivers of persons with Alzheimer's Disease.

Null and Alternative Hypotheses

In everyday usage, if something is null it amounts to nothing; that is, it is invalid. In hypothesis testing, *the **null hypothesis** is one that may potentially be shown to be invalid.* That is, we form a hypothesis that our test *may* nullify, or reject. The **alternative hypothesis** is logically the opposite of the null hypothesis. If the null hypothesis is invalidated, or rejected, then the alternative hypothesis must be true or accepted. On the other

---

4. It is not actually necessary to carry out steps 3 and 4 before steps 5 through 7. Once you have specified *a*, the critical value of *t* cannot be influenced, either consciously or unconsciously, by what you see in the data. In fact, you may not be able to determine *df* before you calculate the mean because you may not know how many valid cases (*n*) there are until the mean is calculated. It is useful to keep steps 2 through 4 together, however, because it helps to emphasize that these steps are all part of determining the criterion (critical *t*) that will be used to make a decision about H_0.

BOX 6.3   Two-Tailed *t*-Test for the Mean

1. State null and alternative hypotheses.
2. Specify α.
3. Determine degrees of freedom.
4. Find $t_{\alpha/2}$.

5. Calculate sample mean.
6. Calculate standard error of mean.
7. Calculate the test statistic t.
8. Make decision and arrive at conclusion.

---

hand, if our test cannot reject the null hypothesis, that does not prove it is true. Although this may appear paradoxical, the null hypothesis is never accepted.

In a two-tailed hypothesis test about the mean, the null hypothesis (H_0) specifies a specific value *a* for the population mean μ that we want to test. The alternative hypothesis (H_1), as the logical alternative to H_0, states that the population mean is not equal to *a*:

$$\text{Null hypothesis, } H_0: \mu = a$$

$$\text{Alternative hypothesis, } H_1: \mu \neq a$$

This means that the alternative hypothesis is specifying that the population mean may be either greater than or less than *a*. That fact suggests why the test is called a two-tailed *t* test. We may reject the null hypothesis if the sample mean is significantly greater than *a* (i.e., it is in the upper tail of the *t* distribution) or if the sample mean is significantly less than *a* (i.e., it is in the lower tail of the *t* distribution).

Let's take the example of caregiver depression as measured by the CES-D scale (Box 6.1). Because a score of 16 on this scale is the minimum score that indicates significant depression, we want to know if the mean level of caregiver depression is greater than 16 or less than 16. Therefore, we specify the following null and alternative hypotheses:

$$\textit{Step 1.} \quad H_0: \mu = 16$$

$$H_1: \mu \neq 16$$

Scores other than 16 can be tested, depending upon the investigator's main research interest. For example, studies have administered the CES-D scale to representative samples of the adult population in which only a small minority of persons could be expected to be clinically depressed, and the mean CES-D score was identified as about 9 (Radloff 1977). An interesting null hypothesis would be that the mean CES-D of AD caregivers is the same as the mean for all other persons. The test of this hypothesis would indicate whether AD caregivers are more depressed, or less depressed, than the average person.

Alpha, Degrees of Freedom, and Critical *t*

When we covered confidence intervals, α was defined as the probability that the population mean is not in the $100P\%$ confidence interval—that is, $\alpha = 1 - P$. Thus, α equaled the probability of making an error. Alpha has an analogous meaning in hypothesis testing: $\boldsymbol{\alpha}$ *equals the maximum probability of incorrectly rejecting the null hypothesis that we are willing to risk*—that is, the probability of rejecting a true H_0. This type of decision-making error is called a **Type I error.** A **Type II error** is incorrectly failing to reject the null hypothesis—that is, of failing to reject a false H_0. *The probability of a Type II error is called β (beta).* To sum up,

$$\alpha = \text{probability of a Type I error}$$

$$\beta = \text{probability of a Type II error}$$

The validities of the various outcomes of a hypothesis test are summarized in Figure 6.6.

Ideally, we would like to minimize both α and β. Unfortunately, there is a trade-off between reducing one probability and restricting the size of the other probability. The smaller we set α, for example, the larger β will be. By trying to protect ourselves against rejecting a true null hypothesis, we increase the probability of failing to reject a false null hypothesis. For example, if we choose an α of .10, β may equal .15, but if we choose a smaller α of .05, β will increase to perhaps .20. We use the words *may* and *perhaps* because the relationship between α and β is rather complex, depending on sample size and how different the true population mean is from the mean specified by the null hypothesis; the larger n and the greater the difference between μ and μ_0, the smaller β will be for any given value of α.

Social scientists have focused more on preventing Type I errors than avoiding Type II errors. As we will see, the standard method for testing hypotheses emphasizes the use of a specified α level when making a decision about H_0. On the other hand, β does not directly enter the equation of hypothesis testing. Therefore, the probability of a Type II error often lurks vaguely in the background or is ignored or entirely forgotten.[5]

We will be using t (Equation 6.6) as the test statistic for deciding whether or not to reject the null hypothesis. The logic of the t test of the null hypothesis requires us first to assume that the null hypothesis is true. Operationally, that means we will enter the mean specified by the null hypothesis (μ_0) into Equation 6.6 to obtain the following test statistic:

$$t = \frac{\overline{X} - \mu_0}{s_{\overline{X}}} \tag{6.9}$$

Essentially, we are asking: If the null hypothesis is true, how far away from the mean of the sampling distribution (μ_0) is our sample mean in units of t? If the sample mean is so far from μ_0 that it is unlikely to deviate much due to random sampling error, then we will no longer continue to assume the null hypothesis is true. That is, we will reject the null hypothesis. We start out by assuming the null hypothesis is true, and we may or may not decide to abandon that assumption.

The α we have chosen is our operational definition of *unlikely to occur by chance.* Now, we need to determine what values of t are so large (in absolute value) that they have a probability of only α *or less* of occurring by chance if the null hypothesis is true. Figure 6.7 illustrates the issue. If our sample mean equals the mean specified by the null hypothesis ($\overline{X} = \mu_0$), then $t = 0$, which is the mean of the t distribution. Because the alternative hypothesis states simply that $\mu \neq \mu_0$, we want to allow for both unlikely pos-

---

5. The probability of a Type II error is considered more formally when a study is being designed. For example, the investigator may choose α and then design the study to collect a sufficiently large sample size n to ensure a desired β. Once the data have been collected, and thus the sample size fixed, we cannot do anything to reduce β after α has been specified.

Decision	Null hypothesis true	Null hypothesis false
Do not reject null hypothesis	Correct decision	Type II error Probability $= \beta$
Reject null hypothesis	Type I error Probability $= \alpha$	Correct decision

FIGURE 6.6 Validities of Hypothesis-Test Decisions

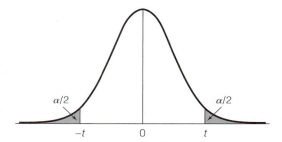

FIGURE 6.7 Critical Values of t for Two-Tailed α

itive values of t and unlikely negative values of t. Therefore, we divide α in half and put $\alpha/2$ under both the upper and lower tails of the t distribution (Figure 6.7). We need to find the positive value of t associated $\alpha/2$ in Table A.2. That is, we need to determine the value of $t_{\alpha/2}$. This value is called the **critical t.** If our calculated t, to be determined shortly, is greater than or equal to $t_{\alpha/2}$ or less than or equal to $-t_{\alpha/2}$, then we know our sample mean lies in the area of the t distribution that is very unlikely to occur by chance if the null hypothesis is true. Therefore, if our calculated $t \geq t_{\alpha/2}$ or $\leq -t_{\alpha/2}$, we will reject the null hypothesis as sufficiently unlikely to be true. That is why $t_{\alpha/2}$ is called the *critical t;* it is necessary for the absolute value of our calculated t to be greater than or equal to the critical t—that is, $|t| \geq t_{\alpha/2}$—in order to reject H_0.

To find the critical values of t, we must determine the degrees of freedom, $df = n - 1$, because there is a different t distribution for each df. Let $n = 36$, and thus $df = 35$. If $\alpha = .05$, then the critical t from Table A.2 is $t_{A,df} = t_{.025,35} = 2.03$. The critical values of t are shown in Figure 6.8. If, on the other hand, we had selected a more liberal $\alpha = .10$, the critical t would be $t_{A,df} = t_{.05,35} = 1.69$ (Figure 6.8). You can see that the region of rejection of the null hypothesis for $\alpha = .10$ is much larger ($t \leq -1.69$ plus $t \geq 1.69$) than for $\alpha = .05$ because there is a greater chance of making the Type I error of rejecting a true null hypothesis.

We are now ready to complete steps 2 through 4 in our hypothesis test of the mean level of depression among AD caregivers:

Step 2. $\alpha = .05$

Step 3. $df = 362 - 1 = 361$

Step 4. $t_{.025,361} = 1.967$

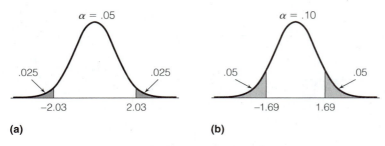

FIGURE 6.8 Critical Values of t for Two-Tailed α, $df = 35$

Notice that the critical value of $t = 1.967$ is the same value we used in the 95–percent confidence interval for the mean level of depression. As we discussed earlier, confidence intervals and hypothesis tests are closely related types of inferential statistics.

Mean, Standard Error, and Calculated t

Steps 1 through 4 should be carried out prior to analysis of the sample observations. The first four steps consist of the specification of the hypotheses and the criterion that will be used to decide between the null and alternative hypotheses. The next three steps consist of analysis of the data—that is, computing the necessary statistics to evaluate the hypotheses. We already computed the sample mean and standard error for the CES–D depression scale when we constructed a confidence interval for the mean. We list them below as steps 5 and 6 along with the **calculated t** (step 7), a statistic that is not involved in confidence intervals:

Step 5. $\overline{X} = 12.5$

Step 6. $s_{\overline{X}} = \dfrac{s}{\sqrt{n-1}} = \dfrac{9.5}{\sqrt{362-1}} = \dfrac{9.5}{19} = .5$

Step 7. $t = \dfrac{\overline{X} - \mu_0}{s_{\overline{X}}} = \dfrac{12.5 - 16}{.5} = \dfrac{-3.5}{.5} = -7.00$

Step 7 consists of dividing the difference between the sample mean and the mean of the null hypothesis by the estimated standard error of the mean to obtain $t = -7$. The negative sign indicates, of course, that the sample mean is less than the mean specified by the null hypothesis. But what does the value 7 represent? As noted before, t is an estimated standard score for the sample mean. As such, it indicates that the sample mean obtained in this study is about 7 standard deviations or standard errors below the hypothesized mean of the sampling distribution. Because we know that t is almost normally distributed in a large sample like ours ($n = 362$), we can appreciate that if the null hypothesis is true, we have obtained a very extreme value of the sample mean.

Decision and Conclusion

Making the decision about the null hypothesis is a straightforward but important final step in hypothesis testing. Reaching the decision is simple, but some subtle points about the conclusion require a little more discussion. The **decision rule** can be stated as follows:

Decision rule $\begin{cases} \text{If } |t| \geq t_{\alpha/2}, \text{ reject } H_0 \\ \text{If } |t| < t_{\alpha/2}, \text{ do not reject } H_0 \end{cases}$

In other words, we compare the absolute value of the calculated t to the critical t (always positive) and reject the null hypothesis if the former is greater than or equal to the latter. The decision if the absolute value of t is less than the critical t is to *not reject* the null hypothesis. This is not the same as *accepting* the null hypothesis. This is a subtle but important point that we will return to later.

Now, let's apply this decision rule to the calculated and critical ts for the mean caregiver depression:

$$\text{Step 8. } (|t| = |-7| = 7) > (1.967 = t_{a/2}), \therefore \text{ reject } H_0$$

The absolute value of t for the mean depression score is greater than the critical t so we reject the null hypothesis that $\mu = 16$. Rejecting the null hypothesis is logically the same as accepting the alternate hypothesis that $\mu \neq 16$. So, technically, our decision is that the mean level of depression on the CES-D scale among AD caregivers is not 16!

This may not sound like much of a conclusion because if the mean is not 16, it could be greater than or less than 16. Remember that in conducting a two-tailed t test we wanted to keep the door open to the possibility that our sample mean might be either greater than or less than μ_0. Therefore, the alternate hypothesis is simply that $\mu \neq 16$, and this is the hypothesis we have accepted. But we can *conclude* more than this because we know whether the sample mean is greater than or less than μ_0. To formalize how we use this information, let us offer the conclusion rule:

$$\text{Conclusion rule } \begin{cases} \text{If } t \leq -t_{\alpha/2}, \text{conclude } \mu < \mu_0 \\ \text{If } t \geq t_{\alpha/2}, \text{conclude } \mu > \mu_0 \end{cases}$$

If our calculated t is less than or equal to the negative of the critical t, we conclude that the population mean is less than the value specified by the null hypothesis. Alternatively, if the calculated t is greater than the critical t, the conclusion is that the population mean is greater than the null hypothesis value. This step is entirely logical but one that can be overlooked because of the wording of the two-tailed alternate hypothesis. For our depression example,

$$t = -7 < -1.967 = -t_{\alpha/2}, \therefore \mu < 16$$

Thus, we conclude that the mean level of depression among AD caregivers is less than 16, the minimum score for significant depression. In somewhat less precise words, we conclude that the average AD caregiver is not depressed.

What is the probability that we have incorrectly rejected the null hypothesis—that is, the probability that we have made a Type I error? We will use the symbol p for this probability. Our α of .05 is the maximum probability that we are willing to risk of making a Type I error, and we used it to determine our critical t of 1.967. If our calculated t had been exactly 1.967 or -1.967, we would have just rejected the null hypothesis, and the probability that we were in error in doing so would be exactly $p = .05$, our α level. But our calculated $t = -7.00$, which is far greater in absolute value than the critical t. Thus, the probability of a Type I error is much smaller than .05—that is, $p < .05$. Looking in Table A.2 in the right-hand column of the rows for $df = 300$ and $df = 400$, we find that $t_{.00005,361} \approx 3.935$. This means that only $2 \times .00005 = .0001$ of all possible samples will have an absolute value of t that is greater than or equal to 3.935. Our $|t| = 7$ is much larger than 3.935, so $p < .0001$. That is, fewer than 1 out of every 10,000 samples would have a mean with a calculated $|t| \geq 3.935$. This is as close as we can get in Table A.2 to the actual value of p. Computer programs that calculate t statistics can also calculate the exact value of p associated with t. In our case, the two-tailed probability for $t = -7$ is $p = .00000000001$! In words, only about 1 out of every 100 billion sample means would deviate this far from a population mean of 16 by chance (i.e., due to random sampling). Thus, the probability that we incorrectly rejected the null hypothesis is virtually 0.

The probability p associated with the calculated t is called the **level of significance.** This is somewhat of a misnomer because it does not indicate the importance of the results. Furthermore, the smaller the level of significance the better, because the level of significance (p) indicates the probability of a Type I error. The probability of incorrectly rejecting the null hypothesis is actually the probability that the null hypothesis is true. So the smaller the level of significance p, the more "significant" is our result because it is less likely to have occurred by chance. When the null hypothesis is rejected, we say that the sample mean is *significantly* different from that of the null hypothesis. In our example, we can say that the mean CES-D score of the AD caregivers is *significantly* less than 16.

We have said there is a close relationship between hypothesis tests and confidence intervals. A $100P$ percent confidence interval, where $P = 1 - \alpha$, will include μ_0 when our t test does not reject H_0; μ_0 will not be included in the confidence interval when the t test leads to the rejection of H_0. In the caregiver depression example, $\mu_0 = 16$ is rejected and the 95% CI = 11.5–13.5 does not include 16. The two methods are entirely compatible. One method, the t test, places the emphasis on making a decision about a specific value of μ, whereas the other method specifies a range of values in which μ is likely to lie. It is arguably the case that the confidence interval is preferable. For example, our t test concluded that $\mu < 16$, whereas the confidence interval told us this and more, namely, that there is a 95-percent probability that $11.5 < \mu < 13.5$.

The confidence interval also helps us to understand the meaning of failing to reject the null hypothesis. Remember that it is not valid to conclude that the null hypothesis is *accepted*. If H_0 is not rejected at the α level, we already noted that μ_0 must lie in the $100(1 - \alpha)$ percent confidence interval. But μ_0 is only one of many values that may lie in that interval. We have no way of knowing which of these values in the interval is μ. Therefore, we have to be content to say that H_0 *is not rejected*. We can never prove or say with any confidence that μ equals any specific value, including μ_0. The null hypothesis may be shown to be invalid, but it can never be shown to be valid.

Another Example

In Chapter 2 we noted that a fertility rate of 2.1 children per woman is needed for replacement of the population. In some countries the fertility rate is below the replacement rate. One factor that may contribute to the fertility rate is the average number of children that women prefer to have. In the 1998 GSS, respondents were asked how many children is the ideal number for a family to have. A subsample of respondents consisting of 161 women aged 20–29 who responded to this question was selected for analysis. The eight-step procedure for conducting a hypothesis test for the mean ideal number of children is presented in Table 6.2.

The null hypothesis is that the mean ideal number of children among young women in the United States equals 2.1, the replacement fertility rate. The alternative hypothesis is that the mean does not equal 2.1, which is a two-tailed test. This will allow us to test whether young American women prefer more children or fewer children, on the

Table 6.2 *t* Test for Mean Ideal Number of Children

1. H_0: $\mu = 2.1$
   H_1: $\mu \neq 2.1$

2. $\alpha = .05$

3. $df = n - 1 = 161 - 1 = 160$

4. $t_{\alpha/2, df} = t_{.025, 160} = 1.975$

5. $\bar{X} = 2.323$

6. $s_{\bar{x}} = \dfrac{s}{\sqrt{n-1}} = \dfrac{.853}{\sqrt{161-1}} = \dfrac{.853}{12.649} = .0674$

7. $t = \dfrac{\bar{X} - \mu_0}{s_{\bar{x}}} = \dfrac{2.323 - 2.1}{.0674} = \dfrac{.223}{.0674} = 3.31$

8. $(|t| = 3.31) > (1.975 = t_{.025})$, $\therefore$ reject H_0
   $(t = 3.31) > (1.975 = t_{.025})$, $\therefore \mu > 2.1$
   $p \approx .001$

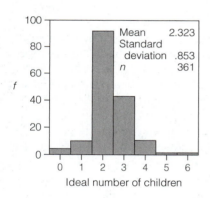

FIGURE 6.9 Ideal Number of Children, Females Aged 20–29, 1998 GSS

average, than the number needed to replace the population. Using the conventional α and the degrees of freedom, we obtain the critical value of $t = 1.975$ from Table A.2.

Now we turn to our data to compute the mean ideal number of children (Figure 6.9), which is slightly greater than the replacement rate. We need to determine, however, whether this result is unlikely to have occurred by chance. Using the standard deviation from Figure 6.9, we calculate that the standard error of the mean equals .0674. The next and final statistic to calculate is t itself, which indicates that our sample mean would be about 3.3 standard deviations above the mean of the sampling distribution if the null hypothesis were true.

In the final step we reject the null hypothesis because the absolute value of the calculated t is greater than the critical t. Furthermore, because the calculated t is positive (the sample mean is greater than μ_0), we conclude that the average woman in her 20s today believes that the ideal number of children is greater than the replacement fertility rate. Finally, we can estimate from Table A.2 that $\alpha/2$ for our *calculated* t is approximately .0005, and thus the two-tailed level of significance is $p \approx .001$. This result indicates that only about 1 out of 1,000 samples would have a mean that deviates from μ_0 as much as our mean if the null hypothesis is true. Thus, the actual risk of a Type I error is much smaller than the risk we were willing to take ($\alpha = .05$).

One-Tailed Hypothesis Tests

In a two-tailed test of the mean, the investigator hypothesizes that the population mean equals some value of interest (the null hypothesis), calculates the difference between the sample mean and the hypothesized mean, and rejects the hypothesis if there is only a small probability that the difference is due to chance. The alternative hypothesis in a two-tailed test does not predict whether the population mean is greater than or less than some specified value. The alternative hypothesis simply says the mean is not equal to the value specified by the null hypothesis—that is, μ_0. This strategy gives the researcher a chance of rejecting the null hypothesis if $\mu < \mu_0$ or $\mu > \mu_0$. This type of hypothesis testing does not use any theory to construct the alternative hypothesis. It may be that the investigator has a hypothesis that $\mu > \mu_0$ (e.g., mean ideal number of children is greater than 2.1) but recognizes that there is a possibility that he or she may be wrong. Thus, the investigator buys protection against the unexpected by putting half of α in the upper tail and half in the lower tail. This is the strategy used most often in research.

There are occasions, however, when a researcher may have a strong theory about the direction of the difference between the population mean and the value indicated by H_0. Or sometimes a difference in one direction may be of little or no interest. For example, there may be good reasons for believing that American women have reduced their preferences for children below the replacement rate, on average, and also it may not be of much interest that fertility preferences are greater than the replacement rate because this has been the status quo throughout history. Thus, the researcher's alternative hypothesis may be that the population mean is less than 2.1. This is called a **one-tailed test.** One-tailed t tests may take either of the two forms in Table 6.3. In both cases, H_0 is the hypothesis the researcher wants to reject and H_1 is the hypothesis (s)he hopes to accept. Notice that now the null hypothesis specifies a range of values rather than a single value as in a two-tailed test.

What are the advantages and disadvantages of the one-tailed test? Figure 6.10a shows a t distribution and a one-tailed critical t for null and alternate hypotheses of form (a) in Table 6.3. The critical t is selected by placing the entire α under the upper tail of the distribution. The critical t is positive and smaller in value than the critical t in a two-tailed test that places only $\alpha/2$ under the upper tail (i.e., $t_{\alpha,df} < t_{\alpha/2,df}$). Because the critical t is smaller for a one-tailed test, it will be easier for the calculated t to be greater than the critical t, assuming the calculated t is positive as hypothesized. In Figure 6.10a, all values of t greater than or equal to t_α will result in rejection of the null hypothesis. All values of t to the left of t_α, including negative ts way out in the left tail, will not result in the rejection of the null hypothesis. The opposite applies to Figure 6.10b, where values of t less than or equal to the negative critical t result in rejecting

Table 6.3 Hypothesis Forms for One-Tailed Tests

(a)		(b)
$H_0: \mu \leq a$	*or*	$H_0: \mu \geq a$
$H_1: \mu > a$		$H_1: \mu < a$

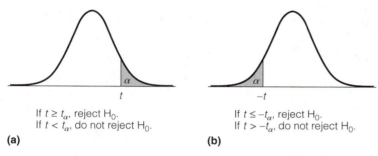

If $t \geq t_\alpha$, reject H_0.
If $t < t_\alpha$, do not reject H_0.

If $t \leq -t_\alpha$, reject H_0.
If $t > -t_\alpha$, do not reject H_0.

(a) **(b)**

FIGURE 6.10 Critical Values of t for One-Tailed α

the null hypothesis and values of t greater than the critical t, including large positive ts in the right tail, do not result in rejection. In a nutshell, if t has the sign predicted by H_1, the probability of rejecting H_0 is greater than with a two-tailed test, but if t has the unexpected sign, it is impossible to reject H_0.

To illustrate a one-tailed t test, we will conduct a test of the mean number of hours worked per week by married women with children at home. Years ago, married women with children at home were not expected to work for pay outside the home, and if they did work they were much more likely to work part time than men were. These traditional roles, however, have changed dramatically in recent decades. We will conduct a test of the hypothesis that when married women with children work, they work full time on the average. Full time will be operationally defined as 35 or more hours per week.[6] Respondents in the 1988 GSS were asked how many hours they worked in the week prior to the survey interview. Married women with babies, preteens, or teenagers at home who were employed (but not self-employed) during the previous week were selected for the analysis. There were $n = 207$ such persons.

The eight-step hypothesis test for the mean is outlined in Table 6.4. The key differences from the two-tailed test occur in steps 1, 4, and 8. The null hypothesis is that the mean hours worked per week by the married mothers is less than or equal to 35.0. The alternate hypothesis is that the mean is greater than 35.0 (this hypothesis includes any fractional amount over a mean of exactly 35). We use the conventional $\alpha = .05$, but in

6. The length of the average work week has been falling, and many full-time jobs are now less than 40 hours per week.

Table 6.4 _t_ Test for Mean Hours Worked per Week by Married Women with Children at Home

1. $H_0: \mu \leq 35.0$
   $H_1: \mu > 35.0$

2. $\alpha = .05$

3. $df = n - 1 = 207 - 1 = 206$

4. $t_{\alpha, df} = t_{.05, 206} = 1.652$

5. $\bar{X} = 36.580$

6. $s_{\bar{x}} = \dfrac{s}{\sqrt{n-1}} = \dfrac{12.497}{\sqrt{207-1}} = \dfrac{12.497}{14.353} = .871$

7. $t = \dfrac{\bar{X} - \mu_0}{s_{\bar{x}}} = \dfrac{36.580 - 35.0}{.871} = \dfrac{1.580}{.871} = 1.814$

8. $(t = 1.814) > (1.652 = t_{.05})$, $\therefore$ reject H_0
   $p \approx .036$

this one-tailed test, it will all be included under the upper right-hand tail. In step 4, the critical $t = 1.652$ is found in the column A $= .05$ because in the one-tailed test, A $= \alpha$ instead of A $= \alpha/2$. Note that the critical t for a two-tailed test would have been 1.972. The mean and standard deviation for steps 5 and 6 are found in Figure 6.11. Notice that the sample mean is 36.58, or 1.58 hours greater than 35. Because the calculated $t = 1.814$ is greater than the critical $t = 1.652$, the null hypothesis is rejected and the alternate hypothesis that married women with children at home work an average of more than 35 hours per week is accepted. Observe that the decision to reject the null hypothesis and the conclusion about the direction of the mean are one and the same in the one-tailed test. Returning to Table A.2, we find that the level of significance for $t = 1.814$ is approximately $p = .036$, just a little smaller than $\alpha = .05$.

The test of the mean hours worked per week provides a good example of the advantages of a one-tailed test. If a two-tailed test had been conducted, the calculated t (1.814) would not have been greater than the critical t (1.972) and thus the null hypothesis would not have been rejected. By daring to put the entire α under the upper tail, we found support for our hypothesis. Remember, however, that if the sample mean had been less than 35.0, we could not have claimed that the mean hours worked by married women with children was less than full time, even if $t \leq -1.972$.

_t_ Test for the Mean of a Difference Score

The study of differences is an important part of social research. Social scientists try to determine how similar or different are pairs of related persons such as spouses, neighbors, best friends, daughters and mothers, and so on. In such research the pair or dyad is the unit of analysis. When studying such dyads, we will use a variable that equals the difference in some attribute, such as personality or socioeconomic status, between each pair. To obtain such a difference, we first obtain a measure of the attribute for both

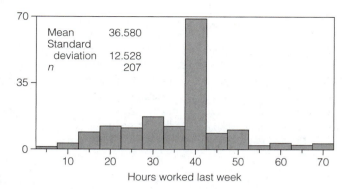

FIGURE 6.11 Hours Worked in Prior Week by Married Women with Children at Home, Not Self-Employed, 1998 GSS

members of the pair, which results in two variables, and then we create a new variable that equals the difference between the first two—that is, a difference score.

The study of change makes use of difference scores. In this case we measure some attribute twice for each person or other unit of analysis, allowing an appropriate amount of time to pass between the measurements. For example, we can study changes in personality (allowing a long enough time interval), financial status, and health. Studies such as this are a special type of *longitudinal analysis* called *panel studies*. In such cases, we have repeated measures of the same variable. Thus, we can subtract the first measure from the second to obtain a difference score that measures change in the variable.

Differences between Members of Pairs

The most important change in conducting a *t* test for the mean difference between members of pairs is simply the construction of the difference-score variable. Let X_1 be some trait for one member of a pair, and let X_2 be the same trait for the other member of a pair. The **difference score** (D) is

$$D = X_2 - X_1 \qquad (6.10)$$

A positive score indicates that person 2's trait is greater than that of person 1; a negative score indicates that person 2's trait is less than person 1's. We will be testing hypotheses about the mean of D, which equals

$$\overline{D} = \frac{\Sigma D}{n} = \frac{\Sigma(X_2 - X_1)}{n} = \frac{\Sigma X_2 - \Sigma X_1}{n} = \frac{\Sigma X_2}{n} - \frac{\Sigma X_1}{n} = \overline{X}_2 - \overline{X}_1 \qquad (6.11)$$

Equation 6.11 first indicates that the mean of D is computed, just as the mean of any X, by summing the scores and dividing by n. After substituting $X_2 - X_1$ for D, Equation 6.11 goes on to indicate that the mean of D has a very special meaning, namely, it equals the mean of X_2 minus the mean of X_1. The mean of the differences equals the difference of the means. Thus, when we test a hypothesis about D we will be testing whether one member of the pair has a greater mean than the other.

Generally speaking, it does not make any difference which person's trait is designated by X_1 and which by X_2. If, however, you have a hypothesis that one member will have a larger mean than the other, you should designate the former person's trait as X_2 so that positive differences will be consistent with your hypothesis. For example, if each pair consisted of a daughter and her mother, you might hypothesize that mothers would have more children than their daughters, on the average. You would thus designate X_2 as the number of children of the mother and X_1 as the daughter's children.

Once the member of the pair to be designated as X_2 has been decided and the difference scores have been computed for all pairs, the **t test for the mean of D** is carried out just like the one-tailed and two-tailed tests that we have covered. This includes the calculation of the standard deviation of D, the standard error of the mean of D, and t, whose formulas are found by substituting D for X in the usual formulas for these statistics:

$$s_D = \sqrt{\frac{\Sigma(D - \overline{D})^2}{n}}$$

$$s_{\overline{D}} = \frac{s_D}{\sqrt{n - 1}}$$

$$t = \frac{\overline{D} - \mu_0}{s_{\overline{D}}}$$

Let's take the education of husbands and wives as an example of a t test for a difference score. Persons tend to marry people with similar social characteristics (but not necessarily with similar personalities). That is, husbands and wives should be more similar to each other on many social variables than random pairs of men and women. Thus, we might reasonably expect little or no difference in the education of husbands and wives. On the other hand, traditional social norms may still tend to work in favor of men in terms of educational attainment, which would suggest that husbands may have more education than their wives, even if spouses are more similar than random pairs of men and women. There is probably no reason, however, to expect that wives have more education than their husbands. So, on the assumption that if there is any significant difference between spouses' educational attainment it should favor husbands, we designate:

$$X_2 = \text{husband's education}$$

$$X_1 = \text{wife's education}$$

To test the difference between X_2 and X_1 we will use years of school completed by married respondents and their spouses from the 1998 GSS. Married respondents reported both their own education and that of their spouse. To illustrate the calculation of $D = X_2 - X_1$ and the mean and standard deviation of D, twenty spouses were randomly selected from the 1998 GSS (Table 6.5). Looking at the column of Ds, we can count that there are seven cases in which the husband has greater education (positive D), four cases with no difference ($D = 0$), and nine cases in which the wife has more education (negative D), including one couple in which the wife has 8 more years of schooling! The mean of the differences is $\overline{D} = -5/20 = -.25$. That is, in this small sample, the husbands averaged one-fourth of a year less education than their wives did. Notice that the difference of means gives the same result; $\overline{X}_2 - \overline{X}_1 = 12.65 - 12.90 = -.25$. This unexpected difference, however, can easily happen by chance in a small sample of twenty pairs.

Table 6.5 Difference in Education (*D*) of Husbands (*X₂*) and Wives (*X₁*), Random Sample of 20 Spouses from 1998 GSS

Pair	X_2	X_1	D	$D - \bar{D}$	$(D - \bar{D})^2$
1	10	12	−2	−1.75	3.0625
2	8	3	5	5.25	27.5625
3	11	12	−1	−.75	.5625
4	12	14	−2	−1.75	3.0625
5	16	16	0	.25	.0625
6	16	18	−2	−1.75	3.0625
7	8	9	−1	−.75	.5625
8	16	14	2	2.25	5.0625
9	12	13	−1	−.75	.5625
10	12	12	0	.25	.0625
11	14	12	2	2.25	5.0625
12	14	12	2	2.25	5.0625
13	15	16	−1	−.75	.5625
14	12	10	2	2.25	5.0625
15	16	16	0	.25	.0625
16	7	15	−8	−7.75	60.0625
17	13	13	0	.25	.0625
18	16	15	1	1.25	1.5625
19	16	15	1	1.25	1.5625
20	9	11	−2	−1.75	3.0625
Sum	253	258	−5	.00	125.7500
Mean	12.65	12.90	−.25	.00	6.2875

We will now turn to the full sample of 1,306 respondents and their spouses from the GSS to conduct the *t* test for the mean difference in their educational attainment. Table 6.6 shows the eight-step *t* test. The null hypothesis specifies that the mean difference in the population (μ_D) equals 0. This is a common null hypothesis when testing differences. Thus, we will be testing to see if the null hypothesis of no difference between husbands and wives can be rejected in favor of the alternate hypothesis that the mean difference is not equal to 0. In other words, we are conducting a two-tailed test to leave the door open to the possibility that the wives may have more education. Note that these hypotheses can be stated in terms of the mean difference (μ_D) or the difference of means ($\mu_2 - \mu_1$). The conventional $\alpha = .05$ is used and the critical value of $t = 1.962$ is retrieved from Table A.2 using $df = 1305$ and A $= \alpha/2 = .025$.

Figure 6.12 shows the distribution of difference scores. It is obvious that the modal difference is 0 years of education (no difference). The small positive mean of .070 is consistent with this result. There are, however, substantial numbers of couples with either positive differences or negative differences, which is reflected in the standard deviation from the mean difference of 2.6 years. This standard deviation is used in step 6 to

Table 6.6 *t* Test for Mean Difference in Husband-Wife Education

1. H_0: $\mu_D = 0$, or $\mu_2 - \mu_1 = 0$
   H_1: $\mu_D \neq 0$, or $\mu_2 - \mu_1 \neq 0$

2. $\alpha = .05$

3. $df = n - 1 = 1306 - 1 = 1305$

4. $t_{\alpha/2,df} = t_{.025,1305} = 1.962$

5. $\overline{D} = .070$

6. $s_{\overline{D}} = \dfrac{s_D}{\sqrt{n-1}} = \dfrac{2.596}{\sqrt{1306-1}} = \dfrac{2.596}{36.125} = .0719$

7. $t = \dfrac{\overline{D} - \mu_0}{s_{\overline{D}}} = \dfrac{.070 - 0}{.0719} = \dfrac{.070}{.0719} = .974$

8. $(|t| = .974) < (1.962 = t_{.025})$, $\therefore$ do not reject H_0
   $p < .25$

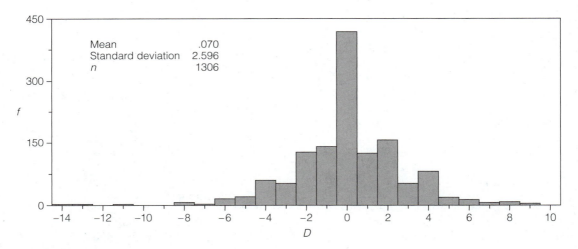

FIGURE 6.12 Husband–Wife Education Differences (*D*), *n* = 1,306

compute the standard error of the mean difference, which equals .072. This small value means that the standard deviation of the sampling distribution of $\overline{D}$ is only about .072, which coincidentally is also approximately the value of the mean. Because of this, the value of the calculated *t* in step 7 is approximately unity (.974). Because this value is less than the critical *t*, we cannot reject the null hypothesis of no educational difference between husbands and wives, which tells us that the significance level $p > .05$. If you look up the calculated *t* in Table A.2, you will find that it is smaller than the *t* of 1.151 listed for $df = 1500$ and A $= .125$, which is the smallest in the row. Thus, we can actually say that $p > (2 \times .125)$ (i.e., $p > .25$).

Table 6.7 t Test for Mean Change in Depression of AD Caregivers

1. H_0: $\mu_D \le 0$, or $\mu_2 - \mu_1 \le 0$
   H_1: $\mu_D > 0$, or $\mu_2 - \mu_1 > 0$

2. $\alpha = .05$

3. $df = n - 1 = 177 - 1 = 176$

4. $t_{\alpha,df} = t_{.05,176} = 1.654$

5. $\overline{D} = 1.921$

6. $s_{\overline{D}} = \dfrac{s_D}{\sqrt{n-1}} = \dfrac{7.071}{\sqrt{177-1}} = \dfrac{7.071}{13.266} = .533$

7. $t = \dfrac{\overline{D} - \mu_0}{s_{\overline{D}}} = \dfrac{1.921 - 0}{.533} = \dfrac{1.921}{.533} = 3.604$

8. $(|t| = 3.604) > (1.654 = t_{.025})$, $\therefore$ reject H_0
   $p < .0005$

Differences between Repeated Measures

We can also use difference scores to study change in traits. Let $X_1 =$ some trait as measured at time 1 and let X_2 be the same trait measured at time 2. Then $D = X_2 - X_1$ measures the amount of change in the trait from time 1 to time 2. A positive D means that X has increased over that time span, and a negative D means it has decreased. Because $\overline{D} = \overline{X}_2 - \overline{X}_1$, a positive $\overline{D}$ indicates that the mean of X at time 2 is greater than its mean at time 1. That is, for the population being investigated, the trait has increased. A negative $\overline{D}$ means the trait has decreased in the population.

Aside from a difference in the interpretation of $\overline{D}$ that involves the time dimension, there is no difference between testing this $\overline{D}$ and testing the $\overline{D}$ for members of pairs. We will return to the example of the level of depression among caregivers of persons with AD. There is currently no way to reverse the progression of AD; the decline in the mental capacities of the afflicted person is steady and noticeable. Thus, a reasonable hypothesis is that the depression of the caregivers will increase under the strain of caring for their increasingly dependent loved one. This hypothesis will be tested for 177 of the caregivers to whom the CES-D depression scale was administered approximately 1 year after the first measurement [(time 2 − time 1) ≈ 1 year]. The eight-step hypothesis test is shown in Table 6.7.

As specified in step 1, this is a one-tailed test for the strong alternative hypothesis that caregiver depression will increase. Because the entire α is placed in the upper tail of the t distribution, the critical t of 1.654 is much smaller than it would be for a two-tailed test, thus making it easier to reject the null hypothesis if the mean change is positive as predicted. As we can see in Figure 6.13, although some caregivers' depression decreased, there were considerably more for whom depression increased over a year's time, resulting in a mean increase in depression of 1.9 points on the CES-D scale. After computing the standard error of D, we calculate $t = 3.604$, which is more than twice as great as the critical t. Thus, we reject the null hypothesis and accept the alternative hypothesis that depression increased for the AD caregivers over a period of a year. The one-tailed level of significance ($p < .0005$) is much less than α. In this case, as it turns out, we

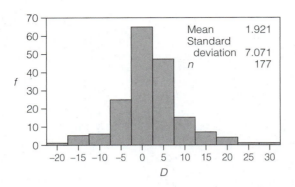

FIGURE 6.13 Change in Depression (*D*) of AD Caregivers Over a Year

could easily have rejected the null hypothesis with a two-tailed test, but one never knows that before conducting the test.

INFERENCES ABOUT PROPORTIONS

A proportion equals the number of cases with some attribute *a* divided by the total number of cases—that is, $p = f_a/n$. We often want to use a sample proportion to make an inference about the proportion in a population that has a given attribute. Although a proportion may appear to be different from a mean, we can use basically the same methods for proportions that we used to make inferences about means. We can do this because a sample proportion is a descriptive statistic that is analogous to a mean and thus has a sampling distribution that is much like the sampling distribution of the mean.

Sampling Distribution of *p*

Let the population proportion be symbolized by the Greek letter π, or pi (pronounced "pie"). We want to make an inference about π based on our sample p. The **sampling distribution of *p*** has the following characteristics:

$$\mu_p = \pi \tag{6.12}$$

$$\sigma_p = \sqrt{\frac{\pi(1 - \pi)}{n}} \tag{6.13}$$

$$p \sim \text{binomial} \tag{6.14}$$

Equation 6.12 states that the mean of the sampling distribution of *p* equals the population proportion π. This means that p is an unbiased estimator of π. This characteristic is analogous to the fact that μ is the mean of the sampling distribution of $\overline{X}$.

Equation 6.13 gives, without derivation, the standard deviation of the sampling distribution of *p*. This is the *standard error of p*. In the numerator is $\pi(1 - \pi)$. Remember that because π is a proportion, its range is .0 to 1.0. Because of this, $1 - \pi$ is also a pro-

Table 6.8 Selected Values of $\pi(1 - \pi)$

π	$\pi(1 - \pi)$
.0	$.0(1 - .0) = .0(1) = .00$
.1	$.1(1 - .1) = .1(.9) = .09$
.2	$.2(1 - .2) = .2(.8) = .16$
.3	$.3(1 - .3) = .3(.7) = .21$
.4	$.4(1 - .4) = .4(.6) = .24$
.5	$.5(1 - .5) = .5(.5) = .25$
.6	$.6(1 - .6) = .6(.4) = .24$
.7	$.7(1 - .7) = .7(.3) = .21$
.8	$.8(1 - .8) = .8(.2) = .16$
.9	$.9(1 - .9) = .9(.1) = .09$
1.0	$1.(1 - 1.) = 1.(.0) = .00$

portion, which gives the proportion of population cases that do not have the attribute *a*. If $\pi = .35$, for example, $1 - \pi = .65$. In words, this means that if .35 of the population has attribute *a*, .65 of the population does not have attribute *a*. The product $\pi(1 - \pi)$ is maximized when $\pi = .5$, as illustrated in Table 6.8. When $\pi = .5$, there are the same number of persons with the attribute as without the attribute. This situation represents the maximum variance in the attribute. Remember that the index of qualitative variation (IQV) discussed in Chapter 4 is maximized when there are the same number of persons in each category of a nominal variable. When $\pi = .1$ or $\pi = .9$, there is very little variance; either the vast majority have *a* or do not have *a*. When $\pi = .0$ or $\pi = 1.0$, there is no variation in the attribute, which is indicated by $\pi(1 - \pi) = 0$.

What are the consequences of this relationship for the standard error of *p*? Because $\pi(1 - \pi)$ is in the numerator of Equation 6.13, it means that the larger this product is, the greater will be the standard error of the sample proportion. This makes sense. When $\pi = .5$, a sample proportion can be both quite a bit higher and quite a bit lower than π by chance. When $\pi = .1$, however, the sample proportion cannot be very much smaller than .1 and, although it can be very much larger than .1 by chance, the probability is extremely small that it will deviate more than in the case of $\pi = .5$. Thus, the greater the variance of the attribute in the population, as indicated by $\pi(1 - \pi)$, the greater the standard error (σ_p). This variance is the same as in the formula for the standard error of the mean, which has the population standard deviation (σ_X) in the numerator.

The other factor that affects the standard error is sample size *n*. Because *n* is in the denominator of Equation 6.13, the greater the sample size, the smaller the standard error of *p*.

Equation 6.14 states that the sample proportion has a **binomial distribution.** Just as there is a family of *t* distributions, there is a family of binomial distributions. The binomial family, however, is larger than the *t* family. Not only is there a different binomial distribution for each value of *n*, but also for each value of *n* there is a different binomial distribution for each value of π. Figure 6.14 shows two binomial distributions for $n = 10$. These distributions are not smooth because only a limited number of discrete values

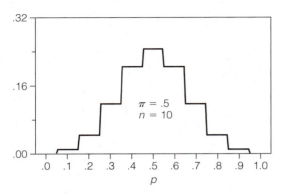
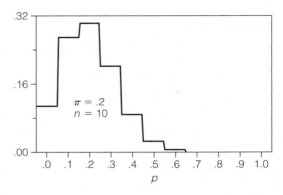

FIGURE 6.14 Binomial Sampling Distributions of the Proportion (p), $n = 10$

of p may occur. Because there are only ten cases in the sample, the number of cases that may have attribute $a(f_a)$ can only be 0, 1, 2, 3, 4, . . . , 10. Therefore, the sample proportion (f/n) can take on only eleven values, namely, .0, .1, .2, .3, .4, .5, .6, .7, .8, .9, and 1.0. The number of possible values for a sample p is $n + 1$, in this case, $10 + 1 = 11$. Thus, although the binomial distribution is a *discrete distribution,* as opposed to a continuous distribution, the larger the n, the more similar the binomial is to a continuous distribution.

Now, although it is possible for p to take on $n + 1$ distinct values, it does not follow that each of those values is likely to occur. You can see in Figure 6.14 that when $\pi = .5$, there is virtually a zero probability that $p = .0$ or $p = 1.0$. That is, if half of the people in a population have a, it will be virtually impossible in a random sample of ten persons that none or all of them will have a. You can also see that when $\pi = .2$, it will be extremely unlikely that $p \geq .7$.

Equation 6.12 states that the mean of the sampling distribution equals π. The graph for $\pi = .5$ clearly shows the mode is $p = .5$. This distribution can also be seen to be symmetrical, and thus the mean equals the mode of .5. The distribution for $\pi = .2$ clearly has a mode of $p = .2$, but it is not symmetrical. Because, however, there are more cases below the mode but greater extreme deviations above the mode, the mean equals the mode of .2 (without proof).

Eyeballing Figure 6.14 suggests that there is greater variance for $\pi = .5$ than for $\pi = .2$; the former is less peaked and is spread out over a wider range of values. Using Equation 6.13 to compute the standard deviation of p confirms this supposition:

$$\sigma_p = \sqrt{\frac{\pi(1 - \pi)}{n}} = \sqrt{\frac{.5(1 - .5)}{10}} = \sqrt{\frac{.25}{10}} = \sqrt{.025} = .158$$

$$\sigma_p = \sqrt{\frac{\pi(1 - \pi)}{n}} = \sqrt{\frac{.2(1 - .2)}{10}} = \sqrt{\frac{.16}{10}} = \sqrt{.016} = .126$$

These results indicate that the standard amount that a sample proportion will deviate from a population proportion of $\pi = .5$ when $n = 10$ is .158. The standard error (.126) is smaller for $\pi = .2$.

What will these sampling distributions be like if we enlarge the sample size by a factor of 10 to $n = 100$? We know that the standard error should get smaller and that the

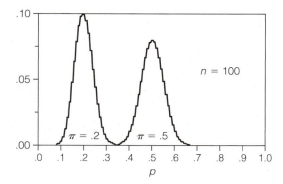

FIGURE 6.15 Binomial Sampling Distributions of the Proportion (*p*)

number of distinct values of *p* should increase from 11 to $n + 1 = 101$. A quick look at Figure 6.15 confirms this. Both distributions are now much less spread out than they were in Figure 6.14. The distribution of $\pi = .5$ is more dispersed, however, than that of $\pi = .2$. The fact that *p* may take on a greater number of distinct values results in the curves being much smoother than for $n = 10$, although only a minority of the potential values occurs frequently enough to be noticed. Also notice that although the distribution for $\pi = .2$ was not symmetrical when $n = 10$, it is now symmetrical. Because its tails were both pulled in, it no longer jams up against the zero value on the scale.

The most important point to note about Figure 6.15 is that both distributions are approximately normally distributed, which suggests that

$$\text{As } n \to \infty, p \stackrel{.}{\sim} \text{N}$$

This says that as *n* goes to infinity (or simply, as *n* increases), *p* becomes approximately normally distributed. In other words, the normal distribution is the **limiting case of the binomial distribution.** This was also the case, you will remember, for the *t* distribution. Actually, *n* doesn't have to approach infinity before this happens. You can see in Figure 6.15 that these distributions look normal at $n = 100$. Here is a useful guideline for what is a sufficiently large *n*:

$$\text{If } n > \frac{10}{\min(\pi, \, 1 - \pi)}, n \stackrel{.}{\sim} \text{N}$$

If sample size is greater than 10 divided by the minimum of π or $1 - \pi$, we can treat the sampling distribution of *p* as having a normal distribution. The necessary sample sizes for $\pi = .5$ and $\pi = .2$ are

$$\frac{10}{\min(.5, \, 1 - .5)} = \frac{10}{.5} = 20$$

$$\frac{10}{\min(.2, \, 1 - .2)} = \frac{10}{.2} = 50$$

This indicates that sample sizes of 20 and 50 are sufficiently large for $\pi = .5$ and $\pi = .2$, respectively, to use the normal distribution in hypothesis tests and confidence inter-

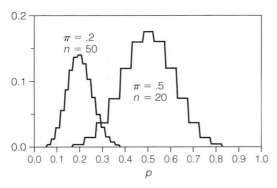

FIGURE 6.16 Binomial Distributions
Approximating the Normal Distribution

vals (see Figure 6.16).[7] Notice that it takes longer if π is small or large for p to become nearly normally distributed. A sample size of 50, however, is not particularly large. If $\pi = .1$ or $\pi = .9$, we need an n of 100.

Hypothesis Tests about *p*

When n is sufficiently large, we will use the standard normal distribution (Z) to conduct hypothesis tests about the population proportion.[8] To use the standard normal distribution (Z), we need to compute a **Z statistic for *p*** as follows:

---

7. The distribution for $\pi = .2$ in Figure 6.16 appears smaller than that for $\pi = .50$ because continuous lines are used to describe these discrete variables. Although there are only 21 distinct values for $\pi = .50$, they are spread out more because of the smaller sample size, and thus it appears that there is more area under the curve than under the curve for $\pi = .2$, which has 51 distinct values.

8. Ideally, we might want to use the binomial distribution itself. Because the binomial is discrete, however, critical values of p may not exist for the desired α. For example, in the binomial with $n = 10$ and $\pi = .5$, the nearest cumulative proportions to $\alpha/2 = .025$ are .0107 and .0547 (see the following table). To keep the probability of a Type I error below .05, we have to use as critical values the ps associated with .0107 and $1 -$ .0107 = .9893 (which happen to be $p = .10$ and $p = .90$). In effect, that means we are forced to use an α of .0107 + .0107 = .0214. This results in a conservative test, meaning we might fail to reject a false null hypothesis. This conservativeness, especially in small samples, is why the binomial test is not used very often.

**Cumulative Proportions of Binomial
Distribution for $n = 10$ and $\pi = .5$**

f_a	p_a	cp	f_a	p_a	cp
0	.00	.0010	6	.60	.8281
1	.10	.0107	7	.70	.9453
2	.20	.0547	8	.80	.9893
3	.30	.1719	9	.90	.9990
4	.40	.3770	10	1.00	1.0000
5	.50	.6230			

Table 6.9 Z **Test for Proportion Who Approve of Abortion for Any Reason, Respondents Aged 18–29, 1998 GSS**

1. H_0: $\pi = .50$

   H_1: $\pi \neq .50$

2. $\alpha = .05$

3. $Z_{\alpha/2} = Z_{.025} = 1.96$

4. $p = .4625$

5. $\sigma_p = \sqrt{\dfrac{\pi_0(1 - \pi_0)}{n}} = \sqrt{\dfrac{.50(1 - .50)}{333}}$

   $= \sqrt{\dfrac{.50(.50)}{333}} = \sqrt{\dfrac{.25}{333}} = \sqrt{.00075} = .0274$

6. $Z = \dfrac{p - \pi_0}{\sigma_p} = \dfrac{.4625 - .50}{.0274} = \dfrac{-.0375}{.0274} = -1.37$

7. $(|Z| = 1.37) < (1.96 = Z_{.025})$, $\therefore$ do not reject H_0

   $p = .1707$

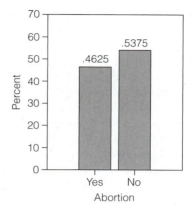

FIGURE 6.17 Abortion If Woman Wants It for Any Reason, $n = 333$; Respondents 18–29 Years of Age, 1998 GSS

$$Z = \frac{p - \pi_0}{\sigma_p} \qquad (6.15)$$

We subtract the population proportion stated in the null hypothesis from the sample proportion and divide the result by the standard error of the proportion as given by Equation 6.13.

To illustrate, we will test the opinions on abortion of young adults in the United States. Using the 1998 GSS again, let young adults be operationalized as respondents under 30 years of age (i.e., 18–29). The amount of support for abortion varies to some extent according to the reason for which a woman wants an abortion (support is much higher, for example, if the woman is a victim of rape or if her health is endangered by the pregnancy). We will use a question from the GSS that is perhaps the most liberal measure of support for abortion: Should it be possible for a pregnant woman to obtain a legal abortion if the woman wants it for any reason? Thus, support for abortion for any reason is the attribute, and we want to estimate what proportion of young adults have this attribute. The hypothesis test is outlined in Table 6.9.

The null hypothesis stated in step 1 says that the proportion of 18- to 29-year-olds in the population supporting abortion for any reason is .50. If we reject this hypothesis, then we can reach a conclusion of whether a majority or only a minority of persons in this age bracket approve of abortion for any reason, because this is a two-tailed test. A null hypothesis that $\pi = .50$ is common in tests for proportions.

We use the conventional $\alpha = .05$. Find the critical value of Z in Table A.1 by locating the standardized score associated with A = $\alpha/2 = .025$, which is 1.96. Notice that there is no step for computing degrees of freedom. There is only one normal distribution, unlike the family of t distributions associated with different degrees of freedom.

Figure 6.17 shows the results of the tabulations. Less than 50 percent of these persons in the sample support abortion for any reason ($p = .4625$). Next, we calculate the standard error of the proportion. We do this under the assumption that the null hypothesis is true. Thus, we use $\pi_0 = .50$ in the formula for the standard error. This is exactly what the standard error of the proportion will equal if the null hypothesis is true. It is different from the result for the t test where we had to use the sample observations to make an estimate of the standard error of the mean. The result means that if the true proportion in the population that favor abortion equals .50, the standard amount that a random sample of 333 respondents will deviate from .50 equals .0274. It would not be surprising to find a sample proportion that is .0274 or more above .50 or below .50.

Now we calculate our standard score—that is, $Z = -1.37$. This means that if the null hypothesis is true, our sample proportion is 1.37 standard deviations below the mean of the sampling distribution. Finally, we compare the calculated Z to the critical Z and find that we cannot reject the null hypothesis. We do not literally accept the hypothesis that exactly 50 percent of the population of young adults support abortion for any reason, but we can't claim to know whether a minority or a majority favor abortion. This result is too close to call. We can look up $Z = 1.37$ in Table A.1 and determine that the exact level of significance is $p = .1701$. This is not a sample proportion—it is the probability that our sample proportion would deviate from a population proportion of .50 by chance.

Confidence Intervals for *p*

We construct confidence intervals for proportions in much the same way we did for means. Equation 6.3 gives the formula for the true confidence interval of the mean. If we substitute p for $\overline{X}$ in that equation, we get the following formula for the **confidence interval for *p*:**

$$100P\% \ \mathrm{CI}_\pi = p \pm Z_{\alpha/2}\sigma_p \tag{6.16}$$

This formula is valid when we have a sufficiently large n to assume that p has approximately a normal distribution.

We cannot apply Equation 6.16, however, because we do not know the true standard error of p, $\sigma_p = \sqrt{\pi(1 - \pi)/n}$ (Equation 6.13), because we do not know π. When we conduct hypothesis tests for p, we use the value of π specified by the null hypothesis to compute the standard error of p. When constructing confidence intervals, however, we do not have a null hypothesis. Therefore, we have to estimate the standard error of p as follows:

$$s_p = \sqrt{\frac{p(1 - p)}{n - 1}} \tag{6.17}$$

The Roman s is used to indicate that this is an estimate of the standard error. Equation 6.17 contains p, the unbiased estimate of π (Equation 6.12), instead of π itself. The denominator contains $n - 1$ instead of n, just as in the case of the estimated standard error of the mean (Equation 6.5). Now we can substitute s_p for σ_p in Equation 6.16 to get a formula for the confidence interval that we can calculate:

$$100P\% \ \mathrm{CI}_\pi = p \pm Z_{\alpha/2}s_p \tag{6.18}$$

We will use the abortion data from the last section (Figure 6.17) to construct a 95-percent confidence interval for the proportion of persons aged 18 to 29 who favor abortion for any reason a pregnant woman might have for wanting it. First we compute the **estimated standard error of p:**

$$s_p = \sqrt{\frac{p(1-p)}{n-1}} = \sqrt{\frac{.4625(1-.4625)}{333-1}} = \sqrt{\frac{.2486}{332}} = \sqrt{.00075} = .0274$$

Next we need to determine the critical value of $Z_{a/2}$. From Table A.1 we retrieve $Z_{a/2} = 1.96$. It is appropriate to use the normal distribution because $n = 333$ is sufficiently large for p to be normally distributed. Now we enter these statistics into Equation 6.18:

$$100(.95)\% \text{ CI}_\pi = p \pm Z_{.025}s_p = .4625 \pm 1.96(.0274)$$
$$= .4625 \pm .0536 = .4089\text{--}.5161$$

This indicates that there is a .95 probability that the population proportion lies somewhere between .41 and .52. Notice that .50, the null hypothesis in our earlier hypothesis test, is within this interval. Thus, the confidence interval is telling us the same thing the hypothesis test did, namely, that we do not know whether a majority or a minority of persons aged 18–29 favor abortion for any reason a pregnant woman might have for wanting it.

SUMMARY

We covered two types of inferences about the mean in this chapter: confidence intervals and hypothesis tests. A confidence interval specifies a range of values within which the population mean has a certain probability of lying. To determine a confidence interval, we must first specify the desired confidence level, which is the probability that the mean lies in the interval. One minus this probability is α, the probability that the mean is outside the interval. We then use the t distribution with degrees of freedom $n-1$ to determine the value of t above which $\alpha/2$ of the distribution lies ($t_{\alpha/2}$). Next, we estimate the standard error of the mean to equal the sample standard deviation divided by the square root of n minus 1. Finally, the calculated confidence interval for the mean equals the sample mean plus or minus $t_{\alpha/2}$ times the estimated standard error.

In hypothesis testing, a specific hypothesis about the population mean is specified before conducting a test to determine whether to reject the hypothesis. There are two types of tests: one-tailed tests and two-tailed tests. In a two-tailed test a null hypothesis specifies that the mean equals a specific value of interest and an alternative hypothesis states the logical alternative, namely, that the mean is greater than or less than this value. In a one-tailed test, the null hypothesis says the mean is in a specific range of values (e.g., less than or equal to a specific value, or greater than or equal to a value), and the alternative hypothesis states that the mean is outside that range (e.g., greater than the value, or less than the value, respectively). After specifying the α level, or the probability of incorrectly rejecting the null hypothesis and determining the degrees of freedom ($n-1$), we can determine the critical value of t, which is $t_{\alpha/2}$ in a two-tailed test and t_α in a one-tailed test. After computing the mean and the estimated standard error of the mean, we calculate the t statistic to equal the sample mean minus the value specified in the null hypothesis with the difference divided by the standard error of the mean. The final step is to decide whether to reject or not reject the null hypothesis. In a two–tailed test, if the absolute value of the computed t is greater than the critical t, we reject the null hypothesis. In a one-tailed test, we reject the null hypothesis if the absolute value of the calculated t is greater than the critical t and the sample mean is as specified in the alternative hypothesis. If the sample mean is not as specified in the alternative hypothesis, the null hypothesis may not be rejected, no matter how large the absolute value of the calculated t. If the sample mean is in the predicted direction, we have a greater chance of rejecting the null hypothesis in a one-tailed test than in a two-tailed test.

We can also conduct hypothesis tests about mean differences between two paired groups or between two points in time. To do so, we must compute a new variable, a difference score, that measures how different a pair of persons (e.g., husbands and wives) are on a variable of interest, or how much change there is in a variable between two points in time for each person. We can then conduct a one-tailed or two-tailed test on the difference score in exactly the same manner as for an ordinary variable.

Hypothesis tests and confidence intervals for the proportion of cases having some specific attribute can also be accomplished using the same logic as for means. Instead of using t as the test statistic, however, we use Z when n is large enough so that the sampling distribution of the proportion is approximately normal. The standard error of the proportion is computed in slightly different ways for hypothesis tests and confidence intervals.

KEY TERMS

confidence interval	t-score confidence interval	one-tailed test
confidence level	hypothesis test	difference score
α	two-tailed test	t test for the mean of D
theoretical confidence interval of the mean	null hypothesis	sampling distribution of p
	alternative hypothesis	binomial distribution
estimated standard error of mean	Type I error	limiting case of binomial distribution
	Type II error	
t statistic	critical t	Z statistic for p
t distribution	calculated t	estimated standard error of p
degrees of freedom	decision rule	confidence interval for p
limiting case of t distribution	level of significance	

SPSS PROCEDURES

To construct confidence intervals for a mean, we can obtain the needed standard error of the mean from either the *Frequencies* procedure or the *Descriptives* procedure. Alternatively, the standard error of the mean *and* the desired confidence interval can be computed with the *One-Sample T Test* procedure (click on *Analyze, Compare Means,* and *One-Sample T Test*). If we want the confidence interval for the mean depression of

Alzheimer's Disease caregivers, we enter the name of the variable in the *Test Variable(s)* box (i.e., *cesd1*) and use the default value of 0 in the *Test Value* box at the end of this paragraph. After you click *OK*, the standard error of the mean will appear in the output window in the table named *One-Sample Statistics* and the 95–percent confidence interval will be in the following table named *One-Sample Test*. This table also contains the hypothesis test for whether the mean is different from 0, which is the test value in this case. More than likely, we are not interested in this *t* test. We simply specify a test value of 0 to get the desired confidence interval. If a confidence interval of other than 95 percent is desired, click on *Options* in the *One-Sample T Test* window and specify the desired confidence interval.

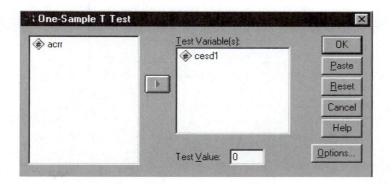

One-Sample Statistics

	N	Mean	Std. Deviation	Std. Error Mean
CESD1	362	12.50829	9.517221	.500214

One-Sample Test

	Test Value = 0					
					95% Confidence Interval of the Difference	
	t	df	Sig. (2-tailed)	Mean Difference	Lower	Upper
CESD1	25.006	361	.000	12.50829	11.52459	13.49199

We can use the *One-Sample T Test* procedure to conduct a two–tailed hypothesis test that the mean caregiver depression score is 16, the value indicating significant depression on the CES-D scale. Simply enter 16 in the *Test Value* box in the *One-Sample T Test* window. The results of the test will appear in the following *One-Sample Test* table in the output window. Note that the 95–percent confidence interval is not the confidence interval for the mean. Instead, it is the confidence interval for the *difference* between the mean and the test value (i.e., 16).

One-Sample Test

	Test Value = 16				95% Confidence Interval of the Difference	
	t	df	Sig. (2-tailed)	Mean Difference	Lower	Upper
CESD1	-6.980	361	.000	-3.49171	-4.47541	-2.50801

Notice that the level of significance given under *Sig. (2-tailed)* in the *One-Sample Test* table rounded to three decimal places is .000. Because the level of significance cannot be exactly 0, this means that $p < .0005$. If you want to know p more precisely, double click on the table to put it in edit mode, right click on the cell containing .000 to get a menu for editing the cell, click on *Cell Properties . . .* in the menu, change the number of decimals to 15 in the *Decimals:* box of the *Cell Properties* window, and click *OK*. With the *One-Sample Test* table still in edit mode, you can widen the *Sig. (2-tailed)* cell so that it will show all 15 decimal places by clicking on the right border of the cell and dragging it to the right. The result follows:

One-Sample Test

	Test Value = 16				95% Confidence Interval of the Difference	
	t	df	Sig. (2-tailed)	Mean Difference	Lower	Upper
CESD1	-6.980	361	.000000000014180	-3.49171	-4.47541	-2.50801

You can also get the level of significance for any t by finding the area under the t distribution that lies beyond the value of t by using the CDF.T(q,df) function in the *Compute Variable* window (click on *Transform* and *Compute*). Enter the computed value of t in place of q in the function and enter the degrees of freedom in place of df. After you have entered a name such as cp in the *Target Variable* box and clicked on *OK*, the cumulative proportion corresponding to the values of t and df will appear in the data editor window under the variable name cp. After computing $1 - cp$, you will have the area that lies beyond the specified t—that is, the one-tailed level of significance. Double $1 - cp$ to get the two-tailed level of significance.

You may also compute critical values of t corresponding to α and df by using the IDF.T(p,df) function in the following *Compute Variable* window. This may be a useful alternative to using Table A.2 when your α and/or df are not listed in the table. If you desire the critical t for a two-tailed test, compute the cumulative proportion $1 - \alpha/2$. In the *Compute Variable* window, move the IDF.T(p,df) function to the *Numeric Expression* box, substitute the value of $1 - \alpha/2$ (e.g., .975) for p in the function, substitute the degrees of freedom for df in the function (e.g., 65), enter a variable name (e.g., t) in the *Target Variable* box, and click *OK*. The critical value of t will appear in the data editor window under the variable name t.

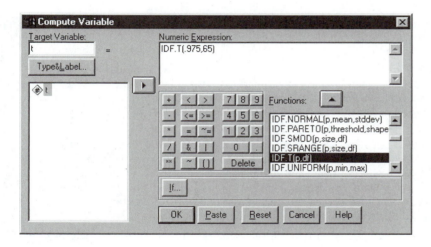

EXERCISES

1. The GSS 2000 (*gss2000b*) asked people who use a computer how many hours per week they spend sending and answering electronic mail (email). For the 1,321 persons who responded, the mean was 3.34 and the standard deviation was 6.278.

   a. What is the standard error of the mean?
   b. What value of $t_{\alpha/2}$ is needed for a 95-percent confidence interval?
   c. What is the 95-percent confidence interval?

2. For 292 persons in GSS 2000 aged 18–29, the mean hours per week using email was 3.83, and the standard deviation was 7.684.

   a. What is the standard error of the mean?
   b. What value of $t_{\alpha/2}$ is needed for a 99-percent confidence interval?
   c. What is the 99-percent confidence interval?

3. The GSS 2000 (*gss2000a*) asked respondents to estimate what percentage of the U.S. population is Black (*usblk*). According to the 2000 U.S. Census (www.census.gov/statab/www/poppart.html), people who identify themselves as being Black and no other race make up 12.3 percent of the population. Conduct a two-tailed test at $\alpha = .05$ of the null hypothesis that 12.3 percent is the mean estimated percentage in the population from which the GSS respondents are selected. For 1,310 responses in GSS 2000, the mean is 31.5 percent and the standard deviation is 17.45. Present your test in the eight-step format.

4. The GSS 2000 (*gss2000a*) asked respondents how much income an unskilled worker should earn per year (*givunskl*). The poverty level for a family of four in 1999 was $17,029 (*Statistical Abstract of the United States 2001,* Table 681). Conduct a two-tailed test at $\alpha = .01$ of the null hypothesis that the poverty level is the mean amount that people believe an unskilled worker should earn. For 1,003 respondents who answered the question in GSS 2000, the mean was $26,008.32 and the standard deviation was $18,119.65. Present your test in the eight-step format.

5. For the GSS statistics given in Exercise 3, conduct a one-tailed test at $\alpha = .05$ to determine if you can conclude that Blacks are believed to be a minority group—that is, less than 50 percent of the population. Present your test in the eight-step format.

6. For the GSS statistics given in Exercise 4, conduct a one-tailed test at $\alpha = .05$ to determine if you can conclude that the earnings that respondents believe unskilled workers should receive are less than those of the average full-time, year-round worker in 1999, which was $33,408 (*Statistical Abstract of the U.S. 2001,* Table 677). Present your test in the eight-step format.

7. The GSS 2000 asked respondents how many years of education were completed by their fathers (*paeduc*) and their mothers (*maeduc*). Conduct a two-tailed test at $\alpha = .05$ of the null hypothesis that there is no difference between fathers' and mothers' education. The variable *educdiff* equals *paeduc* minus *maeduc*. For the 1,805 respondents who reported both *paeduc* and *maeduc,* the mean of *educdiff* is $-.139$ year and the standard deviation is 3.242. Present your test in the eight-step format.

8. Among 333 persons aged 60 or older in the GSS 2000, the mean *educdiff* described in Exercise 7 is $-.784$ year and the standard deviation is 3.747. Conduct a two-tailed test at $\alpha = .05$ of the null hypothesis that there is no difference between fathers' and mothers' education. Present your test in the eight-step format.

9. The GSS 2000 (*gss2000a* and *gss2000b*) asked the following question: "On the average, Blacks/African-Americans have worse jobs, income, and housing than White people. Do you think these differences are because most Blacks/African-Americans just don't have the motivation or will power to pull themselves up out of poverty?" (*racdif4*). Conduct a two-tailed test at $\alpha = .10$ of the null hypothesis that Whites are evenly split on this issue—that is, that 50 percent of the White population would say yes. For 1,346 Whites who responded yes or no, 50.8 percent said yes. Present your test in the seven-step format.

10. Conduct a two-tailed test at $\alpha = .10$ of the null hypothesis that Blacks are evenly split on the issue in Exercise 9—that is, that 50 percent of the Black population would say yes. For 294 Blacks who responded yes or no, 39.5 percent said yes. Present your test in the seven-step format.

11. The following SPSS tables show descriptive statistics for the variable *usblkhsp* from GSS 2000, which equals the estimated percentage of the U.S. population that is Black (*usblk*)

minus the estimated percentage that is Hispanic (*ushisp*). The mean of *usblkhsp* indicates that the average respondent estimated that the Black population is 6.68 percent larger than the Hispanic population. The fact is, however, that the two groups are equal in size (www.census.gov/statab/www/poppart.html). Calculate the missing statistics (?) in the tables. What do you conclude about the mean difference in the population?

One-Sample Statistics

	N	Mean	Std. Deviation	Std. Error Mean
USBLKHSP usblk - ushisp	1276	6.6803	15.06138	?

One-Sample Test

	Test Value = 0					
					95% Confidence Interval of the Difference	
	t	df	Sig. (2-tailed)	Mean Difference	Lower	Upper
USBLKHSP usblk - ushisp	?	?	?	6.6803	?	?

12. The GSS 2000 (*gss2000b*) asked persons who use a computer how many hours per week they spend on the Internet or World Wide Web. First, use SPSS to construct a 95-percent confidence interval for *wwwhr*. Second, use SPSS to conduct a two-tailed test at $\alpha = .05$ of the null hypothesis that the mean hours per week spent on the World Wide Web is 7 hours. Present the test in the eight-step format.

13. SPSS has a *Weight Cases* feature that can be used, among other things, to change the sample size of your data. For example, if you assign a weight of 2 to each case, the sample size will be doubled, and if you assign a weight of .5 to each case, the sample size will be cut in half. If you do not weight the cases, which is the default, each case is in effect treated as having a weight of 1, which gives the actual sample size. You can use *Weight Cases* to examine the effect of different sample sizes on hypothesis tests and confidence intervals. To use *Weight Cases,* it is first necessary to create a variable that equals the desired weight. For example, use *Transform* and *Compute* to compute a variable that may be named *weight* that equals .1 for each case. Then click on *Data* and *Weight Cases* to get the *Weight Cases* window, where you click on *Weight cases by* and then move the variable named *weight* into the *Frequency Variable* box. After each case has been weighted by .1, the sample size will be treated as being only .1 of the actual size by all statistical procedures, such as the *One-Sample T Test*. For this exercise, examine the effect of weights of .1, .25, .5, .75, and 1 on the standard error, *t*, and significance of the variable *wwwhr* when conducting the hypothesis test specified in Exercise 12. Show your results in a table listing *n*, the standard error, *t*, and the significance for each weight. Interpret the effect of the differing weights on these statistics.

14. The GSS 2000 (*gss2000a*) asked respondents to rate the intelligence levels of members of several groups on a scale from 1 to 7, where 1 means almost everyone is not intelligent and 7 means almost everyone is intelligent. Assume these ratings constitute

an interval variable. The variable *intlwhts* is the rating given to Whites and *intlblks* is the rating for Blacks. Create a new variable *intldiff* that equals *intlwhts* minus *intlblks*. Use SPSS to conduct a two-tailed test of whether Blacks believe there is a difference in intelligence between Whites and Blacks. To select only Blacks for this test, click on *Data* and *Select Cases* to open the *Select Cases* window, click on *If condition is satisfied* and the *If* button, and enter "race = 2" in the box of the *Select Cases: If* window. What do you conclude from the results of this test?

7

Associations

BIRDS OF A FEATHER FLOCK TOGETHER
AND OPPOSITES ATTRACT

They had been together for 40 years, first as sweethearts and later as best friends. Surely they were made for each other. But what was the secret of their success?

In so many ways they were exactly alike. They had met in college, and both graduated with degrees in botany. They were good students, if not outstanding ones, who enjoyed learning new things. Their love of learning continued throughout their lives. They liked to discuss and debate political and social issues with each other. They almost always reached a consensus in the end, however, because they were both liberal Democrats who believed you had to change things to make the world better. One thing they did not want to change, however, was the environment. They were staunch conservationists who loved the outdoors and wanted to preserve its beauty and wildness for their children and their children's children. Thus, it is no surprise that they shared a passion for gardening, although she was partial to hostas whereas he loved roses. And when it came to movies, music, and food, yes, their tastes were very much alike. The two of them were alike in so many ways. Surely, that was the secret.

And yet, they also were so different. When they discussed things, he quickly reached a conclusion based on intuition and a deeply felt sense of the right thing to do. She, on the other hand, preferred to carefully examine the pros and cons of an issue before reaching a decision. He was very gregarious and enjoyed parties with lots of people, whereas she, being more introverted, preferred small dinner parties with close friends. Spontaneous and uninhibited, he quickly broke the ice with new acquaintances. A cordial, but cautious, style made her harder to get to know. Deeper down, however, she had a greater sense of self-confidence than he, who often doubted his ability to succeed. Consequently, in the big decisions in their life together, he usually trusted her to take the lead. The two of them were different in so many ways. Surely, that was the secret.

What do you think is the secret of this couple's long and successful life together? Their likenesses or their differences? Do birds of a feather flock together, or do opposites attract? There is reason to believe that both alikeness and difference are important. Couples who share the same values but have complementary personalities, as do this couple, may have a greater chance of success. This conclusion, however, is not beyond dispute in the social science community. What kind of data would you need, and what type of statistical analysis should you use, to test this hypothesis?

First, it would be interesting to find out how alike or different couples are on different types of variables. A number of variables, such as inquisitiveness and political orientation, are used to describe the similarity of the couple in the preceding vignette. These variables consist of values and attitudes. Other variables, such as extroversion and self-esteem, show how different they are. These are personality variables. To test our hypothesis, we would need, ideally, a sample of several hundred couples, with measures of at least one value/attitude variable and at least one personality variable for each member of the couple. Let's say we had a measure, for example, of how politically liberal or conservative is each member of each couple. Thus, we would have two variables, the political orientation of the woman and the political orientation of the man (assuming we are studying heterosexual couples). We expect the couples to be similar on this variable. What kind of pattern should we look for? We expect to find that women with liberal orientations have partners with liberal orientations and women with conservative orientations have partners with conservative orientations. At least, we suspect this would be the predominant pattern. Not every couple would fit this mold. We might find some liberal women with conservative men and some conservative women with liberal men, but these patterns would probably be the exceptions to the rule.

Now, assume we also have a measure of how extroverted or introverted each partner is. What kind of pattern would we expect to find? If opposites attract, we would expect extroverted men to pair up with introverted women and introverted men to be with extroverted women. Again, there may be exceptions, and how many exceptions there are is an important issue, but this is the predominant pattern we would expect to find.

In contrast to our work in Chapters 2 through 6, we are now analyzing two variables at a time. In the example under consideration, one variable is the male partner's personality (or attitude/value), and the other variable is the female partner's personality (or attitude/value). More specifically, we are analyzing the **association,** or *relationship,* between the two variables. *There is an association between two variables if knowledge of the attribute of a case on one variable helps predict the attribute of that case on the other variable.*

There are two kinds of associations between variables: positive and negative. A **positive association** means that if a case is above average on one variable, it is expected to be above average on the other variable. For example, if a woman is more liberal than average, we expect her partner to also be more liberal than average. Also, if a case is below average on one variable, we expect it to be below average on the second variable if there is a positive association. Therefore, a positive association between couples' attributes means we *expect* both partners to be above average or both partners to be below average. In other words, birds of a feather tend to flock together. We say *tend* and *expect* because there will undoubtedly be exceptions to the dominant pattern.

A **negative association** means that a case that is above average on one variable will tend to be below average on the other variable, and vice versa. If there is a negative association between the personalities of partners, then men who are extroverted, for example, would tend to have partners who are introverted, and introverted men would tend to pair up with extroverted women. That is, opposites attract.

Both positive and negative associations may vary in the **strength of association.** We have mentioned exceptions to the rule of an association between variables. If there are just a few exceptions, then the association is strong. If there are many exceptions, the association is weak. And if the number of exceptions is somewhere between these extremes, there is a moderate association. When the exceptions to the rule become so numerous that they are no longer exceptions, we have the case of **no association** or a **zero association.** If there is no association, knowing whether a case is above or below average on the first variable does not help us to predict whether it will be above or below average on the second. If there are zero associations between the attributes of a couple, then birds of a feather do not flock together *and* opposites do not attract.

Here's a final word about the theory suggested by the vignette. The theory predicts that successful couples have positive associations between attitude/value variables and negative associations between personality variables. One way to test this theory is to divide couples into two groups based on the success of their relationship. Success may be measured by whether or not they broke up or divorced after a given number of years (you will have to wait a while to find out). Or success may be operationalized by whether the couple is above average or below average with respect to how happy or satisfied they are with their relationship. Then, we would predict that the more successful couples have stronger positive associations between values *and* stronger negative associations between personalities than the less successful couples. The less successful couples might actually have *zero association* or associations opposite those of those who were more successful.

VISUALIZING ASSOCIATIONS

In the previous section, ordinary words were used to introduce the concept of an association between variables. In Chapter 2 we used tables—which are numerical descriptions of distributions—and graphs—which are visual descriptions of distributions—to examine the characteristics of a distribution. In this chapter we will again use both tables and graphs to describe the association between two variables. This chapter and the next four chapters will focus on *associations between interval or ratio variables*. We will begin with visual descriptions, perhaps the most intuitive method of understanding associations.

Scatterplots

In Chapter 2 the distributions of variables were displayed in bar charts, histograms, and frequency polygons. Each of these graphs consisted of two perpendicular axes. The attributes of the variable being described were marked or scaled along the horizontal axis. The frequencies or percentages of the attributes were scaled on the vertical axis. The graph for the association between two variables also consists of two perpendicular axes. However, the attributes of one variable (call it X) are scaled along the horizontal axis and the attributes of the other variable (call it Y) are scaled along the vertical axis (Figure 7.1). For now, it does not matter which of the two variables is scaled on the X axis and which is scaled on the Y axis.

The graph we are about to construct is a **scatterplot.** Instead of displaying the distribution of a single variable, it shows the **joint distribution** of two variables. The scatterplot is a member of a family of graphs based on the *Cartesian coordinate system,* which is named after French mathematician René Descartes. In this system, each pair of values of X and Y is represented by a single point in the graph. The values of X and Y corresponding to the point are called the *coordinates* of the point and are symbolized as (X, Y). Let one pair of coordinates be $X = 8$ and $Y = 14$, which is symbolized as $(8, 14)$. The point in the graph for this pair would lie at the intersection of a vertical line drawn through the graph at $X = 8$ and a horizontal line drawn at $Y = 14$ (Figure 7.1).

Now, let's examine a scatterplot for three cases based on the following values of X and Y as shown in Table 7.1. The three pairs of coordinates are plotted in Figure 7.2. The coordinates (X, Y) are displayed near each point for clarification. Normally they would not be shown in the scatterplot. Instead, we would look down from the point along a line perpendicular to the X axis to find the value of the X coordinate and look across along a line perpendicular to the Y axis to find the Y coordinate, as indicated by the dashed lines in Figure 7.2 (which normally are not included in the plot).

To make it easier to see the nature of the association in a scatterplot, it is useful to draw a vertical line through the graph at the mean of X and a horizontal line through the graph at the mean of Y. These lines create four **scatterplot quadrants** (Figure 7.3). The NE quadrant contains points representing coordinates that are above the mean on both X and Y, $(+, +)$. If, for example, X is a female's trait and Y is her partner's trait, a point in the NE quadrant would indicate that both partners are above average. The SW quadrant, on the other hand, indicates cases that are below average on both variables. Thus, points in both the NE and SW quadrants are indicative of a positive association. If the points represent couples, they are birds of a feather. The SE and NW quadrants, on the other hand, represent cases that are above the mean on one variable and below the mean on the other variable. Cases in these quadrants are indicative of a negative association between X and Y. If the cases represent couples, these quadrants support the saying that opposites attract.

Now, let's think about how to use these quadrants to interpret the association, if any, that is shown in a scatterplot. If there are more cases in the NE and SW quadrants than in the SE and NW quadrants, there is a positive association between X and Y. The greater the percentage of cases in the NE–SW quadrants, and the smaller the percentage in the SE–NW quadrants, the greater will be the **strength of the positive asso-**

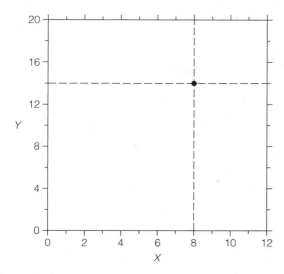

FIGURE 7.1 Locating a Point in the Cartesian Coordinate System

Table 7.1 Values of X and Y	
X	Y
2	4
6	6
10	16

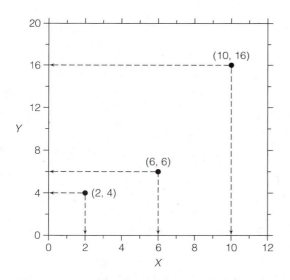

FIGURE 7.2 Scatterplot for a Sample of Three Cases

ciation. If all, or virtually all, of the cases are in the NE–SW quadrants, there will be a very strong association, although just how strong it is depends on where the cases are located in these quadrants. In Figure 7.3, for example, two out of three points are in the NE–SW quadrants and one is right on the line between SW and SE (average on X and below average on Y), so there is a fairly strong positive association in this scatterplot. A

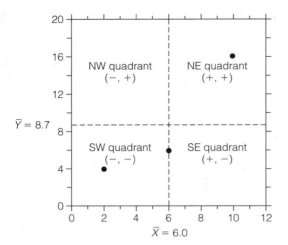

FIGURE 7.3 Quadrants of a Scatterplot

negative association is represented by a preponderance of cases in the SE–NW quadrants, and the greater this preponderance, the stronger the negative association. Finally, if the cases are pretty evenly dispersed across all four quadrants, there is no association between X and Y. A scatterplot must always have some cases in each of two opposite quadrants, either the NE–SW or the SE–NW. Otherwise, it would imply that all of the cases are either above the mean or below the mean on at least one variable—a mathematical impossibility.

What does the strongest possible association between X and Y look like in a scatterplot? Figure 7.4a shows a positive association in which all the points lie exactly on a straight line running through the NE–SW quadrants, and Figure 7.4b shows a negative association in which all the points form a straight line in the SE–NW quadrants. These are **perfect linear associations.** There is no pattern of association that is any stronger than these.[1] A perfect linear association, either positive or negative, will be our model for the strongest possible association. However, it is only a model that is useful for comparison purposes. A perfect linear association between two observed variables will never occur in the social sciences. The world is too complex and our measurements are too imprecise for a perfect linear association to be observed.

Instead of thinking about the strength of associations in simple terms such as weak, moderate, or strong, which are *vague quantifiers,* we will use numbers to represent the strength of association in order to be more precise and accurate. It is fairly common today to hear people rating things on a 10-point scale. For example, you might rate a person's attractiveness on a scale from 1 to 10: "He's an 8," or "She's a 10." We will rate associations between two variables in much the same way. We will, however, specify not only how strong the association is but also whether it is negative or positive. We also

---

1. A perfect nonlinear association, in which all the points lie on a mathematically defined curve, is an equally strong association. In this book, however, we will restrict ourselves to the simpler case of linear associations.

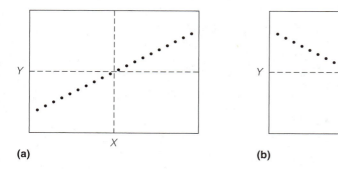

(a) **(b)**

FIGURE 7.4 Scatterplots of Two Perfect Linear Associations

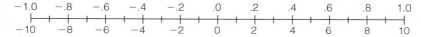

FIGURE 7.5 Rating Scales for Strength and Direction of Association

will add a 0 point to stand for no association. Consequently, we end up with a 21-point scale ranging from −10 to +10, with 0 lying exactly in the middle (Figure 7.5). One final modification that statisticians make to the commonplace 10-point scale is to divide by 10, which results in a rating scale ranging from −1, a perfect negative association, to +1, a perfect positive association (Figure 7.5). The sign of the association is called the **direction of association** because it specifies whether the association is in the direction of the negative pole of the scale or the positive pole.

To gain practice in visually judging the direction and strength of association, examine the scatterplots in Figure 7.6. For each of the ten plots, use the rating scale in Figure 7.5 to assign the association a score between −1 and 1. Although we can theoretically use numbers with two or three decimal places to rate the associations, an eyeball rating cannot be that precise. Thus, choose values on the scale to only a single decimal place to rate the associations. You may, for example, say that one plot is a −.6, another is a .1, and so on. These plots use empty circles instead of solid black circles to represent each point. The empty circles make it easier for the eye to distinguish between points when many points are close together or overlap one another. Keep in mind that this is simply a visual exercise in seeing the association. Don't try to use any mathematical means for making your judgment, such as counting the number of points in each quadrant. Allow your eye to make an overall rating by observing which parts of the plot are the most dense and which are the most sparse.

The correct numerical ratings for the associations in Figure 7.6 are determined by mathematical formulas that you will learn later in this chapter and the next. The correct numeric ratings for the scatterplots in Figure 7.6 are

(a)	(b)	(c)	(d)	(e)	(f)	(g)	(h)	(i)	(j)
.8	.5	−.4	.0	−.6	.7	.9	.1	.3	−.2

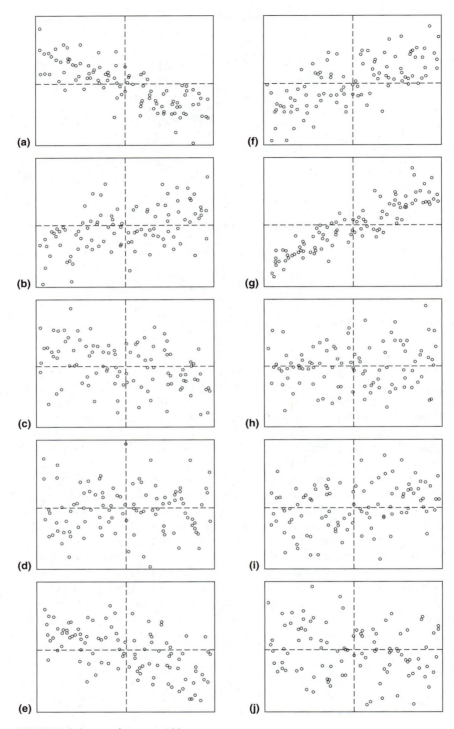

FIGURE 7.6 Scatterplots, $n = 100$

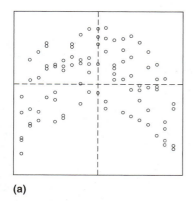 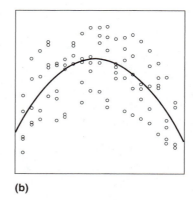

(a)   (b)

FIGURE 7.7 A Nonlinear Association

Don't worry if you weren't right on target with all of your ratings. It is sometimes difficult to judge the strength of associations by visual observation alone. That is why mathematical measures of association were developed. However, it is valuable to have in your mind an image of the scatterplot on which a particular numerical measure of association is based.

Another reason for examining a scatterplot is to see whether a **nonlinear association** may be present. An examination of Figure 7.7a may suggest that there is no association between X and Y because the joint distribution in the NE–SW quadrants is very similar to that in the NW–SE quadrants. Although it is true that there is no *linear* association, there is a nonlinear association that is described by the curve in Figure 7.7b. If you calculated a measure of linear association for these data without examining the scatterplot, you might draw the erroneous conclusion that X and Y are not related to one another.

Jittered Scatterplots

When we plot variables that have a relatively small number of discrete values, there are often multiple cases with exactly the same values of X and Y. Consequently, these cases will be located at the same point in the scatterplot, and their plotting symbols will lie on top of one another. Because of this, when we look at a point shared by multiple cases, we will be able to see only one plotting symbol. Due to this stacking of cases at various locations, we cannot always tell which areas are densely populated with cases and which are only sparsely settled.

To illustrate this problem, Table 7.2 shows the values of X and Y for eight cases. A regular scatterplot for these cases is shown in Figure 7.8a. Although there are eight cases in the plot, only five plotting symbols are visible due to stacking.

To unstack the cases, we will add a small random number to each value of X and Y. This is called **jittering.** Formulas for jittering X and Y are

$$X^\star = X + r - .5$$

$$Y^\star = Y + r - .5$$

Table 7.2 Jittering *X* and *Y*

X	Y	r_X	r_Y	X*	Y*
5	10	.5	.6	5.0	10.1
10	20	.6	.6	10.1	20.1
15	20	.0	.9	14.5	20.4
10	10	.3	.5	9.8	10.0
15	20	.1	.7	14.6	20.2
10	15	.9	.8	10.4	15.3
5	10	.8	.2	5.3	9.7
10	15	.5	.4	10.0	14.9

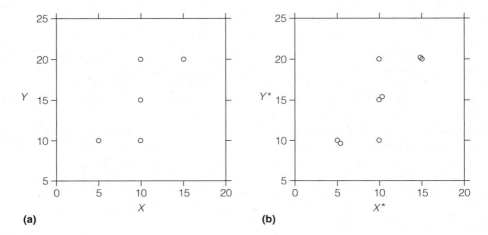

(a)   (b)

FIGURE 7.8 Regular and Jittered Scatterplots

where *r* is a random number from a uniform distribution between 0 and 1. The logic for this procedure is that we want to jitter each variable within its *real limits*. The real limits for each value of *X* and *Y* are $X \pm .5$ and $Y \pm .5$, respectively. The reason for subtracting .5, which is one-half the range of the random numbers, is to make some jittered values as much as .5 smaller than the original value and others as much as .5 greater than the original value. Because there are eight cases and two variables, we will need sixteen random numbers (*r*). We will use the first sixteen digits in the last row of Table 5.1 as our random numbers. It is all right to use the same random number more than once. Because we need random numbers between 0 and 1, we place a decimal point in front of each random value (i.e., divide by 10). These sixteen random numbers are assigned to *X* and then *Y*, one case after another, as shown in columns r_X and r_Y in Table 7.2. The jittered values of the first pair of coordinates are

$$X^\star = 5 + .5 - .5 = 5.0$$

$$Y^\star = 10 + .6 - .5 = 10.1$$

The jittered values $X^\star$ and $Y^\star$ are plotted in Figure 7.8b. You can now see eight plotting symbols. After the cases have been unstacked, the association between X and Y appears stronger than it did in the regular scatterplot. This will not always be the outcome of jittering. Whether the association appears stronger or weaker after jittering, the point of jittering is to let us see all of the points and thereby make a better visual assessment of the true association. The jittered values of X and Y are used only in graphs, not for calculating statistics that provide a numerical measure of association.

Examples of Associations in Scatterplots

A potentially important issue about which couples may agree or disagree is whether men and women should occupy different roles in society. A traditional gender-role attitude maintains that women should not perform the same roles as men. In a traditional family structure, for example, women were expected to keep house and raise the children while men were expected to provide for the family's financial needs. Although traditional gender-role expectations have been challenged for several decades, significant numbers of men and women still hold on to such standards. We will look at data on husbands' and wives' gender-role attitudes to see how closely spouses agree with one another and to determine whether the degree to which they agree may significantly affect the success of their marriage.

Box 7.1 shows five attitude statements that were used to measure traditional gender-role attitudes in a 1987 telephone survey conducted in the Akron, Ohio, metropolitan area (1987 Akron Area Survey). Agreement with the statements indicates traditional attitudes and disagreement indicates nontraditional attitudes. Responses to each statement were scored from 0 (strongly disagree) to 3 (strongly agree). A gender-role scale was created by summing the scores of the five items. The resulting scores on the gender-role scale ranged from 0 to 15. We will analyze the gender-role attitudes of 328 married respondents and their spouses.

Dashed lines are drawn through the scatterplot (Figure 7.9) at the wife's mean (6.26) and the husband's mean (7.35). These means indicate that the husbands are somewhat more traditional than their wives. Because there are only fifteen discrete attitude scores for each spouse, we expect considerable stacking in a scatterplot of the joint distribution of the husbands' and wives' scores (Figure 7.9a). After the coordinates are jittered, it is clear that most of the stacking occurred on or near a diagonal line running from the lower-left corner to the upper-right corner of the plot. Points along this diagonal represent couples in which each spouse has the same attitude. Looking at the quadrants formed by the mean lines, the NE quadrant (husbands and wives who are more traditional than average) and the SW quadrant (both spouses are less traditional than average) clearly contain many more spouses than the SE and NW quadrants. This means that there is a positive association between the wife's gender-role attitude and her husband's attitude. Numerically speaking, how strong do you think the association is?[2]

If holding similar attitudes promotes a better marriage, then the association between spouses' gender-role attitudes should be stronger among those couples who are happiest with their marriage. Marital satisfaction was rated on a 7-point scale, with 7 indicating

---

2. The association is approximately .8.

BOX 7.1   Gender-Role Attitude Statements

1. Women should take care of running their homes and leave running the country up to men.[a]
2. Most men are better suited emotionally for politics than are most women.[a]
3. It is much better for everyone if the man is the main provider and the woman takes care of the home and family.[a]

4. It is more important for a wife to help her husband's career than to have one herself.[a]
5. There is some work that is men's and some that is women's, and they should not be doing each other's.[a]

[a]Strongly disagree = 0; disagree = 1; agree = 2; strongly agree = 3.

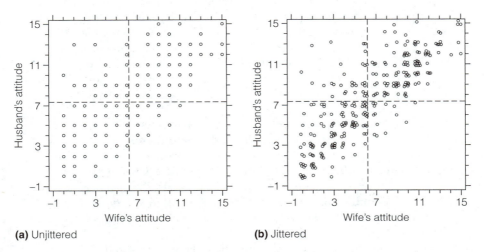

(a) Unjittered   **(b)** Jittered

FIGURE 7.9 Wife's and Husband's Gender-Role Attitudes

the greatest satisfaction. In each household, either the husband or the wife was randomly chosen to rate marital satisfaction. We will examine whether wives who were more satisfied had closer agreement with their husbands on gender roles than wives who were less satisfied.[3] Because marital satisfaction is relatively high, wives are divided into two groups based on whether they rated their marriage 7 (Figure 7.10b) or less than 7 (Figure 7.10a). Do you think the association between husbands' and wives' attitudes is stronger among the more satisfied wives (Figure 7.10b) or among the less satisfied wives?[4]

Thus far, we have been examining associations between two variables that measure the same trait for both members of a couple (husband's gender-role attitude and wife's gender-role attitude). We can also examine associations between variables that represent

---

3. The results for husbands (not shown) are very similar to those for wives.

4. The association is .85 in Figure 7.10b and .67 in Figure 7.10a. Thus, the difference in associations supports the hypothesis, although the associations are not dramatically different in the two scatterplots.

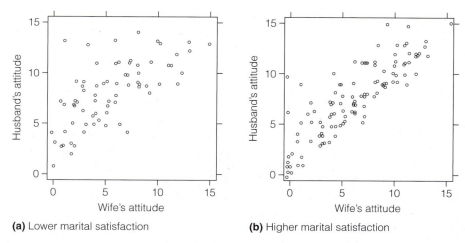

(a) Lower marital satisfaction   **(b)** Higher marital satisfaction

FIGURE 7.10 Wife's and Husband's Gender-Role Attitudes by Wife's Marital Satisfaction

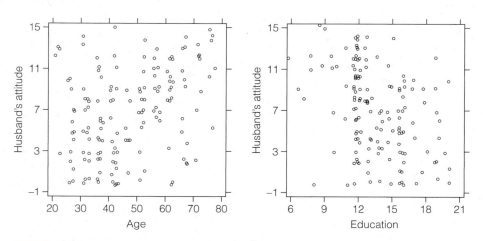

FIGURE 7.11 Association of Husband's Gender-Role Attitude with Age and Education

different traits for the same person. For example, we can ask which attributes of wives are associated with their gender-role attitudes and/or which attributes of husbands are associated with their gender-role attitudes. Two potentially important attributes may be age and education. We might expect older people (that is, older cohorts) to be more traditional and people with more years of schooling to be less traditional. How would you rate the association between husband's age and his gender-role attitude (Figure 7.11a) and how would you rate the association between husband's education and his gender-role attitude (Figure 7.11b)? These relationships, if any, are clearly less strong than those in Figure 7.10.[5]

QUANTIFYING ASSOCIATIONS

Although scatterplots provide a complete picture of any association between X and Y, quantitative measures of association give a more precise description of important aspects of association. In this section, we will examine the basic building block for measures of association between interval/ratio variables, the *covariance*. Whereas the *variance* (and standard deviation) measures how much a single variable varies, the covariance indicates how closely two variables vary together (i.e., covary). As we will see, there is a close correspondence between the way the covariance is computed and the rationale for using the means of X and Y to divide a scatterplot into four quandrants.

Cross-Products

Remember that if most cases are in either the NE quadrant (above the mean on both X and Y) or the SW quadrant (below the mean on both variables), there is a positive association between X and Y. A positive association means that cases tend to be either above the mean on both X and Y or below the mean on both X and Y. With variables that are measured at the interval or ratio level, however, we are able to take into account not only whether both variables are on the *same side of the mean*, but whether they both deviate from the mean by similar amounts or by different amounts. If a case is far above or far below the mean on both X and Y, it indicates a *stronger* positive association than if a case deviates greatly from the mean on one variable but only slightly on the other.

The way that we measure simultaneously both the direction of association (positive or negative) and the strength of association is by multiplying each case's deviation score on X times its deviation score on Y. The product of deviation scores on X and Y is called the **cross-product** (CP):

$$CP = (X - \overline{X})(Y - \overline{Y}) \tag{7.1}$$

You remember that the deviation score $X - \overline{X}$ indicates how far above or below the mean is a score on X. The deviation score $Y - \overline{Y}$ indicates the same thing for Y. The product of the two deviation scores indicates the contribution that each case makes to the direction and strength of association. The reason why the *product* carries this information may be illustrated nicely with the small data set in Table 7.3.

The means of both X and Y are $9/3 = 3$. The sums of deviation scores equal 0 for both X and Y, which you will remember is always the case for deviation scores. The first case in Table 7.3 is below the mean for both X and Y. Thus, it is located in the SW corner of the scatterplot (Figure 7.12). The cross-product for this case equals $(-2)(-1) = 2$. Because the case is below the mean on both variables, it has negative deviation scores

---

5. The association is .34 for age and gender role and $-.31$ for education and gender role.

Table 7.3 Cross-Products of X and Y

X	Y	$X - \bar{X}$	$Y - \bar{Y}$	$(X - \bar{X})(Y - \bar{Y})$
1	2	−2	−1	2
3	2	0	−1	0
5	5	2	2	4
$\overline{9}$	$\overline{9}$	$\overline{0}$	$\overline{0}$	$\overline{6}$

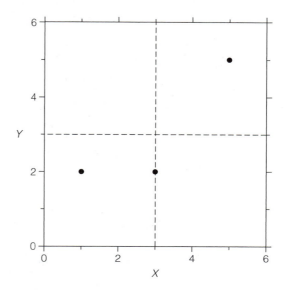

FIGURE 7.12 Scatterplot of X and Y

on both. Because the product of two negative numbers is positive, the cross-product contributes to a positive association.

The second case in Table 7.3 is on the line between the SW and SE quadrants because it is at the mean of X. Thus, $X - \bar{X} = 0$. It contributes to neither a positive nor a negative association. Whenever a case is at the mean of either X, Y, or both, it is not in any of the quadrants and contributes 0 to the association between X and Y.

The third case is two units above the mean on both X and Y and thus has a positive cross-product, $(2)(2) = 4$. Consequently, it contributes to a positive association. Because the third case is further above the mean of Y than the first case is below the mean of Y (-1), its cross-product is greater than that of the first case. Therefore, although the first and third cases both contribute to a positive association, the third case contributes more.

If a case fell in either the SE or NW quadrants, it would have a positive deviation score on one variable and a negative deviation score on the other variable. Thus, it would have a negative cross-product and would make a contribution to a negative association. There are, however, no such cases in Table 7.3.

The **sum of the cross-products** (SCP) of all the cases represents the aggregate association between X and Y. If the sum of all the positive cross-products is larger than the sum of all the negative cross-products, the total sum will be positive and we will say there is a positive association. The sum of cross-products is symbolized as

$$\text{SCP} = \Sigma(X - \bar{X})(Y - \bar{Y}) \tag{7.2}$$

$\text{SCP} = 2 + 0 + 4 = 6$ in Table 7.3 and, thus, there is a positive association between X and Y.

To illustrate how cross-products capture other types of associations, *what if* the values of Y for the second and third cases in Table 7.3 were interchanged? This would not

**Table 7.4 Types of Association
between X and Y**

X	Y	$X - \bar{X}$	$Y - \bar{Y}$	$(X - \bar{X})(Y - \bar{Y})$
		a. Positive association		
1	2	−2	−1	2
3	2	0	−1	0
5	5	2	2	4
$\overline{9}$	$\overline{9}$	$\overline{0}$	$\overline{0}$	$\overline{6}$
		b. Zero association		
1	2	−2	−1	2
3	5	0	2	0
5	2	2	−1	−2
$\overline{9}$	$\overline{9}$	$\overline{0}$	$\overline{0}$	$\overline{0}$
		c. Negative association		
1	5	−2	2	−4
3	2	0	−1	0
5	2	2	−1	−2
$\overline{9}$	$\overline{9}$	$\overline{0}$	$\overline{0}$	$\overline{-6}$

affect the deviation scores of Y for these cases because the mean of Y would not be changed (Table 7.4b). It would, however, affect the cross-products because it would change the correspondence between the deviation scores of X and Y for these cases. Now, the highest score on Y (5) is paired with the mean score on X, resulting in a cross-product of 0. Also, the highest score on X (5) now corresponds with a score of 2 on Y, resulting in a cross-product of −2. Consequently, the sum of cross-products equals 0, indicating that there is *no association between X and Y.*

Now, *what if* the first case had the highest value of Y (5) and the second and third cases each had scores of 2? The fact that the first case is two units below the mean on X and two above the mean on Y results in a large negative cross-product of −4. The third case, on the other hand, is above the mean on X and below the mean on Y, resulting in a negative cross-product (−2). As a result, the sum of cross-products equals −6, which means there is a *negative association between X and Y.*

Covariance

If the sum of cross-products equals 0, there is *no association* between X and Y. If SCP does not equal 0, the sign of SCP indicates the **direction of association**—that is, whether it is a positive or negative association. The absolute value of SCP, however, cannot be used to judge the *strength of association* because it is influenced by more than the signs and magnitudes of the cross-products. It is also influenced by the number of cross-products. Thus, the larger the n, the greater the sum of cross-products, all other things being equal.

It is undesirable for a measure of association to be influenced by the number of observations. There is, of course, a simple way to eliminate this influence, namely, *divide by*

n. The sum of X was divided by n to get the mean, and the sum of squared deviations around the mean was divided by n to get the variance. Dividing the sum of cross-products by n gives the **covariance** (COV) or s_{XY}:

$$\text{Covariance} = \text{COV}(X, Y) = s_{XY} = \frac{\Sigma(X - \bar{X})(Y - \bar{Y})}{n} \tag{7.3}$$

In Table 7.4a, for example, $s_{XY} = 6/3 = 2$. The covariance between X and Y equals 2.

Why is this statistic called the *covariance?* A comparison with the variance may be instructive. The variance may be expressed as

$$s^2 = \frac{\Sigma(X - \bar{X})^2}{n} = \frac{\Sigma(X - \bar{X})(X - \bar{X})}{n}$$

The variance involves the product of $X - \bar{X}$ times itself. If $Y - \bar{Y}$ is substituted for one $X - \bar{X}$, we get the formula for the covariance. That is why the variance is sometimes referred to as the covariance of X with itself. It also suggests that $\Sigma(X - \bar{X})(Y - \bar{Y})/n$ indicates how closely X varies with Y—that is, how closely they *covary*.

This analogy between the variance and the covariance indicates why s_{XY} is often used as a symbol for the covariance. This symbol is somewhat ambiguous, however, because s is the symbol for the standard deviation, not the variance. It would be misleading to use s_{XY}^2, though, because the covariance is not the square of the standard deviation or any other statistic. This may account for the reason that COV(X, Y) is also used often.

Still, there is ambiguity in how to interpret the number to which the covariance is equal. In Table 7.4a, again, the covariance equals 2. But 2 of what? The covariance is a mixture of the units of X and the units of Y. What if X equals the number of siblings that each person has and Y equals the number of children they have? Then the covariance, which is based on products of X and Y, is some combination of siblings and children. It mixes together apples and oranges. Literally, *the covariance is the mean product of the deviation scores of X and Y*. Thus, a covariance of 2 between siblings and children means that the average person has a product of two sibs–children above the means of both variables or two sibs–children below the means of both variables.

There is another difficulty with interpreting the statistic. We know that the larger the absolute value of the covariance, the stronger the association. But what is a large value? What is considered a large covariance depends on how large or small are the standard deviations of X and Y. To illustrate, if we multiply each value of Y in Table 7.4 by 100, we will have $Y' = 100 \cdot Y$ (Table 7.5). All the values of Y have been increased proportionately. If Y is dollars of income, for example, Y' would equal income in cents. The standard deviation of Y' equals 141.42 (calculations not shown), which is 100 times greater than the standard deviation of Y (1.4142). As a consequence of the increase in the standard deviation, which was a result of a change in the units of measurement (e.g., from dollars to cents), the covariance increased by a hundredfold to 600 (Table 7.5). If X had been multiplied by 100 instead of Y, the covariance would have increased by the same amount. This means that the size of the covariance is directly related to the sizes of the standard deviations of X and Y. Consequently, there are no upper or lower limits on the value of the covariance. Thus, you cannot judge the *strength of association* by looking at the covariance alone. You have to use additional information to adjust the covariance. This adjustment will be taken up in Chapter 8.

Table 7.5 Sum of Cross-Products of X and $Y' = 100 \cdot Y$

X	Y'	$X - \bar{X}$	$Y' - \bar{Y}'$	$(X - \bar{X})(Y' - \bar{Y}')$
1	200	−2	−100	200
3	200	0	−100	0
5	500	2	200	400
9	900	0	0	600

Despite the problems with the covariance as a measure of association—ambiguity of interpretation and lack of upper and lower limits—the covariance is a basic building block in the construction of more easily interpretable measures of association. Thus, an understanding of the covariance is essential for interpreting these more advanced measures.

Examples of Covariances

We examined scatterplots in Figures 7.9 through 7.11 in an attempt to visually judge the associations between husbands' and wives' gender-role attitudes, as well as the associations of age and education with husbands' gender-role attitudes. The quantitative measures of association that were just discussed will be used now to describe those same associations (Table 7.6).

There is a positive sum of cross-products for the attitudes of the 328 married couples plotted in Figure 7.9. This positive sum indicates that the more traditional the attitude of the wife (or husband), the more traditional the attitude of the husband (or wife) tends to be. The SCP cannot be used to judge the strength of this association, however, because it is generally strongly affected by n. Dividing by $n = 328$ gives a covariance of 11.97. Because the covariance has no upper or lower limit, however, it cannot be used to assess the strength of this association. The best guide to judging this strength with the tools we have developed thus far is to look at the scatterplot in Figure 7.9.[6]

If we cannot use SCP and the covariance to assess the strength of one association, perhaps we can use them to compare the association between one pair of variables with the association between another pair. Instead of asking how strong an association is, we can ask which association is *stronger.* Figure 7.10b shows the relationship between spouses' attitudes for wives who were the most satisfied with their marriages, and Figure 7.10a shows the relationship for wives who were less satisfied. We cannot use the SCP to compare these two scatterplots, however, because there were many more women in the most satisfied group (113) than the less satisfied one (72). Because both scatterplots involve associations between variables that have the same scales of measurement (the gender-role scale with a range of 0–15), we may be able to validly compare the covariances to determine which association is stronger. The covariance is substantially larger among the most satisfied wives (11.7138) than among the less satisfied (7.4877). Thus, we may be inclined to conclude that the former wives' attitudes are more like their

---

6. Footnote 2 indicated that on the scale of −1 to +1, the association is .8.

Table 7.6 Covariances for Gender-Role Attitude Examples[a]

Figure	X	Y	n	SCP	Cov
7.9	Wife's attitude	Husband's attitude	328	3927.2835	11.9734
7.10.a	Wife's attitude	Husband's attitude	72	539.1111	7.4877
7.10.b	Wife's attitude	Husband's attitude	113	1323.6549	11.7138
7.11.a	Age	Husband's attitude	154	3224.0260	20.9352
7.11.b	Education	Husband's attitude	154	−628.2597	−4.0796

[a]1987 Akron Area Survey

husbands' (a more positive association) than are the latter wives' attitudes, as expected. Before jumping to this conclusion, however, we should check the standard deviations of the variables in the two groups. It turns out that there is more variance in both the husbands' and wives' gender-role attitudes in the group of women who are the most satisfied with their marriages (the standard deviations for the husbands and wives are 3.65 and 3.91, respectively) than in the group who are less satisfied (the standard deviations are 3.14 and 3.64). Because variables that have greater standard deviations will have greater cross-products, all other things being equal, we should be very cautious about drawing any conclusions based on a comparison of the covariances.[7]

The associations in the last two lines of Table 7.6 both involve 154 cases. Thus, can we use their SCPs to determine which is more closely associated with the husbands' attitudes, their ages or their educations? The sign of SCP for age and attitude indicates that older men are more traditional, whereas the sign of SCP for education and attitude shows that more educated husbands are less traditional. Although the directions of the associations are opposite, it is still valid to raise the question of which association is stronger. The answer is that despite the fact that *n* is the same for both associations, we should not use either SCP or the covariance to compare these associations. Age has a much greater standard deviation (15.0) than years of school completed (3.2). The magnitude of this difference may well account for the fact that the covariance of age and attitude (20.9) is much greater than the absolute value of the covariance of education and attitude (−4.1).[8]

SUMMARY

In this chapter we examined methods of displaying and measuring associations between interval/ratio variables. Both visual methods (graphs) and numerical methods were explored. *Linear* relationships between variables were emphasized. Two aspects of associations were discussed: the *direction* of association and the *strength* of association. The visual

---

7. Based on the scale of −1 to +1, footnote 4 indicated that the association *is* stronger among the most satisfied wives (.85) than among the less satisfied (.67).

8. Again, based on the −1 to +1 scale, footnote 5 indicates that the association is about the same strength for age and attitude (.34) as for education and attitude (−.31).

and numerical approaches to interpreting associations share a common logic. Graphically, scatterplots are divided into four quadrants, with each quadrant containing cases that are in the same relative positions with respect to the means of X and Y. Quantitatively, cross-products are computed for deviation scores of X and Y, indicating whether a case is on the same side of the mean of both X and Y or on opposite sides of the means of X and Y. The advantage of the scatterplot is that it lets us see at a glance all of the major features of the association. The advantage of the quantitative measures, such as the sum of cross-products and the covariance, is that they are more precise. Also, as we will see later, they can be used in more complex analyses involving more than two variables (multivariate statistics). Despite their precision, however, the quantitative measures developed up to this point do not provide an unambiguous *interpretation* of the linear relationship between the two variables nor the *strength* of that relationship. Statistics such as the regression equation and the correlation coefficient, both of which build on the sum of cross-products and the covariance, provide better descriptions of associations between interval/ratio variables. These statistics will be introduced in the next chapter.

KEY TERMS

association	joint distribution	sum of cross-products
positive association	scatterplot quadrants	direction of association
negative association	perfect linear associations	covariance
strength of association	jittering	
scatterplot	cross-product	

SPSS PROCEDURES

We will make a scatterplot for husband's gender-role attitude (*hsexrole*) and wife's gender-role attitude (*wsexrole*). First, jitter the two variables to unstack them. Click on *Transform* and *Compute* in the data editor window. In the *Compute Variable* window, select RV.UNIFORM(min,max) from the *Functions* list and move it to the *Numeric Expression* box. Substitute −.5 for *min* and .5 for *max*. Move the variable (e.g., *hsexrole*) you want to jitter to the *Numeric Expression* box and create the expression: hsexrole + RV.UNIFORM(−.5,.5). Type the name you choose for the jittered variable in the *Target Variable* box (e.g., *jithsexr*). After you click *OK*, the *jithsexr* variable will be computed and placed in the data editor window. Follow the same procedure to jitter the other variable (e.g., *wsexrole*).

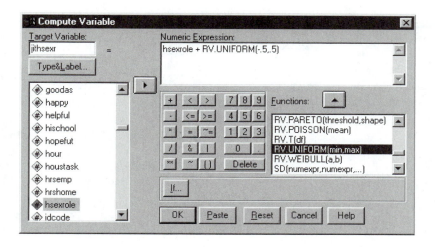

To make the scatterplot, click on *Graphs* and *Scatter* in the main menu, click on *Simple* in the *Scatterplot* window, and move the variables you want on the vertical and horizontal axes into the *Y Axis* and *X Axis* boxes, respectively, in the *Simple Scatterplot* window that follows. Click *OK* and then double click on the scatterplot in the output window to edit it. In the SPSS Chart Editor, click on *Format* and *Marker*, click on the circle in the *Markers* window, and click on *Apply All* to change the data–point symbol from a square to a circle. To edit the scales of the axes, click on *Chart, Axis,* and *X scale* (or *Y scale*). In the following *X Scale Axis* window, enter −1 in the *Minimum Displayed* box and enter 15 in the *Maximum Displayed* box; then set the *Major Division Increment* to 4 and the *Minor Division Increment* to 1 to finish scaling the axis. Select *Center* for the *Title Justification* and enter the desired title in the *Axis Title* box. To divide the scatterplot into four quadrants, click on *Chart, Reference Line,* and *X scale* (or *Y scale*), type the mean of the variable in the *Position of Line* box, and click on *Add*. To change the reference lines from solid to dashed lines, click on one of the reference lines, click on *Format,* click on *Line Style,* click on the type of dashed or dotted line you want in the *Style* box, and click on *Apply*. After repeating these steps for the other reference line, you will have the scatterplot shown next.

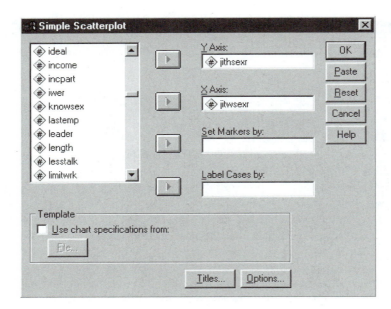

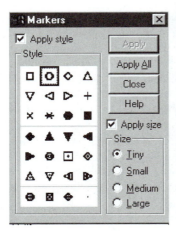

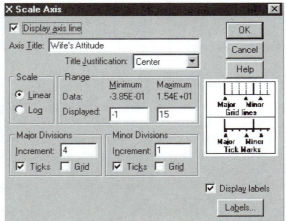

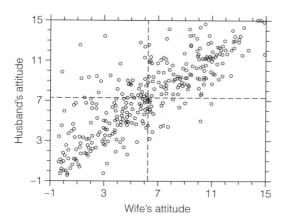

Now we will compute the numerical measures of association, namely, the sum of cross–products and the covariance. Click on *Analyze, Correlate,* and *Bivariate*. In the following *Bivariate Correlations* window, move the two variables to the *Variables* window and do not change the default selections *Pearson Correlation Coefficient* and *Two-Tailed Test of Significance*. Click on *Options* and click on *Cross-product deviations and covariances*. After you have clicked on *Continue* and *OK*, the following correlation table will appear in the output window. The sum of cross-products and the covariance appear in the upper-right cell and lower-left cell of the table (see the shaded areas). The covariance is slightly larger than that given in Table 7.6 because SPSS divides the sum of cross-products by $n - 1$ to obtain the unbiased estimate of the population covariance. You may ignore the Pearson correlation, which will be covered in the next chapter.

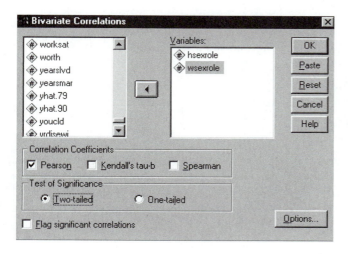

Correlations

		HSEXROLE	WSEXROLE
HSEXROLE	Pearson Correlation	1	.817
	Sig. (2-tailed)	.	.000
	Sum of Squares and Cross-products	5032.858	3927.284
	Covariance	14.630	12.010
	N	345	328
WSEXROLE	Pearson Correlation	.817	1
	Sig. (2-tailed)	.000	.
	Sum of Squares and Cross-products	3927.284	5188.385
	Covariance	12.010	15.171
	N	328	343

EXERCISES

1. Below is a sample of six cases for the variables *emailhr* and *wwwhr* from GSS 2000. Jitter the values using the random numbers in the 11th column of Table 5.1. Construct a scatterplot divided into quadrants for the jittered values.

   emailhr: 3   3   0   5   0   5

   wwwhr: 2   5   2   4   2   4

2. Below is a sample of six cases for the variables *age* and *emailhr* from GSS 2000. Construct a scatterplot divided into quadrants.

   age: 40   44   24   37   39   29

   emailhr: 0   9   2   2   4   5

3. On a scale of −1.0 to +1.0, rate the direction and strength of association of the following scatterplots.

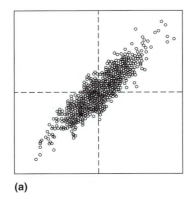 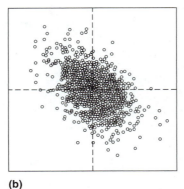 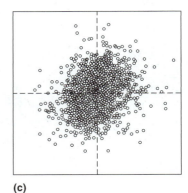

(a)   (b)   (c)

4. Which of the following values best describes the direction and strength of association of these scatterplots? $-1.0, -.7, -.3, .0, .2, .6,$ or $.9$

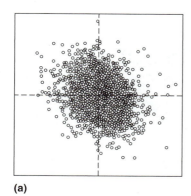
(a)

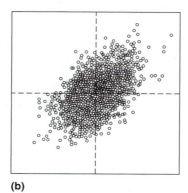
(b)

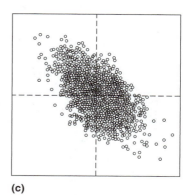
(c)

5. The following five cases are representative of husband's gender-role attitude (*hsexrole*) and wife's gender-role attitude (*wsexrole*) in the 1987 Akron Area Survey. High scores indicate a preference for traditional gender roles.

    hsexrole: 13   9   6   5   2

    wsexrole: 12   5   9   3   1

    a. Calculate the sum of cross-products.
    b. Calculate the covariance.
    c. Interpret the covariance.

6. The following five cases are representative of husband's gender-role attitude (*hsexrole*) and education in the 1987 Akron Area Survey. High scores on *hsexrole* indicate a preference for traditional gender roles.

    educ: 13   16   12   11   18

    hsexrole: 13   9   6   5   2

    a. Calculate the sum of cross-products.
    b. Calculate the covariance.
    c. Interpret the covariance.

7. The GSS 2000 (*gss2000a*) asked respondents to estimate what percentage of the U.S. population is Black (*usblk*). Use SPSS to study the relationship between years of schooling (*educ*) and *usblk*. Use jittered values of both variables to construct a scatterplot divided into quadrants; use the unjittered values to compute the sum of cross-products and the covariance. Interpret your results.

8. The GSS 2000 (*gss2000a*) asked respondents to rate the intelligence of members of several groups on a scale from 1 to 7, where 1 means almost everyone is not intelligent and 7 means almost everyone is intelligent. Assume these ratings constitute an interval variable. The variable *intlwhts* is the rating given to Whites and *intlblks* is the rating for Blacks. Create a new variable *intldiff* that equals *intlwhts* minus *intlblks*. Use SPSS to study the relationship between *age* and *intldiff*: use jittered values of both variables to construct a scatterplot divided into quadrants; use the unjittered values to compute the sum of cross-products and the covariance. Interpret your results.

8

Bivariate Regression
Analysis

SWEET-PEA REGRESSION

"You don't eat sweet peas, James. You grow them for the beauty and fragrance of their flowers," Professor Peabody told me. She had been discussing the class project we would be doing later this spring in Biostatistics 101 when I asked if we could eat the sweet peas after we grew them. She said we were going to attempt to *replicate* a famous study by the English scientist Francis Galton in which he discovered the statistical principles that laid the foundation for modern *regression analysis* (Galton, 1877). The study consisted of growing sweet peas from seeds of various weights and examining the relationship between the weights of the parent seeds and the seeds of their offspring, or progeny. Professor Peabody speculated that Galton might have chosen sweet peas because they are one of the most popular flowers in England. At any rate, Galton carefully weighed sweet-pea seeds and divided them into seven groups of differing weights. Then he enlisted the help of seven friends and gave each of them 70 seeds to grow, ten of each of the seven weights.

Professor Peabody said our class of ten students would do essentially the same thing as Galton and his friends. Each of us would grow 7 seeds × 7 weights = 49 seeds, for a total of 490 seeds. Dr. Peabody obtained permission for us to grow the sweet peas in an Ag-School greenhouse, with the assistance of a horticulture graduate student. She said that in the greenhouse environment, we would be able to collect the seeds from the pods of the mature plants in mid-April if the seeds were sown at the beginning of March. After weighing the offspring seeds from each of the seven weight groups of parents, we would be ready to see if we could replicate Galton's results.

"What did Galton find that was so important?" Professor Peabody asked rhetorically before proceeding to describe the highlights. First, and not surprisingly, he found that there was a positive association between the weights of the offspring and those of their parents. The offspring of seeds that were above average in weight also tended to be above average, whereas the offspring of below average parents tended to be below average themselves. What was more surprising, however, was that on the average, the offspring of seeds that were above average in weight were not nearly as much above average as their parents, and the offspring of below average parents were not nearly as far below average as their parents. Galton referred to this as *reversion* and as *regression toward mediocrity*. Furthermore, this reversion or regression followed the simplest possible pattern: It was linear. In particular, no matter what the weight of the parent seed, the offspring's weight deviated from the average by only one-third as much as the parent. For example, parents who were 21 milligrams above average had offspring who were only 7 milligrams above average, and parents who were 15 milligrams below average had offspring who were 5 milligrams below average.

"Wouldn't this reversion toward averageness lead to less dispersion among the progeny than among the parents?" asked Professor Peabody. "And, if regression continued from generation to generation, wouldn't all peas ultimately be the same weight?" she asked. She then indicated that the answer to both questions was no, and she described how Galton's other findings explained this. Galton found that the offspring in each of the seven groups were normally distributed about *their* group's average value. Furthermore, the dispersion of weights within each group was quite high and just sufficient to offset the tendency of reversion to reduce diversity in the new generation of sweet peas. In other words, although the offspring of seeds of a given weight had an average weight that was closer to the center of the entire distribution than the parent's weight (i.e., regression), they were distributed widely enough around that average to prevent a reduction in diversity from the first generation to the second. "This is the regression paradox, as I see it," stated Professor Peabody. "Reversion toward averageness does not beget averageness. This is Galton's great contribution to statistics. OK, see you in the stat lab on Thursday, and don't forget to smell the sweet peas," she concluded.

Although Francis Galton was not a mathematician, he nevertheless made important conceptual contributions to what is now arguably the most useful statistical method, **regression analysis,** which acquired its name from Galton's work. Although Galton made these discoveries in his studies of inheritance in both plants and human beings, modern regression analysis is used to study associations between any kinds of interval/ratio variables, not just those representing the traits of parents and offspring. The widespread applicability of regression analysis has also broadened the meaning of the terms *reversion* and *regression*. Although Galton spoke of *regression toward mediocrity,* the correct term today is *regression toward the mean*. Galton favored the median as his measure of central tendency and mediocrity (or averageness). It is now known, however, that the mean is the more valid measure for estimating the rate of linear regression. We will carefully consider why this is so and also look closely at the concept of regression toward the mean. Our purpose in this chapter is to learn how regression analysis is used to provide the best mathematical description of the linear association between any two interval variables X and Y.

THE LINEAR MODEL

Linear Functions

Linear regression analysis, or simply regression analysis, involves finding the mathematical formula for the linear association or relationship between X and Y. Thus, before examining how regression determines the equation for this relationship, we will describe the characteristics of linear functions. In Chapter 7 we looked at various patterns of positive and negative linear associations. We now examine both graphically and mathematically the nature of positive and negative linear equations.

Positive Functions

To illustrate the nature of a positive linear relationship, let X equal the Celsius temperature scale and let Y equal the Fahrenheit temperature scale. Figure 8.1 shows the linear relationship between Celsius and Fahrenheit temperatures. For the ranges of Celsius temperatures shown, you can determine the Celsius temperature that corresponds to any particular Fahrenheit temperature and you can determine the Fahrenheit temperature that corresponds to any Celsius temperature. For example, if you want to know the Celsius temperature when it is 50° F, you draw a horizontal line from 50 on the Y axis to the conversion line, drop a vertical line down to the X axis, and read off the value 10° C. To find the Fahrenheit temperature when it is 30° C, draw a vertical line from 30 on the X axis to the conversion line, draw a horizontal line over to the Y axis, and read the corresponding Fahrenheit temperature of 86° F.

We have found two pairs of coordinates that fall on the conversion line: (10, 50) and (30, 86). Because two points define a straight line, we can find the formula for converting Celsius temperatures into Fahrenheit temperatures. The general formula for a line is

$$Y = a + bX \tag{8.1}$$

In this equation a and b are constants. The constant a is named the **Y intercept** because it is the value of Y at the point where the line intercepts or crosses the Y axis. In Figure 8.1 the line intercepts the Y axis at 32° F, so $a = 32$. When Y equals 32° F, the value of X is 0° C. This illustrates another way of defining a:

a is the value of Y when X equals 0

This statement will always be true. In point of fact, a is equal to the Y intercept only when 0 is the origin of the X axis, as it is in Figure 8.1. This is often the case, but not always. In the case of the Celsius scale, temperatures in many places are frequently be-

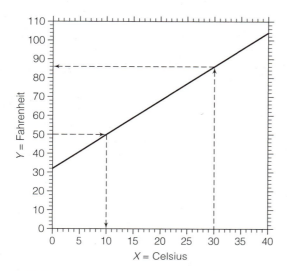

FIGURE 8.1 Linear Association between Celsius and Fahrenheit Temperatures

low $0°$ C in the winter. Consequently, it would be more useful to make the origin of the X axis some value significantly less than 0. In Figure 8.2, the origin of the Celsius scale is $-20°$ C. The line now intercepts the Y axis at $-4°$ F; at $-20°$ C, it is $-4°$ F. However, a does not equal -4. The constant a still equals $32°$ F, the value of the Fahrenheit scale when the Celsius scale equals 0. No matter what the origin of the X axis is, a always equals the value of Y associated with $X = 0$. Nevertheless, even though a is not always equal to the Y intercept, it is conventional to call it the Y intercept. Therefore, we will do so in this book.

The second constant in $Y = a + bX$ (i.e., b) is the **slope** of the line. The idea of b as the *slope of a line* is based on a graph of the line (Figure 8.1 and Figure 8.2), just as naming a the Y *intercept* comes from a graph of the line. The slope indicates the steepness of the line—that is, it tells us how much the line rises or falls as we move from left to right on the horizontal axis.

The slope may be calculated as

$$b = \frac{dY}{dX} = \frac{Y_2 - Y_1}{X_2 - X_1} = \frac{\text{rise}}{\text{run}} \qquad (8.2)$$

In Equation 8.2, dY and dX are the *difference in Y* and the *difference in X*, respectively, between two points (X_2, Y_2) and (X_1, Y_1) on the line. Specifically, $dY = Y_2 - Y_1$ and $dX = X_2 - X_1$. The difference in Y may be thought of as the rise, and the difference in X may be thought of as the run. Thus, the slope of a line equals the difference in Y divided by the difference in X between two points on the line, or the rise divided by the run. Using the two points that were read off the Celsius–Fahrenheit line in Figure 8.1, let $(X_2, Y_2) = (30, 86)$ and $(X_1, Y_1) = (10, 50)$. Figure 8.3 illustrates the difference

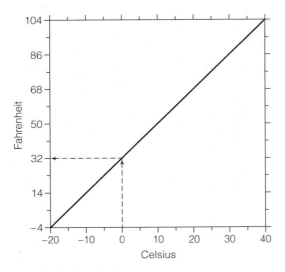

FIGURE 8.2 Non-Zero Origin for the Celsius Scale

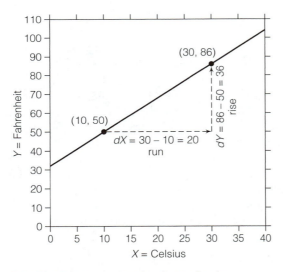

FIGURE 8.3 Determining the Slope of a Line

in Y and the difference in X for these two points. In the figure, imagine that you are moving from the left-hand point to the right-hand point, which corresponds to an increase in the value of X. Then, dY is the amount that the line changes vertically (its rise), and dX is the amount that the line increases horizontally (its run). Thus, we may also refer to the slope of the line as the *change in Y divided by the increase in X* between two points.

The slope of the line in Figure 8.3 equals

$$b = \frac{86 - 50}{30 - 10} = \frac{36}{20} = 1.8$$

Although the slope indicates the steepness of the line, what precisely does $b = 1.8$ mean? When the difference (change) in Y is divided by the difference (change) in X, the result indicates the difference or change in Y associated with *a 1-unit increase* in X. The sign of b (i.e., positive or negative) indicates whether the change in Y is an increase or a decrease. In this case, b is positive, so the slope says that a unit increase in Celsius temperature is associated with a 1.8 unit *increase* in Fahrenheit temperature. In other words, there is an increase of 1.8° F for each increase of 1° C.

Now, we can give a more precise definition of the slope of a line.

b equals the change in Y per unit increase in X

This tells us that if X increases by one unit, Y will change by b units. Because it is a straight line, no matter where on the scale of X a unit increase occurs, there will always be a change of b units in Y. Figure 8.4 illustrates this for the change in F temperature per unit increase in C temperature, for 0 to 5° C. The positive slope can be thought of as describing a series of discrete steps for climbing the line. Thus, a line can be viewed as a staircase with the height of each step specified by the slope.

Although $a = 32$ was originally determined by reading the Y intercept off Figure 8.1, once b has been calculated, we can calculate a with the following formula:

$$a = Y - bX \tag{8.3}$$

By entering the values of b and any pair of (X, Y) coordinates on the line into Equation 8.3, we can calculate a. For example, with $b = 1.8$ and either $(10, 50)$ or $(30, 86)$, we get

$$a = 50 - 1.8(10) = 50 - 18 = 32$$

$$a = 86 - 1.8(30) = 86 - 54 = 32$$

We have determined that the slope of Fahrenheit temperature as a function of Celsius temperature equals 1.8 and the Y intercept equals 32. The linear equation for Fahrenheit (F) as a function of Celsius (C) is thus

$$F = 32 + 1.8C$$

If Celsius equals 0,

$$F = 32 + 1.8(0) = 32 + 0 = 32$$

That is, when Celsius equals 0, Fahrenheit is equal to the Y intercept. If Celsius equals 30,

$$F = 32 + 1.8(30) = 32 + 54 = 86$$

Using the formula gives us the same Fahrenheit temperature that we read off Figure 8.1.

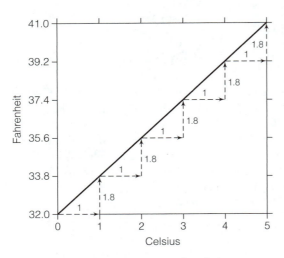

FIGURE 8.4 Step-Up in Degrees Fahrenheit per Degree Increase in Celsius

If we want an equation for converting Fahrenheit temperatures into Celsius temperatures, we need to find the slope and intercept of the following equation:

$$C = a' + b'F$$

In this equation, b' equals the change in C per unit increase in F, and a' equals C when F equals 0. To find b', we can use the same two pairs of coordinates that we used to find b, namely, (10, 50) and (30, 86). However, we now need to divide the difference in C by the difference in F. In terms of X and Y, where, as before, $X = C$ and $Y = F$, the equation for b' is

$$b' = \frac{dX}{dY} = \frac{X_2 - X_1}{Y_2 - Y_1} \qquad (8.4)$$

Entering coordinates (10, 50) and (30, 86) into Equation 8.4 gives

$$b' = \frac{30 - 10}{86 - 50} = \frac{20}{36} = \frac{5}{9} = .5556$$

The change in Celsius temperature per unit increase in Fahrenheit equals $5/9° = .5556°$ C. Notice that $b' = 1/b = 1/1.8 = .5556$. Thus, you would not have to use Equation 8.4 to find b'. Equation 8.4 is presented to make it clear that b' represents the change in X per unit increase in Y.

We can find a' with the following formula:

$$a' = X - b'Y \qquad (8.5)$$

Entering $b' = .5556$ and $(X, Y) = (30, 86)$ into Equation 8.5 gives

$$a' = 30 - .5556(86) = 30 - 47.7778 = -17.7778$$

This result indicates that when the temperature is 0° F, the Celsius temperature is −17.8° C.

We can now write the equation for expressing Celsius as a function of Fahrenheit:

$$C = -17.7778 + .5556F$$

Notice that the slope and intercept for expressing Celsius as a function of Fahrenheit are not the same as the slope and intercept when Fahrenheit is expressed as a function of Celsius. In general, $b \neq b'$ and $a \neq a'$.

Negative Functions

The Fahrenheit–Celsius conversion function has provided an example of a *positive* linear relationship. We will now look at a different example, the earned income tax credit (EITC), which illustrates a negative linear association as well as positive and zero associations. The EITC is a form of income transfer from the federal government to poor families and individuals. It provides a subsidy to low-income workers that is designed to create an incentive for them to work more and earn more. If the amount of the credit exceeds what a taxpayer owes, he or she will receive a direct payment from the U.S. Treasury for the difference.

The amount of the tax credit/refund depends on the earnings of an individual or family, and the relationship between earnings and EITC depends on whether the workers' earnings place them into the subsidy, plateau, or phase-out stages of the EITC. Figure 8.5 illustrates the three stages.

In the subsidy stage, which is defined by earnings of $0 to $9,800, the more income the family earns, the greater their income tax credit. Thus, there is a double benefit to working more in this income range; the worker earns more and also receives a larger tax credit. Because a family of four with earnings of $9,800 would not owe any federal income tax, they would receive all of their tax credit in the form of a check from the U.S. Treasury. If a family's earnings are $9,800, their EITC equals $3,556 (Figure 8.5), and their total income will thus be $9,800 + 3,556 = \$13,356$.

The exact relationship between EITC and earnings in the subsidy stage is

$$\text{Subsidy stage: EITC} = 0 + .4 \text{ earnings} = .4 \text{ earnings}$$

This indicates that for each additional dollar of earnings, the family's EITC increases by 40 cents. Thus, the slope for this positive linear relationship is $b = .4$. The EITC equals 0 when earnings are 0. That is, the Y intercept equals 0. This provides the incentive for the unemployed to enter the labor force. The positive slope of .4 provides the incentive to work more hours.

The association between EITC and earnings in the plateau stage, namely, $9,800 to $11,610 of earnings, is

$$\text{Plateau stage: EITC} = 3,556 + 0 \cdot \text{earnings} = 3,556$$

In the plateau stage, the line in Figure 8.5 is flat or horizontal; that is, $b = 0$. The Y intercept is $a = 3,556$. So, for earnings from $9,800 to $11,610, the amount of earnings has no effect on EITC. The EITC provides no incentive for working more and earning more in this earnings interval. On the other hand, there is no disincentive either. The workers' families do not lose any benefits from the EITC by earning a little more.

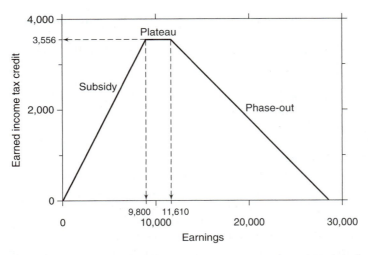

FIGURE 8.5 Earned Income Tax Credit as a Function of Earnings (Family of Four, 1996)

During the final stage, EITC benefits are phased out. That is, above $11,610 the tax credits are reduced by a fixed fraction of each additional dollar earned. In 1996, $11,610 was just over what a full-time worker would earn at the minimum wage. The EITC reaches $0 for a family of four when earnings reach $28,543. The relationship between EITC and earnings in the phase-out stage is

Phase-out stage: EITC = 5,994 − .21 earnings

The phasing out of benefits is reflected in the negative slope of $b = -.21$. In this earnings range ($11,610 to $28,543), for each additional dollar earned there is a reduction of 21 cents in the EITC. Thus, the benefits are not dropped all at once; instead, they are gradually phased out. Because a worker loses $.21 in EITC for each additional $1.00 earned, the worker's income increases by only $1.00 − .21 = $.79 per increase of $1 in earnings. Figure 8.6 illustrates the slope $b = -.21$ as a series of steps down in EITC, in thousands of dollars, per $1,000 increase in earnings. Thus, the negative slope describes a staircase for descending the linear function.

The Y intercept in the phase-out stage is $a = 5,994$. Normally, the intercept equals the value of Y when X equals 0. Thus, in this case, the intercept implies that EITC equals $5,994 when earnings equal $0. We know, however, from the equation for the subsidy stage that EITC equals 0 when earnings equal 0. The reason that we cannot interpret a in the usual fashion is that the equation applies only to the earnings range of $10,610 to $28,543. Thus, the intercept is merely a number that is needed to calculate EITC over the specified range of earnings. It has no intrinsic meaning.

We have examined positive, zero, and negative linear relationships between EITC and earnings that apply in three different ranges of earnings. We should add that above $28,543 in earnings there is a fourth linear function, namely, EITC = 0. The four linear relationships mean that over the full range of earnings there is a nonlinear association, as can be seen in Figure 8.5. Sometimes it is possible to describe a nonlinear

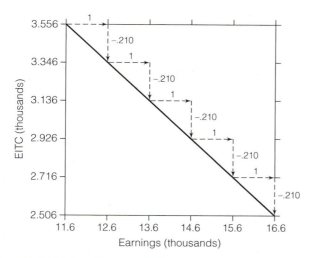

FIGURE 8.6 Step Down in EITC for Each Additional $1,000 of Earnings (Phase-Out Stage)

relationship with a single equation, albeit one that is more complex than a linear function. (We will not discuss nonlinear equations in this text.) At other times, it is better, or even necessary, to describe nonlinear associations with a series of linear relationships, as is done for EITC and earnings.

Independent and Dependent Variables

We have used Y and X as symbols for two variables that are associated. As shown in Equation 8.1, $Y = a + bX$, it is conventional to place Y to the left of the equal sign and X to the right. What determines which of the variables is specified as Y and placed to the left of the equal sign? If there is a potential *cause-and-effect relationship* between the two variables, the one that is believed to be the effect is designated as the Y variable, and the one that is thought to be the cause becomes the X variable.

A cause can be defined as something that produces another thing as an effect, result, or consequence. You undoubtedly already have a commonsense understanding of the meaning of *cause* and *causality*. If your teacher asks you *why* you did so badly on an examination, the teacher is trying to discover the *cause* of the poor performance. If you explain that you did poorly *because* the material was so hard, you are saying that the material was the cause of the performance. That is, level of difficulty of the test is a cause and performance on the test is the effect. Thus, exam difficulty is the X variable and exam performance is the Y variable.

The variable in the X–Y relationship that is believed to be the cause is called the **independent variable,** and the variable that is specified as the effect is the **dependent variable.** We use this terminology because the level of the effect (Y) *depends* on the level of the cause (X) and the cause does not depend on the effect. Independent does not mean that X is not associated with Y. If there is a causal relationship between X and

Y, they will be associated. Furthermore, specifying X as the independent variable does not mean that it is not dependent on some other variable(s), but only that it does not depend on Y.

Not every linear equation is intended to represent a causal relationship between X and Y. In the equation $F = 32 + 1.8C$ that we previously examined, there is no implication that Celsius temperature (X) is a cause of Fahrenheit temperature (Y). The equation simply shows how to convert Celsius into Fahrenheit. If you want to translate Fahrenheit into Celsius, you use the equation $C = -17.7778 + .5556F$. Thus, in the Celsius–Fahrenheit relationship, there is not an independent variable or a dependent variable. In the example of the earned income tax credit (EITC), however, there is a causal relationship. The level of earned income affects the amount of EITC. Or, the amount of EITC to which a person is entitled depends on the person's earned income. Thus, EITC is the dependent variable Y and earned income is the independent variable X.

How do we know if X is a cause of Y? There are three **criteria for a causal association:**

1. X precedes Y in time.
2. There is an association between X and Y.
3. The association is not produced by a common cause of X and Y.

The first criterion states that the cause must exist before the effect. For example, the difficulty of an exam is established *before* the students' scores on the exam are determined, and earned income exists *prior* to the amount of EITC that a worker receives. Because the effect follows the cause, it is probably no accident that Y, which follows X in the alphabet, is the symbol for the effect and X is the symbol for the cause. The time-order criterion means that when there is a clear-cut temporal sequence between two variables, the variable that occurs earlier in time is ruled out as a potential dependent variable. For example, the attributes of some variables, such as race and gender, are established at birth and thus cannot be dependent on variables that are established later in life, such as education, gender-role attitudes, and number of children. Thus, the later variables are potentially dependent on the earlier variables, but not vice versa.

The second criterion states that there must be an association between the independent and dependent variables. If X is a cause of Y, an observable statistical association will result from the effect. This association can be established with bivariate regression analysis, the subject of this chapter. Regression analysis is an important technique for measuring linear relationships between interval/ratio variables because the slope of X in the linear regression equation provides a meaningful measure of the effect of X on Y. If the slope is statistically significant, we can conclude that the two variables are associated.[1] The slope also indicates the direction of the association and gives a clear numerical description of the association.

The third criterion addresses the fact that finding an association between X and Y does not necessarily mean that the association resulted from the effect of X on Y. Both

---

1. If bivariate regression does not reveal an association, one may still be found in the results of a multiple regression analysis, as we will see in Chapter 10.

variables might be dependent on a third variable, and this common cause may be the reason they are associated. For example, in Chapter 7 we found an association between a married man's education and his gender-role attitude; that is, more educated men have less traditional gender-role attitudes. Does this mean that there is a causal relationship between these two variables? Not necessarily. It is possible that education and gender-role attitude are both dependent upon age, which causes education and attitude to be associated. As age increases, men tend to have less education and as age increases, they tend to have more traditional gender-role attitudes. This type of common cause will result in less educated men having more traditional gender-role attitudes, a negative association between education and traditional attitudes. If this is the reason for the association between education and gender-role attitude, the association is **spurious,** which means that it does not represent a genuine casual relationship.

Multiple regression (Chapter 10), which expresses a dependent variable as a function of two or more independent variables, is an excellent statistical technique for addressing the third criterion. With multiple regression, we are able to control a second independent variable, or several other independent variables, that are potential causes of a spurious association between our target dependent and independent variables. If there is no association between the target X and Y variables when these potential common causes of X and Y are controlled, or held constant, then the association between X and Y is spurious—that is, the third criterion has not been satisfied. If there is an association between X and Y after controlling for potential common causes, the association gains validity as representing a genuine causal relationship.

Bivariate Linear Model

We have described the characteristics of linear functions. In the linear equations that we have examined, all of the points fall exactly on a straight line. Thus, these are examples of *perfect linear associations.* In the world of social research, however, we cannot find a perfect linear relationship between X and Y. Why not? There are at least three reasons. First, the dependent variable almost always, if not always, has more than one cause. That is, there are likely to be two or more independent variables. The presence of a second cause means that if we examine the relationship between the dependent variable and only one of the independent variables, a perfect linear relationship cannot exist. The effects of the second independent variable will disturb the perfect relationship. To compensate for this, we will learn how to use regression models with more than one independent variable in Chapter 10.

Second, even if there is only one cause of the dependent variable, social variables appear to be characterized by a certain amount of random variation. That is, a certain amount of the variance of social phenomena is seemingly unrelated to any systematic causes. Instead, it is the product of chance events. This apparently random variance may be truly random or it may be a function of many small, offsetting causes that in the aggregate behave as if their effect was purely random. A third reason for the lack of a perfect linear relationship is random measurement error. Because of the unreliability of our measurements, additional random variance is added to our observations. This additional variance is artificial noise that is created by our efforts to study natural phenomena.

As a result of the natural and unnatural random variance in our data, our linear analyses must incorporate a variable that represents this chance, unpredictable aspect of our data. This may be accomplished by adding a *random variable* to the basic linear equation to obtain the **bivariate linear model:**

$$Y = \alpha + \beta X + \varepsilon \qquad (8.6)$$

This equation should be recognized as one that involves population parameters symbolized by the Greek letters α, β, and ε. The Y intercept in the population is represented by α (alpha); β (beta) stands for the population slope; and ε (epsilon) is the new variable that represents the random part of Y. As before, population parameters are unknowns that we attempt to estimate with our sample observations. Although α and β are unknown constants, ε is a new kind of unknown, an *unobservable variable*. It is unobservable because it is impossible to measure random phenomena. As you will see, we have to infer and estimate the values of ε on the basis of other variables that are measurable or observable.

The manner in which ε upsets a perfect linear relationship between X and Y will be illustrated with a hypothetical data set. The data set is hypothetical because it is desirable to keep it simple for illustrative purposes. It must also be hypothetical because to illustrate the nature and effect of ε, we must pretend that ε is observable. In the example, X represents students' true knowledge of the material on which they are to be examined. It is the percentage of the examination material that they know. Y is the students' scores on the examination, expressed as percentages of the total points that they answered correctly. Ideally, there would be a perfect one-to-one relationship between how much students know and their scores on the test. As you know, however, such a perfect relationship does not exist in reality. The relationship between knowledge and test scores in this example is described by

$$Y = 10 + .9X + \varepsilon \qquad (8.7)$$

In this equation, $\alpha = 10$ and $\beta = .9$. Assume that α indicates the level of difficulty of the test. In this case, the test is relatively easy because the intercept means that all students, including those who have zero knowledge of the material, will get ten free percentage points correct. That is, ten percent of the points are giveaways that everyone will get. If the exam were 100 percent valid, β would equal 1. In this case, β is less than 1, which means that students are not converting all of their knowledge into correct answers. The difference between 1 and .9 may be due to test anxiety, trick questions, and other difficulties of testing. $\beta = .9$, however, indicates a pretty strong relationship, which means that this exam has high validity.

The random variable ε indicates that there is a certain amount of luck involved in the scores of the students. In other words, ε represents a guessing factor. Because of guessing, some students have good luck and some have bad luck. Good luck is represented by positive values of ε and bad luck is indicated by negative values. A score of 0 means that luck was not a factor. Because ε is random, the mean should equal 0. Because ε is luck, ε (the lowercase Greek letter *e*) literally represents *errors* in the test scores. Some students get more points than they should, some get less, and some get just about the right number.

Equation 8.7 describes the effects of X and ε on Y. The scores on X (knowledge) and ε (luck), coupled with the values of α and β, determine or cause the values of Y. X and ε are input variables and Y is the outcome variable. The scores on X and ε are given in Table 8.1 for ten students. You are asked to imagine that we know the values

Table 8.1 The Effects of Knowledge (X) and Guessing (ε) on Text Score (Y)

X	ε	α	+	βX	+	ε	=	(α + βX)	+	ε	=	Y
55	−7.5	10	+	.9(55)	−	7.5	=	59.5	−	7.5	=	52.0
59	5.3	10	+	.9(59)	+	5.3	=	63.1	+	5.3	=	68.4
63	8.7	10	+	.9(63)	+	8.7	=	66.7	+	8.7	=	75.4
71	3.2	10	+	.9(71)	+	3.2	=	73.9	+	3.2	=	77.1
74	−5.9	10	+	.9(74)	−	5.9	=	76.6	−	5.9	=	70.7
76	5.5	10	+	.9(76)	+	5.5	=	78.4	+	5.5	=	83.9
80	−8.7	10	+	.9(80)	−	8.7	=	82.0	−	8.7	=	73.3
86	.5	10	+	.9(86)	+	.5	=	87.4	+	.5	=	87.9
90	−3.7	10	+	.9(90)	−	3.7	=	91.0	−	3.7	=	87.3
96	.5	10	+	.9(96)	+	.5	=	96.4	+	.5	=	96.9

of these variables. The mean and standard deviation of X are 75 and 12.8, respectively. X is roughly normally distributed. The values of ε were created by randomly selecting numbers between 0 and 20 from a uniform distribution and subtracting 10 from each random number. In effect, this means that ε is randomly selected from the interval -10 to 10. You can see that there are four negative values (bad luck), two values equal to .5 (little or no luck), and four additional positive numbers (good luck). The mean of ε equals $-.21$ (almost 0), and the standard deviation is 5.7.

Table 8.1 shows how the outcome Y is computed from the inputs X and ε. α is added to the product of β and X to get $\alpha + \beta X = 10 + .9X$. These are the test scores that would occur if there was no guessing (ε), or the test scores that would be expected on the basis of the students' knowledge (X) alone. As can be seen in Figure 8.7a, the points formed by these expected scores and the values of X form a straight line. The slope of this line is .9, which indicates that for each percentage point increase in knowledge, there is a .9 percentage point increase in expected test score.

Next, the guessing scores (ε) are added to the expected scores to get the students' actual test scores (Y). This disturbs the perfect linear relationship (Figure 8.7b). Some students who would be close on the basis of their knowledge become widely separated due to chance (see the cases with scores of 55 and 59 on X). On the other hand, students who would differ greatly on the basis of knowledge wind up close together due to luck (see X equals 63 and 80). Consequently, there is not a perfect linear association between Y and X. In Chapter 7 you were asked to estimate the strength of association (from -1 to 1) between X and Y by examining scatterplots of relationships. How strong is the association in Figure 8.7b?

The bivariate linear model represents the way values of the dependent variable are caused by the linear effect of a single independent variable and a random error term. Although it is a very simple model, it represents how random phenomena disrupt perfect linear relationships in the social world and consequently complicate considerably the task of finding the true linear association between X and Y. The task is complicated because we cannot directly observe ε. If ε could be measured, we could compute the

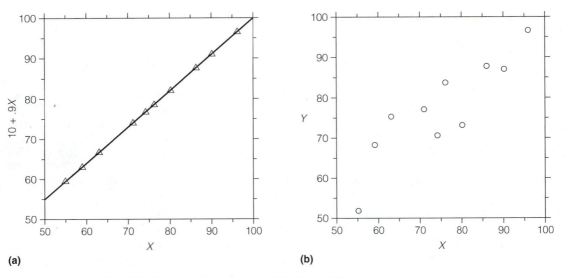

FIGURE 8.7 Knowledge (X), Expected Test Score, and Test Score (Y)

**Table 8.2
Knowledge (X) and
Text Score (Y)**

X	Y
55	52.0
59	68.4
63	75.4
71	77.1
74	70.7
76	83.9
80	73.3
86	87.9
90	87.3
96	96.9

linear relationship between $(Y - \varepsilon)$ and X (e.g., Figure 8.7a). In real social research, however, we have measures of only X and Y to work with. In our example, we would have the values of X and Y in Table 8.2 to work with (we are imagining that we actually know X). The scatterplot for the values of X and Y will not show an exact linear relationship (e.g., Figure 8.7b). If we were analyzing actual research data, how could we uncover, or, more accurately, estimate the true linear relationship underlying the scatter of points in the plot? That is, how could we estimate α and β? These parameters are estimated with the statistical methods of modern regression analysis, the culmination of the groundbreaking work that Francis Galton conducted with his sweet peas more than 125 years ago.

THE BIVARIATE REGRESSION EQUATION

The Best-Fitting Line

The objective of regression analysis is to estimate the true linear relationship underlying a bivariate distribution (i.e., scatterplot) of X and Y. We do this by determining the line that has the best fit to the data—that is, the best-fitting line. This objective is very similar to the issue addressed in Chapter 4 concerning which measure of central tendency best fits the distribution of a single variable. Before we can determine the best-fitting line, however, we must specify some criterion for assessing goodness-of-fit. Although there are many possibilities, two important alternative methods will be illustrated. These methods are analogous to the criteria considered for determining the best-fitting measure of central tendency, the *sum of absolute deviations* and the *sum of squared deviations*.

The small data set from Table 7.2 that was used to illustrate the covariance statistic will serve as an example of how to find the best-fitting line. As you can see in Figure 8.8, we do not find a perfect linear relationship. What line fits this scatter of points best?

Figure 8.9 shows an educated guess about a line that appears to fit the data reasonably well. Fitting the data well involves drawing a line that comes as close as possible to all the data points. The educated guess was not based on any rigorous mathematical criterion; it was simply an eyeball guess that has some nice features. It passes exactly through the highest point and splits the first two in half; that is, it comes equally close to the first two points. The Y intercept for the line is 0, which establishes (0, 0) as one point that the line passes through. A second point on the line is (5, 5). With these two points we can compute the slope as $b = (5 - 0)/(5 - 0) = 1$. The equation for this line is $Y = 0 + 1X$, or simply $Y = X$.

Because the eyeball line was created to fit the observed values of X and Y, we are going to place a *circumflex* over Y (i.e., $\hat{Y}$) to acknowledge that fact. Thus, the formula for the eyeball line is $\hat{Y} = X$. The variable $\hat{Y}$ is called "Y-hat." Whenever you see a circumflex over a variable—that is, a variable with a hat on it—it stands for **predicted values** of the variable. $\hat{Y}$ means predicted Y. The predicted values of Y are determined by inserting the observed values of X into the formula for $\hat{Y}$ as a function of X; for example, $\hat{Y} = X = 1$ for the first case in Table 8.3.

To make a quantitative assessment of how well the eyeball line fits the data, we will compare each predicted value $\hat{Y}$ with the observed value. The difference between Y and the predicted value of Y is the error of prediction. This error is called a **residual:**

$$\text{Residual} = Y - \hat{Y}$$

The residual of Y is the amount that is left over after the predicted value is subtracted from Y. Because $\hat{Y}$ is subtracted from Y, a positive residual means that the predicted value is too small and a negative value means the prediction was too large. For the first case in Table 8.3, $Y - \hat{Y} = 2 - 1 = 1$, which is an underestimate of Y. In Figure 8.9, *the residuals are the vertical distances between each point and the line.*

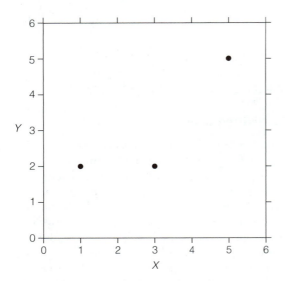

FIGURE 8.8 Observed *X–Y* Scores

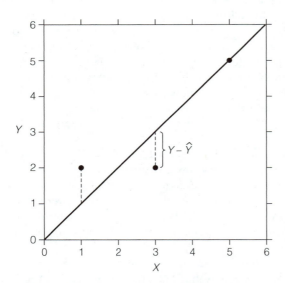

FIGURE 8.9 Residuals of the Eyeball Line

Now, how should we use these residuals to measure how well the eyeball line fits the data points? Our first thought may be to sum the residuals to get the total amount of error. However, because the prediction line will always pass somewhere through the middle of the scatterplot, some points will be above the line, resulting in positive residuals, and some will be below the line, creating negative residuals. Thus, the negative

Table 8.3 Residuals for the Eyeball Line

X	Y	$\hat{Y}$	$Y - \hat{Y}$	$|Y - \hat{Y}|$	$(Y - \hat{Y})^2$
1	2	1	1	1	1
3	2	3	−1	1	1
5	5	5	0	0	0
				2	2

residuals will tend to cancel out the positive residuals. In Table 8.3, the sum of the residuals turns out to be exactly 0. To avoid this problem, we can ignore the signs of the residuals and compute the sum of the absolute values:

$$\text{Sum of absolute residuals} = \Sigma |Y - \hat{Y}|$$

This measure of fit is like the *sum of absolute deviations* $(\Sigma |X - M|)$ that was used to assess the fit of the mean and the median in Chapter 4. The sum of absolute residuals for the eyeball line equals 2 (Table 8.3).

A second measure of the goodness-of-fit of a line is the **sum of squared residuals:**

$$\text{Sum of squared residuals} = \Sigma (Y - \hat{Y})^2$$

This measure of fit is like the *sum of squared deviations* $\Sigma (X - M)^2$ that was used to assess the fit of the mean and the median in Chapter 4. The sum of squared residuals for the eyeball line equals 2 (Table 8.3). It is merely a coincidence that the sum of squared residuals equals the sum of absolute residuals. Generally, because squaring a residual makes it larger, unless it is a fraction, the sum of squared residuals should be the larger of the two measures. The objective, however, is not to compare the two measures of fit for a single line. Instead, we want to compare two lines, the eyeball line and a second line to be introduced next, on each of the two measures.

The formula for the second line is $\hat{Y} = .75 + .75X$, and it is plotted in Figure 8.10. Both the slope and the Y intercept equal .75. This line is called the **least-squares line** because it minimizes the sum of squared residuals. That is, the sum of squared residuals for the small example data set will be smaller for this line than for any other line. We cannot find any other slope and intercept that would give more accurate predictions of Y when accuracy is measured by the sum of squared residuals. Therefore, this is the least-squares line for the example data. In the next section we will see how the values of the slope and intercept of the least-squares line were determined.

To illustrate the principle of least squares, we will compare the fit of the least-squares line to that of the eyeball line. Table 8.4 shows the predicted values of Y for the least-squares line. The predicted value for the third case, for example, is $\hat{Y} = .75 + .75(5) =$ .75 + 3.75 = 4.5. You can see from the predicted values, and also from Figure 8.10, that unlike the eyeball line, the least-squares line does not pass through any of the points in the scatterplot.

When the residuals of the least-squares line are calculated (Table 8.4), you can see that their sum equals 0. That is, the sum of the negative errors of prediction exactly cancels

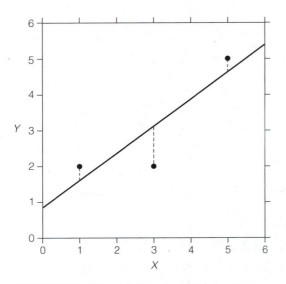

FIGURE 8.10 Residuals of the Least-Squares Line

Table 8.4 Residuals for the Least-Squares Line

X	Y	$\hat{Y}$	$Y - \hat{Y}$	$|Y - \hat{Y}|$	$(Y - \hat{Y})^2$
1	2	1.5	.5	.5	.25
3	2	3.0	−1.0	1.0	1.00
5	5	4.5	.5	.5	.25
			.0	2.0	1.50

out the sum of the positive errors of prediction. Consequently, the mean residual equals 0. This will be true for the least-squares line for any pair of variables. It will not always be true for any eyeball line that you might draw through the scatterplot.

Now, we compare the measures of fit of the least-squares line with those of the eyeball line. In terms of the sum of absolute residuals, both lines have identical goodness-of-fit scores (2). Thus, neither line is superior to the other according to this criterion. The sum of squared residuals, however, is smaller for the least-squares line. Thus, it has a better fit than the eyeball line. Of course, this is what we would expect, because it is named the *least-squares line,* which means that its sum of squared residuals will be smaller than that of any other line, including the eyeball line.

Why is the sum of squared errors smaller for the least-squares line? Notice that the eyeball line has two big residuals (1 and −1) and one small residual (0). The least-squares line has one big residual (−1) and two medium-sized residuals (.5 and .5). Thus, the least-squares line reduces the number of large errors by increasing the number of moderate errors. The reason that the least-squares solution results in the reduction of large residuals is that the squaring of residuals disproportionately increases large values relative to smaller values.

So, $\hat{Y} = .75 + .75X$ has a better fit than $\hat{Y} = 0 + 1X$ according to the least-squares criterion, but both lines have equally good fits according to the least-absolute-residuals criterion. Should we choose the least-squares line as providing the best estimate of the true linear association, should we flip a coin to choose between the two lines, or should we choose some other line? The answer is that *the least-squares line provides the best estimate of the true linear function* because it minimizes sampling error, the difference between a sample slope and the population slope. To be more precise, the slope of the least-squares line will have a smaller standard error than the slope of any other line. We studied the standard error of the mean in Chapters 5 and 6. It was used to construct confidence intervals around the sample mean and to conduct hypothesis tests about the population mean. Thus, minimizing the standard error is an essential aspect of inferential statistics. The regression slope also has a standard error. The formula for the standard error of the slope and its use in hypothesis tests will be taken up in Chapter 9. For now, the important point is that the least–squares slope is used in regression analysis because it provides the most efficient sample estimate of the population slope.

Least-Squares Regression Coefficients

The Slope

The values of the slope and intercept of the least-squares line were given above without describing how they were determined. The formula that defines the least-squares slope is

$$b_{YX} = \frac{\Sigma(X - \bar{X})(Y - \bar{Y})}{\Sigma(X - \bar{X})^2} \tag{8.8}$$

Subscripts have been attached to b to indicate which variable is the dependent variable. The first letter in the subscript is the dependent variable, and the second letter indicates the independent variable. It is important to note this because the regression slope is an *asymmetric* measure of association. If X is used as the dependent variable and Y is the independent variable, the value of the slope will be different, as we shall see.

If you study Equation 8.8, you should recognize the quantities that are in the numerator and denominator. The numerator is the *sum of cross-products* that is the main component of the *covariance* statistic, the measure of association introduced in Chapter 7. The denominator is the *sum-of-squares* of X that is the basis of the *variance* statistic. Thus, *the **least–squares slope** equals the sum of cross-products divided by the sum-of-squares of the independent variable.* The calculation of the sum–of–squares for X and Y and the sum of cross-products is shown in Table 8.5 for our example data. If the values of these statistics are entered into Equation 8.8, the least-squares slope is

$$b_{YX} = \frac{6}{8} = .75$$

This indicates that for every unit increase in X there is a .75 increase in Y.

Table 8.5 Squared Deviations and Cross-Products of X and Y

X	Y	$X - \bar{X}$	$(X - \bar{X})^2$	$Y - \bar{Y}$	$(Y - \bar{Y})^2$	$(X - \bar{X})(Y - \bar{Y})$
1	2	−2	4	−1	1	2
3	2	0	0	−1	1	0
5	5	2	4	2	4	4
9	9	0	8	0	6	6

$$s_X = \sqrt{\frac{8}{3}} = 2.67 \qquad s_Y = \sqrt{\frac{6}{3}} = 1.414 \qquad b_{YX} = \frac{6}{8} = .75$$

If we divide both the sum-of-squares and the sum of cross-products by *n*, Equation 8.8 becomes

$$b_{YX} = \frac{[\Sigma(X - \bar{X})(Y - \bar{Y})]/n}{[\Sigma(X - \bar{X})^2]/n} = \frac{s_{XY}}{s_X^2} \qquad (8.9)$$

This equation shows that *the least-squares slope equals the covariance divided by the variance of the independent variable.* Thus, the covariance and the variance—two statistical building blocks—are used to construct the least-squares estimate of the linear effect of X on Y. Intuitively, this may make sense when you remember that a slope is the difference in Y between any two points on a line divided by the difference in X between the same two points. Because all the points in the scatterplot do not fall on a line, however, we cannot determine the best-fitting slope so easily. Instead, we divide the amount that X and Y vary together overall (the covariance) by the amount that X itself varies overall (the variance).

The Intercept

Having determined the least-squares slope, we can now find the Y intercept of the equation as follows:

$$a_{YX} = \bar{Y} - b_{YX}\bar{X} \qquad (8.10)$$

Because the formula indicates that *the **least-squares intercept** equals the mean of Y minus the product of the slope and the mean of X*, we must calculate the slope before we can find the intercept. After entering the values of these statistics into Equation 8.10, we find the intercept equals

$$a_{YX} = 3 - .75(3) = 3 - 2.25 = .75$$

The reason Equation 8.10 gives the correct value of the intercept is that *the least-squares line always passes through the pair of coordinates corresponding to the means of X and Y; i.e., $(\bar{X}, \bar{Y})$.* Therefore,

$$\bar{Y} = a_{YX} + b_{YX}\bar{X} \qquad (8.11)$$

That is, when X is at its mean, the best prediction of Y is always $\bar{Y}$. This means that people who are average on X are predicted to be average on Y. You can see in Figure 8.10

that the line passes through $(3, 3)$, the means of X and Y. If we enter the mean of X into the least-squares equation, the prediction is the mean of Y:

$$\hat{Y} = .75 + .75(3) = .75 + 2.25 = 3 = \overline{Y}$$

Because this relationship holds, you should be able to see that Equation 8.10 is a simple rearrangement of Equation 8.11.

Examples of Least–Squares Lines

To illustrate more realistically the least-squares regression line, we return to the data on gender-role attitudes that were displayed in scatterplots and quantified with covariances in Chapter 7. The gender-role attitudes were measured on a scale that could range from 0 (nontraditional attitude) to 15 (traditional attitude). For our regression analysis, the husband's attitude will be the dependent variable Y and the wife's attitude will be the independent variable X. That is, we will be using the wife's attitude to predict the husband's attitude. This does not mean that we believe the wife's attitude affects or causes the husband's attitude, or that the reverse influence does not occur. Each spouse probably has some influence on the other. We simply are using the regression equation to describe how differences in wives' attitudes are associated with differences in husbands' attitudes. We will examine later the regression equation when the wife's attitude is treated as the dependent variable.

The top panel of Table 8.6 (labeled Slopes) shows the standard deviations and covariances that can be used to compute the slopes of the regression equations. The bottom panel (Intercepts) gives the means and slopes that are used to determine the intercepts. The first line of each panel provides these data for the total sample of 328 spouses. Using Equations 8.9 and 8.10, the regression equation can be determined as follows:

$$b_{YX} = \frac{s_{XY}}{s_X^2} = \frac{11.973}{3.861^2} = .80$$

$$a_{YX} = \overline{Y} - b_{YX}\overline{X} = 7.387 - .80(6.210) = 2.4$$

$$\hat{Y} = 2.4 + .8X$$

The slope indicates that for each unit increase in the wife's traditional gender-role attitude, there is a .8 increase in the husband's attitude. The more traditional the wife's attitude, the more traditional the husband's attitude tends to be. The intercept indicates that wives with a gender-role score of 0 (the most nontraditional) are predicted to have husbands with a gender role score of 2.4, which is somewhat less nontraditional than the wives' scores.

Next we look at the husband's gender-role attitude as a function of the wife's attitude separately for wives who are very satisfied with their marriages and for wives who are less satisfied with their marriages (lines 2 and 3 in each panel of Table 8.6). The slope for the less satisfied wives is $b = 7.4777/(3.586)^2 = .582$, and the slope for the more satisfied wives is $b = 11.7138/(3.809)^2 = .807$. This indicates that a unit difference in the wife's attitude is associated with a greater difference in the husband's attitude for wives who are more satisfied with their marriages than for those who are less satisfied. In other words, wives who have greater marital satisfaction have husbands who are more in agreement with them on gender roles than is the case for wives with less marital satisfaction.

Table 8.6 Least-Squares Regression Coefficients for Gender-Role Attitude Examples[a]

Figure	X	Y	n	s_X	s_Y	s_{YX}	b_{YX}
			a. Slopes				
7.9	Wife's attitude	Husband's attitude	328	3.861	3.797	11.9734	.803
7.10a	Wife's attitude	Husband's attitude	72	3.586	3.134	7.4877	.582
7.10b	Wife's attitude	Husband's attitude	113	3.809	3.606	11.7138	.807
7.11a	Age	Husband's attitude	154	14.973	4.153	20.9352	.093
7.11b	Education	Husband's attitude	154	3.149	4.153	−4.0796	−.411

Figure	X	Y	n	$\overline{X}$	$\overline{Y}$	b_{YX}	a_{YX}
			b. Intercepts				
7.9	Wife's attitude	Husband's attitude	328	6.210	7.387	.803	2.400
7.10a	Wife's attitude	Husband's attitude	72	5.486	8.111	.582	4.917
7.10b	Wife's attitude	Husband's attitude	113	6.142	7.584	.807	2.626
7.11a	Age	Husband's attitude	154	47.390	6.766	.093	2.341
7.11b	Education	Husband's attitude	154	13.604	6.766	−.411	12.364

[a]1987 Akron Area Survey

Lines 4 and 5 in Table 8.6 show two independent variables, age and education, that may help to explain why some husbands are more traditional than others. The slope for age as a predictor of the husband's attitude is $b = 20.9352/(14.973)^2 = .093$. For each additional year of age there is a .093 increase in traditional gender-role attitudes. Thus, there is a positive association between how old the husband is and how traditional his attitudes are. Although a slope of .093 may not look very substantial, a unit difference in age, which ranges from 22 to 77, is not as important as a unit difference in gender-role attitudes, which range only from 0 to 15. When husband's education is used to predict his attitude, the slope is $b = -4.0796/(3.149)^2 = -.411$. Each additional year of education is associated with a .411 point decrease in traditional gender-role attitude. In other words, the more educated the husband is, the less traditional are his attitudes. You should not compare the slopes for age and education, however, because years of age and years of education are two different kinds of attributes.

No Association

An important point follows from Equations 8.10 and 8.11. If there is no linear association between X and Y—that is, if b equals 0—the intercept will equal

$$a_{YX} = \overline{Y} - 0 \cdot \overline{X} = \overline{Y}$$

and the least-squares equation will be

$$\hat{Y} = a_{YX} + 0 \cdot X = \overline{Y}$$

This means that the best prediction of Y when the slope equals 0 is the mean of Y, for all values of X. For instance, a scatterplot of the data in Table 7.4b is shown in Figure 8.11. These values of X and Y were shown to have a covariance of 0; thus, they also have

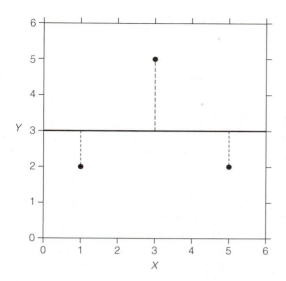

FIGURE 8.11 Least-Squares Regression Line When $b = 0$

a slope of 0. As shown by the plot of the line $\hat{Y} = \overline{Y} = 3$, all of the residuals equal the deviation scores of Y. That is, $Y - \hat{Y} = Y - \overline{Y}$. Because this is the case, the sum of squared residuals also equals the sum of squared deviation scores:

$$\Sigma(Y - \hat{Y})^2 = \Sigma(Y - \overline{Y})^2$$

This result verifies again (as shown in Chapter 4) that according to the least-squares criterion, the mean has a better goodness-of-fit than any other value. That is, *the mean is a least-squares statistic;* the sum of squared deviations around the mean is smaller than the sum of squared deviations around any other value. Therefore, if you had to make the same prediction of Y for every case, the mean of Y would be the best prediction.

Regression of X on Y

In regression analysis, we typically designate the dependent variable as Y and the independent variable as X when estimating the slope and intercept for the equation $\hat{Y} = a + bX$. This process is called *regressing Y on X;* the variable to the left of the equal sign is *regressed* on the variable to the right of the equal sign. It is also possible to calculate the regression coefficients for the equation that uses the variable designated Y as a predictor of the variable labeled X—that is, $\hat{X} = a_{XY} + b_{XY}Y$. In other words, we can determine the slope and intercept for X regressed on Y. Two researchers, for example, may disagree about the direction of causality. It is thus instructive to examine how the regression coefficients differ in the two cases.

The formula for the slope of X regressed on Y, when X is the dependent variable and Y is the independent variable, is

$$b_{XY} = \frac{\Sigma(X - \overline{X})(Y - \overline{Y})}{\Sigma(Y - \overline{Y})^2} \tag{8.12}$$

This formula has the sum of cross-products in the numerator and the sum-of-squares of Y, which is now the independent variable, in the denominator. Returning to our simple example with three cases in Table 8.5,

$$b_{XY} = \frac{6}{6} = 1$$

For each unit increase in Y, there is a one-unit increase in X. When X was used as the independent variable, the slope was .75. This illustrates the fact that, in general, the slope of Y regressed on X is not equal to the slope of X regressed on Y:

$$b_{YX} \neq b_{XY}$$

Because the slope of the line relating X and Y differs depending on which variable is designated as dependent, b *is an asymmetric measure of association.* We will discuss shortly the research implications of the **asymmetry of the least–squares slope.**

The intercept, called the X intercept, for X regressed on Y equals

$$a_{XY} = \overline{X} - b_{XY}\overline{Y} \qquad\qquad (8.13)$$

The X intercept equals the mean of the dependent variable, which is now X, minus the slope times the mean of the independent variable, which is now Y. For the example data in Table 8.5,

$$a_{XY} = 3 - (1 \cdot 3) = 0$$

Thus, the intercept for X regressed on Y does not equal the intercept for Y regressed on X, which was .75. The least-squares equation for X regressed on Y is

$$\hat{X} = 0 + 1Y = Y$$

There is a one-to-one relationship between Y and the predicted value of X.

Let's look at a couple of examples from the gender-role data in Table 8.6. First, we will regress the wife's attitude on the husband's attitude (line 1). The slope is $b = 11.9734/(3.797)^2 = .830$. For each unit increase in husband's attitude there is an increase of .83 unit in the wife's attitude. Although $b = .830$ is not identical to the slope for husband's attitude regressed on wife's attitude ($b = .803$), the two slopes are very similar because the variances of wives' and husbands' attitudes [$(3.861)^2$ and $(3.797)^2$, respectively], which are the denominators in the formulas for b_{XY} and b_{YX}, are approximately equal.

Second, let's regress the husband's age on the husband's gender-role attitude (line 3). The slope is $b = 20.9352/(4.153)^2 = 1.214$. That is, for each unit increase in husband's gender-role traditionalism, there is a 1.2-year increase in his age. This slope is much larger than the slope for gender-role attitude regressed on age ($b = .093$) because the variances of the two variables are so different [$(4.153)^2$ and $(14.973)^2$ for attitude and age, respectively]. The reason for this difference can be seen in the formula for the ratio of b_{XY} to b_{YX}:

$$\frac{b_{XY}}{b_{YX}} = \frac{\cancel{s}_{XY}/s_Y^2}{\cancel{s}_{YX}/s_X^2} = \frac{s_X^2}{s_Y^2} = \frac{(14.973)^2}{(4.153)^2} = 13.0$$

The slope for age regressed on attitude is thirteen times as large as the slope for attitude regressed on age because the variance of age is thirteen times the variance of attitude.

The preceding formula may be used to express b_{XY} as a function of b_{YX}:

$$b_{XY} = \frac{s_X^2}{s_Y^2} \ b_{YX}$$

The ratio of the variances (s_X^2/s_Y^2) times the slope for Y regressed on X (b_{YX}) equals the slope for X regressed on Y (b_{XY}). For the husband's gender-role attitude and age,

$$b_{XY} = \frac{(14.973)^2}{(4.153)^2} \ (.0934) = 13.0(.0934) = 1.214$$

This formula can be used as an alternative to Equation 8.12 for computing b_{XY}. The reason for introducing the formula, however, is to show how the ratio of the variances (s_X^2/s_Y^2) influences the size of b_{XY} relative to that of b_{YX}.

Finally, we have another word about selecting the most appropriate variable as the dependent variable. A researcher would not compute both of the regression equations ($\hat{Y} = a_{YX} + b_{YX}X$ and $\hat{X} = a_{XY} + b_{XY}Y$) for a pair of variables. One variable would be selected as the dependent variable and only that regression would be carried out. Sometimes it is hard to choose. Should the husband's attitude or the wife's attitude be the dependent variable? At other times it is clear. It seems quite logical that being older will tend to cause a man to be more traditional. Men and women who were born longer ago were socialized at a time when traditional attitudes were the norm (this is called an age-cohort effect). It is also possible that as a man grows older he may become more traditional or conservative (this is an aging effect). On the other side of the coin, it is clearly unreasonable to think that age is dependent on gender-role attitude. It is implausible that an increase in traditionalism would cause a man to live longer or would cause a man to have been born a longer time ago. So you must use common sense, logic, and/or theory to decide which variable is dependent and which is independent, if either. Once you have specified the dependent variable, you will know which of the two regression equations to compute. The formulas for the two regression equations have been introduced for the sole purpose of demonstrating that you get different slopes and intercepts depending on which variable you choose as the dependent variable.

b Is Unstandardized

When we discussed the covariance, we emphasized that its magnitude was directly related to the size of the standard deviations of both X and Y; the greater each standard deviation, the greater the absolute value of the covariance. This means that there are no fixed upper or lower limits on the covariance.

There are also no fixed limits on the least-squares slope. The first panel in Table 8.7 shows the result of doubling the values of Y to get Y'. First, the mean of Y' is doubled from 3 to 6. Second, the standard deviation of Y' is doubled from 1.414 to 2.828. As a consequence of the doubling of the standard deviation, the slope is doubled from .75 to 1.50. This illustrates the principle that *the greater is the standard deviation of Y, the greater will be the slope, all other things being equal*. When the standard deviation (and variance) of

Table 8.7 Effects of Doubling Y and Doubling X on the Slope

X	$2Y = Y'$	$X - \bar{X}$	$(X - \bar{X})^2$	$Y' - \bar{Y}'$	$(Y' - \bar{Y}')^2$	$(X - \bar{X})(Y' - \bar{Y}')$
			a. Y doubled			
1	$2 \cdot 2 = 4$	-2	4	-2	4	4
3	$2 \cdot 2 = 4$	0	0	-2	4	0
5	$2 \cdot 5 = 10$	2	4	4	16	8
9	$2 \cdot 9 = 18$	0	8	0	24	12

$$\bar{Y}' = \frac{18}{3} = 6 \qquad s_{Y'} = \sqrt{\frac{24}{3}} = 2.828 \qquad b_{Y'X} = \frac{12}{8} = 1.50$$

$2X = X'$	Y	$X' - \bar{X}'$	$(X' - \bar{X}')^2$	$Y - \bar{Y}$	$(Y - \bar{Y})^2$	$(X' - \bar{X}')(Y - \bar{Y})$
			b. X doubled			
$2 \cdot 1 = 2$	2	-4	16	-1	1	4
$2 \cdot 3 = 6$	2	0	0	-1	1	0
$2 \cdot 5 = 10$	5	4	16	2	4	8
$2 \cdot 9 = 18$	9	0	32	0	6	12

$$\bar{X}' = \frac{18}{3} = 6 \qquad s_{X'} = \sqrt{\frac{32}{3}} = 3.266 \qquad b_{YX'} = \frac{12}{32} = .375$$

Y is greater, a unit difference in X will be associated with a greater difference in Y. To illustrate this fact, let c be any constant. In general,

$$\text{If } s_{Y'} = c \cdot s_Y, \text{ then } b_{Y'X} = c \cdot b_{YX}$$

If the standard deviation of Y' is c times as great as the standard deviation of Y, then the regression slope for Y' and X will be c times as great as the slope for Y and X. This is true because when the standard deviation is greater, the sum of cross–products will be greater, which increases the numerator of the formula for the least-squares slope. The denominator of the formula, which is the sum-of-squares of the independent variable, does not change. Consequently, an increase in the numerator coupled with no change in the denominator results in an increase in the slope. Incidentally, the intercept for Y' and X will be c times as great as the intercept for Y and X ($a_{Y'X} = c \cdot a_{YX}$).

Now, how does the size of the standard deviation of X affect the slope? The second panel of Table 8.7 shows the consequence of doubling the standard deviation by doubling the values of X. The increase in the standard deviation increases the sum of cross–products, which will tend to increase the slope. However, the increase in the standard deviation of X also involves an increase in the sum-of-squares of X, which is the denominator of the formula for the slope. The increase in the denominator will be greater than the increase in the numerator, and consequently the slope will become smaller. In this case doubling the standard deviation of X has the consequence of cutting the slope in half, from .75 to .375. In general,

$$\text{If } s_{X'} = c \cdot s_X, \text{ then } b_{YX'} = \frac{1}{c} b_{YX}$$

That is, if the standard deviation of X' is c times as great as the standard deviation of X, then the slope for Y and X' will only be $1/c$ as large as the slope for Y and X. The principle is that *the greater is the standard deviation of the independent variable, the smaller the slope will be, all other things being equal.*

An example from the gender-role data will help to clarify this principle. Two regression equations were calculated for the variables husband's age and husband's gender-role attitude:

$$\widehat{\text{Attitude}} = 2.34 + .093\text{age}$$

$$\widehat{\text{Age}} = 39.18 + 1.214\text{attitude}$$

The slope for age regressed on attitude (1.214) is more than ten times as large as the slope for attitude regressed on age (.093). These slopes are different because the standard deviations of age and attitude (4.153 and 14.973, respectively) are different. When attitude is regressed on age, the slope is smaller because the standard deviation of the dependent variable attitude (4.153) is much smaller than the standard deviation of the independent variable age (14.973). But when age is regressed on attitude, the slope is much larger because the standard deviation of the dependent variable age is much larger than that of the independent variable attitude.

The point of these illustrations is to show that the magnitude of the least-squares slope is heavily dependent on the values of the standard deviations of X and Y. When s_Y is large and s_X is small, the slope will tend to be very large, and when s_Y is small and s_X is large, the slope will tend to be very small. There is no upper or lower limit on the value of b_{YX}. Remembering that b_{YX} may be positive or negative, the range of the least-squares slope is

$$-\infty < b_{YX} < +\infty$$

Because there are no fixed boundaries on b_{YX}, it is called an **unstandardized slope.** The consequence is that you cannot judge the strength of association by the value of b_{YX}. If b_{YX} seems to be large by some intuitive standard, it may be large because the standard deviation of Y is very large relative to the standard deviation of X. There may be a widely dispersed scatter of points around the regression line, indicating that there are large errors of prediction, despite the fact that the slope appears to be large.

Regression toward the Mean

In this section we will return to the topic with which this chapter began, namely, Francis Galton's experiment with sweet peas. In his study he discovered the statistical phenomenon called regression toward the mean. You will remember that in the opening vignette a class of ten biostatistics students planned to conduct a class project to replicate Galton's results. Professor Peabody acquired 70 seeds of each of the following weights: 70, 80, 90, 100, 110, 120, and 130 milligrams (mg). Each of the students received 7 seeds of each weight, for a total of 49 seeds per student. Their assignment was to plant the seeds, nurture them to full growth, harvest and weigh the seeds of each plant (the offspring or progeny), and conduct a regression analysis of the parents' and their offsprings' weights. To simplify the analysis, the students randomly selected only one offspring seed per parent. We will now review the results of a simulation of the student project.

The means and standard deviations of the weights of the offspring seeds, for each of the 7 weights of parent seeds, are shown in Table 8.8. Notice that the mean of all the offspring (100.7) is essentially equal to the mean of the parents (100). If you compare the mean weight of each group of offspring with the weight of their parents, you will see that each offspring mean is closer to the overall mean than is their parents' weight. The mean of the offspring of parents that weighed 70 mg is 90.74 mg, which is 20.74 mg heavier and closer to the mean than their parents' weight. The mean offspring weight of 130-mg parents is 110.33, which is 19.67 mg lighter than their parents. Parents whose weight equaled the mean (100) had offspring whose mean weight (101.34) was approximately equal to the overall mean of the offspring (100.7). Thus, the offspring of heavier-than-average parents were lighter than their parents, although still above average in weight, and the offspring of lighter-than-average parents were heavier than their parents even though they were below average in weight. In other words, the mean offspring of parents who were below the overall mean were not as far below the overall mean as their parents. The reverse was true of the offspring of parents who were above the overall mean; their offspring were not as far above average as the parents. Galton called this phenomenon reversion or regression toward mediocrity. Today, it is called **regression toward the mean** (RTM).

Regression toward the mean does not necessarily indicate that dispersion is shrinking among the offspring or that ultimately, over many generations, all offspring will be at the mean. You can read in Table 8.8 that the offsprings' standard deviation is large within each parental group. Figure 8.12 also shows how much spread or diversity there is in the offspring of each parental weight. For example, one offspring of the 70-mg parents actually weighed almost twice as much as the parent (about 135 mg). This seed actually "regressed through the mean." The lightest-weight parents also had an offspring that weighed about 20 mg less than they did. This seed regressed away from the mean. The term *regression,* however, does not apply (strictly speaking) to the behavior of individuals. It refers to the movement of the mean of individuals whose parents have a specific value on the variable.

The least-squares regression line is plotted in Figure 8.12. You can see that it has a positive slope. Although the phrase *regression toward the mean* sounds negative, it is not incompatible with a best-fitting line that has a *positive* slope. The best-fitting line is

$$\widehat{\text{Progeny}} = 68.15 + .326 \text{parent}$$

The slope indicates that each additional milligram of parental weight is associated with a .326-mg increase in progeny weight. So, the heavier the parent, the heavier will be the offspring on the average. The slope of .326 is very close to the slope of Galton's best-fitting line (1/3), although Galton did not use the least-squares criterion that we use today.

If parent's and progeny's weights are expressed as deviations from their means (i.e., $X - \overline{X}$ and $Y - \overline{Y}$), the regression equation is

$$\widehat{Y - \overline{Y}} = b(X - \overline{X}) = .326(X - \overline{X})$$

The intercept equals 0 because the means of the deviation scores are equal to 0. Expressed this way, the equation indicates that progeny are predicted to deviate from the

Table 8.8 Weights (mg) of Parent and Progeny Sweet Peas

Parent	n	Progeny Mean	Progeny Std. Dev.
70	70	90.74	16.67
80	70	93.88	17.39
90	70	97.33	19.63
100	70	101.34	19.11
110	70	105.13	17.27
120	70	106.17	16.30
130	70	110.33	19.46
Total	490	100.70	19.07

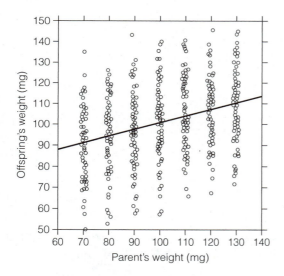

FIGURE 8.12 Weights of Parent (X) and Progeny (Y) Sweet Peas

mean only about one-third as much as their parents deviate from the mean. So, if the parent had a score of 130, the predicted deviation from the mean of the offspring would equal

$$\widehat{Y - \overline{Y}} = .326(130 - 100) = .326(30) = 9.78$$

The progeny of a parent that is 30 mg above the mean is predicted to be only 9.78 mg above the mean. A parent that has a weight of 70 mg, or 30 mg below the mean, is predicted to have an offspring that deviates from the mean by

$$\widehat{Y - \overline{Y}} = .326(70 - 100) = .326(-30) = -9.78$$

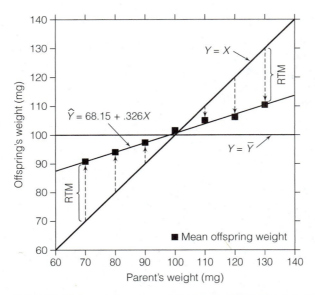

FIGURE 8.13 Regression toward the Mean (RTM) Weight of Progeny Sweet Peas

Deviations away from the mean, however, do not directly express the amount of regression *toward* the mean. Regression toward the mean equals

$$\widehat{RTM} = (b - 1)(X - \overline{X}) \qquad (8.14)$$

If the parent has a weight of 130, the offspring's predicted regression toward the mean equals

$$\widehat{RTM} = (.326 - 1)(130 - 100) = -.674(30) = -20.22$$

The offspring are expected to regress downward toward the mean by -20.22 mg. If the parent is 30 mg below the mean, the offspring will regress upward toward the mean by $-.674(-30) = 20.22$ mg.

The amount of regression toward the mean is shown in Figure 8.13. The mean weights of the seven groups of offspring (indicated by the small black squares) lie approximately on the regression line ($\hat{Y} = 68.15 + .326X$). The line $Y = X$ is the weight the offspring would have if their weights were equal to their parents' weights. Thus, the vertical distance between each group's mean and the line $Y = X$ equals the amount of RTM for that group. The predicted regression toward the mean ($\widehat{RTM}$) equals the vertical distance between the regression line and $Y = X$.

In a later study, Galton (1886) analyzed the relationship between the heights of human parents and their children. His diagram of the relationship is shown in Figure 8.14. "Mid-parent" refers to the mean of the heights of the mother and father. The diagram is very similar to Figure 8.13. The line labeled MID-PARENTS in Figure 8.14 is the same as the line $Y = X$ in Figure 8.13. The statement "The Deviates of the Children Are to Those of Their Mid-Parents as 2 to 3" means that Galton's estimate of the slope of the

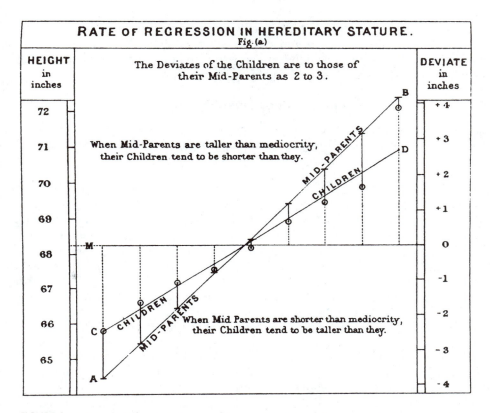

FIGURE 8.14 Francis Galton's Diagram of Regression in Hereditary Height (Galton, 1886)

regression line is two-thirds, or .667. This result indicates that the rate of regression toward the mean for human height is only about one-half as much as for sweet-pea weight.

MEASURES OF STRENGTH OF ASSOCIATION

We know that the least-squares equation is the best-fitting line in terms of minimizing the sum of squared residuals. The best-fitting line, however, is not necessarily a good-fitting line. What we need is a statistic that describes, for a fixed range of values, just how closely the regression line fits the data. Such a statistic is called a measure of the strength of association. In Chapter 7, you were asked to rate the direction and strength of associations displayed in scatterplots on a scale of -1 to $+1$. We have seen, however, that there is no upper or lower limit on the values of the covariance or the regression slope. Thus, these statistics cannot be used to judge the strength of association. In the following sections we will examine two statistics that measure the association on a fixed scale. These are the *coefficient of determination,* which has a scale of 0 to 1, and the *standardized slope,* which has a scale of -1 to $+1$. Not only do these statistics have fixed scales, but they also have very useful interpretations. Although we will see that these statistics are closely related, they provide two different perspectives on the nature of association that will enhance our understanding of this phenomenon.

Coefficient of Determination

To understand the coefficient of determination, you should recognize that a deviation score of Y can be expressed as the sum of the residual of Y and the regression deviation of Y:

$$Y - \overline{Y} = (\hat{Y} - \overline{Y}) + (Y - \hat{Y}) \tag{8.15}$$

This equation holds for every case. Although we have studied the deviation scores $(Y - \overline{Y})$ and residuals $(Y - \hat{Y})$, the regression deviation $(\hat{Y} - \overline{Y})$ is a new type of score. It indicates how far above or below the mean of Y is the predicted score. It tells us the direction and distance from the mean that is predicted for a case based on that case's value of X. The first two columns of Table 8.9 give the observed and predicted values of Y for the example data we have been using. The mean of Y equals $9/3 = 3$. With this information and Equation 8.15, we can decompose each deviation score into the two components shown in Table 8.10.

The first case is predicted to be 1.5 points below the mean, but the observed value is actually .5 point higher than predicted, resulting in a deviation score of 1 point below the mean. The third case is predicted to be 1.5 points above the mean and the observed value is .5 higher than predicted, resulting in a deviation score of 2 points above the mean. The regression deviations and the residuals are shown graphically in Figure 8.15. The regression deviation is the vertical distance between the regression line and the mean line, and the residual is the vertical distance between the observed point and the regression line. Note that the sum of deviations, regression deviations, and residuals all equal 0 (Table 8.9). Thus, the mean of all three variables equals 0.

Equation 8.15, which expresses each deviation score as the sum of the regression deviation and the residual, can be used to derive the equation that divides the sum-of-squares into the following components:

$$\Sigma(Y - \overline{Y})^2 = \Sigma(\hat{Y} - \overline{Y})^2 + \Sigma(Y - \hat{Y})^2 \tag{8.16}$$

This equation states that the total sum-of-squares (SS_Y) equals the **regression sum-of-squares** ($\text{SS}_{\hat{Y}}$) plus the **residual sum-of-squares** ($\text{SS}_{Y-\hat{Y}}$). This relationship also may be displayed as

$$\text{Regression sum-of-squares} = \text{SS}_{\hat{Y}} = \Sigma(\hat{Y} - \overline{Y})^2$$
$$\underline{+ \text{ Residual sum-of-squares} = \text{SS}_{Y-\hat{Y}} = \Sigma(Y - \hat{Y})^2}$$
$$\text{Total sum-of-squares} = \text{SS}_Y = \Sigma(Y - \overline{Y})^2$$

Table 8.9 shows the calculation of the total, regression, and residual sums-of-squares. Inserting the regression and residual sums-of-squares into Equation 8.16 gives the total sum-of-squares:

$$\Sigma(Y - \overline{Y})^2 = \Sigma(\hat{Y} - \overline{Y})^2 + \Sigma(Y - \hat{Y})^2 = 4.5 + 1.5 = 6.0$$

Table 8.9 Total, Residual, and Regression Sum-of-Squares of Y for Example Data

		Total		Residual		Regression	
Y	$\hat{Y}$	$Y - \bar{Y}$	$(Y - \bar{Y})^2$	$Y - \hat{Y}$	$(Y - \hat{Y})^2$	$\hat{Y} - \bar{Y}$	$(\hat{Y} - \bar{Y})^2$
2	1.5	−1	1	.5	.25	−1.5	2.25
2	3.0	−1	1	−1.0	1.00	0.0	0.00
5	4.5	2	4	.5	1.25	1.5	2.25
9	9.0	0	6	.0	1.50	0.0	4.50

Table 8.10 Decomposition of $Y - \bar{Y}$

$Y - \bar{Y}$	=	$(\hat{Y} - \bar{Y})$	+	$(Y - \hat{Y})$				
2 − 3	=	(1.5 − 3.0)	+	(2.0 − 1.5)	=	−1.5	+	.5	=	−1
2 − 3	=	(3.0 − 3.0)	+	(2.0 − 3.0)	=	.0	−	1.0	=	−1
5 − 3	=	(4.5 − 3.0)	+	(5.0 − 4.5)	=	1.5	+	.5	=	2
						.0		.0		0

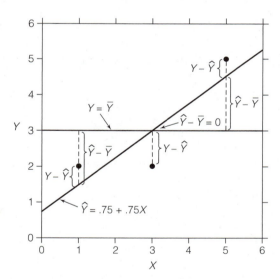

FIGURE 8.15 Residuals $(Y - \hat{Y})$ and Regression Deviations $(\hat{Y} - \bar{Y})$ for the Example Data

This shows that the total sum-of-squares, 6.0, can be divided into two meaningful components, the regression sum-of-squares (4.5) and the residual sum-of-squares (1.5).

We know that if we divide the total sum-of-squares by n, we get the *variance:*

$$\frac{\text{Total sum-of-squares}}{n} = \frac{\text{SS}_Y}{n} = \frac{\Sigma(Y - \bar{Y})^2}{n} = s_Y^2$$

What do we get if we divide the regression sum-of-squares by n? Table 8.9 shows that the sum of the predicted values equals 9, and thus the mean of the predicted values equals $9/3 = 3$. This is the same value as the mean of Y. In general, $\hat{Y} = \overline{Y}$. As a result, dividing the regression sum-of-squares by n gives the variance of the predicted scores, or the **regression variance:**

$$\frac{\text{Regression sum-of-squares}}{n} = \frac{SS_{\hat{Y}-\overline{Y}}}{n} = \frac{\Sigma(\hat{Y} - \overline{Y})^2}{n} = s_{\hat{Y}}^2$$

What does the residual sum-of-squares divided by n equal? Because the sum of the residuals equals 0, the mean of the residuals is also $0(Y - \hat{Y} = 0)$. Thus, the residual sum-of-squares divided by n equals the **residual variance:**

$$\frac{\text{Residual sum-of-squares}}{n} = \frac{SS_{Y-\hat{Y}}}{n} = \frac{\Sigma(Y - \hat{Y})^2}{n} = s_{Y-\hat{Y}}^2$$

The importance of the results of dividing the sums-of-squares by n is that dividing Equation 8.16 by n gives

$$s_Y^2 = s_{\hat{Y}}^2 + s_{Y-\hat{Y}}^2 \tag{8.17}$$

In words, the variance of Y equals the variance of $\hat{Y}$ plus the variance of $Y - \hat{Y}$. Or, the total variance equals the regression variance plus the residual variance. Using the data in Table 8.9 to illustrate, we have

$$s_Y^2 = s_{\hat{Y}}^2 + s_{Y-\hat{Y}}^2 = \frac{4.5}{3} + \frac{1.5}{3} = 1.5 + .5 = 2.0$$

Thus, we can decompose both the sum-of-squares of Y and the variance of Y into regression and residual components.

Now, let's consider the meaning of the regression sum-of-squares and variance and the meaning of the residual sum-of-squares and variance. Remember that investigators often consider the independent variable X to be a cause of Y, and thus, the dependent variable Y to be an effect of X. If this is the case, then $\hat{Y}$ is the value of Y that is caused by X because $\hat{Y}$ is an exact linear function of X (i.e., $\hat{Y} = a + bX$). And if $\hat{Y}$ is caused by X, then the sum-of-squares and the variance of X cause the sum-of-squares and the variance of $\hat{Y}$. If you find a cause of the variance of a variable, you are *explaining* part of the variance of that variable. Therefore, we can say that $\hat{Y}$ is *explained* by X, its cause, and the sum-of-squares and variance of $\hat{Y}$ are the *explained sum-of-squares and variance:*

$$\text{Regression sum-of-squares} = \Sigma(\hat{Y} - \overline{Y})^2 = \text{explained sum-of-squares}$$
$$\text{Regression variance} = \frac{\Sigma(\hat{Y} - \overline{Y})^2}{n} = \text{explained variance}$$

If $\hat{Y}$ is the part of Y that is explained by X, then the residual $Y - \hat{Y}$ is the part of Y that is not explained by X. Thus, the residual sum-of-squares and variance are called the *unexplained sum-of-squares and variance:*

$$\text{Residual sum-of-squares} = \Sigma(Y - \hat{Y})^2 = \text{unexplained sum-of-squares}$$

$$\text{Residual variance} = \frac{\Sigma(Y - \hat{Y})^2}{n} = \text{unexplained variance}$$

Because the total sum-of-squares equals the regression sum-of-squares plus the residual sum-of-squares (Equation 8.16) and the total variance equals the sum of the regression and residual variances (Equation 8.17), we can also write

$$\text{Total sum-of-squares} = \text{explained sum-of-squares} + \text{unexplained sum-of-squares}$$

$$\text{Total variance} = \text{explained variance} + \text{unexplained variance}$$

With the preceding results in hand, we are now ready to construct a measure of the strength of association, the **coefficient of determination** (CD). *The coefficient of determination equals the proportion of the variance of Y that is explained by X.* Although we prefer to define the CD in terms of variances because the term *variance* is more intuitively meaningful than the term *sum-of-squares,* you should recognize also that the CD *equals the proportion of the sum-of-squares of Y that is explained by X.* To get this proportion, you divide the explained variance by the total variance or the explained sum-of-squares by the total sum-of-squares:

$$\text{Coefficient of determination} = \frac{\text{explained variance}}{\text{total variance}} = \frac{\text{explained sum-of-squares}}{\text{total sum-of-squares}}$$

Remembering that the explained component is either the variance of $\hat{Y}$ or the sum-of-squares of $\hat{Y}$, the formula for the coefficient of determination is

$$\text{CD} = \frac{s_{\hat{Y}}^2}{s_Y^2} = \frac{\Sigma(\hat{Y} - \overline{Y})^2}{\Sigma(Y - \overline{Y})^2} \tag{8.18}$$

To illustrate, we use the sums-of-squares from the data in Table 8.9 (computationally, it is more efficient to use the sums-of-squares than the variances unless the variances have already been computed):

$$\text{CD} = \frac{\Sigma(\hat{Y} - \overline{Y})^2}{\Sigma(Y - \overline{Y})^2} = \frac{4.5}{6.0} = .75$$

How do we interpret this value of .75? Remembering that the CD is a proportion, we can say that .75 of the variance of Y is explained by X. Because a percent equals 100 times a proportion, the CD is often interpreted in terms of percentages. Thus, in our example, X explains 75 percent of the variance of Y.

As a proportion, the coefficient of determination has a fixed range of values, namely, 0 to 1. The greater the regression variance and the smaller the residual variance, the closer to 1 will be the CD. When the regression slope equals 0, there will be no regression variance because $\hat{Y} = \overline{Y}$ for every case. Consequently, the CD will equal 0:

$$\text{If } b = 0, \text{CD} = \frac{\Sigma(\overline{Y} - \overline{Y})^2}{\Sigma(Y - \overline{Y})^2} = \frac{0}{\Sigma(Y - \overline{Y})^2} = .0$$

When the observed Y equals the predicted Y ($Y = \hat{Y}$) for every case (i.e., the residuals are all 0), the coefficient of determination will equal 1:

$$\text{If } Y = \hat{Y}, \text{CD} = \frac{\Sigma(Y - \overline{Y})^2}{\Sigma(Y - \overline{Y})^2} = 1.0$$

Alternate Formula for CD

You remember that the variance of Y equals the explained variance plus the unexplained variance. It follows that the explained variance equals the total variance *minus* the unexplained variance:

$$\text{Explained variance} = \text{total variance} - \text{unexplained variance}$$

This leads to an alternative formula for the coefficient of determination:

$$\text{CD} = \frac{s_Y^2 - s_{Y-\hat{Y}}^2}{s_Y^2} = \frac{\Sigma(Y - \overline{Y})^2 - \Sigma(Y - \hat{Y})^2}{\Sigma(Y - \overline{Y})^2} \tag{8.19}$$

Again, using the sums-of-squares in Table 8.9:

$$\text{CD} = \frac{\Sigma(Y - \overline{Y})^2 - \Sigma(Y - \hat{Y})^2}{\Sigma(Y - \overline{Y})^2} = \frac{6.0 - 1.5}{6.0} = \frac{4.5}{6.0} = .75$$

This equation is often given in statistics books. The only advantage to this formula is that if the sum-of-squares of the residuals had been calculated but the sum of squared regression deviations had not, Equation 8.19 would be more efficient than Equation 8.18. This alternative formula for the CD, however, is not as simple in appearance nor as directly meaningful as the first formula.

Standard Error of Estimate

We saw earlier that the sum of the squared residuals divided by n is the variance of the residuals. Thus, the standard deviation of the residuals is

$$s_{Y-\hat{Y}} = \sqrt{\frac{\Sigma(Y - \hat{Y})^2}{n}} = \sqrt{\frac{1.5}{3}} = .707$$

The standard amount that the observed values of Y deviate from the predicted values is .707. This is the standard deviation of the residuals *in the sample*. We have seen, however, that s, the sample standard deviation of X, is a biased estimate of the population standard deviation (σ). Specifically, s is an underestimate of σ. To correct this bias, we divide the sum-of-squares by $n - 1$ instead of n when we want to estimate the population standard deviation. The sample standard deviation of the residuals is also an underestimate of the standard deviation of the residuals in the population ($\sigma_{Y-\hat{Y}}$). Instead of dividing by $n - 1$ to correct this bias, however, we divide by $n - 2$. We have 1 less degree of freedom because we had to estimate two parameters (α and β) to calculate the predicted values of Y. Thus, the estimated population standard deviation of the residuals is

$$\hat{s}_{Y-\hat{Y}} = \sqrt{\frac{\Sigma(Y - \hat{Y})^2}{n - 2}} = \sqrt{\frac{1.5}{1}} = 1.225 \tag{8.20}$$

Because we divided by $n - 2$ instead of n, $\hat{s}_{Y-\hat{Y}}$ is greater than $s_{Y-\hat{Y}}$. The name of $\hat{s}_{Y-\hat{Y}}$ is the **standard error of estimate.** It is not to be confused with the standard deviation of a sampling distribution such as the standard error of the mean.

Coefficient of Determination for X

What would the coefficient of determination equal if X had been chosen as the dependent variable and Y was the independent variable? What proportion of the variance of X is explained by Y when the regression equation is $\hat{X} = a_{XY} + b_{XY}Y$? To get the formula for the CD, simply substitute X for Y in Equation 8.18:

$$\mathrm{CD}_{XY} = \frac{s_{\hat{X}}^2}{s_X^2} = \frac{\Sigma(\hat{X} - \overline{X})^2}{\Sigma(X - \overline{X})^2} \tag{8.21}$$

We have placed a subscript on CD to indicate that X (the first variable in the subscript) is the dependent variable and Y (second variable in subscript) is the independent variable.

You may remember that the equation for regressing X on Y is $\hat{X} = 0 + 1Y = Y$. Table 8.11 gives the predicted values of X along with the total sum-of-squares and the regression sum-of-squares. Substituting the appropriate sums-of-squares into Equation 8.21 gives

$$\mathrm{CD}_{XY} = \frac{\Sigma(\hat{X} - \overline{X})^2}{\Sigma(X - \overline{X})^2} = \frac{6}{8} = .75$$

This means that Y explains .75 (75%) of the variance in X, which is the same as the proportion of variance in Y explained by X. In general, the coefficient of determination for Y regressed on X will equal the coefficient for X regressed on Y:

$$\mathrm{CD}_{YX} = \mathrm{CD}_{XY}$$

In other words, *the coefficient of determination is a symmetrical measure of association.* A symmetrical measure of association gives the same value no matter which one of two variables is specified as the dependent variable. You will remember that the regression slope is an *asymmetrical* measure because $b_{YX} \neq b_{XY}$.

Examples of Coefficients of Determination

We now return to our gender-role examples to look at the strength of association as measured by the coefficient of determination. Table 8.12 gives the regression slope b_{YX}, our previous measure of association, and the variances of Y and $\hat{Y}$, which we will use to calculate the coefficients of determination. For the total sample of husbands and wives (line 1), the coefficient of determination equals

$$\mathrm{CD} = \frac{s_{\hat{Y}}^2}{s_Y^2} = \frac{9.612}{14.417} = .667$$

The CD indicates that .667, or 66.7 percent of the variance of husband's gender-role attitude, is "explained" by his wife's gender-role attitude. We put the word *explained* in quotation marks because we do not necessarily believe that the husbands' attitudes are caused by the wives' attitudes. We are merely computing the CD to measure how closely the husband's and wife's attitudes are associated. Instead of saying explained, we might say

Table 8.11 Total and Regression Sum-of-Squares of X for Example Data

		Total		Regression	
X	$\hat{X}$	$X - \bar{X}$	$(X - \bar{X})^2$	$\hat{X} - \bar{X}$	$(\hat{X} - \bar{X})^2$
1	2	−2	4	−1	1
3	2	0	0	−1	1
5	5	2	4	2	4
9	9	0	8	0	6

Table 8.12 Coefficients of Determination (CD) for Gender-Role Attitude Examples[a]

Figure	X	Y	n	b_{YX}	s_Y^2	$s_{\hat{Y}}^2$	CD
7.9	Wife's attitude	Husband's attitude	328	.803	14.417	9.612	.667
7.10a	Wife's attitude	Husband's attitude	72	.582	9.822	4.356	.443
7.10b	Wife's attitude	Husband's attitude	113	.807	13.003	9.449	.727
7.11a	Age	Husband's attitude	154	.093	17.247	1.939	.112
7.11b	Education	Husband's attitude	154	−.411	17.247	1.675	.097

[a]1987 Akron Area Survey

instead that 66.7 percent of the variance in husbands' attitudes is associated with the variance in wives' attitudes. Remember that we would get the same value for CD if we regressed wives' attitudes on husbands' attitudes. Thus, we could also say that .667 of the variance in wives' attitudes is associated with the variance in husbands' attitudes.

Line 2 in Table 8.12 indicates the relationship between husband's and wife's attitudes for wives who are not highly satisfied with their marriages. For these women, CD = 4.356/9.822 = .443. That is, a little less than half the variance in husband's gender-role attitude is shared with the variance in wife's attitude. In contrast, CD = .727 for women who are highly satisfied with their marriages (line 3). Clearly there is a much higher association between spouses' attitudes for the more satisfied wives than for wives who are less satisfied with their marriages. This suggests that the closeness of the relationship between spouses' gender-role attitudes is a factor in how satisfied wives are with their marriages.

Lines 4 and 5 in Table 8.12 show how closely the husband's gender-role attitude is related to his age and years of education. The regression slope is much larger for education (−.411) than for age (.093). The CD for age and attitude (.112 ≈ .11), however, is approximately equal to the CD for education and attitude (.097 ≈ .10), which indicates that age and education explain approximately the same percent of the variance in husband's attitude. Thus, age and education are equally good predictors of gender-role attitude. The CDs for these variables do indicate that neither is as good a predictor of a husband's attitude as is his wife's attitude; that is, the CDs in lines 1 to 3 are much greater than the CDs in lines 4 to 5.

Finally, the coefficient of determination for the simulation of Galton's sweet-pea research is CD = .193. This indicates that 19.3 percent of the variance in the weights of the offspring sweet peas is explained by the variance in their parents' weights. Inheritance does not appear to be a strong factor in determining the weight of a sweet pea.

Standardized Regression Slope

The second important approach to measuring the strength of association is based directly on the regression equation itself. We saw that the size of the regression slope was directly related to the standard deviation of Y and inversely related to the standard deviation of X. The greater the variance of Y and the less the variance of X, the greater will be the regression slope. That is why the slope is called an unstandardized slope. The logic behind the standardized slope is to transform or standardize X and Y so that they have equal standard deviations and then to calculate the regression equation on these standardized variables. The resulting slope will be a standardized slope that is a valid measure of the strength of association.

In Chapter 4 you learned that a standardized score (z) equals the deviation score divided by the standard deviation. The standardized scores for X and Y are

$$z_X = \frac{X - \overline{X}}{s_X}$$

$$z_Y = \frac{Y - \overline{Y}}{s_Y}$$

These formulas are used to transform each raw score of X and Y into a z score. The reason the z scores are called standardized scores is that the mean of z-score distribution equals 0 and its standard deviation equals 1. The interpretation of a z score is that it indicates how many standard deviations the raw score is above or below the mean. To illustrate, Table 8.13 gives the standardized scores for the example data. The z scores for the first case indicate that it is 1.225 standard deviations below the mean of X and .707 standard deviation below the mean of Y. What we are going to do next is to find the least-squares regression slope for z_X and z_Y.

The general formula for the least-squares slope for any pair of variables is

$$b = \frac{\text{covariance}}{\text{variance of IV}} = \frac{\text{sum of cross-products}}{\text{sum-of-squares of IV}}$$

In the above formula, IV means independent variable (DV would be the dependent variable). This formula applies for raw X and Y scores, for standardized X and Y scores, or for any other transformation of X and Y. When applied to the raw scores, the general formula becomes the equation we first learned for the regression slope:

$$b_{YX} = \frac{\Sigma(X - \overline{X})(Y - \overline{Y})}{\Sigma(X - \overline{X})^2}$$

If the general formula is applied to the z scores (substitute z_X for X and z_Y for Y in the above raw score formula), the result is

$$b_{z_Y z_X} = \frac{\Sigma(z_X - \bar{z}_X)(z_Y - \bar{z}_Y)}{\Sigma(z_X - \bar{z}_X)^2} = \frac{\Sigma(z_X - 0)(z_Y - 0)}{\Sigma(z_X - 0)^2} = \frac{\Sigma z_X z_Y}{\Sigma z_X^2} = \frac{\Sigma z_X z_Y}{n} \qquad (8.22)$$

Table 8.13 Standardized Scores for the Example Data

z_X	z_Y	z_X^2	z_Y^2	$z_X z_Y$
−1.225	−.707	1.50	.50	.866
.000	−.707	.00	.50	.000
1.225	1.414	1.50	2.00	1.732
.000	.000	3.00	3.00	2.598

Because the means $\bar{z}_X$ and $\bar{z}_Y$ equal 0 (note that the sums of the z scores in Table 8.13 equal 0), $\Sigma(z_X - \bar{z}_X)(z_Y - \bar{z}_Y)/\Sigma(z_X - \bar{z}_X)^2$ simplifies to $\Sigma z_X z_Y/\Sigma z_X^2$. And because the sum of the squared z scores always equals n (see Table 8.13 and Chapter 4), $\Sigma z_X z_Y/\Sigma z_X^2$ simplifies to $\Sigma z_X z_Y/n$. Thus, the initial complex formula for the slope for standardized scores reduces to a very simple equation. We will simplify the symbol for this standardized slope from $b_{z_Y z_X}$ to B_{YX} and rewrite the formula for the standardized slope as

$$B_{YX} = \frac{\Sigma z_X z_Y}{n} \qquad (8.23)$$

In this notation, the capital italic B signifies that this is a standardized slope, and we need only put YX in the subscript of B.

Notice that the numerator of Equation 8.23 contains the sum of the products of z_X and z_Y. Because the means of the z scores equal 0, $\Sigma z_X z_Y$ *is the sum of cross-products of the standardized scores.* When the sum of cross-products is divided by n, the result is the covariance. Thus, *the* **standardized slope** *equals the covariance of the standardized scores for X and Y:*

$$B_{YX} = s_{z_X z_Y}$$

You will remember that the covariance was the first measure of association presented in this text. It described some valuable properties of associations, but it was somewhat difficult to interpret, and it had no upper or lower limits. The covariance of the standardized scores, however, is a completely different story. It has none of the limitations of the covariance of the raw X and Y scores. As we will see, it has fixed limits and a clear interpretation.

What are the limits of B_{YX}? In a perfect positive association, each case would have equal values for z_X and z_Y. If $z_Y = z_X$, the residuals of prediction would all equal 0. In a perfect negative association, z_X and z_Y would have the same absolute values for each case, but the sign of z_Y would be the opposite of the sign of z_X ($z_Y = -z_X$). The values of B_{YX} for perfect positive and negative associations would be as follows:

$$\text{If } z_Y = z_X\text{: } B_{YX} = \frac{\Sigma z_X z_X}{n} = \frac{\Sigma z_X^2}{n} = \frac{n}{n} = 1$$

$$\text{If } z_Y = -z_X\text{: } B_{YX} = \frac{\Sigma - z_X z_X}{n} = \frac{\Sigma - z_X^2}{n} = \frac{-n}{n} = -1$$

In other words, $B_{YX} = 1$ when there is a perfect positive association between X and Y, and $B_{YX} = -1$ when there is a perfect negative association. The range of values of B_{YX} is -1 to $+1$:

$$-1 \leq B_{YX} \leq +1$$

Because B_{YX} is a covariance, the standardized slope equals 0 when there is no association. Because B_{YX} has a fixed range of values and can be either positive or negative, we can use it to judge both the direction of association and the strength of association. The closer B_{YX} is to either $+1$ or -1, the stronger the association.

Because the standardized slope is a covariance, *B is a symmetrical measure of association.* This can also be seen from the formula for z_X regressed on z_Y:

$$B_{XY} = \frac{\Sigma z_X z_Y}{\Sigma z_Y^2} = \frac{\Sigma z_X z_Y}{n}$$

Because z_Y is the independent variable, the sum-of-squares of z_Y goes in the denominator. And because $\Sigma z_Y^2 = n$, the formula for B_{XY} reduces to the same formula as B_{YX}. Thus,

$$B_{YX} = B_{XY}$$

No matter which variable we choose as the dependent variable, we will get the same value for the standardized slope. Thus, we can use B without a subscript as the formula for the standardized slope.

Now, what is the formula for the intercept of the standardized regression equation? We will use the capital letter A for the intercept of the standardized regression equation. The general formula for the intercept of a least-squares regression equation is

$$\text{Intercept} = \text{mean of DV} - \text{slope} \cdot \text{mean of IV}$$

Substituting standardized terms into this general equation, we get

$$A_{YX} = \bar{z}_Y - B_{YX}\bar{z}_X = 0 - B_{YX} \cdot 0 = 0$$

This formula indicates that the standardized intercept equals 0 because the means of z_X and z_Y equal 0. The intercept A_{XY} equals 0 for the same reason. Because $A_{YX} = A_{XY} = 0$ and B is symmetrical, the formulas for the standardized regression equation are

$$\hat{z}_Y = Bz_X$$

$$\hat{z}_X = Bz_Y$$

Let's return to our example data (Table 8.13) and calculate the standardized slope:

$$B = \frac{\Sigma z_Y z_X}{n} = \frac{2.598}{3} = .866$$

We know by the sign that the association is positive. The absolute value of B tells us that the association is .866 as large as a perfect association of 1, which is a pretty strong association.

Now, aside from the fact that B indicates the strength of association, how do we interpret what .866 means? To interpret the value of B, we must remember that although

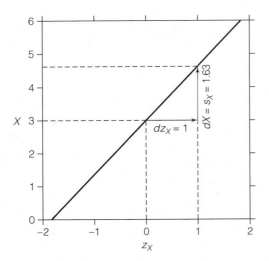

FIGURE 8.16 Change in X Corresponding to a Unit Increase in z_X

it is a special kind of covariance, it is also a special kind of slope. Remember that in general, *a slope equals the change in the dependent variable per unit increase in the independent variable.* So, a standardized slope also can be interpreted this way, as long as it is expressed in terms of the standardized scores z_X and z_Y. Thus, *B equals the change in z_Y per unit increase in z_X.* The example value of B indicates that if z_X increases by 1 unit, z_Y is expected to increase by .866 unit.

What does a 1–unit difference in z represent? Remember that a z score indicates the number of standard deviations that a score deviates from the mean. For our example data, Figure 8.16 shows the conversion line for changing z scores into X or for changing X into z. The arrows indicate that a 1–unit increase in z_X corresponds to a 1.63–unit increase in X. This number, 1.63, is the value of the standard deviation of X. Thus, we can say that a 1–unit increase in z_X is equivalent to a 1 standard deviation increase in X. Likewise, a change in z_Y of $B = .866$ is equivalent to a change in Y of .866 standard deviation. Thus, we can interpret B in terms of changes in standard deviations of X and Y: *B equals the number of standard deviations that Y changes per standard deviation increase in X.* For our example data, if X increases by one standard deviation, Y increases by .866 standard deviation.

Although Equation 8.23 is the defining formula for the standardized slope, there are two additional formulas that are both insightful and useful:

$$b_{YX} = \frac{s_Y}{s_X} B \tag{8.24}$$

$$B = \frac{s_X}{s_Y} b_{YX} \tag{8.25}$$

Equation 8.24 shows that the unstandardized regression slope is a function of the strength of the relationship between X and Y, as indicated by B, and the ratio of the standard deviations of Y and X. The latter influence was discussed several times previously. Now we see both factors contained in a single formula such that the greater the standardized slope and the greater the standard deviation of Y relative to that of X, the greater will be the change in Y caused by a unit increase in X. Equation 8.25, on the other hand, shows that you can remove the influence of the standard deviations on the unstandardized slope by multiplying $s_X/s_Y \cdot b_{YX}$. The result is the standardized slope. Both of these formulas may also be useful in efficiently computing one slope from the other. For instance, you can compute the standardized slope with Equation 8.25 without having to go through the process of converting raw scores to standard scores in order to use Equation 8.23.

Examples of Standardized Slope

Now we will examine the strength of association for the gender-role variables as measured by the standardized slope. We will use Equation 8.25 to calculate the standardized slopes directly from the unstandardized slope and the standard deviations (Table 8.14). For the total sample of husbands' and wives' gender-role attitudes (line 1), the standardized slope equals

$$B = \frac{s_X}{s_Y} b_{YX} = \frac{3.861}{3.797} (.803) = 1.017(.803) = .817$$

In this case, the standardized slope is not much different from the unstandardized slope because the husbands' and wives' standard deviations are about equal. The positive sign indicates a positive association (the more nontraditional the wife, the more nontraditional the husband). The absolute value of B indicates that the association is .817 of the maximum possible association. The slope also indicates that for each increase of 1 standardized unit in the wife's attitude there is an increase of .817 standardized unit in the husband's attitude. Or, for a standard deviation increase in the wife's attitude there is a .817 standard deviation increase in the husband's attitude.

The standardized slope is greater for wives who are the most satisfied with their marriages ($B = .852$) than for wives who are less satisfied ($B = .666$). This indicates that there is a stronger association between spouses' attitudes for wives who are the most satisfied with their marriages than for the less-satisfied wives. The last two lines indicate that there is little difference between the strength of association between the husband's age and attitude ($B = .335$) and the strength of association between his education and attitude ($B = -.312$). These associations are much weaker, however, than those for husband's and wife's attitudes. In other words, a husband's gender-role attitude can be predicted more accurately by knowing his wife's attitude than by knowing his age or his education.

Pearson Product–Moment Correlation

The **Pearson product–moment correlation** (symbolized by r) is also a symmetrical measure of association with a fixed range of values ($-1 \leq r \leq +1$). Equation 8.23 for the standardized slope (B) often is given for the correlation coefficient. Thus, these two statistics are one and the same:

$$r = B$$

Table 8.14 Standardized Regression Slopes (B) for Gender-Role Attitude Examples[a]

Figure	X	Y	n	s_X	s_Y	b_{YX}	B
7.9	Wife's attitude	Husband's attitude	328	3.861	3.797	.803	.817
7.10a	Wife's attitude	Husband's attitude	72	3.586	3.134	.582	.666
7.10b	Wife's attitude	Husband's attitude	113	3.809	3.606	.807	.852
7.11a	Age	Husband's attitude	154	14.973	4.153	.093	.335
7.11b	Education	Husband's attitude	154	3.149	4.153	−.411	−.312

[a]1987 Akron Area Survey

The importance of Equation 8.23 for the correlation is indicated by the term product-moment. In mathematics, taking the "moment" of a distribution of scores is another way of saying that we divide by n to get the arithmetic average. Thus, product-moment correlation means the arithmetic average of the products of the z scores for X and Y. The identity of the standardized slope and the Pearson correlation means that the correlation can be given a very explicit interpretation in terms of a slope. Thus, the correlation not only tells us the strength of association between X and Y, it also indicates the change in standardized Y associated with a unit increase in standardized X, and vice versa.

The Pearson correlation does not tell us anything more or different about the association between X and Y than the standardized slope. It is just another name for the same statistic. Because the correlation coefficient is so frequently used, however, it is important that you know these two statistics are identical. This fact often is not recognized. Two formulas for the correlation that you may encounter in other books are

$$r = \frac{s_{XY}}{s_X s_Y} = \frac{\Sigma(X - \overline{X})(Y - \overline{Y})}{\sqrt{[\Sigma(X - \overline{X})^2][\Sigma(Y - \overline{Y})^2]}} \tag{8.26}$$

Both of these formulas can be derived readily from Equation 8.23 (and vice versa). The first formula indicates that if you divide the covariance by the product of the standard deviations of X and Y you get the correlation coefficient (and B). Because the size of the covariance is directly related to the size of the standard deviations of X and Y, as discussed earlier, dividing by the product of s_Y and s_X removes their influences. Consequently, *the correlation coefficient and the standardized slope are standardized covariances.*

Relationship between CD and B

We have covered two measures of association, the coefficient of determination (CD) and the standardized slope (B). The coefficient of determination was defined as the ratio of the variance of the predicted values Y to the total variance of Y. The standardized slope equals the least-squares slope for the standardized scores of Y regressed on the standardized scores of X. Now we will compare the two measures. The square of the standardized slope for our three-case example data is $B^2 = (.866)^2 = .750$. The coefficient of determination for the same data is CD = .75. It is no coincidence that the square of the standardized slope equals the coefficient of determination. In general,

$$CD = B^2$$

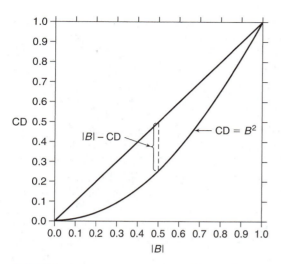

FIGURE 8.17 Relationship between CD and B

You may verify this by squaring the values of B in Table 8.14 to see if they equal the corresponding values of CD in Table 8.12.

Because B is a fraction (unless it is exactly equal to 0 or 1), its square will always be less than itself. Thus, $CD < B$. And although B itself may be negative, the square of a negative number is positive. So, $CD > 0$. In general, $0 \leq CD \leq |B| \leq 1$. The size of the difference between CD and $|B|$ is shown in Figure 8.17 as the vertical distance between the diagonal line and the curve $CD = B^2$. The largest gap is when $|B| = .5$. As the association gets stronger than .5 or weaker than .5, the difference gets smaller. Thus, the gap is small for very strong and very weak associations, and the gap is large for moderate associations.

Is there a reason for preferring one measure to the other? Many investigators appear to prefer CD because of the intuitive appeal of its interpretation: CD equals the proportion of variance in one variable explained by the other. Because $CD = B^2 = r^2$, this statistical interpretation is often referred to as "r^2." The interpretation of B, although not as well known, is just as meaningful as that of CD: B equals the standard deviation change in Y per standard deviation increase in X. We do not prefer CD because the explained variance and the total variance consist of *squared* deviations from the mean. The standard deviations involved in the interpretation of B, however, are the square roots of the variances and thus are expressed in terms of the measurement scales of X and Y. There is, however, no right or wrong answer to the question of which of these two statistics is preferable. Whichever measure you choose, remember that the association will appear to be weaker if you choose CD.

Regression toward the Mean, Revisited

When Galton discussed the concept of regression toward the mean, he was analyzing the relationship between a parent's trait and the same trait of its offspring. In this case, both variables are measures of the same trait (e.g., weight) for each of a pair of units (e.g., parents and offspring). When we are examining the association between a single attribute for a pair of units (e.g., gender-role attitudes of wives and husbands), b will usually be positive and will almost always be less than 1. When this is the case, $(b - 1)$ equals the rate of regression toward the mean (RTM) and $(b - 1)(X - \bar{X})$ equals the amount of RTM (Equation 8.14). For example, $b = .8$ for wife's attitude (X) and husband's attitude (Y). If the wife's attitude is 5 points above the mean, RTM $= (.8 - 1)(5) = -.2(5) = -1$. That is, the husband is expected to be 1 point less above the mean than his wife.

Is it appropriate to speak of regression toward the mean when X and Y represent different traits, such as husband's age (X) and husband's attitude (Y)? The slope for age and attitude is $b = .093$. Thus, for each year of age that the husband is above the mean, he is .093 attitude point above the mean gender-role attitude. Does this indicate that he is closer to the attitude mean than the age mean? It is difficult to answer this question, because we can't directly compare attitude points with years of age.

If, however, we standardize attitude and age by converting them to z scores, we can compare standardized attitude points with standardized years of age. In this case, the standardized regression slope (B) will indicate the amount of regression toward the mean in terms of standardized units. The standardized slope for attitude and age is $B = .335$. This indicates that if the husband is 1 standard deviation above the mean of age, we expect him to be only .335 standard deviation above the mean gender-role attitude. If B is positive, the amount of regression toward the mean equals

$$\text{If } B > 0: \text{RTM} = (B - 1)z_X \tag{8.27}$$

If a husband is 2 standard deviations above the mean of age (i.e., $z_X = 2$), he will regress toward the mean attitude by $(.335 - 1)(2) = -.665(2) = -1.33$ standard deviations. If he were 2 standard deviations below the mean age, the husband would regress toward the mean by $(.335 - 1)(-2) = -.665(-2) = 1.33$ standard deviations. These regressions toward the mean are shown in Figure 8.18 as the vertical distances between the line $z_Y = z_X$ and the line $\hat{z}_Y = .335z_X$.

If the standardized regression slope is negative,

$$\text{If } B < 0: \text{RTM} = (B + 1)z_X \tag{8.28}$$

For example, $B = -.312$ for husband's education and attitude. In this case, if a husband is 2 standard deviations above the mean of education, he will regress toward the attitude mean by $(-.312 + 1)(2) = (.688)(2) = 1.376$ standard deviations. To understand this, remember that if there were a perfect negative association, the man would be 2 standard deviations below the attitude mean. He is predicted, however, to be $Bz_X = -.312(2) = -.624$ standard deviation from the mean ($.624s_Y$ below the mean). The difference between the predicted z_Y and the perfect association z_Y equals $-.624 - (-2) = -.624 + 2 = 1.376$, which is a regression upward toward the mean of 1.376 standard deviations. Regression toward the mean in the case of a negative association indicates, for example, that if a case is above the mean on X, it will be "regressing" upward toward the mean of Y. This means that the case will not be as many standard deviations below the mean of Y as it is above the mean of X. If, on the other hand, a case is below the mean of X, it will not be as many standard deviations above the mean of Y as it is below the mean of X.

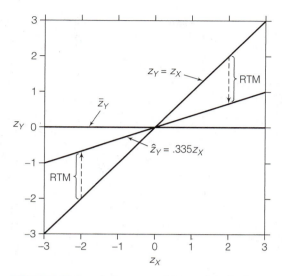

FIGURE 8.18 Regression toward Mean (RTM) of Husband's Attitude (Y) on Age (Y)

When Galton introduced the notion of regression toward the mean, he was applying it to a very specific type of regression analysis, namely, the analysis of hereditary traits. Modern regression analysis is not limited to studies of parents and offspring. It is used to analyze all types of Xs and Ys. Regression toward the mean *in standardized units* exists whenever there is less than a perfect association $(-1 < B < 1)$. Because there is never a perfect association in real research, there will always be some regression toward the mean. The strength of regression toward the mean is indicated by $1 - |B|$; or, the smaller the absolute value of B, the greater the regression toward the mean.

SUMMARY

Regression analysis is a statistical technique that is used to study linear associations between interval/ratio variables. A linear relationship between X and Y is defined by the slope and the Y intercept of the line. The slope is the change in Y that is associated with a 1-unit increase in X. The slope may be either positive, indicating that Y increases when X increases, or negative, indicating that Y decreases when X increases. The Y intercept is the value of Y when X equals 0. In regression analysis we usually assume that Y is the dependent variable (i.e., the effect) and X is the independent variable (i.e., the cause). Even if X has a linear effect on Y, however, we will never observe a perfect linear relationship between the observed values of X and Y because there may be other causes of Y and because of random disturbances that disrupt the perfect linear relationship. In this chapter we took up the case of a single independent variable.

Least-squares regression analysis is a method for estimating the underlying linear relationship (i.e., the slope and intercept) that is being masked by other causes and random disturbances. The least-squares line is the one that minimizes the sum of squared residuals, or squared differences between the observed Y and the predicted Y. This criterion is used because the least-squares line minimizes sampling error. The least-squares slope equals the sum of cross-products of X and Y divided by the sum-of-squares of X, or, equivalently, the covariance divided by the variance of X. The intercept equals the mean of Y minus the product of the slope and the mean of X. The least-squares slope is an asymmetrical measure of association, which means that the slope for Y regressed on X is not equal to the slope for X regressed on Y. It is also unstandardized because there are no maximum or minimum limits on the value it can take. In general, the greater the standard deviation of Y relative to that of X, the greater will be the slope of Y regressed on X. When X and Y are measured on the same scale, such as parent's and offspring's weights, the mean of Y for cases with any given value of X will be closer to the overall mean of Y than X is to its mean, which is called regression toward the mean. The greater the slope, the less the regression toward the mean.

Because the least-squares slope is unstandardized, we considered two measures of the strength of association, namely, the coefficient of determination and the standardized slope. To understand the coefficient of determination, we described how the sum-of-squares of Y equals the residual sum-of-squares plus the regression sum-of-squares, which is the sum of squared deviations of the predicted values of Y from the mean of Y. The regression sum-of-squares, or regression variance, is called the explained variance, and the residual sum-of-squares, or residual variance, is called the unexplained variance. The coefficient of determination equals the regression sum-of-squares divided by the total sum-of-squares, which indicates the proportion of the variance of Y that is explained by, or shared with, X. Because it has a fixed range of values (0 to 1), it is a standardized measure of association. The coefficient of determination is also a symmetrical measure of association that gives the same value whether Y or X is the dependent variable.

The second measure of strength of association is the standardized slope. Because standardized scores have a standard deviation of 1, the least-squares regression of the standardized Y on the standardized X gives a standardized slope whose values range from -1 to 1. Thus, it indicates both the direction and strength of association. The standardized slope indicates the number of standard deviations that Y increases or decreases (depending on the sign) per standard deviation increase in X. It is also a symmetrical measure of association. Although the coefficient of determination and the standardized slope have different interpretations, the coefficient of determination equals the square of the standardized slope. Thus, the coefficient of determination is always smaller than the standardized slope.

In general, we can speak of regression toward the mean whenever there is less than a perfect association between X and Y. When X and Y are not measured on the same scale, or when they have different standard deviations, we must speak of regression toward the mean in terms of standardized units. One minus the absolute value of the standardized slope indicates the strength of regression toward the mean; the greater the standardized slope, the less will be the regression toward the mean.

KEY TERMS

regression analysis

Y intercept

slope

independent variable

dependent variable

criteria for a causal association

spurious association

bivariate linear model

predicted values

residuals

sum of squared residuals

least-squares line

least-squares slope

least-squares intercept

asymmetry of least-squares slope

unstandardized slope

regression toward the mean

regression sum-of-squares

residual sum-of-squares

coefficient of determination

standard error of estimate

standardized slope (B)

Pearson product-moment correlation

relationship between CD and B

SPSS PROCEDURES

To compute the regression of husband's gender-role attitude on wife's gender-role attitude, click on *Analyze, Regression,* and *Linear.* In the *Linear Regression* window, move the husband's gender-role variable (*hsexrole*) into the *Dependent* box and move the wife's gender-role variable (*wsexrole*) into the following *Independent(s)* box. After clicking on *OK,* you will see the tables shown next in the output window. In the *Model Summary* table, *R* is the Pearson correlation coefficient and *R Square* is the coefficient of determination. The value of the coefficient of determination is the same as that given in Table 8.12. In the *ANOVA* table are found the regression, residual, and total sum-of-squares, which can be used to calculate the coefficient of determination. In the *Coefficients* table under *B* are found the unstandardized slope for *wsexrole* and the *Constant,* which is the *Y* intercept. The values of these statistics equal those given in Table 8.6. The standardized slope for *wsexrole* is found under *Beta* in the *Coefficients* table, the value of which equals that given in Table 8.14.

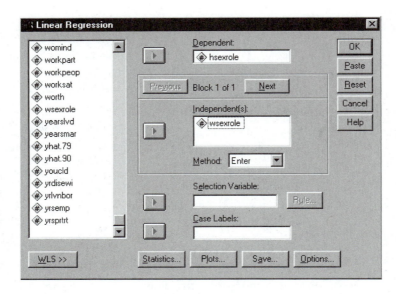

Model Summary

Model	R	R Square	Adjusted R Square	Std. Error of the Estimate
1	.817[a]	.667	.666	2.1987430

a. Predictors: (Constant), WSEXROLE

ANOVA[b]

Model		Sum of Squares	df	Mean Square	F	Sig.
1	Regression	3153.789	1	3153.789	652.355	.000[a]
	Residual	1576.037	326	4.834		
	Total	4729.826	327			

a. Predictors: (Constant), WSEXROLE

b. Dependent Variable: HSEXROLE

Coefficients [a]

Model		Unstandardized Coefficients		Standardized Coefficients	t	Sig.
		B	Std. Error	Beta		
1	(Constant)	2.400	.230		10.438	.000
	WSEXROLE	.803	.031	.817	25.541	.000

a. Dependent Variable: HSEXROLE

If you would like descriptive statistics (e.g., means, standard deviations, correlations, covariances, and sums of cross-products) for the variables in the regression analysis, click on *Statistics...* in the *Linear Regression* window, click on *Descriptives* in the following *Linear Regression: Statistics* window, and click on *Continue.*

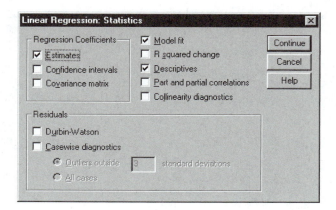

The descriptive statistics available through the menu selections do not include the covariances and sums of cross-products. To obtain these statistics, click on *Paste* instead of *OK* in the *Linear Regression* window. This will translate the menu selections you have made into SPSS syntax and paste the syntax into the following window named the *SPSS Syntax Editor.* Edit the "/DESCRIPTIVES . . ." line by deleting "SIG" and "N" and entering XPROD and COV. Click on *Run* in the main menu and *All* in the submenu and the regression analysis will be executed. The descriptive statistics follow. In the *Covariance* panel of the *Correlations* table, 12.010 is the covariance between *hsexrole* and *wsexrole,* 14.464 is the variance of *hsexrole,* and 14.956 is the variance of *wsexrole.* In the *Sum of Squares and Cross-products* panel of the *Correlations* table, 3927.284 is the sum of cross-products, 4729.826 is the sum-of-squares of *hsexrole,* and 4890.485 is the sum-of-squares of *wsexrole.*

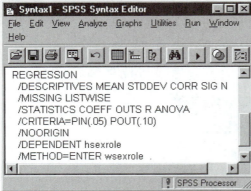

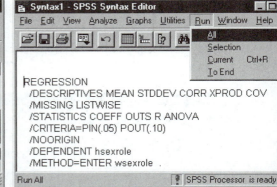

Descriptive Statistics

	Mean	Std. Deviation	N
HSEXROLE	7.387195	3.8031961	328
WSEXROLE	6.210366	3.8672485	328

Correlations

		HSEXROLE	WSEXROLE
Pearson Correlation	HSEXROLE	1.000	.817
	WSEXROLE	.817	1.000
Covariance	HSEXROLE	14.464	12.010
	WSEXROLE	12.010	14.956
Sum of Squares and	HSEXROLE	4729.826	3927.284
Cross-products	WSEXROLE	3927.284	4890.485

EXERCISES

1. Let X be the percentage of children living in single-parent households in each state, and let Y be the number of homicides per 100,000 population in each state. In 1993, the coordinates for Louisiana and Kansas were (32, 20.3) and (18, 6.4), respectively. What are the slope and the Y intercept of the line relating homicides to children in single-parent households for Louisiana and Kansas?

2. Let X be the percentage of the population below the poverty line in each state, and let Y be the percentage of homicides that are knife-related in each state. In 1993, the coordinates for Mississippi and Iowa were (20.6, 14.7) and (9.6, 28.9), respectively. What are the slope and the Y intercept of the line relating knife-related homicides to poverty for Mississippi and Iowa?

3. The plot shows the regression line relating homicides per 100,000 to percentage of children living in single-parent households for the 50 states. What are the slope and Y intercept?

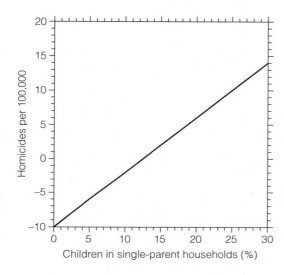

4. The plot shows the regression line relating percentage of homicides that are knife-related to percentage of persons living in poverty for the 50 states. What are the slope and Y intercept?

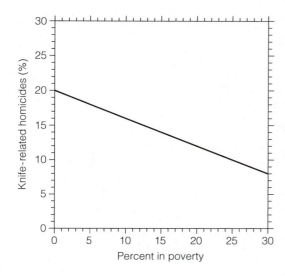

5. The SPSS table reports the homicide rate (homicides per 100,000) and percentage of children in single-parent households for a random sample of five states. Let the homicide rate be Y.

	STATENAM	RATHOMS Rate of Homicides	SINGLE Percentage of Children in Single-parent households
1	Florida	8.9	27
2	Idaho	2.9	15
3	Maryland	12.7	26
4	Nebraska	3.9	17
5	Texas	11.9	23

   a. Calculate the least-squares slope and intercept. Interpret.
   b. Calculate the coefficient of determination. Interpret.
   c. Calculate the standardized slope (using z scores). Interpret.
   d. Calculate the standard error of estimate. Interpret.

6. The SPSS table reports the percentage of homicides that are knife-related and the percentage of the population living in poverty for a random sample of five states. Let percentage in poverty be X.

	STATENAM	PERKNIFE Percent of Homicides Knife-Related	POVERTY percent below poverty level
1	Colorado	15.5	10.6
2	Florida	11.7	14.2
3	Mass.	27.1	10.1
4	Minnesot	22.1	9.8
5	Miss.	14.7	20.6

   a. Calculate the least-squares slope and intercept. Interpret.
   b. Calculate the coefficient of determination. Interpret.
   c. Calculate the standardized slope (using z scores). Interpret.
   d. Calculate the standard error of estimate. Interpret.

7. In the following SPSS tables, calculate the missing statistics (?). Use definitional formulas except in the case of Beta, where you may use Equation 8.25.

Descriptive Statistics

	Mean	Std. Deviation	Variance	N
RATHOMS Rate of Homicides	7.330	3.9809	15.8474	50
SINGLE Percentage of Children in Single-parent households	22.14	4.170	17.388	50

Correlations

		RATHOMS Rate of Homicides	SINGLE Percentage of Children in Single-parent households
Covariance	RATHOMS Rate of Homicides	15.847	13.773
	SINGLE Percentage of Children in Single-parent households	13.773	17.388
Sum of Squares and Cross-products	RATHOMS Rate of Homicides	776.525	674.890
	SINGLE Percentage of Children in Single-parent households	674.890	852.020

Model Summary

Model	R	R Square	Adjusted R Square	Std. Error of the Estimate
1	?[a]	?	.682	?

a. Predictors: (Constant), SINGLE Percentage of Children in Single-parent households

ANOVA[b]

Model		Sum of Squares	df	Mean Square	F	Sig.
1	Regression	534.584	1	534.584	106.059	.000[a]
	Residual	241.941	48	5.040		
	Total	776.525	49			

a. Predictors: (Constant), SINGLE Percentage of Children in Single-parent households

b. Dependent Variable: RATHOMS Rate of Homicides

Coefficients[a]

Model		Unstandardized Coefficients		Standardized Coefficients	t	Sig.
		B	Std. Error	Beta		
1	(Constant)	?	1.732		-5.893	.000
	SINGLE Percentage of Children in Single-parent households	?	.077	?	10.299	.000

a. Dependent Variable: RATHOMS Rate of Homicides

8. In the following SPSS tables, calculate the missing statistics (?). Use definitional formulas except in the case of Beta where you may use Equation 8.25. *Note:* Three states had missing values for the variable PERKNIFE.

Descriptive Statistics

	Mean	Std. Deviation	Variance	N
PERKNIFE Percent of Homicides Knife- Related	15.170	6.9908	48.8713	47
POVERTY percent below poverty level	12.809	4.0329	16.2643	47

Correlations

		PERKNIFE Percent of Homicides Knife- Related	POVERTY percent below poverty level
Covariance	PERKNIFE Percent of Homicides Knife- Related	48.871	-6.187
	POVERTY percent below poverty level	-6.187	16.264
Sum of Squares and Cross-products	PERKNIFE Percent of Homicides Knife- Related	2248.078	-284.618
	POVERTY percent below poverty level	-284.618	748.157

Model Summary

Model	R	R Square	Adjusted R Square	Std. Error of the Estimate
1	?a	?	.027	?

a. Predictors: (Constant), POVERTY percent below poverty level

ANOVA[b]

Model		Sum of Squares	df	Mean Square	F	Sig.
1	Regression	108.276	1	108.276	2.277	.138a
	Residual	2139.802	45	47.551		
	Total	2248.078	46			

a. Predictors: (Constant), POVERTY percent below poverty level
b. Dependent Variable: PERKNIFE Percent of Homicides Knife- Related

Coefficients[a]

Model		Unstandardized Coefficients		Standardized Coefficients		
		B	Std. Error	Beta	t	Sig.
1	(Constant)	?	3.382		5.926	.000
	POVERTY percent below poverty level	?	.252	?	-1.509	.138

a. Dependent Variable: PERKNIFE Percent of Homicides Knife- Related

9. The following tables report the regression of respondent's education (*educ*) on father's education (*paeduc*) for males 25+ years of age in the GSS 2000.

Descriptive Statistics

	Mean	Std. Deviation	N
EDUC	13.96	3.106	784
PAEDUC	11.17	4.325	784

Coefficients[a]

	Unstandardized Coefficients		Standardized Coefficients		
	B	Std. Error	Beta	t	Sig.
(Constant)	10.319	.274		37.637	.000
PAEDUC	.326	.023	.454	14.234	.000

a. Dependent Variable: EDUC

a. Interpret the unstandardized slope.
b. What is the regression toward the mean in raw years of education for sons whose fathers had 8 years of education, 12 years of education, and 16 years of education?
c. If father's education had been treated as the dependent variable (an absurdity), what would be the slope and intercept of the regression line?

10. The following tables report the regression of respondent's education (*educ*) on mother's education (*paeduc*) for females 25+ years of age in the GSS 2000.

Descriptive Statistics

	Mean	Std. Deviation	N
MAEDUC	10.96	3.628	1193
EDUC	13.54	2.652	1193

Coefficients[a]

	Unstandardized Coefficients		Standardized Coefficients		
	B	Std. Error	Beta	t	Sig.
(Constant)	9.976	.219		45.545	.000
MAEDUC	.325	.019	.445	17.126	.000

a. Dependent Variable: EDUC

a. Interpret the unstandardized slope.
b. What is the regression toward the mean in raw years of education for daughters whose mothers had 8 years of education, 12 years of education, and 16 years of education?
c. If mother's education had been treated as the dependent variable (an absurdity) what would be the slope and intercept of the regression line?

11. Using the GSS 2000 (*gss2000a* or *gss2000b*) and women aged 40 or older, use SPSS to regress the number of children a woman has had (*childs*) on the age at which her first child was born (*agekdbrn*). Construct a jittered scatterplot and draw the regression line on the plot. Interpret the unstandardized slope, the intercept, the standardized slope, and the coefficient of determination. What is the amount of regression toward the mean in standardized scores for women who were 18 at first birth and for women who were 25 at first birth?

12. Using the GSS 2000 (*gss2000a* or *gss2000b*) and persons 25 or older, use SPSS to regress the number of hours of TV watched per day (*tvhours*) on years of education (*educ*). Construct a jittered scatterplot and draw the regression line on the plot. Interpret the unstandardized slope, the intercept, the standardized slope, and the coefficient of determination. What is the amount of regression toward the mean in standardized scores for persons with 8 years of education and for those with 16 years of education?

9

Inferences about the Regression Slope

IS COLLEGE WORTH IT?

Carmen was worried. She had carefully estimated all the costs of going to college, and now, despite what almost everyone had told her, she was asking herself seriously if going to college was worth it. Carmen had graduated from Madison High five weeks ago, and in just forty-seven days it would be time to head for college. She had been admitted to three schools and had decided on Gates College because of its strong programs in computer science and business. She had always been goal-oriented and practical about things. Therefore, Carmen had applied herself in school and earned good grades. If truth be known, however, she would rather get a job than go to college. If she was going to subject herself to four more years of school, it had better pay off in the long run.

Carmen figured that the true expenses of college were tuition and books, plus the money she would have earned if she had gone to work instead of to college. Room and board, transportation, and personal expenses would have to be paid for anyway. Her tuition and books at Gates would come to $23,700 the first year. After subtracting $8,000 in need-based grants that wouldn't have to be paid back and $3,000 in summer earnings and work-study income, she calculated that it would cost her $12,700 the first year for tuition and books. Taking tuition inflation into account, that would come to $55,500 over the course of four years. Then Carmen estimated what she would earn if she weren't in school. She figured she could get a job that initially paid $23,000 a year. With annual raises of 4 percent, she calculated that she could earn about $97,700 over four years if she didn't go to college. Thus, her total costs of college, according to her thinking, would total approximately $153,200!

"Is a college education worth $153,200 to me?" she asked herself. "How much more per year should I expect to earn if I get my four-year college degree?" Carmen estimated off the top of her head that she might earn about $10,000 a year more if she went to college. That sounded like a pretty conservative estimate to her. If she worked for forty years before retiring, she calculated that she would earn $400,000 more over her work career than she would earn with only a high school degree. Subtracting her estimated college costs of $153,200, her net gain would be $246,800. College certainly would be worth that, she thought. But that conclusion assumes her estimated college-earnings advantage of $10,000, or $2,500 per year of college, is accurate. What she really needs to know, Carmen reasoned, is the minimum-earnings advantage that she would be almost certain to get. If, for example, she could be very sure to earn at least $4,000 more per year due to her college education, she could be quite certain that her total advantage (4,000 × 40 = $160,000) would exceed her college costs ($153,200). "I need to get some more information," she thought, "before I can decide to go ahead with my college plans."

Did Carmen do her math right? Did she take the right factors into consideration? What kind of statistical information does she need to answer the crucial question about the minimum-earnings advantage of a college education?

Estimating the financial return of a college education is no easy task. Individuals vary greatly in terms of how much monetary benefit they receive from their college education. The payoff depends on many factors, including the student's major, college or university, grade-point average, personal attributes, and just plain luck. Because incoming students cannot know with much certainty what they will end up majoring in, what college or university they will actually graduate from, what their grades will be, whether they possess the relevant personal attributes or will be in the right place at the right time, they cannot know how much they will individually benefit from a college education. Consequently, the safest method is to examine the financial gain that the average person realizes from college.

One way to accomplish this task is to take a relatively large sample of workers and conduct a regression analysis in which earnings is the dependent variable and the number of years of college completed (e.g., 0–4) is the independent variable. The regression slope will indicate how much each additional year of college increases earnings, on the average. The value of the sample slope will be used to make an inference about the average financial advantage from education in the population of all workers. Inferential statistics involve both a margin of error and a likelihood of being in error. One form of inferential statistics is a confidence interval. A confidence interval may be calculated for a sample slope in a similar manner to that for the sample mean. The 95-percent confidence interval for the slope of earnings regressed on education would help Carmen to decide whether to attend college. The lower limit of this confidence interval would tell her the minimum financial return that she can be very confident of receiving from a college education. This value can then be used to calculate her lifetime-earnings advantage from college so she can compare it with her estimate of the costs of attending college for four years. Thus, regression analysis and inferences about the regression slope are directly relevant to the question, "Is college worth it?"

SAMPLING DISTRIBUTION OF THE SLOPE

Whether we are making inferences about the mean or inferences about the regression slope, knowledge about the sampling distribution of the statistic is a crucial component of inferential statistics. *The sampling distribution of a statistic is the distribution of that statistic across all possible unique samples of a given size (n).* Chapter 5 illustrated and described the sampling distribution of the mean. We saw that the mean of the sampling distribution equals the population mean ($\mu_{\overline{X}} = \mu_X$), which indicates that the sample mean is an unbiased estimate of the population mean. We learned that the standard deviation of the sampling distribution, called the standard error, equals the population standard deviation divided by the square root of the sample size ($\sigma_{\overline{X}} = \sigma_X/\sqrt{n}$). And we learned that the sampling distribution is a normal distribution ($\overline{X} \sim N$). In this section you will learn about the characteristics of the sampling distribution of the slope. First, we will illustrate some of its characteristics with data for a small population. Second, we will describe its theoretical characteristics, followed by the assumptions on which they are based.

Illustrative Sampling Distribution of *b*

The population equation for the bivariate linear model given in Equation 8.6 is

$$Y = \alpha + \beta X + \varepsilon$$

β is the unknown regression slope in the population. We will be estimating it with the sample least-squares slope *b*. The other two unknowns are the population intercept α and the random disturbance ε.

We will illustrate the sampling distribution of *b* by using a small population consisting of five cases. Imagine that the linear population equation is

$$Y = 23 + 2X + \varepsilon$$

Let *Y* be yearly earnings in thousands of dollars, and let *X* be the number of years of college completed (0–4). The intercept equals 23 ($23,000), and the slope equals 2 ($2,000). Thus, a person with zero years of college ($X = 0$) is expected to earn $23,000, and each year of college results in $2,000 of additional earnings, all other things being equal. All other things are not equal, however, because the random variable ε disturbs this relationship.

The values of *X* and ε are given in Table 9.1 for each of five persons in the population. Table 9.1 also shows how these values are converted into earnings by the intercept and slope of the linear model. Looking at case number 5, for example, $Y = 23 + 2X + \varepsilon = 23 + (2 \cdot 4) - 1 = 30$. This person would have earned $31,000 had it not been for some random event(s) ($\varepsilon = -1$) that caused him/her to earn $1,000 less.

Now, what if a researcher selected a simple random sample of size $n = 2$ in order to estimate the population slope β? After selecting the two cases, the researcher would determine the values of *X* and *Y* for each case. *X* and *Y* are observable variables that the

Table 9.1 The Effects of X and Ramdom Error (ε) on Y

Case	X	ε	α	+	βX	+	ε	=	$(\alpha + \beta X)$	+	ε	=	Y
1	0	-1	23	+	2(0)	$-$	1	=	23	$-$	1	=	22
2	1	2	23	+	2(1)	+	2	=	25	+	2	=	27
3	2	-2	23	+	2(2)	$-$	2	=	27	$-$	2	=	25
4	3	2	23	+	2(3)	+	2	=	29	+	2	=	31
5	4	-1	23	+	2(4)	$-$	1	=	31	$-$	1	=	30

researcher can measure. Remember, however, that ε is an unobservable variable and thus is unknown to the investigator. The investigator then has to enter the observed values of X and Y for the selected cases into the least-squares formula for b to calculate an estimate of β. What will the estimate equal?

Well, that depends on which two cases were randomly selected. In a population of $N = 5$, there are $[N(N - 1)]/2 = (5 \times 4)/2 = 10$ unique samples of $n = 2$ that can be selected by the investigator. Table 9.2 shows the ten different pairs of cases. For each pair, the sample slope is computed as $b = dY/dX$. Because there are just two pairs of coordinates in a sample of two, the slope of the line passing through the two points is equal to the least-squares slope. You can verify this by using the least-squares formula $b = \Sigma(X - \bar{X})(Y - \bar{Y})/\Sigma(X - \bar{X})^2$ to calculate one of the slopes.

As you can see by examining the values of b for the ten different samples, the estimated value of β depends greatly on which sample is selected. The values of b are plotted in Figure 9.1 to assist you in seeing the dispersion in the estimates of β. Six of the sample slopes are centered fairly closely around $\beta = 2$. This is a good thing. However, two sample slopes are negative, indicating that earnings decrease as years of college increase! Also, two extremely large slopes (5 and 6) greatly overestimate the monetary value of a year of college.

The distribution of the sample bs shown in Table 9.2 and Figure 9.1 is the **sampling distribution of b.** The mean of the sampling distribution of b equals the sum of the slopes (Table 9.2) divided by the number of samples (m):

$$\mu_b = \frac{\Sigma b}{m} = \frac{20}{10} = 2$$

The value of μ_b (2) is the same as the value of the population slope. Thus, the mean of the ten sample slopes equals the population parameter β. To compute the standard deviation of the sampling distribution of b, we must compute the squared deviations of b from the mean of b $[(b - \mu_b)^2]$ (Table 9.2). The standard deviation of b equals

$$\sigma_b = \sqrt{\frac{\Sigma(b - \mu_b)^2}{m}} = \sqrt{\frac{52.5}{10}} = \sqrt{5.25} = 2.29$$

Thus, the standard amount that sample slopes deviate from the mean of the slopes (the population slope) is 2.29. Because deviations of sample statistics from population parameters are sampling errors, *the standard deviation of the sampling distribution of b is called* the **standard error of b.**

Table 9.2 Slopes (*b*) for All Unique Samples of *n* = 2

Sample	Cases	dY / dX	= b	b − μ_b	(b − μ_b)²
1	1, 2	(27 − 22) / (1 − 0) =	5.0	5.0 − 2.0 = 3.0	9.00
2	1, 3	(25 − 22) / (2 − 0) =	1.5	1.5 − 2.0 = −.5	.25
3	1, 4	(31 − 22) / (3 − 0) =	3.0	3.0 − 2.0 = 1.0	1.00
4	1, 5	(30 − 22) / (4 − 0) =	2.0	2.0 − 2.0 = .0	.00
5	2, 3	(25 − 27) / (2 − 1) =	−2.0	−2.0 − 2.0 = −4.0	16.00
6	2, 4	(31 − 27) / (3 − 1) =	2.0	2.0 − 2.0 = .0	.00
7	2, 5	(30 − 27) / (4 − 1) =	1.0	1.0 − 2.0 = −1.0	1.00
8	3, 4	(31 − 25) / (3 − 2) =	6.0	6.0 − 2.0 = 4.0	16.00
9	3, 5	(30 − 25) / (4 − 2) =	2.5	2.5 − 2.0 = .5	.25
10	4, 5	(30 − 31) / (4 − 3) =	−1.0	−1.0 − 2.0 = −3.0	9.00
Sum			20.0	0.0	52.50

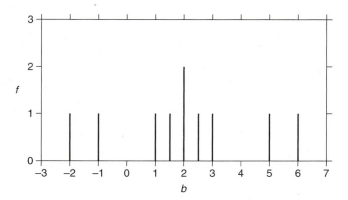

FIGURE 9.1 Sampling Distribution of *b* for *n* = 2

Now, what if a researcher selected a simple random sample of size *n* = 3 from the population in Table 9.1 to estimate the population slope *β*? Table 9.3 uses the least–squares formula to calculate the slope if the selected sample consists of the first three cases (*b* = 1.5).

There are also ten unique samples of size *n* = 3. Table 9.4 shows the value of the least–squares slope for each of these samples. This is the sampling distribution of *b* for *n* = 3. The mean of this sampling distribution is μ_b = 19.9999/10 = 2.00. This result indicates that the mean of the *b*s equals the population slope, just as was the case for *n* = 2.

How does the dispersion of the sampling distribution of *b* for *n* = 3 compare to that for *n* = 2? Comparing Figure 9.2 to Figure 9.1 shows clearly that there is much less dispersion of the *b*s for *n* = 3 than for *n* = 2. All of the *b*s for *n* = 3 fall between

Table 9.3 Calculation of Slope for Sample 1, $n = 3$

Case	X	Y	$X - \bar{X}$	$(X - \bar{X})^2$	$Y - \bar{Y}$	$(X - \bar{X})(Y - \bar{Y})$
1	0	22	−1	1	−2.6667	2.6667
2	1	27	0	0	2.3333	0
3	2	25	1	1	.3333	.3333
Sum	$\overline{3}$	74	$\overline{0}$	$\overline{2}$	$\overline{-.0001}$	$\overline{3.0000}$

$$b = \frac{\Sigma(X - \bar{X})(Y - \bar{Y})}{\Sigma(X - \bar{X})^2} = \frac{3}{2} = 1.5$$

Table 9.4 Slopes (b) for All Unique Samples of $n = 3$

Sample	Cases	b	$b - \mu_b$	$(b - \mu_b)^2$
1	1,2,3	1.5000	−.5000	.2500
2	1,2,4	2.8570	.8570	.7344
3	1,2,5	1.7692	−.2308	.0533
4	1,3,4	2.7857	.7857	.6173
5	1,3,5	2.0000	.0000	.0000
6	1,4,5	2.2308	.2308	.0533
7	2,3,4	2.0000	.0000	.0000
8	2,3,5	1.2143	−.7857	.6173
9	2,4,5	1.1429	−.8571	.7346
10	3,4,5	2.5000	.5000	.2500
Sum		19.9999	−.0001	3.3103

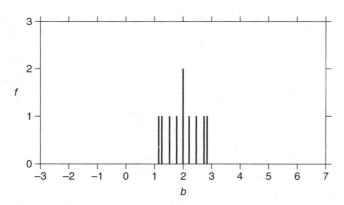

FIGURE 9.2 Sampling Distribution of b for $n = 3$

1 and 3, whereas for $n = 2$ they range from -2 to 6. The standard deviation of the sampling distribution for $n = 3$ is

$$\sigma_b = \sqrt{\frac{\Sigma(b - \mu_b)^2}{m}} = \sqrt{\frac{3.3103}{10}} = \sqrt{.33103} = .575$$

The standard error is .575, which is much smaller than the standard error for $n = 2$ ($\sigma_b = 2.29$).

Theoretical Sampling Distribution of *b*

The preceding illustrations of sampling distributions of *b* were based on an unrealistically small population ($N = 5$) so that we could easily construct and view a sampling distribution. The exercise was also unrealistic because it assumed that we knew the population values of X and Y. If we had this information, however, we would not need to take a sample.

Although we can't observe a realistic sampling distribution of the slope, statistical theory can tell us what the characteristics of the sampling distribution will be *under certain specified conditions.* This section will review certain characteristics of the theoretical sampling distribution of the slope, and the following section will discuss the conditions under which they hold.

Investigators often select a simple random sample of n cases and compute the least-squares slope (*b*) to serve as an estimate of β in the population linear model:

$$Y = \alpha + \beta X + \varepsilon$$

To evaluate how well *b* estimates β, the investigator must know the characteristics of the sampling distribution of *b*.

The discussion of the sampling distribution of *b* will parallel that of the sampling distribution of the mean given in Chapter 5. We saw that the sampling distribution of the mean can be described by formulas for its mean, standard deviation, and shape (Equations 5.3, 5.4, and 5.5, respectively). We also learned the specific conditions under which these equations are true. Statistical theory tells us very precisely what will be the mean, the standard deviation, and the shape of the sampling distribution of *b*, under certain conditions:

$$\mu_b = \beta \tag{9.1}$$

$$\sigma_b = \frac{\sigma_\varepsilon}{s_X \sqrt{n}} \tag{9.2}$$

$$b \sim N \tag{9.3}$$

Equation 9.1 says that the mean of the sampling distribution of *b* (i.e., μ_b) equals the slope of the regression line in the population (β). Although some random samples will produce *b*s that are too high and some will have *b*s that are too low, on the average the *b*s will be right on target—neither too high nor too low. This means that *the least-squares b is an* **unbiased estimate of the population slope β.** This characteristic of *b* is analogous to Equation 5.3, which indicated that the mean of the sampling distribution of $\overline{X}$ equals the population mean ($\mu_{\overline{X}} = \mu_X$). Thus, the mean of an unbiased sampling distribution equals the population parameter (μ_X or β) that we wish to estimate.

Equation 9.2 provides a formula for the standard deviation of b (i.e., σ_b), called the **standard error of b.**[1] This formula is analogous to Equation 5.4 for the standard deviation of $\overline{X}$ ($\sigma_{\overline{X}} = \sigma_X/\sqrt{n}$). Equation 9.2 points to three factors that determine the size of the standard error (i.e., the sampling error). First, σ_ε (the numerator) is the standard deviation of the random disturbance in the population. Because it is in the numerator, the greater this standard deviation, the greater the standard error. In other words, the greater the standard deviation of the disturbances, the greater the sampling error. The disturbance stands for other causes of Y that disrupt the linear relationship between X and Y. When these other factors are contributing a lot to Y, as indicated by a large standard deviation of ε, we will have more difficulty getting a precise sample estimate of the linear relationship of X and Y (β). This is illustrated in Figure 9.3, which shows scatterplots for two hypothetical populations. The solid lines are population regression lines, and the dashed lines represent sample regression lines that could occur by chance in some samples. The deviations of the sample lines from the population line are greater in Figure 9.3a than in Figure 9.3b. The deviations of Y from the population line (i.e., ε) are greater in Figure 9.3a than in Figure 9.3b, making it possible for larger deviations of the sample regression line (i.e., sampling errors) to occur in the former than in the latter. In other words, because the standard deviation of ε is greater in Figure 9.3a than it is in Figure 9.3b, the standard error of the slope will be larger in Figure 9.3a than it is in Figure 9.3b.

The second factor that determines the size of the standard error is the fact that the sample standard deviation of X is in the denominator of Equation 9.2.[2] This means that the greater the dispersion of X, the smaller will be the standard error. In other words, if X has a wide range of values with which to estimate β, the estimates of β will bounce around less from sample to sample than when we must use a narrow band of X values to make an estimate. This is illustrated in Figure 9.4, which shows a greater range and standard deviation of X in Figure 9.4a than in Figure 9.4b. Because of the restricted range of X in Figure 9.4b, it is possible for sample regression lines to deviate more from the population line than in Figure 9.4a. In sum, because X has a greater standard deviation in Figure 9.4a, the standard error of the slope will be less in Figure 9.4a.

The third influence on the size of the standard error is the fact that the square root of n is in the denominator of Equation 9.2. Thus, the larger the n, the smaller the standard error of b. Because it is the square root of n that is involved, however, there are diminishing returns to increases in n. To cut the standard error in half, for example, we must quadruple the size of n. This effect of n on the standard error of b is exactly the same as its effect on the standard error of the mean (Equation 5.4).

To summarize, Equation 9.2 shows that the standard error of the sampling distribution is positively related to the standard deviation of ε and inversely related to both the standard deviation of X and the sample size. Although the sample size and the standard deviation of X are known, the standard deviation of ε is not known because σ_ε is

---

1. Although our emphasis is on the slope, the parameters of the sampling distribution of the intercept a are $\mu_a = \alpha$ and $\sigma_a = \sigma_\varepsilon\sqrt{1/n + \overline{X}^2/\Sigma(X - \overline{X})^2}$.

2. Equation 9.2 is correct when X is a *fixed* variable—that is, when the values of X are under the control of the investigator so that the same values would be observed in each sample (i.e., s_X would be constant from sample to sample). In the social sciences, however, it is more often the case that *the investigator does not control* X. In this case, X is a random variable; its values are a random sample of those occurring naturally in the population from which the sample is drawn. When X is random, the standard error of b is $\sigma_b = \sigma_\varepsilon/(\sigma_X\sqrt{n - 2})$. Whether X is fixed or stochastic, however, the formula that will be used for estimating σ_b is the same.

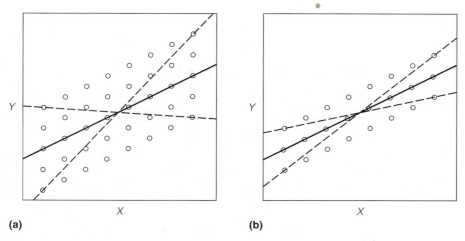

FIGURE 9.3 Influence of Standard Deviation of ε on Standard Error of Slope: Because σ_ε in (a) > σ_ε in (b), σ_b in (a) > σ_b in (b) (— population line; - - - sample line)

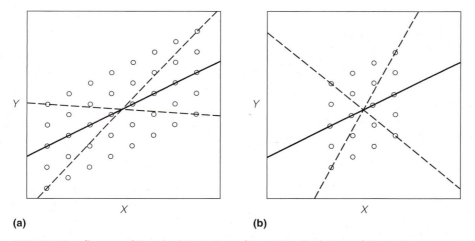

FIGURE 9.4 Influence of Standard Deviation of X on Standard Error of Slope: Because s_X in (a) > s_X in (b), σ_b in (a) < σ_b in (b) (— population line; - - - sample line)

a population parameter. Thus, although the identity indicated by Equation 9.2 is true (given certain conditions described later), we can't use the formula to compute the standard error of b directly. Equation 9.2 can, however, be modified slightly to allow us to estimate the standard error of b with a single sample of cases from the population.

The third characteristic of the sampling distribution of b (Equation 9.3) is the fact that b has a normal distribution. This was also seen to be true of the sampling distribution of $\overline{X}$ (Equation 5.5). Figure 9.5 shows some characteristics of the **normal distribution of b,** which is analogous to the normal distribution of $\overline{X}$ shown in Figure 5.5. β is shown at the center of the normal distribution of b because the mean of the

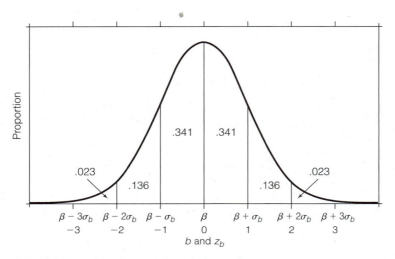

FIGURE 9.5 Normal Sampling Distribution of the Slope (*b*)

Table 9.5 Characteristics of the Sampling Distributions of the Mean and the Least-Squares Slope

Characteristic	$\bar{X}$	b
Mean	$\mu_{\bar{X}} = \mu_X$	$\mu_b = \beta$
Standard error	$\sigma_{\bar{X}} = \dfrac{\sigma_X}{\sqrt{n}}$	$\sigma_b = \dfrac{\sigma_\varepsilon}{s_X\sqrt{n}}$
Shape	$\bar{X} \sim N$	$b \sim N$

sample *b*s equals β. Also, about two-thirds (68.2%) of the sample *b*s will be within 1 standard deviation (standard error) of β (i.e., $\beta \pm \sigma_b$), and about 95 percent (95.4% to be exact) will be within 2 standard deviations of β (i.e., $\beta \pm 2\sigma_b$). The *Z* scores shown under the horizontal axis are equal to $Z_b = (b - \beta)/\sigma_b$. This is a standardized *b* because β, which is the mean of the *b*s, is subtracted from *b* and the result is divided by the standard deviation of *b*. Because *b* is normally distributed, Z_b is also normally distributed.

A comparison of the characteristics of the sampling distributions of the mean and the least-squares slope is shown in Table 9.5. Both statistics are unbiased and normally distributed. Also, the standard errors of both are inversely related to the square root of the sample size.

Gauss–Markov Theorem

We have examined the characteristics of the least-squares slope. The **Gauss–Markov theorem** provides the justification for using the least-squares slope. It states that the least-squares *b* provides the *best linear unbiased estimate* (BLUE) of the population slope. *Best* means that the least-squares slope will be a more efficient estimator of β than any other unbiased linear estimator of β. The *efficiency* of a sample statistic (i.e., an estimator) refers to the size of its standard error; the smaller the standard error, the greater the efficiency. More efficient estimators are those whose values vary less from sample to sample. The fact that the least-squares estimator is more efficient than any other unbiased method of estimating β is the major justification for choosing the line that minimizes the sum of squared errors of prediction (the squared residuals). Remember that in Chapter 8 we compared the fit of the least-squares line with that of an eyeball line. Both lines fit the example data equally well in terms of the sum of the absolute values of the residuals. It was claimed, however, that the least-squares line was superior because it is less susceptible to sampling errors. We now know that this claim is based on the Gauss-Markov theorem, which states that it minimizes the standard deviation of the sampling distribution—that is, the standard error. Thus, the least-squares b will have a smaller standard error than an estimator of β that minimizes the absolute values of the residuals.

Assumptions Underlying Least-Squares Regression

The characteristics of the sampling distribution of b that have been described depend on the existence of several statistical conditions. These conditions are called **least-squares regression assumptions.** We must assume that certain statistical conditions exist in order for the specified characteristics of the sampling distribution of b to be accurate. The conditions consist of six assumptions:

$$\text{SRS} \tag{A9.1}$$

$$n/N \approx 0 \tag{A9.2}$$

$$\varepsilon \text{ is uncorrelated with } X \tag{A9.3}$$

$$\varepsilon \text{ has the same variance for all values of } X \tag{A9.4}$$

$$\varepsilon\text{'s values are independent of one another} \tag{A9.5}$$

$$\varepsilon \sim N \tag{A9.6}$$

The first two assumptions are identical to those pertaining to the sampling distribution of the mean described in Chapter 5. The assumption of simple random sampling (A9.1) means that each case has an equal probability of selection. Random sampling with equal probability of selection is necessary for b to be an unbiased estimator of β (Equation 9.1). The second assumption (A9.2) states that the sampling fraction (n/N) is approximately equal to 0. This assumption is necessary for the formula for the standard error of b (Equation 9.2) to be correct. If the sampling fraction is not approximately 0, Equation 9.2 must be multiplied by the finite population correction (Chapter 5) to get the correct value of the standard error:

$$\sigma_b = \sqrt{\frac{N-n}{N-1}} \frac{\sigma_\varepsilon}{s_X\sqrt{n}} \tag{9.4}$$

The last four assumptions are about the error term ε in the population equation $Y = \alpha + \beta X + \varepsilon$. We will describe each of these assumptions so that you will understand the circumstances under which least-squares regression analysis is most valid.

ε Is Uncorrelated with X

To ensure that b is an unbiased estimator, we must also assume that ε is uncorrelated with X. This assumption is necessary in nonexperimental research. If X and ε were correlated, that would mean that ε was not entirely random, because random variance cannot correlate with anything. Thus, ε would represent one or more unmeasured variables that are causes of Y that are not included in the equation and that are correlated with the measured variable X. In that case, we should be using multiple regression (Chapter 10) with two or more Xs in the equation instead of bivariate regression. Because we have failed to include some other causes of Y in the equation, we are unable to take them into account when estimating the relationship between the measured X and Y. Thus, because omitted causes of Y are correlated with X, b is a biased estimate of β. That is, β is not the mean of the sampling distribution of b. This means that our sample b is more likely to be too high (in some cases) or too low (in other cases) than right on target. This is a potentially serious problem. However, the seriousness of the bias depends on how strongly X is correlated with ε. The stronger the correlation, the greater the bias. Conversely, the weaker the correlation, the less the bias. A little bias is not necessarily a serious problem.

ε Has the Same Variance for All Values of X

To justify Equation 9.2 for the standard error of b, we must assume that ε has the same variance for all values of X. The variance of any variable such as ε under certain specified conditions (e.g., $X = 1$) is called a *conditional variance*. Thus, the conditional variance of ε is assumed to be constant across all values of X. If ε has a constant variance for all values of X, it is homoskedastic (Figure 9.6a). **Homoskedasticity** means that the predicted values of Y are equally good (or poor) at all levels of X; that is, the average squared error of prediction does not vary as X changes. If the variance of ε is greater for some values of X than for others, **heteroskedasticity** exists. For example, heteroskedasticity would exist if the equation made poorer predictions of Y (large variance of ε) for high values of X than for low values of X (Figure 9.6b); in this case there is a positive correlation between X and the variance of ε. The greater is X, the greater the variance of ε. Or the variance of ε might be greater for low levels of X than for high levels—that is, a negative correlation between X and the variance of ε. Other patterns of heteroskedasticity also are possible.

Notice that in Figure 9.6 the slope of the line is the same for the heteroskedastic pattern of ε as it is for the homoskedastic pattern. In both plots $Y = 1 + 1X + \varepsilon$. This is intended to illustrate the point that even if heteroskedasticity exists, the least-squares b will still be an unbiased estimate of β. That is, the mean of the sampling distribution of b will still equal β even if the homoskedasticity assumption is violated.

One problem created by heteroskedasticity is that Equation 9.2 no longer correctly describes the standard error of b. The true standard error of b may be either larger or smaller than the magnitude indicated by Equation 9.2; or, looked at from the opposite direction, Equation 9.2 will either overstate or understate the true standard deviation of the sampling distribution. If, for example, there is a positive correlation between X and the variance of ε, the value of the standard error as given by Equation 9.2 will be too small; that is, the sample estimator of β will not be as efficient as Equation 9.2 would indicate. On the other hand, if there is a negative correlation between X and ε, the vari-

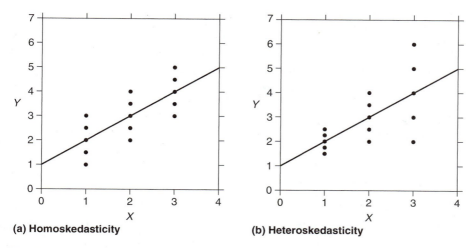

FIGURE 9.6 Illustrative Patterns of the Variance of ε

ance of the standard error given by Equation 9.2 will be too large. In sum, Equation 9.2 is a biased formula for the standard error of the least-squares b when heteroskedasticity is present. The direction of the bias depends on the direction of the correlation between X and ε.

Another problem also results from a violation of the homoskedasticity assumption. When heteroskedasticity is present, the Gauss–Markov theorem is no longer valid. That is, the least-squares estimate of b is no longer BLUE. There may be other linear unbiased methods of estimating b that provide more efficient estimates—that is, that have smaller standard errors. Even if we correct Equation 9.2 for the standard error, other estimation techniques, such as weighted least-squares (McClendon, 1994), provide more efficient estimates than ordinary least-squares. To summarize the difficulties created by heteroskedasticity, the usual formula for the least-squares standard error is wrong, and the least-squares slope is no longer the most efficient (i.e., best) unbiased estimate of β.

ε's Values Are Independent of One Another

To justify Equation 9.2 for the standard error of b, we must assume that the values of ε are independent of one another. When the independence assumption is not true, **autocorrelation** exists. Autocorrelation occurs when there is a systematic pattern in which certain pairs of cases have similar values of ε (positive autocorrelation) or dissimilar values of ε (negative autocorrelation). Temporal, physical, and social proximity are common factors that cause autocorrelation. Longitudinal research, in which repeated measures of the same variable are taken over time, is an example of temporal proximity that causes autocorrelation. Observations taken adjacent to one another in time are likely to have similar εs because the various unmeasured causes of Y that are summarized by ε tend to change slowly from one point in time to the next. Cluster sampling is an example of physical proximity of cases that also results in autocorrelation. People living in the same cluster, such as a neighborhood or block, are likely to have similar values of ε because of the existence of residential segregation based on socioeconomic status, race, and ethnicity. Pairs of persons in kin relationships or social relationships, such as married couples, are examples of social proximity that lead to autocorrelation. Many sources of similarity other than X between husbands and wives cause ε to be correlated for couples. The key factor producing autocorrelation is nonindependence between some observations as a result of some type of proximity.

The consequences of violating the independence assumption are the same as for violation of the homoskedasticity assumption. When autocorrelation is present, Equation 9.2 no longer correctly describes the standard error of b. The true standard error of b may either be larger or smaller than the magnitude indicated by Equation 9.2, depending on the nature of the autocorrelation. Also, the least-squares slope will not be the most efficient linear unbiased estimator.

ε Is Normally Distributed

The third characteristic of the sampling distribution (Equation 9.3) is that b is normally distributed. It is based on the assumption that ε is normally distributed. This assumption is analogous to the assumption that X is normally distributed that we made about the sampling distribution of the mean in Chapter 5. If ε is not normally distributed, b will not, in general, be normally distributed. However, b will still be the most efficient linear unbiased estimator of β (i.e., BLUE) if the assumptions of simple random sampling and homoskedasticity are true. Equation 4.3 for the standard error of b will also remain valid. Moreover, if ε is not normally distributed, the *central limit theorem* (Chapter 5) assures us that the sampling distribution of b will become increasingly similar to a normal distribution as sample size increases, as long as the other assumptions are true. That is, the sampling distribution of b will be approximately normal for large samples when ε is not normally distributed. Thus, when we are working with the relatively large samples that are often characteristic of social science research (where there may be several hundred cases or more), the sampling distribution of b will be approximately normal, regardless of the shape of the distribution of ε.

Violations of Assumptions

Methods for detecting violations of assumptions and for making corrections when violations are found are beyond the scope of this book (see McClendon, 1994). Assumptions, however, are never completely true, and the greater the violation, the more serious will be the consequent biases. Thus, violations and their consequences are a matter of degree. Furthermore, least-squares regression is robust. The **robustness of b** means that it can tolerate substantial violations of some assumptions without seriously biasing the analyses. In particular, the assumptions of homoskedasticity and normalcy are quite robust. We described how the assumption of a normally distributed ε is robust when the sample size is fairly large. The usual formula for the standard error of b is also robust in the face of sizeable heteroskedasticity. The fact that least-squares regression is robust, however, does not mean that we should take these assumptions lightly. It simply means that we should not be overly timid about using the method when we believe that our data do not exactly conform to the assumptions.

CONFIDENCE INTERVALS FOR THE SLOPE

We have examined the mean, standard deviation, and shape of the sampling distribution of the least-squares slope. These characteristics are important because they allow us to use a single sample slope to make inferences about the population slope. There are two types of statistics that are used to make inferences: confidence intervals and hypothesis tests. Confidence intervals and hypothesis tests for the mean were described in Chapter 6. The methods for constructing confidence intervals and hypothesis tests for the regression slope are nearly identical to those for the mean. We will describe confidence intervals first.

Theoretical Confidence Intervals for β

Because we usually compute a sample slope in order to estimate an unknown population slope, we would like to construct a confidence interval around the sample slope that has a certain likelihood (e.g., $P = .95$) of including the population slope. Because the sampling distribution of b is normal, we can determine from Table A.1 that the slope we have calculated has a 95-percent probability of being within 1.96 standard errors of the population slope. The reverse is also true. There is a 95-percent probability that the population slope is within 1.96 standard errors of our observed sample slope. Therefore, the 95-percent confidence interval (CI) for the population slope equals

$$95\% \text{ CI}_\beta = b \pm 1.96\sigma_b$$

This means that there is a 95-percent probability that the population slope is within plus or minus 1.96 standard errors of the sample slope.

The 95-percent confidence interval is illustrated in Figure 9.1. Although we know that there is a 95-percent probability that the sample slope is within 1.96 standard errors of the population slope, we do not know whether the sample slope is above, equal to, or below the population slope. Figure 9.7 shows two sampling distributions, A and B, whose means form the lower and upper extremities of the 95-percent confidence interval. The mean of sampling distribution A (β_A) is 1.96 standard errors below the sample slope, and the mean of sampling distribution B (β_B) is 1.96 standard errors above the sample slope. We know that there is a 95-percent probability that our sample slope did not come from a sampling distribution that is further away than the sampling distributions A and B. There is a probability of .025 that the sample slope came from a sampling distribution to the left of A and a probability of .025 that it came from a distribution to the right of B. These probabilities, however, are small. Therefore, we can be very confident that the population slope β is within this interval.

In general, confidence intervals for the slope equal

$$100P\% \text{ CI}_\beta = b \pm Z_{\alpha/2}\sigma_b \tag{9.5}$$

where $\alpha = 1 - P$. P is the confidence level or probability that the population slope β lies within the confidence interval and α is the probability that β is outside of the interval. That is, α is the probability that the interval is incorrect. If you want a probability of P that the slope is in a certain interval, which is the $100P$ percent confidence

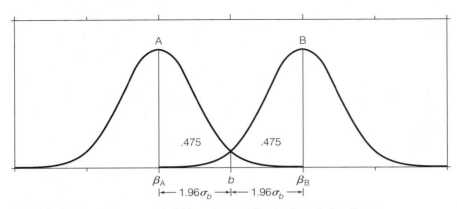

FIGURE 9.7 The 95-Percent Confidence Interval for the Population Slope

Table 9.6 Example Confidence Intervals

100P% CI	$\alpha/2$	$Z_{\alpha/2}$	$b \pm Z_{\alpha/2}\sigma_b$
99	.005	2.575	$b \pm 2.575\sigma_b$
95	.025	1.960	$b \pm 1.960\sigma_b$
90	.050	1.645	$b \pm 1.645\sigma_b$
75	.125	1.150	$b \pm 1.150\sigma_b$

interval, then you would calculate $\alpha/2$, locate A $= \alpha/2$ in Table A.1, find the associated $Z_A = Z_{\alpha/2}$ in the table, and enter $Z_{\alpha/2}$ in Equation 9.5. Examples of the Z scores needed for four confidence intervals are shown in Table 9.6. Notice that the greater the desired confidence (i.e., P) that the interval contains β, the larger will be $Z_{\alpha/2}$, and thus the wider will be the confidence interval.

Remember that we locate these key values of $Z_{\alpha/2}$ in the standard normal table because the sample slope has a normally distributed sampling distribution, and thus

$$z_b = \frac{b - \beta}{\sigma_b} \sim \text{N} \tag{9.6}$$

This equation indicates that the standardized sample slope, which has a mean of 0 and a standard deviation of 1, is normally distributed. It is normally distributed because subtracting the population slope and dividing by the standard error of the slope (both constants) gives a linear transformation of the sample slope. A linear transformation, while changing the mean and standard deviation, does not change the *shape* of the distribution. Therefore, it is appropriate to use a standard normal (Z) table to find the Z score associated with $\alpha/2$.

Calculated Confidence Intervals for β

In a research project we want to use Equation 9.5 to calculate the 95-percent confidence interval. Although we know the sample slope b, we seldom if ever know the standard error of the slope (σ_b). Therefore, although Equation 9.5 is true, we do not have enough information to use this equation to calculate the 95-percent confidence interval. Despite the fact that we do not *know* the true value of σ_b, however, we can *estimate* it with our sample observations. This will allow us to use a formula that is closely related to Equation 9.5 to calculate the desired confidence interval.

Estimated Standard Error of b

According to Equation 9.2, the standard error of the slope equals

$$\sigma_b = \frac{\sigma_\varepsilon}{s_X \sqrt{n}}$$

We know n and s_X, but we need to estimate σ_ε in order to estimate the standard error of b.

The *standard error of estimate* (Equation 8.20) will be used to estimate σ_ε:

$$\hat{s}_{Y-\hat{Y}} = \sqrt{\frac{\Sigma(Y - \hat{Y})^2}{n - 2}}$$

The standard error of estimate is itself an estimate of the standard deviation of the residuals in the population. The sample residuals $Y - \hat{Y}$ provide an estimate of the population disturbances ε. Therefore, $\hat{s}_{Y-\hat{Y}}$ provides an estimate of σ_ε. Substituting $\hat{s}_{Y-\hat{Y}}$ for σ_ε in Equation 9.2 gives the **estimated standard error of b:**

$$s_b = \frac{\hat{s}_{Y-\hat{Y}}}{s_X \sqrt{n}} \tag{9.7}$$

Equation 9.7 indicates that the standard error of the slope can be estimated from our sample data by simply dividing the standard error of estimate by the sample standard deviation of X and the square root of n. We use the symbol s_b because the lowercase italic Roman letter s indicates that it is computed from our sample observations and thus is not a population parameter. Consequently, we aren't talking theoretically anymore because Equation 9.7 is a statistic to which we can actually assign a number.

To illustrate the estimation of the standard error of the slope, we turn to the problem raised in the vignette at the beginning of the chapter and in the illustrative sampling distribution of the slope—namely, the financial worth of a college education. We will use respondents to the 1998 GSS who were 25 to 65 years of age, worked 35 or more hours per week, and had worked 48 or more weeks during the previous year. We use these restrictions so we can focus on full-time workers in the prime years of their careers, allowing us to better examine the optimum rate of return on the investment in education. To focus on the benefits of an undergraduate education, we will include only people who completed 12 to 16 years of education. We will assume that those with 12 years of education finished high school but did not complete any years of college, those with 13 to 15 years of education completed 1 to 3 years of college, and those with 16

Table 9.7 Coding of College

Education	College
12	0
13	1
14	2
15	3
16	4

years of education completed college. The independent variable (X) is named College and is coded as shown in Table 9.7. This variable will enable us to focus not only on the returns of completing college but also on the worth of a partial college education. The dependent variable (Y) is annual earnings from wages, salaries, and self-employment. The scatterplot for earnings and college shown in Figure 9.8 omits three people with earnings of $150,000 to $200,000 to spread out the remaining data points more in the plot.

Table 9.8 shows the means, standard deviations, and various regression statistics for earnings and college. Using the intercept and slope from the table, the best-fitting regression line is

$$\hat{Y} = 29315.40 + 4399.12X$$

This linear model predicts that high school graduates with no college will earn $29,315.40 on the average. The slope indicates that each year of college completed is worth $4,399.12.[3] A person who completes four years of college is expected to earn $4 \cdot 4399.12 = \$17,596.48$ more per year than a high school graduate. If the young woman named Carmen in this chapter's vignette had seen this figure, it might have resolved her concerns about the worth of a college education. On the other hand, these figures are based on a sample of 786 persons. Consequently, the estimated slope may be too high or too low. What Carmen wants to know is the minimum amount that a college education is likely to be worth. In statistical terms, what she needs is a confidence interval for the slope that has a very high probability (e.g., 95%) of containing the population slope.

To construct such a confidence interval, we need to estimate the standard error of the slope. To find the standard error of b we need the standard error of estimate (Table 9.8), which is calculated from the residual sum-of-squares as follows:

$$\hat{s}_{Y-\hat{Y}} = \sqrt{\frac{\Sigma(Y - \hat{Y})^2}{n - 2}} = \sqrt{\frac{400369319520.709}{786 - 2}} = 22598.123$$

This value means that the standard amount by which predicted earnings deviate from observed earnings is $22,598. The size of the standard error of estimate suggests that there is considerable variance in earnings among workers who have the same amount of education. This can be seen in Figure 9.8 by looking at the dispersion of earnings around the regression line for each year of college.

---

3. This linear model makes the perhaps unrealistic assumption that completing the fourth year of college results in the same additional earnings as completing the first, second, and third years of college.

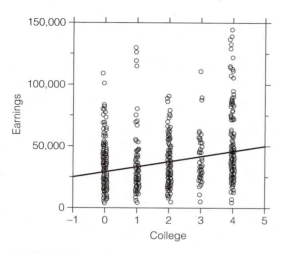

FIGURE 9.8 Regression of Earnings on Years of College for Persons Aged 25–65 Who Worked 35+ Hours per Week and 48+ Weeks per Year, GSS 1998

Table 9.8 Regression of Earnings on Years of College for Persons Aged 25–65 Who Worked 35+ Hours per Week and 48+ Weeks per Year, GSS 1998

Variables	Mean	St. Dev.	n		
Earnings	36647.263	23664.047	786		
College	1.667	1.606	786		
Variables	**b**	**SE**	**B**	**t**	**p**
College	4399.121	501.974	.299	8.764	.000
Intercept	29315.395	1161.745		25.234	.000
Source	**SS**				
Regression	39220584551.745				
Residual	400369319520.709				
Total	439589904072.454				
	R	**R-Squared**	**SE of Est.**		
Goodness-of-fit	.299	.089	22598.123		

The standard error of estimate is inserted in Equation 9.7 along with the standard deviation of X and the sample size to compute the standard error of the slope:

$$s_b = \frac{\hat{s}_{Y-\hat{Y}}}{s_X\sqrt{n}} = \frac{22598.123}{1.606\sqrt{786}} = 501.898$$

The slight discrepancy between the preceding value and the value in Table 9.8 is due to rounding error when computing the standard deviation of college. The value of the estimated standard error of b means that the standard deviation of the sampling distribution of b is about \$502. That is, the worth of a year of college education is estimated to deviate by \$502 from sample to sample. It is important to remember that we are estimating the standard deviation of the sampling distribution of b with a single sample of persons.

t-Score Confidence Intervals for β

Although we now have an estimate of the standard error, we need one more statistic before we can compute a confidence interval for b. The needed statistic is suggested by $Z_{\alpha/2}$ in Equation 9.5, namely, the theoretical formula for the confidence interval of β:

$$100P\% \text{ CI}_\beta = b \pm Z_{\alpha/2}\sigma_b$$

To review, P is the confidence level, $\alpha = 1 - P$ is the probability that β is not in the confidence interval, and $Z_{\alpha/2}$ is the number of standard errors above and below the mean of a normal sampling distribution that contains $100P$ percent of the sample slopes. To find $Z_{\alpha/2}$ we select P, calculate α, and use the standard normal distribution (Table A.1) to find $Z_{\alpha/2}$ associated with α.

As was the case with the sampling distribution of the mean, however, we cannot use the standard normal distribution when we don't know the true value of the standard error of the slope (σ_b). Because we are using s_b to estimate σ_b, we have to use another distribution, the t distribution. The **t statistic for b** is

$$t = \frac{b - \beta}{s_b} \tag{9.8}$$

Because t is an estimate of the number of standard errors by which b deviates from the mean of the sampling distribution (β), the t distribution can be used to determine the number of standard errors ($t_{\alpha/2}$) above and below the mean that contain $100P$ percent of the slopes. You will remember from Chapter 6 that the t distribution is more dispersed than the normal distribution but, as sample size increases, the t distribution approaches the normal distribution.

Substituting $t_{\alpha/2}$ for $Z_{\alpha/2}$ and s_b for σ_b in Equation 9.5 gives the formula for the **calculated confidence interval for β**:

$$100P\% \text{ CI}_\beta = b \pm t_{\alpha/2}s_b \tag{9.9}$$

There is a different t distribution for each sample size. The degrees of freedom for each t distribution are given by

$$df = n - 2 \tag{9.10}$$

The **degrees of freedom of t** for the slope are 1 less than the degrees of freedom for the mean ($df = n - 1$) because two parameters (α and β) are estimated to calculate the standard error of the slope, compared with only one (μ) to calculate the standard error of the mean.

With the degrees of freedom, we can enter Table A.2 to find t for the specified df and $\alpha/2$. We go to the column for A $= \alpha/2$ and go down to the row for $df = n - 2$ to find $t_A = t_{\alpha/2}$. Examples of $t_{\alpha/2}$, for 99- and 95-percent confidence intervals with

Table 9.9 Example Confidence Intervals

100P% CI	df	$t_{\alpha/2}$	$b \pm t_{\alpha/2}s_b$
99	20	2.845	$b \pm 2.845s_b$
99	100	2.626	$b \pm 2.626s_b$
95	20	2.086	$b \pm 2.086s_b$
95	100	1.984	$b \pm 1.984s_b$

degrees of freedom of 20 and 100 for each confidence interval are given in Table 9.9. Notice that for both degrees of freedom $t_{\alpha/2}$ is larger for the 99-percent interval than for the 95-percent interval. Thus, if you want greater confidence that the interval contains β, the interval will have to be wider. Also notice that as the degrees of freedom increase (as n increases), $t_{\alpha/2}$ becomes smaller and thus the confidence interval becomes smaller.

To illustrate, we return to the regression of annual earnings of full-time workers on the number of years of college completed. Let's construct a 95-percent confidence interval for the slope. Table 9.8 shows that the slope is $b = 4399.121$, the standard error of the slope is $s_b = 501.974$, and the sample size is $n = 786$. To find $t_{\alpha/2}$ in Table A.2, we need to determine $\alpha/2$ and the degrees of freedom:

$$\alpha/2 = (1 - P)/2 = (1 - .95)/2 = .025$$

$$df = n - 2 = 786 - 2 = 784$$

Now we go down column A $= .025$ until we reach the row for $df = 784$. Because 784 is not one of the selected values of df given in the table, we look for the two given degrees of freedom between which 784 falls. These values are 750 and 1,000. The values of t for these df are 1.963 and 1.962, respectively. Because our degrees of freedom are much closer to 750 than to 1,000, we choose $t_{.025} = 1.963$.

We are now ready to calculate the confidence interval for b. We enter $b = 4399.121$, $t_{.025} = 1.963$, and $s_b = 501.974$ into Equation 9.9:

$$95\% \ \text{CI}_\beta = b \pm t_{.025}s_b$$

$$= 4399.121 \pm (1.963)(501.974)$$

$$= 4399.121 \pm 985.375$$

$$= 3413.75\text{--}5384.50$$

This means that we are 95-percent confident that the slope for earnings regressed on college is no less than 3,413.75 and no more than 5,384.50. In dollars and cents, the true earnings return for each year of college is between \$3,413.75 and \$5,384.50. If we multiply both the lower bound and the upper bound by 4, the financial return from four years of college is between \$13,655 and \$21,538. If Carmen of this chapter's opening vignette saw this range, would it resolve her concerns about going to college? Remember that she estimated she would have to earn \$4,000/year more as a result of getting a four-year college education to pay off the costs of going to college. Our confidence interval shows that the *minimum* annual advantage she can expect to earn due to four years of college ($13,655) is more than triple what she would need to cover her college expenses. Thus, for financial reasons alone, college is worth the cost to Carmen.

HYPOTHESIS TESTS FOR THE SLOPE

Hypothesis testing is the second type of procedure for making inferences about the slope. The logic and procedures for conducting hypothesis tests for the mean were described in Chapter 6. You may want to review two-tailed and one-tailed tests for the mean before proceeding with this section. The logic and procedures for conducting hypothesis tests of the slope are identical to those for tests of the mean. The eight-step t test for the slope, which may be two-tailed or one-tailed, is outlined in Box 9.1.

The steps are the same as those for the mean except, of course, that we calculate a slope, the standard errors of the slope, and a t score for the slope rather than calculate these statistics for the mean. We will conduct two-tailed and one-tailed tests of b to illustrate the procedure.

Two-Tailed Hypothesis Test for β

We have found that that each year of college completed is associated with higher earnings. Those who complete college are not just paid more for performing the same job as those who did not attend college, although this is sometimes the case. College graduates are usually paid more because they work at different jobs from those who did not attend college. We will conduct a test to see if those who attend college are more satisfied with the jobs that they hold.

Respondents to the GSS who were employed were asked how satisfied they were with their jobs. They were asked to choose one of seven response alternatives shown in Table 9.10 to describe their job satisfaction. The response alternatives are scored from 1 to 7. The higher the score, the greater the job satisfaction. The regression analysis of job satisfaction and college will be conducted on the same types of respondents who were used in our analysis of earnings and college—namely, persons aged 25 to 65 who worked 35 or more hours per week and 48 or more weeks during the previous year (that is, full-time workers). The results of regressing job satisfaction on years of college are shown in Table 9.11. The sample size is only 329 because only about half of the full-time workers were asked the question about job satisfaction.

Using the slope and intercept from Table 9.11, we can write the regression equation as

$$\widehat{\text{JOBSAT}} = 5.2035 + .0967\text{COLLEGE}$$

The intercept (5.203) is the level of job satisfaction on a seven-point scale for those with 0 years of college. The slope indicates that there is a .0967 point increase in job satisfaction for each year of college completed. If we multiply the slope times 4, we get the difference between those who completed 4 years of college and those who completed 0 years, which equals $.0967 \times 4 = .3868$. This is about .4 of a point advantage in job satisfaction, which does not look like a very strong increase in satisfaction. The stan-

---

BOX 9.1   *t* Test for Slope

1. State null and alternative hypotheses.
2. Specify α.
3. Determine degrees of freedom.
4. Find $t_{\alpha/2}$ (two-tailed) or t_{α} (one-tailed).
5. Calculate sample slope.
6. Calculate standard error of slope.
7. Calculate the test statistic t.
8. Reach decision and form conclusion.

Table 9.10 Job Satisfaction Scores

Job Satisfaction	Score
Completely dissatisfied	1
Very dissatisfied	2
Fairly dissatisfied	3
Neither	4
Fairly satisfied	5
Very satisfied	6
Completely satisfied	7

Table 9.11 Regression of Job Satisfaction on Years of College for Persons Aged 25–65 Who Worked 35+ Hours per Week and 48+ Weeks per Year, GSS 1998

Variables	Mean	St. Dev.	*n*		
Job Satisfaction	5.3647	1.1541	329		
College	1.6687	1.6326	329		
Variables	***b***	**SE**	***B***	***t***	***p***
College	.0967	.0387	.137	2.496	.013
Intercept	5.2035	.0904		25.234	.000
Source	**SS**				
Regression	8.192				
Residual	430.039				
Total	438.231				
	R	***R*-Squared**	**SE of Est.**		
Goodness-of-fit	.137	.019	1.147		

**Table 9.12 t Test for the Slope
of Job Satisfaction Regressed
on Years of College**

1. $H_0: \beta = 0$
   $H_1: \beta \neq 0$
2. $\alpha = .05$
3. $df = n - 2 = 329 - 2 = 327$
4. $t_{\alpha/2, df} = t_{.025, 327} = 1.967$
5. $b = .0967$
6. $s_b = \dfrac{\hat{s}_{Y-\hat{y}}}{s_x \sqrt{n}} = \dfrac{1.1468}{1.6326 \sqrt{329}} = .0387$
7. $t = \dfrac{b - \beta_0}{s_b} = \dfrac{.0967 - 0}{.0387} = 2.499$
8. $(|t| = 2.499) > (t_{.025} = 1.967)$, $\therefore$ reject H_0
   $(t = 2.499) > (t_{.025} = 1.967)$, $\therefore \beta > 0$
   $p = .013$

dardized slope ($B = .1367$) confirms this. For a standard deviation increase in college, there is only a .1367 standard deviation increase in job satisfaction. The standardized slope for satisfaction regressed on college is less than half as great as the standardized slope for earnings regressed on college ($B = .299$, Table 9.8). Thus, college pays off better financially than it does in terms of psychological job satisfaction.

Perhaps the relatively weak association between job satisfaction and college may have occurred by chance, when in reality there is actually no association at all. We turn to the t test for the slope to evaluate the likelihood of this possibility. The eight-step **two-tailed t test for β** is shown in Table 9.12.

Step 1 shows that we are conducting a two-tailed test. The null hypothesis states that the slope is 0, and the alternative hypothesis specifies that it is not equal to 0. Thus, we want to have the possibility of rejecting the null hypothesis if the slope is positive or negative. Remember that these hypotheses are to be specified *prior* to computing the sample slope. The two-tailed test of $\beta = 0$ is the most common form of the hypothesis test of the slope. This is a test of whether the independent variable has any association, positive or negative, with the dependent variable.

Steps 2 and 3 specify that $\alpha = .05$ (the maximum probability of committing a Type I error that we are willing to risk) and the degrees of freedom. Because we want to put $\alpha/2 = .025$ in each tail of the t distribution, in step 4 we look up the critical $t_{.025, 327} = 1.967$ in Table A2.

Step 5 shows that the computed slope equals .0967. In step 6 we compute the standard error of the slope from the standard error of estimate, the standard deviation of college, and the sample size, all of which are found in Table 9.11. In step 7 we enter the computed slope, the estimated standard error of the slope, and the value of β specified by the null hypothesis into the formula for the calculation of t. The calculated $t = 2.499$ indicates that if the null hypothesis is true, our sample slope is about 2.5 standard errors above the mean of the sampling distribution.

In step 8 we reject the null hypothesis because the absolute value of the computed t is greater than the critical t. And, because the computed t is positive, we conclude that β is greater than 0. The level of significance associated with the computed t is $p = .013$. Thus, the probability of rejecting a true null hypothesis (a Type I error) is only .013. Our conclusion tells us that each year of college completed increases job satisfaction. Our standardized slope, however, indicates that the increase in satisfaction from 1 year of college education is not particularly strong.

One-Tailed Hypothesis Test for β

If we are interested in only one of the two possible directions of the slope (i.e., positive or negative), we can conduct a one-tailed test of whether the sign of the slope is in the hypothesized direction. The one-tailed hypotheses may take one of the forms shown in Table 9.13. If you are interested in a positive slope, you would use form (a), and if you think the slope is negative, you would use form (b).

To illustrate, we will return to the regression of earnings on college for full-time workers in Table 9.8. We used these results to construct a confidence interval for the population slope. At the beginning of the chapter, Carmen estimated that to cover the expenses of a college education she would have to make $4,000 more per year as a result of graduating from college than she would have made if she had not gone to college at all. This translates to a slope for earnings regressed on college of $\beta = 4000/4 = 1000$. Because she is interested only in whether she will earn more than $1,000 per year of college, the null hypothesis of our test is that β is less than or equal to 1,000, and the alternate hypothesis is that β is greater than 1,000. These hypotheses are specified in step 1 of Table 9.14, which shows the **one-tailed t test for β.**

Step 2 indicates that we will use the conventional $\alpha = .05$. Because the alternative hypothesis is that the slope is greater than 1,000, we want to put the entire α in the upper tail of the t distribution. Consequently, we look up the critical t in Table A.2 for A = .05 and 784 degrees of freedom. Step 4 indicates that this critical value is $t_{.05,784} = 1.647$. This value is much smaller than the critical t (1.963) we would have needed for a two-tailed test.

Steps 5 through 7 show the calculated slope, standard error of the slope, and t score for the slope under the null hypothesis. The standard error of estimate, standard deviation of college, and sample size used in step 6 are found in Table 9.8. Notice that in step 7, we subtract 1,000 from the sample slope when calculating t because we are testing whether the slope is greater than 1,000 (1,000 is the maximum value specified by the null hypothesis). The calculated t equals 6.773.

Finally, in step 8 we reject the null hypothesis because the calculated t (6.773) is much greater than the critical t (1.647). Thus, we conclude that the earnings increase per year of college is greater than $1,000. The level of significance is less than .00005 because the calculated t is greater than the t for A = .00005 in Table A.2 for $df = 750$. Thus, there is only an extremely small probability that we are making a Type I error in rejecting the null hypothesis. This means that Carmen can be extremely confident that her increased earnings from completing four years of college will more than cover her college expenses. Again, we conclude, just as we did with the constructed confidence interval, that college is worth it.

Table 9.13 Hypothesis Forms of One-Tailed Tests

(a)	(b)
$H_0: \beta \leq a$	$H_0: \beta \geq a$
$H_1: \beta > a$	$H_1: \beta < a$

Table 9.14 One-Tailed t Test for the Slope of Earnings Regressed on Years of College

1. $H_0: \beta \leq 1{,}000$
   $H_1: \beta > 1{,}000$
2. $\alpha = .05$
3. $df = n - 2 = 786 - 2 = 784$
4. $t_{\alpha, df} = t_{.05, 784} = 1.647$
5. $b = 4399.12$
6. $s_b = \dfrac{\hat{s}_{Y-\hat{Y}}}{s_X \sqrt{n}} = \dfrac{22598.123}{1.606 \sqrt{786}} = 501.898$
7. $t = \dfrac{b - \beta_0}{s_b} = \dfrac{4399.12 - 1000}{501.898} = 6.773$
8. $(t = 6.773) > (t_{.05} = 1.647)$, $\therefore$ reject H_0
   $p < .00005$

SUMMARY

The material in this chapter on inferences about the regression slope closely parallels the material in Chapter 5 about the sampling distribution of the mean and the material in Chapter 6 on confidence intervals and hypothesis tests for means. You now should be able to see the common logic and theory underlying inferences about the mean and inferences about the slope. In this chapter we started with simple illustrations of what sampling distributions of b actually look like. Then we turned to the theoretical characteristics of the distribution of the least-squares slope that state that b is unbiased, the standard error of b equals the standard error of estimate divided by the product of the standard deviation of X and the square root of n, and b is normally distributed. These characteristics of the sampling distribution are based on the following assumptions: X and ε are uncorrelated; the variance of ε is the same for all values of X (homoskedasticity); the values of ε are independent; and ε is normally distributed. The Gauss–Markov theorem tells us that if these assumptions are true, the least-squares b is the best linear unbiased estimator of β. These characteristics of the sampling distribution of b were then used to define a theoretical formula for the confidence interval of b that is based on the true standard error and the normal distribution. Because we must estimate the true standard error of b, however, it is necessary to use this estimate and the t distribution to calculate the desired confidence interval. Next, we learned how to apply the eight-step procedure for hypothesis tests of the mean to conduct tests of the slope. We examined the two-tailed hypothesis test in which the null hypothesis specifies that β equals 0 as the method for testing whether there is an association, positive or negative, between X and Y. This hypothesis test is the one most commonly used for the slope in the social sciences. We concluded with the one-tailed hypothesis test for β.

KEY TERMS

sampling distribution of b

standard error of b

unbiased estimator of β

normal distribution of b

Gauss–Markov theorem

least-squares regression assumptions

homoskedasticity

heteroskedasticity

autocorrelation

robustness of b

estimated standard error of b

t statistic for b

calculated confidence interval for β

degrees of freedom of t

two-tailed t test for β

one-tailed t test for β

SPSS PROCEDURES

We will regress the earnings of full-time workers, as defined in this chapter, on years of college completed. First, select the full-time workers by clicking on *Data* and *Select Cases* to open the *Select Cases* window, click on *If condition is satisfied* and the *If* button, and enter the criteria for full-time workers in the *Select Cases: If* window, as shown next.

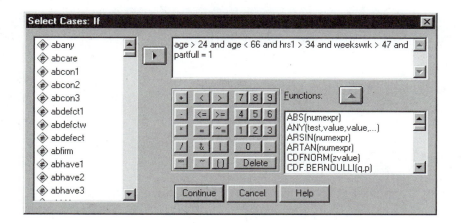

Next, click on *Analyze, Regression,* and *Linear* to get the *Linear Regression* window. Move the variable *earnings* to the *Dependent* box and move the variable *college* to the *Independent* box. To get more statistics than those generated by the default selections, click on the *Statistics* button to open the following *Linear Regression: Statistics* window. You will see that *Regression Coefficients Estimates* is already checked and *Model fit* is already checked. These are the default selections. Click on *Confidence intervals* to get 95-percent confidence intervals for the slope and intercept, and click on *Descriptives* to get means, standard deviations, and correlations for the independent and dependent variables.

After clicking on *Continue* and then *OK,* you will see the Descriptives, Model Fit, ANOVA, and Coefficients tables in the output window. The means and standard deviations are computed for the 786 cases that do not have missing values on either *earnings* or *college,* which is called list-wise deletion of missing values. The ANOVA table contains the regression, residual, and total sum-of-squares that are used to compute the coefficient of determination. The coefficients table contains the standard error, *t* score, two-tailed level of significance, and 95-percent confidence interval for the regression slope. The results in these tables are the same as those in Table 9.8. If you are conducting a one-tailed test and the sign of the slope is as predicted by the alternative hypothesis, divide the two-tailed level of significance by 2 to get the one-tailed significance level.

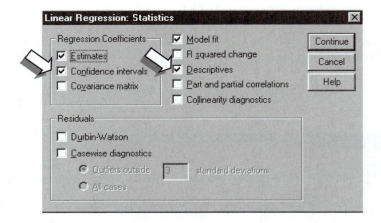

Descriptive Statistics

	Mean	Std. Deviation	N
EARNINGS	36647.262569	23664.0473915	786
COLLEGE	1.666667	1.6067797	786

Model Summary

Model	R	R Square	Adjusted R Square	Std. Error of the Estimate
1	.299[a]	.089	.088	22598.1227638

a. Predictors: (Constant), COLLEGE

ANOVA[b]

Model		Sum of Squares	df	Mean Square	F	Sig.
1	Regression	39220584551.745	1	39220584551.745	76.801	.000[a]
	Residual	400369319520.709	784	510675152.450		
	Total	439589904072.454	785			

a. Predictors: (Constant), COLLEGE
b. Dependent Variable: EARNINGS

Coefficients[a]

Model		Unstandardized Coefficients		Standardized Coefficients	t	Sig.	95% Confidence Interval for B	
		B	Std. Error	Beta			Lower Bound	Upper Bound
1	(Constant)	29315.395	1161.745		25.234	.000	27034.895	31595.895
	COLLEGE	4399.121	501.974	.299	8.764	.000	3413.748	5384.493

a. Dependent Variable: EARNINGS

EXERCISES

1. In the following tables, *sexprtyr* is constructed from GSS 2000 to equal the number of sex partners since age 18 divided by (age − 17). Thus, it is a sex-partner rate—that is, the number of sex partners per year. The data are for females. Calculate the missing statistics (?).

Correlations

		SEXPRTYR	AGE
Sum of Squares and Cross-products	SEXPRTYR	553.359	-3409.364
	AGE	-3409.364	322301.4
N	SEXPRTYR	1088	1088
	AGE	1088	1088

Model Summary

Model	R	R Square	Adjusted R Square	Std. Error of the Estimate
1	.255[a]	.065	.064	.69017

a. Predictors: (Constant), AGE

Coefficients[a]

Model		Unstandardized Coefficients		Standardized Coefficients	t	Sig.
		B	Std. Error	Beta		
1	(Constant)	.779	.059		13.311	.000
	AGE	-.0106	?	-.255	?	?

a. Dependent Variable: SEXPRTYR

2. In the following tables, *sexprtyr* is constructed from GSS 2000 to equal the number of sex partners since age 18 divided by (age − 17). Thus, it is a sex-partner rate—that is, the number of sex partners per year. The data are for males. Calculate the missing statistics (?).

Correlations

		SEXPRTYR	AGE
Sum of Squares and Cross-products	SEXPRTYR	2515.648	-5091.394
	AGE	-5091.394	210861.7
N	SEXPRTYR	854	854
	AGE	854	854

Model Summary

Model	R	R Square	Adjusted R Square	Std. Error of the Estimate
1	.221[a]	.049	.048	1.67581

a. Predictors: (Constant), AGE

Coefficients[a]

Model		Unstandardized Coefficients		Standardized Coefficients	t	Sig.
		B	Std. Error	Beta		
1	(Constant)	1.878	.166		11.307	.000
	AGE	-.0241	?	-.221	?	?

a. Dependent Variable: SEXPRTYR

3. The following table reports the regression of respondent's education (*educ*) on father's education (*paeduc*) for 784 males 25+ years of age in the GSS 2000. Construct the 95-percent confidence interval for the unstandardized slope.

Coefficients[a]

	Unstandardized Coefficients		Standardized Coefficients	t	Sig.
	B	Std. Error	Beta		
(Constant)	10.319	.274		37.637	.000
PAEDUC	.326	.023	.454	14.234	.000

a. Dependent Variable: EDUC

4. The following table reports the regression of respondent's education (*educ*) on mother's education (*maeduc*) for 1,193 females 25+ years of age in the GSS 2000. Construct the 99-percent confidence interval for the unstandardized slope.

Coefficients[a]

	Unstandardized Coefficients		Standardized Coefficients		
	B	Std. Error	Beta	t	Sig.
(Constant)	9.976	.219		45.545	.000
MAEDUC	.325	.019	.445	17.126	.000

a. Dependent Variable: EDUC

5. In the GSS 2000, *sei* is a socioeconomic index of a worker's occupation; the higher the *sei,* the greater the socioeconomic status of the respondent's occupation. Conduct a two-tailed test at $\alpha = .05$ of the null hypothesis that the slope of *emailhrs* (hours/week using email) regressed on *sei* equals 0 for females working full-time. For 434 full-time female workers in GSS 2000, the standard deviation of *sei* is 17.912, the standard error of estimate is 7.483, and the slope is .0478. Present your test in the eight-step format.

6. In the GSS 2000, *sei* is a socioeconomic index of occupations; the higher the *sei,* the greater the socioeconomic status of respondent's occupation. Conduct a one-tailed test at $\alpha = .05$ of the null hypothesis that the slope of *wwwhr* (hours/week on Internet or World Wide Web) regressed on *sei* is less than or equal to 0 for females working full time. For 308 full-time female workers in GSS 2000, the standard deviation of *sei* is 17.201, the standard error of estimate is 7.430, and the slope is .0218. Present your test in the eight-step format.

7. Assume that the 1,380 full-time workers aged 25 or older in the GSS 2000 represent a population. The regression of *sei* (socioeconomic index) on *educ* for these respondents follow. The second table reports the regression of sei on *educ* for a random sample of 30 respondents.

Coefficients, N = 1380[a]

Model		Unstandardized Coefficients		Standardized Coefficients		
		B	Std. Error	Beta	t	Sig.
1	(Constant)	-5.838	2.215		-2.636	.008
	EDUC	4.190	.156	.587	26.940	.000

a. Dependent Variable: SEI

Coefficients, n = 30[a]

	Unstandardized Coefficients		Standardized Coefficients			95% Confidence Interval for B	
	B	Std. Error	Beta	t	Sig.	Lower Bound	Upper Bound
(Constant)	-15.947	9.360		-1.704	.099	-35.119	3.226
EDUC	4.912	.664	.813	7.393	.000	?	?

a. Dependent Variable: SEI

   a. What is the sampling error for the unstandardized slope?
   b. What is the 95-percent confidence interval for the unstandardized slope? Is the
       population parameter within the interval?
   d. What is the 75-percent confidence interval for the unstandardized slope? Is the
       population parameter within the interval?

8. Assume that the 1,380 full-time workers aged 25 or older in the GSS 2000 represent a
   population. The regression of *sei* (socioeconomic index) on *age* for these respondents
   follows. The second table reports the regression of *sei* on *age* for a random sample of
   30 respondents.

Coefficients, N = 1380[a]

Model		Unstandardized Coefficients		Standardized Coefficients		
		B	Std. Error	Beta	t	Sig.
1	(Constant)	52.333	2.122		24.668	.000
	AGE	.0100	.049	.005	.204	.839

a. Dependent Variable: SEI

Coefficients, n = 30[a]

Model		Unstandardized Coefficients		Standardized Coefficients			95% Confidence Interval for B	
		B	Std. Error	Beta	t	Sig.	Lower Bound	Upper Bound
1	(Constant)	67.504	11.217		6.018	.000	44.528	90.480
	AGE	-.376	.258	-.265	-1.453	.157	?	?

a. Dependent Variable: SEI

   a. What is the sampling error?
   b. Because the unstandardized population slope is almost 0, assume it is truly 0. Based
       on the sample test, would you reject the null hypothesis that $\beta = 0$ in favor of a
       two-tailed alternative? Would you be making a Type I error, a Type II error, or
       neither?
   c. What is the 95-percent confidence interval for the unstandardized slope? Is the
       population parameter within the interval?

9. Using persons aged 55 and older in the GSS 2000 (*gss2000b*), regress *wordsum* (number of questions correct on a ten-question vocabulary test) on *age*.

   a. Interpret the unstandardized slope.
   b. Explain why you would or would not reject the two-tailed null hypothesis that
       the slope equals 0 at $\alpha = .01$.
   c. What is the 99-percent confidence interval? Is the value of the slope specified by
       the null hypothesis in the interval?

10. Using females in the GSS 2000 (*gss2000b*), regress *wordsum* (number of questions correct on a ten-question vocabulary test) on *maeduc*.

   a. Interpret the unstandardized slope.
   b. Explain why you would or would not reject the two-tailed null hypothesis that the slope equals 0 at $\alpha = .01$.
   c. What is the 99-percent confidence interval? Is the value of the slope specified by the null hypothesis in the interval?

10

✥

Multiple Regression

THE BOYS OF SUMMER

Ten-year-old Doris adored the "boys of summer." She owned baseball cards for Jackie Robinson, Duke Snider, Gil Hodges, Roy Campanella, and all the rest of the 1953 Brooklyn Dodgers who would go on to win a team-record 105 games that year. And on this day in late July, she could hardly wait until her dad got home from work to tell him all about how the Dodgers had destroyed the cross-town-rival New York Giants 13 to 2 that afternoon. Doris had carefully recorded all the details in her scorebook while listening to the game on the radio, just as she had done for most games every summer since she was six. Now she was ready to share this wealth of information with her father who had taught her how to keep score.

"Why don't you start with the highlights, Doris?" her father said after they had taken their seats on the front porch before dinner. Doris, naturally, started with the Dodger's power hitting, telling her dad about the home runs of Campanella, Hodges, and Snider, and concluding with the most exciting play of the day, the bases-loaded double by Carl Furillo that brought Jackie Robinson home in a cloud of dust all the way from first base. "Our hitting was awesome, Dad," she exclaimed.

"Well, what about the pitching, Doris? The Giants scored only two runs," said her father. Doris dutifully described how Clem Labine had pitched a complete game, striking out five and walking only two. "Labine was never in serious trouble. We staked him to a big lead early and he just coasted the rest of the way," she said.

"You don't sound very excited about the pitching," said her father. "You can't win championships without pitching. Wait until we get in the World Series against the Yankees. Now, they have great pitching, better than any we face in the National League. Our pitching will have to be awfully good if we are going to beat the Yankees, wouldn't you say, Doris?" Doris agreed, but in her heart she felt the Dodger's hitting attack was unbeatable. "Dad, nobody can stop our hitters. They're fantastic," she concluded as they headed in for dinner.[1]

The point of the preceding vignette is not whether the Dodgers would beat the Yankees in the 1953 World Series (they didn't). Instead, this story is included to illustrate the age-old debate about which is more important in competitive sports, offense or defense. In baseball, the debate takes the form of hitting versus pitching. The average spectator gets more excited about hitting—especially home runs—than pitching. Who can blame an impressionable youngster like Doris for being in awe of the Dodgers' power hitting? Some fans, however, get just as much enjoyment out of a well-pitched, low-scoring game. But regardless of what kind of game you prefer, which aspect contributes more to the success of the team—hitting or pitching, offense or defense?

Multiple regression is a statistical technique that is well suited to answer this question. In multiple regression, the dependent variable is regressed on two or more independent variables. In our case, we will choose a measure of team success as the dependent variable and use measures of the teams' offensive and defensive performance as two independent variables. The multiple regression analysis will be executed on a collection of teams that play one another. We will then examine the signs of the regression slopes for the measures of offense and defense to see if they both contribute to success, as indicated by positive slopes. If they are both positive, we can compare the magnitude of their standardized slopes to see which variable—offense or defense—contributes more to team success. (If both variables are measured on the same scale, we can compare their unstandardized slopes.) This comparison is one of the important functions of multiple regression.

In multiple regression, we can validly evaluate and compare the contribution of each independent variable because the slope of each variable indicates the linear relationship between that variable and the dependent variable, with all of the other independent variables *held constant*. This procedure differs greatly from running several bivariate regression analyses, such as regressing team success on offensive performance and then regressing team success on defensive performance. In the two bivariate regressions, we are unable to assess the effect of one variable independent of the other variable. Thus, learning why it is necessary to conduct one multiple regression analysis as opposed to a series of bivariate regressions is an important objective of this chapter. A related important objective is to learn how multiple regression is able to hold constant—that is, *control*—all the other independent variables when describing the effect of one particular independent variable.

A final objective is to learn the way multiple regression measures how well the independent variables as a group predict the dependent variable. For example, we want to

---

1. To learn more about the true stories that inspired this fictional account, read Roger Kahn's *The Boys of Summer* (New York: Harper & Row, 1971) and Doris Kearns Goodwin's *Wait Till Next Year: A Memoir* (New York: Simon & Schuster, 1997).

measure how well offensive performance *and* defensive performance together explain variance in team success. This statistic can tell us whether one or more important variables have been omitted from the multiple regression equation.

MULTIPLE LINEAR FUNCTIONS

We will look first at the case of a dependent variable as a linear function of two independent variables. The wind chill index as a function of temperature and wind velocity will serve as an example. The wind chill index measures the heat loss from exposed human skin surfaces. Heat loss is caused by a combination of high wind velocity and low air temperature. The index is measured in degrees of temperature that are equivalent to heat loss due to a combination of wind and temperature. Risk of frostbite from low wind chill indices makes the wind chill a winter weather hazard.

A simple equation for the wind chill index is

$$WC = -13 + 1.3T - .4V \tag{10.1}$$

where WC is wind chill in degrees Fahrenheit, T is temperature in degrees Fahrenheit, and V is wind velocity in miles per hour.[2] Table 10.1 shows calculations of wind chill scores for different combinations of temperature and wind velocity. The first calculation shows that a temperature of 45° F combined with a wind velocity of 10 miles per hour produces a wind chill of 41.5° F. The two calculations immediately below that one show how much the wind chill changes (dWC) when the temperature remains 45° and the wind increases by 1 mile per hour ($dV = 1$). When the wind increases to 11 miles per hour the wind chill falls to 41.1, and when the wind increases to 12 miles per hour the wind chill falls to 40.7. Thus, we see that when the temperature is held constant, each unit increase in wind velocity causes a drop in the wind chill of $-.4°$. This is the slope for V in Equation 10.1; that is, $dWC/dV = -.4/1 = -.4$.

The next set of calculations in the left-hand column shows the wind chill when the temperature is 15° and the wind velocity is 20, 21, and 22 miles per hour, respectively. With the temperature held constant at 15°, each unit increase in wind velocity produces a .4° drop in the wind chill. Thus, $dWC/dV = -.4/1 = -.4$; that is, the slope once again equals $-.4$. The third set of calculations shows that the slope also equals $-.4$ for a temperature of $-15°$ and wind velocities of 30, 31, and 32 miles per hour, respectively.

The right-hand columns display three sets of calculations showing the change in wind chill per 1° drop in temperature with the wind velocity held constant at 10, 20, and 30 miles per hour, respectively. For each case, $dWC/dT = (-1.3)/(-1) = 1.3$. This is the slope for temperature in Equation 10.1.

The calculations in Table 10.1 are designed to illustrate the meaning of the slopes in a linear function with two independent variables. *The slopes indicate the change in the dependent variable per unit increase in one independent variable when the other independent vari-*

---

2. Equation 10.1 is a simplification of the true equation $WC = 35.74 + .6215T - 35.75V^{.16} + .4275TV^{.16}$. The simplified equation, however, explains 99.5 percent of the variance in true WC for temperatures of $-30°$ F to 45° F and wind velocities of 5 to 50 miles per hour.

Table 10.1 Change in Wind Chill (*dWC*) Per Unit Increase in Wind Velocity (*dV*) and Per Unit Decrease in Temperature (*dT*)

\	dWC per dV = 1	\	\	\	\	\	dWC per dT = −1	\	\	\	\
−13	+1.3T	−.4V =	WC	dV	dWC	−13	+1.3T	−.4V =	WC	dT	dWC
−13	+1.3(45)	−.4(10) =	41.5			−13	+1.3(45)	−.4(10) =	41.5		
				1	−.4					−1	−1.3
−13	+1.3(45)	−.4(11) =	41.1			−13	+1.3(44)	−.4(10) =	40.2		
				1	−.4					−1	−1.3
−13	+1.3(45)	−.4(12) =	40.7			−13	+1.3(43)	−.4(10) =	38.9		
−13	+1.3(15)	−.4(20) =	−1.5			−13	+1.3(15)	−.4(20) =	−1.5		
				1	−.4					−1	−1.3
−13	+1.3(15)	−.4(21) =	−1.9			−13	+1.3(14)	−.4(20) =	−2.8		
				1	−.4					−1	−1.3
−13	+1.3(15)	−.4(22) =	−2.3			−13	+1.3(13)	−.4(20) =	−4.1		
−13	+1.3(−15)	−.4(30) =	−44.5			−13	+1.3(−15)	−.4(30) =	−44.5		
				1	−.4					−1	−1.3
−13	+1.3(−15)	−.4(31) =	−44.9			−13	+1.3(−16)	−.4(30) =	−45.8		
				1	−.4					−1	−1.3
−13	+1.3(−15)	−.4(32) =	−45.3			−13	+1.3(−17)	−.4(30) =	−47.1		

able is held constant. As illustrated here, the other independent variable may be held constant *at any given value of that variable.* Thus, the slope of the first independent variable is the same across the entire range of the second independent variable, and the slope of the second independent variable is the same across the entire range of the first independent variable.

A bivariate linear function defines a line in two-dimensional space. With two independent variables, however, the linear function defines a flat surface (i.e., a plane) in three-dimensional space. A three-dimensional graph of the wind-chill function is displayed in Figure 10.1. Three axes define the graph. There are two horizontal axes representing temperature and wind velocity and a vertical axis representing wind chill. We may think of the three axes as forming a box that contains the plane defined by the wind chill function. The top of the box and the front two panels of the box have been removed so we can see the plane inside. The two horizontal axes (velocity and temperature) outline the bottom of the box. The temperature and wind chill axes outline the left-rear panel, whereas the velocity and wind chill axes form the right-rear panel.

Each point on the wind chill surface represents a unique combination of values of wind chill, wind velocity, and temperature. Parallel lines are drawn on the surface corresponding to each of the six tick-marked values of wind velocity (5, 14, 23, 32, 41, 50). These lines show the rate of change in wind chill as temperature increases with wind velocity held constant. We can see on the back-left panel that the slope of these lines is positive ($dWC/dT = 1.3$). Six parallel lines also have been drawn on the surface for the temperatures tick-marked on the temperature axis (−30, −15, 0, 15, 30, 45). These lines

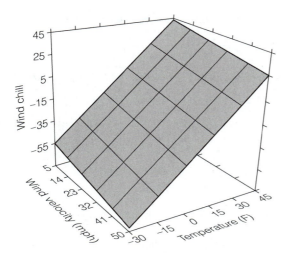

FIGURE 10.1 Wind Chill as a Linear Function of Temperature and Wind Velocity

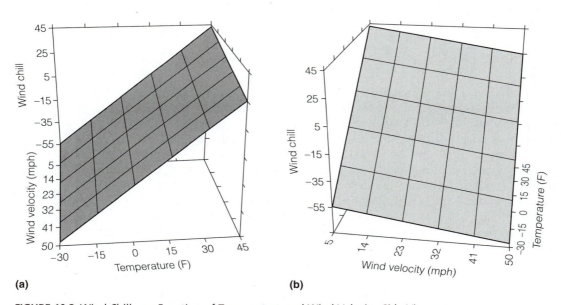

(a)

(b)

FIGURE 10.2 Wind Chill as a Function of Temperature and Wind Velocity, Side Views

show the rate of change in wind chill as wind velocity increases with temperature held constant ($dWC/dV = -.4$).

The relationship between wind chill and each independent variable may be seen more clearly from two side views of the box (Figure 10.2). Whereas these side views are still three-dimensional, they are nearly two-dimensional perspectives. Figure 10.2a emphasizes the linear relationship between temperature and wind chill, whereas Figure

10.2b stresses the relationship between wind velocity and wind chill. From each of these perspectives, the relationship is shown as a series of parallel lines with each successive line being higher than the previous one as we move from front to back. We can also see that the slope of the lines for temperature and wind chill (Figure 10.2a) is greater than the slope of the lines relating wind velocity to wind chill (Figure 10.2b).

THE MULTIPLE REGRESSION EQUATION

The multiple linear function that we have examined is different from a multiple regression equation in two respects. First, the slopes and intercept for the wind chill equation are known. We did not have to estimate them with a sample of measurements of temperature, wind velocity, and wind chill taken at different places or at different times. Second, wind chill is an exact function of temperature and wind velocity. There are no other causes, random or otherwise, to take into account. Consequently, there is no error term (i.e., disturbance term) in the equation.

In comparison, multiple regression analysis is a statistical tool we use to *estimate* the parameters of a multiple regression model that contains a random error term. Thus, we have to use a sample of observations of the variables specified by the model to estimate the slopes and intercept of the equation. We have to learn how to estimate these parameters when there is more than one independent variable. As we will see, the bivariate slopes that we have used will not provide suitable estimates.

The Multiple Regression Model

The symbols and notation used in multiple regression are simple extensions of those used in bivariate regression. In bivariate regression, Y refers to the dependent variable and X is the symbol for the single independent variable. In multiple regression we add subscripts to X to designate different independent variables, as follows:

$$X_1 = \text{1st independent variable}$$

$$X_2 = \text{2nd independent variable}$$

$$X_3 = \text{3rd independent variable}$$

$$\vdots \qquad \vdots$$

$$X_k = k\text{th independent variable}$$

where k indicates the total number of independent variables.

The equation for the simplest **multiple regression model,** one with $k = 2$ independent variables, is

$$Y = \alpha + \beta_1 X_1 + \beta_2 X_2 + \varepsilon \tag{10.2}$$

where

$$\alpha = \text{value of } Y \text{ when } X_1 = X_2 = \varepsilon = 0$$

$$\beta_1 = \text{slope of } Y \text{ on } X_1 \text{ with } X_2 \text{ held constant}$$

$$\beta_2 = \text{slope of } Y \text{ on } X_2 \text{ with } X_1 \text{ held constant}$$

β_1 and β_2 are called partial regression slopes to distinguish them from bivariate slopes. *The **partial regression slope** is the change in the dependent variable per unit increase in one independent variable with the other independent variable(s) held constant.*

The use of the lowercase Greek letters α, β, and ε indicates that Equation 10.2 is a population model containing unknown parameters that must be estimated with sample data. The equation that represents the sample estimates of the parameters in Equation 10.2 is

$$\hat{Y} = a + b_1 X_1 + b_2 X_2 \qquad (10.3)$$

We know that Equation 10.3 is for sample data because the intercept and slopes (a, b_1, and b_2) are indicated by Roman letters. The sample statistics a, b_1, and b_2 are estimates of the population parameters α, β_1, and β_2, respectively. The values of $Y - \hat{Y}$ are estimates of the values of ε.

What criterion should we use to determine the values of the sample slopes and intercept? In the bivariate case we chose values for a and b that minimized the sum of squared residuals $[\Sigma(Y - \hat{Y})^2]$, which is the *least-squares criterion*. We will also use the least-squares criterion in multiple regression to determine the values of a, b_1, and b_2. The Gauss-Markov theorem for the case of multiple regression, which states that the least-squares solution provides the best linear unbiased estimates (BLUE) of α, β_1, and β_2, again provides the justification. As we will see, however, we can't use the bivariate formulas for a and b to calculate the least-squares estimates of the intercept and slopes in the multiple regression equation.

In the social world, differences on one independent variable are usually correlated with differences on another independent variable. That is, there are not only multiple causes of most dependent variables, but these causes tend to be correlated with one another. Thus, if we look at the change in Y related to a unit increase in X_1, it is very likely that X_2 did not stay constant when X_1 increased. A key objective of this chapter is to show how to hold constant one independent variable to get a valid estimate of the slope for a second independent variable.

The Venn diagrams in Figure 10.3 illustrate both the problem created by correlated independent variables and the solution to this problem. Each circle in a Venn diagram represents the variance of a standardized variable (i.e., z), which thus equals 1. The overlap between two circles indicates the proportion of variance in each variable that it shares with the other—that is, the coefficient of determination (CD). Areas a and b in Figure 10.3a indicate the CD for Y with X_1 and the CD for Y with X_2, respectively. The fact that X_1 and X_2 do not overlap means that they are uncorrelated. As a consequence, area a does not overlap with area b, which means that the variance that Y shares with X_1 is independent of the variance that Y shares with X_2. Therefore, in the case of two uncorrelated independent variables, we can compute two separate bivariate regression equations to determine the effects of X_1 and X_2 on Y. Although there are other statistical advantages to using multiple regression when the independent variables are not correlated, it is not absolutely necessary to do so in this case.

Figure 10.3b shows a Venn diagram for two correlated causes of Y. The shared variance between X_1 and Y is represented by area $a + c$, and the shared variance between X_2 and Y is represented by $b + c$. Thus, c is a new area created by the correlation between X_1 and X_2 that represents the proportion of the variance in Y that is shared jointly with X_1 and X_2. To which variable should we assign this jointly shared variance

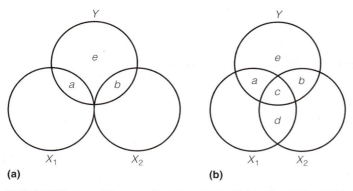

FIGURE 10.3 Venn Diagrams for Uncorrelated (a) and Correlated (b) Independent Variables

c? If we computed two separate bivariate regression equations, area c would be assigned to X_1 in one bivariate regression and to X_2 in the second bivariate regression. Because it would clearly be incorrect to count it twice, which would result in overestimating the effect of each independent variable, bivariate regression is wrong.

Because area c represents variance that is related to both X_1 and X_2, it cannot be uniquely assigned to either variable. Areas a and b, however, are uniquely related to X_1 and X_2, respectively, and thus areas a and b should be used to calculate the partial regression slopes for X_1 and X_2. How can we determine areas a and b? If we remove from X_1 all the variance that it shares with X_2 (i.e., $c + d$), area c will be removed, and the only remaining shared variance between X_1 and Y would be a, the variance in Y uniquely caused by X_1. Thus, if we regress Y on X_1 from which $c + d$ has been removed, we will have a valid estimate of the partial regression slope for X_1. Likewise, if we remove from X_2 all the variance that it shares with X_1 and then regress Y on the remaining variance in X_2, we will have a valid estimate of the partial slope for X_2. We take up the method for removing the shared variance between X_1 and X_2 in the next section.

Multiple Regression Coefficients

A small, hypothetical data set will be used to illustrate how the slopes and intercept in a multiple regression equation are determined. The data provide an example of two variables that affect the number of years of education a person has achieved—namely age and number of siblings. These data have been constructed to simulate the relationships between these variables in the real world. These variables will be designated as follows:

$$X_1 = \text{years of age}$$

$$X_2 = \text{number of siblings}$$

$$Y = \text{years of education}$$

Furthermore, we will imagine that we know the true effects of age and siblings on education, which are

$$Y = 27 - .2X_1 - 2X_2 + \varepsilon \tag{10.4}$$

The slope for age indicates that education decreases by .2 year for each year of increase in age with number of siblings held constant. The slope for siblings means that education decreases by 2 years for each additional sibling with age held constant. So, both age and education have negative effects on education. The intercept of 27 is not meaningful in this case because it represents the expected education when siblings and age equal 0. An age of 0 is far below the range of values for which we want to study educational attainment.

The values of X_1, X_2, and ε are listed in Table 10.2. These values are inserted into Equation 10.4 to determine the values of Y, as follows:

$$Y = 27 - .2X_1 - 2X_2 + \varepsilon$$

$$= 27 - .2(25) - 2(2) - 2 = 16$$

$$= 27 - .2(35) - 2(0) + 2 = 22$$

$$= 27 - .2(45) - 2(3) + 0 = 12$$

$$= 27 - .2(55) - 2(6) + 2 = 6$$

$$= 27 - .2(65) - 2(4) - 2 = 4$$

The unknown regression coefficients (which we are pretending to know), along with the values of X_1, X_2, and ε, have created the distribution of Y.

Now, let's assume that we are researchers who want to estimate these unknown regression coefficients. All we have to work with are the values of X_1, X_2, and Y. These variables are measurable and thus have values that we can observe. The regression coefficients ($\alpha = 27$, $\beta_1 = -.2$, and $\beta_2 = -2$) and the error term ε are unknowns that we must estimate with the observed values of X_1, X_2, and Y. How will we do this?

Let's begin by calculating the bivariate slopes for Y on X_1 and Y on X_2. The first step is to calculate the means from the sums of X_1, X_2, and Y given in Table 10.2:

$$\overline{X}_1 = \frac{225}{5} = 45 \qquad \overline{X}_2 = \frac{15}{5} = 3 \qquad \overline{Y} = \frac{60}{5} = 12$$

The average person is 45 years of age, has 3 siblings, and has completed 12 years of education. Using these means, we can calculate the deviation scores of the three variables—that is, $X_1 - \overline{X}_1$, $X_2 - \overline{X}_2$, and $Y - \overline{Y}$ (see Table 10.2). As always, the deviation scores of each variable sum to 0. Our next step is to calculate the sums of the squared deviation scores of X_1, X_2, and Y, which are needed in the denominators of the formulas for the bivariate slopes. Last, we compute the cross-products—$(X_1 - \overline{X}_1)$ $(Y - \overline{Y})$, $(X_2 - \overline{X}_2)(Y - \overline{Y})$, and $(X_1 - \overline{X}_1)(X_2 - \overline{X}_2)$—and their sums to use in the numerators of the slope formulas. We can now compute the bivariate regression slopes:

$$b_{YX_1} = \frac{\Sigma(X_1 - \overline{X}_1)(Y - \overline{Y})}{\Sigma(X_1 - \overline{X}_1)^2} = \frac{-400}{1000} = -.4$$

$$b_{YX_2} = \frac{\Sigma(X_2 - \overline{X}_2)(Y - \overline{Y})}{\Sigma(X_2 - \overline{X}_2)^2} = \frac{-60}{20} = -3$$

The slopes imply that each additional year of age decreases education by .4 year and each additional sibling decreases education by 3 years. The bivariate slopes overestimate

Table 10.2 Data for the Bivariate Regression Equations for Age (X_1), Siblings (X_2), and Education (Y)

X_1	X_2	ε	Y	$X_1 - \bar{X}_1$	$(X_1 - \bar{X}_1)^2$	$X_2 - \bar{X}_2$	$(X_2 - \bar{X}_2)^2$
25	2	−2	16	−20	400	−1	1
35	0	2	22	−10	100	−3	9
45	3	0	12	0	0	0	0
55	6	2	6	10	100	3	9
65	4	−2	4	20	400	1	1
225	15	0	60	0	1000	0	20

$Y - \bar{Y}$	$(Y - \bar{Y})^2$	$(X_1 - \bar{X}_1)(Y - \bar{Y})$	$(X_2 - \bar{X}_2)(Y - \bar{Y})$	$(X_1 - \bar{X}_1)(X_2 - \bar{X}_2)$
4	16	−80	−4	20
10	100	−100	−30	30
0	0	0	0	0
−6	36	−60	−18	30
−8	64	−160	−8	20
0	216	−400	−60	100

the true negative slopes of age and siblings, which are $-.2$ and 2, respectively. The bivariate slope for age is twice as great ($-.4/-.2 = 2$) as the true slope, and the bivariate slope for siblings is one and a half times as great ($-3/-2 = 1.5$) as the true slope. Thus, the bivariate slopes are seriously biased. Why?

The answer is that the bivariate slope does not indicate the change in the dependent variable associated with a unit increase in one independent variable *with the other independent variable held constant*. A glance at the values of X_1 and X_2 in Table 10.2 and at the scatterplot for X_1 and X_2 in Figure 10.4 shows that there is a positive association between age and siblings. Older persons tend to have more siblings than younger persons. Because both age and siblings affect education, the association between the two independent variables results in biased bivariate slopes. For example, as age increases there is a resulting decrease in education. The increase in age, however, is accompanied by an increase in siblings that causes a further decrease in education. The bivariate slope for education on age is attributing to age both the decrease in education resulting from an increase in age and the decrease in education resulting from the corresponding increase in siblings. Consequently, the bivariate slope overestimates the influence of age on education. What we need is a statistical technique for holding the number of siblings constant to determine the extent to which education changes as a result of an increase in age alone.

The solution to the problem of how to hold X_2 constant is to remove from X_1 that part of it that is correlated with X_2. That is, we want to remove the variance in X_1 that it shares with X_2. This is called residualizing X_1. Because the **residualized X_1** is uncorrelated with X_2, we can use it to find the change in Y that results from an increase in X_1 alone. This procedure can be summarized by the four steps in Box 10.1.

According to step 1, we need to regress X_1 (age) on X_2 (siblings). This means that X_1 is the dependent variable in the regression and X_2 is the independent variable. This

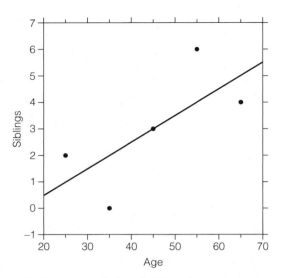

FIGURE 10.4 Age (X_1) and Siblings (X_2) in the Educational Attainment Example

does not mean that siblings are believed to be a cause of age. We are conducting this regression merely for the purpose of statistical control. Using the sum-of-squares of X_2 and the sum of cross-products of X_1 and X_2 from Table 10.2, as well as the means just computed, we get

Age regressed on siblings $\begin{cases} b_{X_1 X_2} = \dfrac{\Sigma(X_1 - \overline{X}_1)(X_2 - \overline{X}_2)}{\Sigma(X_2 - \overline{X}_2)^2} = \dfrac{100}{20} = 5 \\ a_{X_1 X_2} = \overline{X}_1 - b_{X_1 X_2}\overline{X}_2 = 45 - 5(3) = 30 \end{cases}$

The slope indicates that an increase of one sibling is associated with a 5-year increase in age. The intercept means that persons with no siblings are predicted to be 30 years of age.

The second step is to use these regression coefficients to compute predicted ages. We enter the five values of siblings listed in Table 10.2 into the regression equation as follows:

$$\hat{X}_1 = 30 + 5X_2$$
$$= 30 + 5(2) = 40$$
$$= 30 + 5(0) = 30$$
$$= 30 + 5(3) = 45$$
$$= 30 + 5(6) = 60$$
$$= 30 + 5(4) = 50$$

These predicted values of age are entered in Table 10.3, which shows that the sum of $\hat{X}_1$ equals 225. The mean of the predicted ages is $225/5 = 45$. Note that this is the same

---

BOX 10.1   Slope for X_1 Holding Constant X_2

1. Regress X_1 on X_2 to determine $a_{x_1x_2}$ and $b_{x_1x_2}$.
2. Compute $\hat{X}_1 = a_{x_1x_2} + b_{x_1x_2}X_2$.

3. Compute the $X_1 - \hat{X}_1$ residuals.
4. Regress Y on $X_1 - \hat{X}_1$ to determine b_1.

Table 10.3 Data for the Multiple Regression Equation for Age (X_1), Siblings (X_2), and Education (Y)

$\hat{X}_1$	$X_1 - \hat{X}_1$	$(X_1 - \hat{X}_1)(X_2 - \overline{X}_2)$	$(X_1 - \hat{X}_1)^2$	$(X_1 - \hat{X}_1)(Y - \overline{Y})$
40	−15	15	225	−60
30	5	−15	25	50
45	0	0	0	0
60	−5	−15	25	30
50	15	15	225	−120
225	0	0	500	−100

$\hat{X}_2$	$X_2 - \hat{X}_2$	$(X_2 - \hat{X}_2)(X_1 - \overline{X}_1)$	$(X_2 - \hat{X}_2)^2$	$(X_2 - \hat{X}_2)(Y - \overline{Y})$
1	1	−20	1	4
2	−2	20	4	−20
3	0	0	0	0
4	2	20	4	−12
5	−1	−20	1	8
15	0	0	10	−20

as the mean of the observed ages. It is always the case that the mean of the least-squares predictions equals the mean of the dependent variable.

Step 3 is to compute the residual values of X_1 (i.e., $X_1 - \hat{X}_1$). After subtracting the predicted score from each observed value of age, we get the residuals shown in the second column of Table 10.3. Note that they sum to 0 and thus have a mean of 0, as all residuals do. Thus, each residual represents a deviation score from the mean of all residuals. Column 3 shows the cross-products of the residuals of X_1 and the deviation scores of X_2. The sum of cross-products equals 0. This means that the covariance between the two variables is 0. Therefore, there is no association between the residualized X_1 and X_2. This is what we were seeking—namely, a transformation of X_1 that is unrelated to X_2. Thus, *when the residualized X_1 increases, there will be no tendency for X_2 to either increase or decrease. In this way, X_2 is held constant when X_1 changes.* Column 4 shows the squared residuals of X_1. The sum of squared residuals $[\Sigma(X_1 - \hat{X}_1)^2]$ equals 500. This sum is only half as great as the sum-of-squares of X_1, which equals 1,000 (Table 10.2). Thus, half of the variance of X_1 has been removed to produce a residualized variable that is uncorrelated with X_2.

The last step in calculating the slope for X_1 with X_2 held constant is to regress the dependent variable Y on the residuals $X_1 - \hat{X}_1$. That is, $X_1 - \hat{X}_1$ will be the independent variable (IV) in a bivariate least-squares regression. We will use the general formula for a least-squares slope from Chapter 8:

$$b = \frac{\text{sum of cross-products}}{\text{sum-of-squares of IV}}$$

The sum of cross-products (SCP) in our example is

$$\text{SCP} = \Sigma(X_1 - \hat{X}_1)(Y - \overline{Y})$$

Remember that a cross-product is the deviation score for one variable times the deviation score of another. Again, because the residuals have a mean of 0, $X_1 - \hat{X}_1$ is a deviation score. Thus, $(X_1 - \hat{X}_1)(Y - \overline{Y})$ is a product of deviation scores. The sum-of-squares (SS) of the independent variable is

$$\text{SS} = \Sigma(X_1 - \hat{X}_1)^2$$

Now we can enter the formulas for SCP and SS into the general formula for the slope to get the formula for the slope of Y on X_1 with X_2 held constant:

$$b_1 = \frac{\Sigma(X_1 - \hat{X}_1)(Y - \overline{Y})}{\Sigma(X_1 - \hat{X}_1)^2} \tag{10.5}$$

The values of the sum of cross-products and the sum-of-squares from Table 10.3 are entered into Equation 10.5 to get

$$b_1 = \frac{-100}{500} = -.2$$

This result indicates that a 1-year increase in age, with siblings held constant, results in a .2-year decrease in education. This is the true value of the slope for age that was originally used to create the values of education. Thus, the regression of education on residualized age (age from which the effect of siblings has been removed) results in an unbiased value for the slope.

The next task is to calculate the slope of Y on X_2 with X_1 held constant (i.e., b_2). We will use the four-step procedure in Box 10.2 that is analogous to the one we used to get b_1. Step 1 gives

Siblings regressed on age $\begin{cases} b_{X_2X_1} = \dfrac{\Sigma(X_1 - \overline{X}_1)(X_2 - \overline{X}_2)}{\Sigma(X_1 - \overline{X}_1)^2} = \dfrac{100}{1000} = .1 \\ a_{X_2X_1} = \overline{X}_2 - b_{X_2X_1}\overline{X}_1 = 3 - .1(45) = -1.5 \end{cases}$

The slope indicates that for each year of increase in age, the number of siblings increases by .1. The second step is to enter the values of age (25, 35, 45, 55, 65) into the equation $\hat{X}_2 = -1.5 + .1X_1$ to get the predicted values of siblings shown in Table 10.3. After calculating the residuals $X_2 - \hat{X}_2$, we compute the squared residuals $(X_2 - \hat{X}_2)^2$ and cross-products $(X_2 - \hat{X}_2)(Y - \overline{Y})$. (Note that $\Sigma(X_2 - \hat{X}_2)(X_1 - \overline{X}_1) = 0$, which indicates that the residuals of X_2 are not associated with X_1.)

BOX 10.2   Slope for X_2 Holding Constant X_1

1. Regress X_2 on X_1 to determine $a_{x_2 x_1}$ and $b_{x_2 x_1}$.
2. Compute $\hat{X}_2 = a_{x_2 x_1} + b_{x_2 x_1} X_1$.
3. Compute the $X_2 - \hat{X}_2$ residuals.
4. Regress Y on $X_2 - \hat{X}_2$ to determine b_2.

Now we can enter the formula for the sum-of-squares of $X_2 - \hat{X}_2$ and the formula for the sum of cross-products of $X_2 - \hat{X}_2$ and Y into the general formula for the least-squares slope to get the formula for the slope of X_2 with X_1 held constant (i.e., b_2):

$$b_2 = \frac{\Sigma(X_2 - \hat{X}_2)(Y - \overline{Y})}{\Sigma(X_2 - \hat{X}_2)^2} \tag{10.6}$$

Substituting the values of the numerator and denominator from Table 10.3 into Equation 10.6 gives

$$b_2 = \frac{-20}{10} = -2$$

This result indicates that for each additional sibling, education decreases by 2 years, with age held constant. This slope equals the true value of the slope for X_2 that was used to generate the values of Y.

Finally, we must determine the value of the intercept of the multiple regression equation. Because this is a least-squares solution, the predicted value of Y will equal its mean when X_1 and X_2 are equal to their means:

$$\hat{Y} = \overline{Y} = a + b_1\overline{X}_1 + b_2\overline{X}_2$$

When we rearrange this equation to put a alone on the left side of the equals sign, we have the formula for the least-squares intercept:

$$a = \overline{Y} - b_1\overline{X}_1 - b_2\overline{X}_2 \tag{10.7}$$

Entering the slopes and means into Equation 10.7 gives

$$a = 12 - (-.2)(45) - (-2)(3) = 12 + 9 + 6 = 27$$

The value of 27 equals the true value of the intercept used to compute the values of Y.

Combining the slopes and intercept into a single equation gives the following least-squares multiple regression equation:

$$\hat{Y} = 27 - .2X_1 - 2X_2$$

This is the least-squares regression equation. This equation, plus ε, is the equation we originally used to generate the values of Y. By way of two bivariate regressions—namely, Y on residualized X_1 and Y on residualized X_2—we have achieved unbiased measures of the true slopes and intercept.

The slopes b_1 and b_2 are least-squares partial regression slopes. The generic symbol for a partial slope is b_i, where $i = 1, 2, \ldots, k$ and k is the number of independent variables. The name *partial slope* is used because it is the slope between the dependent variable and *part* of an independent variable—namely, the residual part. Thus, it is not the entire slope but instead the part of the slope that results from holding the other independent variable(s) constant. Now we have two types of regression slopes, namely, the *bivariate slope* in a bivariate regression equation and the *partial slopes* in a multiple regression equation.

Goodness-of-Fit

We know that the least-squares equation fits the data better than any other linear, unbiased equation. But how closely do its predictions fit the data? To answer this question, we first calculate the predicted values of education (Y) by entering the values of age (X_1) and siblings (X_2) into the least-squares equation:

$$\hat{Y} = 27 - .2X_1 - 2X_2$$
$$= 27 - .2(25) - 2(2) = 18$$
$$= 27 - .2(35) - 2(0) = 20$$
$$= 27 - .2(45) - 2(3) = 12$$
$$= 27 - .2(55) - 2(6) = 4$$
$$= 27 - .2(65) - 2(4) = 6$$

Next, we calculate the residuals of education ($Y - \hat{Y}$ in Table 10.4). These residuals equal the values of ε (Table 10.2) originally used to create the values of Y.

Standard Error of Estimate

The first statistic used to measure the accuracy of the predictions is

$$\text{Standard error of estimate} = \hat{s}_{Y-\hat{Y}} = \sqrt{\frac{\Sigma(Y - \hat{Y})^2}{n - k - 1}} \qquad (10.8)$$

We encountered the **standard error of estimate** for the bivariate regression in Chapter 8 (Equation 8.20). You will remember that it is an estimate of the standard deviation of the residuals in the population. Thus, the sum of squared residuals was divided by $n - 2$ in the bivariate case. In general, the sum-of-squares of $Y - \hat{Y}$ is divided by $n - k - 1$, where k equals the number of independent variables. In our current example, $k = 2$, so we divide by $n - 2 - 1 = n - 3$. Table 10.4 gives the squared residuals and their sum. Substituting the sum into Equation 10.8 gives

$$\hat{s}_{Y-\hat{Y}} = \sqrt{\frac{16}{5 - 2 - 1}} = \sqrt{\frac{16}{2}} = \sqrt{8} = 2.828$$

This means that the estimated standard error of prediction is 2.828 years of education. You will remember that the standard error of estimate does not have a fixed range of values—that is, it is not standardized. You cannot look at the value of $\hat{s}_{Y-\hat{Y}}$ alone and judge the accuracy of the predictions.

Table 10.4 Accuracy of Predictions for Age (X_1), Siblings (X_2), and Education (Y)

$\hat{Y}$	$Y - \hat{Y}$	$(Y - \hat{Y})^2$	$\hat{Y} - \bar{Y}$	$(\hat{Y} - \bar{Y})^2$	$(\hat{Y} - \bar{Y})(Y - \bar{Y})$
18	−2	4	6	36	24
20	2	4	8	64	80
12	0	0	0	0	0
4	2	4	−8	64	48
6	−2	4	−6	36	48
60	0	16	0	200	200

Coefficient of Multiple Determination

The next measure of goodness-of-fit *is* standardized. It is conceptually the same as the coefficient of determination for bivariate regression from Chapter 8. The **coefficient of multiple determination** (CMD) is

$$\text{CMD} = \frac{s_{\hat{Y}}^2}{s_Y^2} = \frac{\Sigma(\hat{Y} - \bar{Y})^2}{\Sigma(Y - \bar{Y})^2} \tag{10.9}$$

This formula is identical to that for the bivariate coefficient of determination (Equation 8.21). The predicted values in Equation 10.9, however, come from a multiple regression equation:

$$\hat{Y} = a + b_1X_1 + b_2X_2 + \cdots + b_kX_k$$

The numerator of the CMD is the sum of squared deviations of $\hat{Y}$ (Table 10.4), and the denominator contains the sum-of-squares of Y (Table 10.2). Inserting these values in Equation 10.9 gives

$$\text{CMD} = \frac{200}{216} = .9259$$

Remember that the numerator is the regression sum-of-squares (explained SS), and the denominator is the total sum of squares, which value indicates that .926 (92.6%) of the variance of education is explained by age and education together. Because the range of values of CMD is 0 to 1, this is a very high value of the CMD. Therefore, the accuracy of the predictions is very good.

Multiple Correlation

The second standardized measure of goodness-of-fit is the multiple correlation. You will remember that the symbol for a bivariate correlation is r. The symbol for the multiple correlation is R. It is based on the logic that the accuracy of the predicted values of Y can be measured by the strength of their correlation with the observed values of Y. Thus, the **multiple correlation** (R) is equal to the bivariate correlation between Y and $\hat{Y}$, or $r_{Y\hat{Y}}$:

$$R = r_{Y\hat{Y}} = \frac{\Sigma z_Y z_{\hat{Y}}}{n} = \frac{\Sigma(\hat{Y} - \bar{Y})(Y - \bar{Y})}{\sqrt{[\Sigma(\hat{Y} - \bar{Y})^2][\Sigma(Y - \bar{Y})^2]}} \tag{10.10}$$

There are several equivalent formulas for a bivariate correlation. Equation 10.10 contains two of them. In both cases, $\hat{Y}$ is substituted for X in the original defining formulas for the bivariate correlation in Chapter 8. The first formula,

$$r_{Y\hat{Y}} = \frac{\Sigma z_Y z_{\hat{Y}}}{n}$$

is based on the formula for the bivariate standardized slope (Equation 8.23), which you will remember is equal to the correlation coefficient. The second formula,

$$r_{Y\hat{Y}} = \frac{\Sigma(\hat{Y} - \bar{Y})(Y - \bar{Y})}{\sqrt{[\Sigma(\hat{Y} - \bar{Y})^2][\Sigma(Y - \bar{Y})^2]}}$$

is based on that of the Pearson correlation coefficient (Equation 8.26). The numerator contains the sum of cross-products of Y and $\hat{Y}$, which may be found in Table 10.4. The denominator contains the sums-of-squares of $\hat{Y}$ and Y, which are in Table 10.4 and Table 10.2, respectively. When these values are entered into the formula, we get

$$R = \frac{200}{\sqrt{(200)(216)}} = \frac{200}{\sqrt{43200}} = \frac{200}{207.846} = .9623$$

Literally, this means that an increase of 1 standard deviation in $\hat{Y}$ is associated with an increase of .9623 of a standard deviation in Y. This is a very strong positive correlation.

Unlike the bivariate correlation, the multiple correlation can never be negative. A negative correlation would mean that as the predicted scores increase, the observed values decrease! This is not possible. The worst predictions that are possible are that $\hat{Y} = \bar{Y}$ for all cases, which is a correlation of 0. Thus, the range of values of the multiple correlation is

$$0 \leq R \leq 1$$

You will remember that the square of the bivariate correlation (which equals the squared standardized slope) equals the coefficient of determination ($r^2 = CD$). The same equality exists for the two measures of accuracy in multiple regression; that is, the squared multiple correlation equals the coefficient of multiple determination:

$$R^2 = CMD$$

In our example data, $R^2 = (.9623)^2 = .9260$, which is the value of CMD that was calculated by using Equation 10.9. Because of this identity, the coefficient of multiple determination is often called R^2, or R-squared.

Because a squared fraction is always smaller than the fraction itself, the multiple correlation is always greater than the coefficient of multiple determination:

$$R > CMD$$

This relationship, however, does not make one measure better or worse than the other. R and CMD are both equally valid measures of the strength of association between the dependent variable and the set of independent variables used in the multiple regression.

Standardized Partial Slopes

We have developed standardized measures of the accuracy of predictions provided by a multiple regression equation (CMD and R). These measures indicate how closely the dependent variable is related to the independent variables as a set. They do not, however, show the strengths of relationships between Y and the individual Xs.

The partial regression slopes (b_1 and b_2) cannot be used to judge the strength of associations because they are *unstandardized*. That is, they have no fixed maximum and minimum values that would allow us to evaluate the closeness of the relationship. Just as in the bivariate case, the greater is the ratio of the standard deviation of the dependent variable to that of the independent variable (e.g., s_Y/s_{X_1}), the greater the absolute value of the slope. In our example data, for instance, $b_1 = -.2$ and $b_2 = -2$. We cannot, however, conclude that the number of siblings (X_2) has a stronger effect on education than age (X_1) just because the absolute value of its slope is ten times as large as that of age.

We will now develop a **standardized partial slope** to measure the strength of partial association. The logic is the same as that for the standardized bivariate slope. In the bivariate case, we standardized X and Y and then calculated the bivariate slope for z_Y regressed on z_X. Because all standardized variables have a standard deviation equal to 1, the slope for z_Y regressed on $z_X(B_{YX})$ has a maximum value of 1 and a minimum value of -1. Thus, in multiple regression we standardize the dependent variable and each independent variable. The notation for the standardized Xs is

$$z_1 = \frac{X_1 - \overline{X}_1}{s_1}$$

$$z_2 = \frac{X_2 - \overline{X}_2}{s_2}$$

$$\vdots$$

$$z_k = \frac{X_k - \overline{X}_k}{s_k}$$

where $s_1, s_2, \ldots, s_k$ are the standard deviations of $X_1, X_2, \ldots, X_k$. We then use the same procedure we used to calculate the partial slopes (b_1 and b_2), except we apply it to the standardized variables (z_1, z_2, and z_Y). The result is the standardized multiple regression equation

$$\hat{z}_Y = B_1 z_1 + B_2 z_2 \tag{10.11}$$

Capital Roman italic Bs are again used for standardized slopes. There is no intercept in Equation 10.11 because it equals 0, just as in the bivariate case. The steps for calculating B_1 are outlined in Box 10.3. Steps 2 through 5 are the same as those specified in Box 10.1 except for the fact that they are applied to z scores.

First, we must calculate the standard deviations that are needed to standardize X_1, X_2, and Y. Using the sums-of-squares from Table 10.2, the standard deviations are

$$s_1 = \sqrt{\frac{1000}{5}} = 14.1421$$

$$s_2 = \sqrt{\frac{20}{5}} = 2$$

$$s_Y = \sqrt{\frac{216}{5}} = 6.5727$$

BOX 10.3 Standardized Slope for X_1 Holding Constant X_2

1. Compute z_1, z_2, and z_Y.
2. Regress z_1 on z_2 to determine $B_{X_1X_2}$.
3. Compute $\hat{z}_1 = B_{X_1X_2}z_2$.

4. Compute the residuals $z_1 - \hat{z}_1$.
5. Regress z_Y on $z_1 - \hat{z}_1$ to determine B_1.

Now we can specify the following equations for calculating the standardized scores:

$$z_1 = \frac{X_1 - 45}{14.1421}$$

$$z_2 = \frac{X_2 - 3}{2}$$

$$z_Y = \frac{Y - 12}{6.5727}$$

The standardized scores for each of the five cases are given in Table 10.5. For example, $z_2 = (2 - 3)/2 = -.5$ for the first case.

The second step is to regress z_1 on z_2 to get $B_{X_1X_2}$:

$$B_{X_1X_2} = \frac{\Sigma z_1 z_2}{n} = \frac{3.5355}{5} = .7071$$

This standardized slope indicates that age increases by .7071 standard deviation per standard deviation increase in siblings.

The third step is to calculate the predicted z scores for age shown in Table 10.5. For the first case, for example,

$$\hat{z}_1 = .7071z_2 = .7071(-.5) = -.3536$$

Notice that the sum of $\hat{z}_1$ equals 0, and thus the predicted z scores have the same mean as the z scores themselves (0).

Step 4 is to compute the residuals of z_1 shown in Table 10.5. For the first case,

$$z_1 - \hat{z}_1 = -1.4142 - (-.3536) = -1.4142 + .3536 = -1.0607$$

The sum of the residuals for the standardized scores equals 0, just as the sum of any residuals would.

The final step is to regress z_Y on $z_1 - \hat{z}_1$ to determine B_1. Based on the general formula for a least-squares slope, which is $b = \text{SCP}/(\text{SS of IV})$, the formula for the standardized partial slope for X_1 is

$$B_1 = \frac{\Sigma z_Y(z_1 - \hat{z}_1)}{\Sigma(z_1 - \hat{z}_1)^2} \tag{10.12}$$

Table 10.5 Data for Standardized Partial Regression Slope (B_1) for Education (Y) Regressed on Age (X_1) and Siblings (X_2)

z_1	z_2	z_Y	z_1z_2
−1.4142	−.5000	.6086	.7071
−.7071	−1.5000	1.5215	1.0607
.0000	.0000	.0000	.0000
.7071	1.5000	−.9129	1.0607
1.4142	.5000	−1.2172	.7071
.0000	.0000	.0000	3.5355

$\hat{z}_1$	$z_1 - \hat{z}_1$	$(z_1 - \hat{z}_1)^2$	$z_Y(z_1 - \hat{z}_1)$
−.3536	−1.0607	1.1250	−.6455
−1.0607	.3536	.1250	.5379
.0000	.0000	.0000	.0000
1.0607	−.3536	.1250	.3228
.3536	1.0607	1.1250	−1.2910
.0000	.0000	2.5000	−1.0758

This formula says to divide the sum of cross-products for z_Y and $z_1 - \hat{z}_1$ by the sum-of-squares of $z_1 - \hat{z}_1$. When these quantities are entered into Equation 10.12 from Table 10.5, we get

$$B_1 = \frac{-1.0758}{2.5} = -.4303$$

This standardized partial slope means that education decreases by .4303 of a standard deviation for each standard deviation increase in age, *with siblings held constant.*

The formula for the standardized partial slope of X_2 is

$$B_2 = \frac{\Sigma z_Y(z_2 - \hat{z}_2)}{\Sigma(z_2 - \hat{z}_2)^2} \tag{10.13}$$

The residuals $z_2 - \hat{z}_2$ in this equation have not been calculated and shown in Table 10.5. You may want to calculate them and use them to calculate B_2 to insure that you understand this statistic.

Equations 10.12 and 10.13 are the defining equations for B_1 and B_2, respectively. The following formulas, which are analogous to Equation 8.25 for the standardized bivariate slope, are also useful:

$$B_1 = \frac{s_1}{s_Y}b_1$$

$$B_2 = \frac{s_2}{s_Y}b_2 \tag{10.14}$$

Entering the standard deviations calculated earlier into these equations, we get

$$B_1 = \frac{14.1421}{6.5727}(-.2) = -.4303$$

$$B_2 = \frac{2}{6.5727}(-.2) = -.6086$$

Notice that Equation 10.14 gives the same value for B_1 as Equation 10.12. Although B_1 is easier to calculate with Equation 10.14, the objective of using Equation 10.12 is to illustrate the meaning of this slope. With the values of B_1 and B_2 in hand, we can now write the standardized multiple regression equation as

$$\hat{z}_Y = -.43z_1 - .61z_2$$

Do the values of B_1 and B_2 indicate strong partial associations? *Despite the fact that B_1 and B_2 are named standardized partial slopes, they do not have a fixed range of values* because the residuals $z_1 - \hat{z}_1$ and $z_2 - \hat{z}_2$ are not standardized. Notice in Table 10.5 that the sum-of-squares of $z_1 - \hat{z}_1$ equals 2.5. We can compute the standard deviation of these residuals as follows:

$$s_{z_1 - \hat{z}_1} = \sqrt{\frac{2.5}{5}} = \sqrt{.5} = .7071$$

If $z_1 - \hat{z}_1$ was standardized, it would have a standard deviation of 1. Instead, it has a standard deviation of less than 1 because part of its variance was removed to hold constant z_2. Consequently, because z_Y is standardized, the dependent and independent variables (z_Y and $z_1 - \hat{z}_1$) do not have the same standard deviations. Thus, unlike the standardized bivariate slope, B_1 (and B_2) does not have maximum and minimum values of 1 and -1, respectively.

Despite the fact that the standardized partial slope does not have a fixed range of values, it is still commonly used to measure the strength of partial associations in multiple regression analysis.[3] There are two justifications. First, it is quite unusual in social research for its values to be greater than 1 or less than -1. Thus, practically speaking, we may compare the values of B_1 and B_2 to 1 or -1 to evaluate the strength of the partial associations.

Second, we may validly compare the absolute values of B_1 and B_2 to one another to evaluate whether X_1 or X_2 has a stronger partial relationship with Y. This is appropriate because *the standardized partial slopes indicate how many standard deviations the dependent variable changes as a result of a 1-standard-deviation increase in one independent variable, with the other independent variable(s) held constant.* That is, both Bs are interpreted in terms of standardized values of X_1, X_2, and Y. For our example data, therefore, we can conclude that the number of siblings ($B_2 = -.61$) has a stronger effect on education than age ($B_2 = -.43$).

---

3. There are two measures of the strength of partial association that are standardized to range from -1 to 1. They are the semipartial correlation and the partial correlation (McClendon, 1994). The square of the semipartial correlation indicates the proportion of the variance of Y that is uniquely explained by one X. Thus, it could be named a partial coefficient of determination.

You will remember that the square of the standardized bivariate slope equals the coefficient of determination ($B^2 = $ CD). Consequently, B^2 equals the proportion of the variance of Y that is explained by X. There is not, however, an analogous result for the standardized partial slopes in multiple regression. As a result of the fact that the partial Bs do not range from -1 to 1, B_1^2 and B_2^2 do *not* indicate the proportions of variance in Y that are explained by X_1 and X_2, respectively.

Baseball Example

Let's return to the issue raised in the opening section of this chapter, the "boys of summer." Doris and her father disagreed about whether hitting or pitching is more important for winning in major league baseball. We will now use multiple regression to evaluate this issue. The data to be analyzed pertain to the sixteen National League teams during the 2000 season (Table 10.6).

The National League has changed significantly since Doris rooted for the Brooklyn Dodgers in the 1950s. For example, the Dodgers moved to Los Angeles long ago, the New York Giants moved to San Francisco, and the New York Mets, among others, were added to the league. The teams are listed in descending order according to their winning percentage (Win). Post-season games are not included in the statistics. (The New York Mets won the playoffs to become the league champion.)

Winning percentage will be the dependent variable in the regression analysis. Runs batted in (RBI) has been chosen as the measure of a team's hitting ability. RBI is the mean runs scored per game that did not result from defensive errors by the opponents. The measure of team pitching will be the earned run average (ERA) of the pitching staff. ERA is the mean number of runs allowed per game that did not result from defensive errors made by the team. RBI and ERA are the independent variables.

The results of the regression analysis are given in Table 10.7. Based on the unstandardized coefficients (bs) and the intercept listed in the table, we can write the regression equation as

$$\widehat{\text{Win}} = 69.42 + 8.49\text{RBI} - 13.00\text{ERA}$$

The intercept is not meaningful in this case because it represents the expected winning percentage of a team that averaged zero runs batted in per game and zero earned runs given up per game, an impossibility in the major leagues. The slope for RBI indicates that for each additional run batted in per game the team winning percent increases by 8.49 percentage points, with earned run average held constant. The ERA slope shows that team winning percent drops by 13 percentage points for each additional earned run given up by the pitchers, with runs batted in held constant. Which is more important, ERA or RBI? Because ERA and RBI are both measured on the same scale (runs per game), we can compare their unstandardized slopes. Giving up one fewer run per game, in comparison to scoring one more run per game, leads to a $13.00 - 8.49 = 4.51$ greater percentage point increase in winning. Thus, pitching is more important than hitting.

The standardized slopes (Bs) also verify the conclusion that pitching is more important than hitting. An increase of 1 standard deviation in ERA leads to a reduction of .839 of a standard deviation in winning, holding RBI constant. This is substantially greater than the .573-standard-deviation increment in winning that results from a standard

Table 10.6 National League Baseball, 2000

Team	Win	RBI	ERA
1 San Francisco	59.90	5.49	4.21
2 Atlanta	58.60	4.68	4.06
3 St. Louis	58.60	5.19	4.38
4 New York	58.00	4.70	4.16
5 Los Angeles	53.10	4.67	4.10
6 Cincinnati	52.50	4.90	4.33
7 Arizona	52.50	4.67	4.35
8 Colorado	50.60	5.59	5.26
9 Florida	49.10	4.27	4.56
10 San Diego	46.90	4.41	4.52
11 Milwaukee	45.10	4.37	4.63
12 Houston	44.40	5.56	5.41
13 Pittsburgh	42.60	4.62	4.93
14 Montreal	41.40	4.35	5.13
15 Philadelphia	40.10	4.12	4.77
16 Chicago	40.10	4.46	5.25

Table 10.7 Regression of Winning Percentage on Runs Batted in (RBI) and Earned Run Average (ERA) of National League Teams, 2000

Variables	b	B		
RBI	8.485	.573		
ERA	−12.997	−.839		
Intercept	69.422			
		R	R-Squared	SE of Est.
Goodness-of-fit		.955	.913	2.20

deviation increase in runs batted in. Thus, we can validly conclude that pitching is more important than hitting in 2000. We should not conclude, however, that Doris's father was right without first conducting an analysis of the 1953 National League data.

How accurately do these measures of hitting and pitching predict team winning percentage? The multiple correlation ($R = .955$) is close to 1. The coefficient of multiple determination ($R^2 = .913$) indicates that ERA and RBI together explain 91.3 percent of the variance in winning percentage. Thus, our multiple regression model is a very powerful predictor of winning in modern–day baseball.

Three or More Independent Variables

The concepts and formulas for multiple regression with three or more independent variables are straightforward extensions of those for two independent variables. The population regression model for the case of three independent variables is

$$Y = \alpha + \beta_1 X_1 + \beta_2 X_2 + \beta_3 X_3 + \varepsilon$$

The meanings of the coefficients in this equation are

$$\alpha = \text{value of } Y \text{ when } X_1 = X_2 = X_3 = \varepsilon = 0$$

$$\beta_1 = \text{slope of } Y \text{ on } X_1 \text{ with } X_2 \text{ and } X_3 \text{ held constant}$$

$$\beta_2 = \text{slope of } Y \text{ on } X_2 \text{ with } X_1 \text{ and } X_3 \text{ held constant}$$

$$\beta_3 = \text{slope of } Y \text{ on } X_3 \text{ with } X_1 \text{ and } X_2 \text{ held constant}$$

The intercept α is the value of Y when all three Xs and ε equal 0. *The partial slopes (β_is) equal the change in Y per unit increase in one X with the other Xs held constant.*

The sample regression equation for estimating the population model is

$$\hat{Y} = a + b_1X_1 + b_2X_2 + b_3X_3$$

The least-squares coefficients for the preceding equation are

$$b_1 = \frac{\Sigma(X_1 - \hat{X}_1)(Y - \overline{Y})}{\Sigma(X_1 - \hat{X}_1)^2} \quad \text{where } \hat{X}_1 = a' + b_2'X_2 + b_3'X_3$$

$$b_2 = \frac{\Sigma(X_2 - \hat{X}_2)(Y - \overline{Y})}{\Sigma(X_2 - \hat{X}_2)^2} \quad \text{where } \hat{X}_2 = a'' + b_1''X_1 + b_3''X_3$$

$$b_3 = \frac{\Sigma(X_3 - \hat{X}_3)(Y - \overline{Y})}{\Sigma(X_3 - \hat{X}_3)^2} \quad \text{where } \hat{X}_3 = a''' + b_1'''X_1 + b_2'''X_2$$

$$a = \overline{Y} - b_1\overline{X}_1 - b_2\overline{X}_2 - b_3\overline{X}_3$$

The formulas for b_1 and b_2 look identical to Equations 10.5 and 10.6 for two independent variables. The difference is that the predicted values of each independent variable ($\hat{X}_1$, $\hat{X}_2$, $\hat{X}_3$) are obtained by regressing the independent variable on the other *two* independent variables. For example, $\hat{X}_1 = a' + b_2'X_2 + b_3'X_3$. (We use a' and b' to indicate that they are not the same coefficients as those in the equation for $\hat{Y}$.) Consequently, each X is residualized on two other Xs. That is how two independent variables are held constant when Y is regressed on the residuals of the third X. If there were four Xs, each of them would be residualized on the other three Xs.

The procedure specified for computing the appropriate residuals for each independent variable involves conducting a multiple regression with each X dependent on the others. This would be a long and tedious process with three independent variables and virtually impossible with four or more. Matrix algebra solutions can be carried out with a pocket calculator more easily than the residualization procedure that we have learned (Cohen and Cohen, 1983). Even the matrix algebra method, however, is quite demanding. Therefore, we will not carry out any pocket-calculator solutions for multiple regression with three or more independent variables. Instead, we will rely on computer programs that implement the matrix algebra method. We have studied the residualization procedure to understand how multiple regression is able to hold constant independent variables. Such understanding is very difficult to acquire through any other approach.

INFERENCES IN MULTIPLE REGRESSION

In this section we return to the two types of inferential statistics: confidence intervals and hypothesis tests. We now apply them to partial slopes and to the fit of the multiple regression equation. For making inferences about partial slopes, we will again use the t distribution. The procedures are identical to those we followed for the bivariate slope. The formula for the standard error of the partial slope, however, is different in one important way from that for the bivariate slope. The test for the fit of the entire regression equation is quite different from hypothesis tests of slopes. It is based on a new distribution named the F distribution.

Inferences Based on *t* Scores

First, we will use the sample partial slope (b_i) to make inferences about the population partial slope (β_i). The sampling distribution of the partial slope is unbiased; that is, its mean equals the population parameter (β_i), just as in the case of a bivariate slope (b). The partial slope also has a normal distribution, just as the bivariate slope does. Moreover, the partial slope divided by its estimated standard error (s_{b_i}) also has a t distribution:

$$\frac{b_i - \beta_i}{s_{b_i}} \sim t$$

Thus, we can use the t distribution to construct confidence intervals around b_i, and we can compute t scores to test null hypotheses about b_i.

Standard Errors for Partial Slopes

The standard error of partial slope is *not* equal to that of the bivariate slope, although the two are logically related. You will remember from Chapter 9 that the standard error of the bivariate slope equals

$$s_b = \frac{\hat{s}_{Y-\hat{Y}}}{s_X \sqrt{n}}$$

Note that the denominator contains the standard deviation of the raw X values. We have defined the partial slope b_i as the slope of Y regressed on the residuals $X_i - \hat{X}_i$. If we substitute the standard deviation of $X_i - \hat{X}_i$ for the standard deviation of X in the preceding formula, we get a formula for the standard error of the partial slope:

$$s_{b_i} = \frac{\hat{s}_{Y-\hat{Y}}}{s_{X_i-\hat{X}_i} \sqrt{n}} \tag{10.15}$$

where s_{b_i} is the standard error of the partial slope for X_i. The numerator of this formula contains the symbol for the standard error of estimate, which looks identical to the numerator of the formula for the standard error of the bivariate slope. They are, however, not the same quantities. In Equation 10.15, $\hat{s}_{Y-\hat{Y}}$ is the standard error of estimate from a multiple regression equation, not from a bivariate regression.

We can illustrate concretely the standard error of the partial slope by returning to the hypothetical data on education, age, and siblings. The sums-of-squares of $X_i - \hat{X}_i$ and $Y - \hat{Y}$ from Table 10.3 and Table 10.4, respectively, may be entered in Equation 10.15 as follows to compute the standard errors of b_1 and b_2:

$$s_{b_1} = \frac{\hat{s}_{Y-\hat{Y}}}{s_{X_1-\hat{X}_1} \sqrt{n}} = \frac{\sqrt{16/(5-3)}}{\sqrt{500/5}\sqrt{5}} = \frac{2.828}{10(2.236)} = .1265$$

$$s_{b_2} = \frac{\hat{s}_{Y-\hat{Y}}}{s_{X_2-\hat{X}_2} \sqrt{n}} = \frac{\sqrt{16/(5-3)}}{\sqrt{10/5}\sqrt{5}} = \frac{2.828}{1.414(2.236)} = .8944$$

Notice that both equations contain the same standard error of estimate (2.828) and the same square root of n (2.236). They differ in only the standard deviations of $X_1 - \hat{X}_1$ and $X_2 - \hat{X}_2$ (i.e., 10 and 1.414, respectively). Consequently, the standard error of X_1 (.1265) is much smaller than that of X_2 (.8944).

Although Equation 10.15 shows the logical relationship between the standard errors of the partial and bivariate slopes, the following formula shows more clearly the factors that determine the **standard error of the partial slope** b_i:

$$s_{b_i} = \frac{\hat{s}_{Y-\hat{Y}}}{s_i \sqrt{n} \sqrt{(1 - R_i^2)}} \tag{10.16}$$

This formula looks like the one for a bivariate slope except that $\sqrt{(1 - R_i^2)}$ has been added to the denominator. R_i^2 equals the coefficient of multiple determination for X_i. As such, it indicates the proportion of the variance of X_i that is related to the other Xs.[4] If R_i^2 is high, X_i has high **multicollinearity.** High multicollinearity means that one independent variable (e.g., X_1) is highly correlated with the other independent variables (e.g., X_2 and X_3). R_i^2 is not to be confused with the CMD for the dependent variable (R_Y^2). If there are $k = 3$ independent variables, for example, R_1^2 is equal to

$$R_1^2 = \frac{(\hat{X}_1 - \bar{X}_1)^2}{(X_1 - \bar{X}_1)^2}$$

where

$$\hat{X}_1 = a' + b_2' X_2 + b_3' X_3$$

That is, we will regress X_1 on X_2 and X_3 to get the predicted values of X_1. Then we will divide the regression sum-of-squares of X_1 by the total sum-of-squares of X_1 to get R_1^2.

Given the meaning of R_i^2, $1 - R_i^2$ equals the proportion of the variance of X_i that is unrelated to (i.e., independent of) the other Xs. This quantity is called the **tolerance of X_i:**

$$\text{Tolerance of } X_i = 1 - R_i^2$$

The greater the tolerance, the greater the variance of X_i that is free and independent of the other independent variables. High tolerance means low multicollinearity, which is a good thing. Because $\sqrt{(1 - R_i^2)}$ is in the denominator of the standard error of b_i, *the greater the tolerance, the smaller the standard error of b_i.* Small tolerances, or high multicollinearities, cause large standard errors. In other words, the more highly a variable is correlated with the other independent variables, the more difficult it is to reliably estimate its slope separately from those of the other Xs.

When there are just two independent variables, R_i^2 is a squared bivariate correlation:

$$R_i^2 = r_{X_1 X_2}^2 \text{ where } i = 1, 2$$

In our hypothetical data from Table 10.5,

$$r_{X_1 X_2} = \frac{\Sigma z_1 z_2}{n} = \frac{3.5355}{5} = .7071$$

$$r_{X_1 X_2}^2 = (.7071)^2 = .5000$$

---

4. Therefore, $s_i \sqrt{(1 - R_i^2)} = \sqrt{s_i^2 (1 - R_i^2)} = \sqrt{s_{X_i - \hat{X}_i}^2} = s_{X_i - \hat{X}_i}$, which shows that Equations 10.15 and 10.16 are equal.

which indicate that both independent variables (age and siblings) share 50 percent of their variances with one another. Using the standard deviations of X_1 and X_2 calculated earlier ($s_1 = 14.1421$ and $s_2 = 2$), we can use Equation 10.16 to calculate the standard errors of b_1 and b_2:

$$s_{b_1} = \frac{\hat{s}_{Y-\hat{Y}}}{s_1\sqrt{n}\sqrt{(1-R_1^2)}} = \frac{2.828}{14.1421\sqrt{5}\sqrt{1-.5}} = \frac{2.828}{(14.1421)(2.236)(.7071)}$$

$$= .1265$$

$$s_{b_2} = \frac{\hat{s}_{Y-\hat{Y}}}{s_2\sqrt{n}\sqrt{(1-R_2^2)}} = \frac{2.828}{2\sqrt{5}\sqrt{1-.5}} = \frac{2.828}{(2)(2.236)(.7071)}$$

$$= .8943$$

These standard errors equal those calculated earlier using Equation 10.15.

Notice what would happen if $R_i^2 = 0$ and everything else remained the same:

$$s_{b_1} = \frac{2.828}{14.1421\sqrt{5}\sqrt{1-.0}} = \frac{2.828}{(14.1421)(2.236)(1)} = .0894$$

$$s_{b_2} = \frac{2.828}{2\sqrt{5}\sqrt{1-.0}} = \frac{2.828}{(2)(2.236)(1)} = .6324$$

The standard errors would be substantially smaller if the independent variables were uncorrelated. To be precise, as a result of the correlation between the independent variables, the standard errors are increased by a factor of

$$\sqrt{\frac{1}{1-R_i^2}} = \sqrt{\frac{1}{1-.5}} = \sqrt{2} = 1.414$$

which means that the standard error is 1.414 times larger (41.4% larger) than it would be if X_i were uncorrelated with the other Xs. We call this factor the **standard error inflation factor** (SEIF):

$$\text{SEIF} = \sqrt{\frac{1}{1-R_i^2}}$$

The SEIF is always greater than or equal to 1 (SEIF $\geq$ 1). It indicates how much R_i^2 inflates the standard error s_{b_i}. The greater R_i^2, the greater will be s_{b_i}.[5]

Confidence Intervals for Partial Slopes

Now that we know how to estimate the standard error of the partial slope, we may construct a confidence interval for that slope using the same method we used for the bivariate slope. The confidence interval for the partial slope of X_i is

$$100P\%\ \text{CI}_{\beta_i} = b_i \pm t_{\alpha/2}s_{b_i} \tag{10.17}$$

---

5. The quantity under the radical sign is called the **variance inflation factor** (VIF): VIF $= 1/(1-R_i^2)$. It indicates how much the sampling variance of b_i increases due to R_i^2 (thus the name *variance* inflation factor). But the quantity that we really are concerned with is the effect that R_i^2 has on the standard error of the partial slope, which is $\sqrt{\text{VIF}} = \sqrt{1/(1-R_i^2)}$.

As shown in the formula, to establish the confidence interval we must determine the critical value of t for our selected α level ($t_{\alpha/2}$). Because there is a different t distribution for each of the degrees of freedom, we must calculate the degrees of freedom. The degrees of freedom for a partial slope equal

$$df = n - k - 1$$

where, as usual, k equals the number of independent variables.

To illustrate the construction of confidence intervals, we will use the 1998 GSS to estimate a multiple regression model for years of school completed by females. Education will be regressed on four independent variables: age, siblings, father's education, and mother's education:

$$Y = \text{education}$$

$$X_1 = \text{age}$$

$$X_2 = \text{siblings}$$

$$X_3 = \text{father's education}$$

$$X_4 = \text{mother's education}$$

The 875 women who are 30 or more years of age compose the sample. Thus, the degrees of freedom are

$$df = 875 - 4 - 1 = 870$$

We will use the conventional $\alpha = .05$. With α and the degrees of freedom, we can locate the critical value of t in Table A.2. Looking in column $A = \alpha/2 = .025$ and row $df = 750$ (the row closest to our df), we find $t_{\alpha/2} = 1.963$, which is our critical t value.

Table 10.8 shows the computer-generated output of the regression analysis. There are some new statistics in the table associated with the SS and df that we will take up later. We will also defer examining the t scores and p values associated with the partial slopes until the next section. The partial slopes show that both age and number of siblings have negative effects on women's education, as they did in the hypothetical data. As indicated by the standardized partial slopes (Bs), however, these slopes are both close to 0. Thus, they are not very strong effects. The partial slopes for both father's and mother's education are positive, which indicates that as each parent's educational attainment increases, a woman's educational attainment increases if we hold constant age, siblings, and the other parent's education. The partial slope for the father's education is about 50 percent larger than that for the mother's education, indicating that a father's education influences his daughter's education more than does her mother's education. It is valid to compare the two unstandardized slopes because both variables are measured on the same scale (years of school completed). The standardized slopes also indicate that the father's education has a stronger effect than the mother's education. Furthermore, the standardized slopes show that both the mother's and the father's education are more closely related to their daughter's education than is the daughter's age and the number of her brothers and sisters.

How well do these four variables predict women's educational attainment? The multiple correlation is .547, which is slightly above the middle of the range of R, which

Table 10.8 Regression of Education on Age, Siblings, Father's Education, and Mother's Education: Females 30+ Years of Age, GSS 1998

Variables	Mean	St. Dev.	n		
Education	13.655	2.916	875		
Age	50.189	14.591	875		
Siblings	3.976	3.393	875		
Father's Education	10.920	4.154	875		
Mother's Education	11.066	3.548	875		
Variables	**b**	**SE**	**B**	**t**	**p**	**TOL**
Age	−.012	.006	−.062	−2.037	.042	.878
Siblings	−.082	.025	−.095	−3.271	.001	.949
Father's Education	.243	.027	.347	9.171	.000	.564
Mother's Education	.160	.031	.195	5.153	.000	.564
Intercept	10.170	.521		19.528	.000	
Source	**SS**	**df**	**MS**	**F**	**p**	
Regression	2224.343	4	556.086	92.798	.000	
Residual	5213.424	870	5.992			
Total	7437.767	874				
	R	**R-Squared**	**SE of Est.**			
Goodness-of-fit	.547	.299	2.448			

indicates that the predictive power is moderately strong. The coefficient of multiple determination (listed in most computer output as R-squared), however, is .299, which means that about 30 percent of the variance of women's education is explained by the four independent variables. Because the value of R is close to the value at which the greatest discrepancy between R and R^2 occurs (i.e., $R^2 = .5^2 = .25$), the two measures disagree on the goodness-of-fit of the regression equation. Thus, there is some ambiguity in evaluating the fit of the model.

Before constructing the confidence interval, we will use the relevant data in Table 10.8 to illustrate how the standard error of the slope for father's education (X_3) is computed. All the information needed by Equation 10.16 is included in the table. The column labeled TOL contains the tolerance values for each variable. You can see that the father's and mother's educations have much lower tolerances than age and siblings. (It is a coincidence that the tolerances for mother's and father's educations are equal up to three decimal places.) This indicates that R_i^2 is much greater for both parents' educations (primarily because $r = .645$ for this pair of variables [not shown]). The consequence is that their standard errors will be inflated more than those of age and siblings. Entering the appropriate statistics from Table 10.8 into Equation 10.16 gives

$$s_{b_3} = \frac{\hat{s}_{Y-\hat{Y}}}{s_3\sqrt{n}\sqrt{(1 - R_3^2)}} = \frac{2.448}{4.154\sqrt{875}\sqrt{.564}} = .027$$

This result matches the value given in Table 10.8. It indicates that the standard deviation of the sampling distribution for b_3 is estimated to be .027.

Entering the partial slopes and their standard errors from Table 10.8 along with our critical value of t into Equation 10.17, we calculate the 95-percent confidence interval for the four partial slopes as

$$95\% \; CI_{\beta_1} = b_1 \pm t_{\alpha/2}s_{b_1} = -.012 \pm 1.963(.006) = -.012 \pm .0118$$
$$= -.024 \text{ to } -.0002$$

$$95\% \; CI_{\beta_2} = b_2 \pm t_{\alpha/2}s_{b_2} = -.082 \pm 1.963(.025) = -.082 \pm .049$$
$$= -.131 \text{ to } -.033$$

$$95\% \; CI_{\beta_3} = b_3 \pm t_{\alpha/2}s_{b_3} = .243 \pm 1.963(.027) = .243 \pm .053 = .193 \text{ to } .296$$

$$95\% \; CI_{\beta_4} = b_4 \pm t_{\alpha/2}s_{b_4} = .160 \pm 1.963(.031) = .160 \pm .061 = .099 \text{ to } .221$$

Notice that neither the confidence interval for age nor for siblings contains the value 0 (i.e., no association), although it was necessary to carry the upper bound for age to four decimal places to demonstrate this fact. Therefore, we can be very certain that there is a negative association between these variables and women's education. Thus, their small associations with education (as indicated by their Bs) are not likely to have occurred by chance.

Now, let's examine the confidence intervals for mother's education and father's education. The lower end of the interval for father's education equals .193, which is smaller than the upper end of the interval for mother's education (.221), which means that the true slopes for these two variables may be equal or very nearly equal. For example, $\beta_3 = \beta_4 = .21$ is within both confidence intervals. Thus, we cannot be as confident as we might like in drawing the conclusion that the father's education has a stronger effect than the mother's education.[6] Because 0 is not within either interval, however, we can be very certain that each parent's education has an independent influence on the daughter's education.

Hypothesis Tests for Partial Slopes

We discussed hypothesis tests for the mean (Chapter 6) and for the bivariate regression slope (Chapter 9). The procedure for conducting t tests for the partial slope is identical to that for the mean and bivariate slope. Box 10.4 outlines the eight-step procedure.

We already performed steps 2 through 6 when we constructed confidence intervals for partial slopes. Thus, we do not have to review these steps. We will, however, use the regression output for women's educational attainment in Table 10.8 to illustrate the eight-step test. The test will be conducted for the partial slope of education regressed on siblings (β_2). Table 10.9 shows the test.

---

6. The standard error for the difference between two partial slopes equals $s_{b_i - b_j} = \sqrt{s_{b_i}^2 + s_{b_j}^2 - 2s_{b_i b_j}}$, where $s_{b_i}^2$ is the sampling variance for b_i, $s_{b_j}^2$ is the sampling variance for b_j, and $s_{b_i b_j}$ is the covariance between b_i and b_j from sample to sample. $s_{b_i b_j}$ can be requested in the output from regression programs. The covariance between b_3 and b_4 in the education example equals $-.0004879$, which means that as the slope for father's education increases from sample to sample, the slope for mother's education tends to decrease. Thus, if we overestimate the slope for the father's education, we tend to underestimate the slope for the mother's education. The standard error for the difference between the slopes for father's and mother's education is $s_{b_3 - b_4} = \sqrt{(.027)^2 + (.031)^2 - 2(-.0004879)} = .05143$, and the 95% confidence interval for $b_3 - b_4$ is $(.2434 - .1601) \pm 1.963(.05143) = .0833 \pm .10096 = -.0177 \text{ to } .1843$ with $df = 975 - 4 - 1 = 970$. Because the confidence interval contains 0, we cannot be confident that the slopes for father's and mother's education differ.

BOX 10.4 t Test for Partial Slope

1. State null and alternative hypotheses.
2. Specify α.
3. Determine $df = n - k - 1$.
4. Find $t_{\alpha/2}$ (two-tailed) or t_α (one-tailed).
5. Calculate the partial slope.
6. Calculate standard error of the partial slope.
7. Calculate the test statistic t.
8. Make your decision.

Table 10.9 t Test for the Slope of Education Regressed on Siblings with Age, Father's Education, and Mother's Education Held Constant

1. $H_0: \beta_2 = 0$
   $H_1: \beta_2 \neq 0$
2. $\alpha = .05$
3. $df = n - k - 1 = 875 - 4 - 1 = 870$
4. $t_{\alpha/2} = t_{.025} = 1.963$
5. $b_2 = -.082$
6. $s_{b_2} = \dfrac{\hat{s}_{Y-\hat{Y}}}{s_2\sqrt{n}\sqrt{1 - R_2^2}} = \dfrac{2.448}{3.393\sqrt{875}\sqrt{.949}} = .025$
7. $t_2 = \dfrac{b_2 - \beta_2}{s_{b_2}} = \dfrac{-.082 - 0}{.025} = -3.271$
8. $(|t| = 3.271) > (t_{.025} = 1.963)$, $\therefore$ reject H_0
   $(t_2 = -3.271) < (-t_{.025} = -1.963)$, $\therefore \beta_2 < 0$
   $p = .001$

In step 1 the null hypothesis (H_0) states that the population slope for siblings (β_2) equals 0. The alternate hypothesis is that the slope is not equal to 0. This is a two–tailed test. We want to entertain the possibility that the slope may either be positive or negative. Step 2 specifies that the maximum probability of a Type I error that we are willing to risk is the conventional $\alpha = .05$. That is, we will tolerate a probability of .05 of incorrectly rejecting the null hypothesis. Steps 3 and 4 show the same degrees of freedom and critical t that we previously used in constructing the confidence intervals for this example. Step 5 lists the slope for siblings shown in Table 10.8. In step 6, we use the standard error of estimate, the standard deviation of siblings, the sample size, and the tolerance for siblings from Table 10.8 to compute the standard error of the partial slope. The sample t score is calculated in step 7. The value specified by the null hypothesis ($\beta_2 = 0$) is entered into the formula to calculate t. In step 8 we reject the null hypothesis because the absolute value of our calculated sample t is greater than the critical t. Because the sample slope is negative, we conclude that the population slope for siblings is less than 0. The probability that we have made a Type I error (the level of significance) is only .001, which is listed as p in Table 10.8. Thus, we can be very confident that women with more siblings complete fewer years of schooling.

Table 10.10 One-Step Hypothesis Test for β_i

1. If $p \le .05$, reject H_0: $\beta_i = 0$ and conclude β_i has same sign as b_i.

We could go through this eight-step hypothesis test for each of the remaining three independent variables. Practically speaking, however, that is not necessary. For example, assume that our null hypothesis is $\beta_i = 0$, a two-tailed test. This is by far the most common null hypothesis in social research. Most computer programs for regression analysis assume that this is the null hypothesis and calculate $t_i = b_i/s_{b_i}$. That is, they enter $\beta_i = 0$ into the formula for t. The program then calculates the exact two-tailed p value associated with the calculated t and the degrees of freedom. This two-tailed p is then listed as part of the computer output. Consequently, if your null hypothesis is $\beta_i = 0$, all you have to do is compare the α you have chosen to the listed p value. If p is less than α, you reject the null hypothesis. If you use the conventional $\alpha = .05$, you simply compare p to .05. Thus, if you are a conventional researcher (i.e., you use $\alpha = .05$ and the null hypothesis $\beta_i = 0$ without even thinking about it), the eight-step hypothesis test is reduced to one step (Table 10.10). Because the p values for age, father's education, and mother's education in Table 10.8 are all less than .05, the conventional researcher will reject the null hypothesis for each and conclude that each has an effect on educational attainment, holding constant the other three variables in the equation.

What would you do if you wanted to do a one-tailed test of either $\beta_i \le 0$ or $\beta_i \ge 0$? If b_i has the same sign as the one the alternative hypothesis predicts for β_i, you would reject the null hypothesis if $p/2 \le .05$. This would be a two-step test because you would have to make a conscious decision in step 1 that you were performing a one-tailed test before you proceeded to compare $p/2$ with .05 in step 2.

Many researchers routinely use the one-step hypothesis test. There is nothing wrong with it as long as you understand the full eight-step test. It is important that you know the logic and meaning of the eight-step test if you are to understand the conclusions you reach in a one-step test.

F Tests

Now we will develop a test of the coefficient of multiple determination, a measure of the goodness-of-fit of the regression model. This statistic, you will remember, equals R^2. The null and alternative hypotheses are

$$H_0: R^2 = 0$$

$$H_1: R^2 > 0$$

The null hypothesis states that the proportion of variance explained by the variables in the model equals 0. The alternative hypothesis states that the explained variance is greater than 0. Remember, R^2 cannot be less than 0. Thus, the alternative hypothesis is one-tailed. These hypotheses are always tested.

The preceding hypotheses about R^2 are equivalent to

$$H_0: \beta_1 = \beta_2 = \cdots = \beta_k = 0$$

$$H_1: \text{one or more } \beta_i \ne 0$$

In this case, the null hypothesis states that all the partial slopes equal 0. If all the slopes equal 0, $R^2 = 0$. The alternative hypothesis indicates that at least one of the partial slopes is not equal to 0. If $R^2 \neq 0$, then one or more of the slopes must not equal 0. Thus, rejection of the null hypothesis that $R^2 = 0$ does not indicate that all of the independent variables are related to Y. We must conduct t tests of each partial slope to determine which variables are significant.

Unlike the regression slopes, we do not calculate a standard error for R^2. Thus, we will not conduct a t test of the null hypothesis. Instead, we will conduct an F test. Just as with t, the greater the value of F, the more likely it is that we will reject the null hypothesis. A formula for the **F statistic** is

$$F = \frac{R^2/df_1}{(1 - R^2)/df_2} \qquad (10.18)$$

where

$$df_1 = k$$
$$df_2 = n - k - 1$$

The numerator of F contains R^2 divided by $df_1 = k$, the *numerator degrees of freedom*. Again, k is the number of independent variables. Thus, the numerator is the proportion of explained variance per variable. For any given value of R^2, the greater the number of Xs, the smaller will be the numerator and thus the smaller will be F. This should seem valid because the more independent variables we include in the regression equation, the greater the likelihood that R^2 could be greater than 0 by chance.

The denominator of Equation 10.18 contains $1 - R^2$, that is, the proportion of variance of Y that is unexplained by the Xs (i.e., the proportion of residual variance). The greater the unexplained variance, the smaller F will be. The proportion of residual variance, however, is divided by $df_2 = n - k - 1$, the *denominator degrees of freedom,* which is the same degrees of freedom as for the standard error of estimate (Equation 10.8). Remember that the standard error of estimate is the estimated standard deviation of the residuals in the population. The greater df_2, the smaller the denominator of Equation 10.18 will be and thus the larger F will be. The greater n is for any given number of Xs (k), the larger df_2 will be, which means that when the sample size is large, the contribution of $1 - R^2$ will be reduced because R^2 is less likely to deviate from 0 by chance when the sample size is larger.

A second formula that is frequently given for F is

$$F = \frac{\Sigma(\hat{Y} - \bar{Y})^2/df_1}{\Sigma(Y - \hat{Y})^2/df_2} = \frac{\text{regression mean square}}{\text{residual mean square}} \qquad (10.19)$$

The numerator contains the regression sum-of-squares (the explained sum-of-squares), which is analogous to R^2 in the numerator of Equation 10.18. The denominator contains the residual sum-of-squares (the unexplained sum-of-squares), which is analogous to $1 - R^2$ in the denominator of Equation 10.18. The degrees of freedom are the same as for Equation 10.18. The regression SS divided by df_1 is called the *regression mean square*. The residual SS divided by df_2 is called the *residual mean square.*

You will remember that the t distribution is a family of distributions, one for each df. The same is true for the F distribution, except that it consists of a larger family of

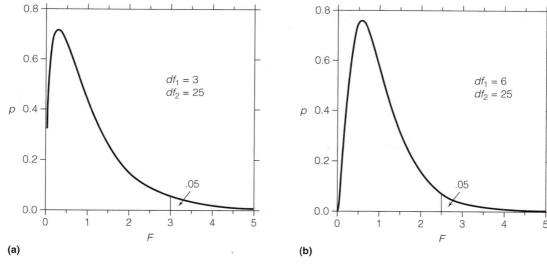

FIGURE 10.5 *F* Distributions with A = .05

distributions. There is a somewhat differently shaped *F* distribution for each pair of numerator and denominator degrees of freedom (df_1, df_2). Figure 10.5 shows two examples. The *F* distribution differs from the *t* (and normal) distribution in two respects. As you can see, the *F* distribution is not symmetrical. It is skewed to the right. Second, the *F* statistic is always positive.

Critical values of the *F* distribution are shown in Tables A.3a, A.3b, and A.3c for A = .05, .01, and .001, respectively. In each table, values of df_1 form the columns and values of df_2 form the rows. At the intersection of each row and column is a critical value of the *F* distribution for the corresponding df_1 and df_2. Table A.3a, for example, shows values of *F* that divide the distribution in two parts, with A = .05 of the distribution lying above *F* and 1 − A = .95 lying below *F*. That is, the table shows the 95th percentile for each pair of df_1 and df_2. For df_1 = 3 and df_2 = 25, for example, $F_{.05}$ = 2.99, which is illustrated in Figure 10.5a. For df_1 = 6 and df_2 = 25, $F_{.05}$ = 2.49, which is illustrated in Figure 10.5b.

Once you have selected your α level, F_{α} will be the critical value of *F* for the appropriate degrees of freedom. This means that the entire α is being placed under one tail—the upper tail—as shown in Figure 10.5. Thus, the *F* test is a one-tailed test.

The steps for conducting the **F test** are outlined in Box 10.5. There are only six steps compared with the eight steps for the *t* test because, first of all, the *F* test always has the same null hypothesis and the same alternative hypothesis. Second, there is no standard error to compute for R^2.

We will use the regression output in Table 10.8 to illustrate the *F* test. In this example, women's education was regressed on age, siblings, father's education, and mother's education. There were *n* = 875 cases and *k* = 4 variables. As listed in Table 10.8, R^2 = .299. The six-step *F* test is shown in Table 10.11. The calculated *F* is far

BOX 10.5   *F*-Test for Regression Equation

1. Specify α.
2. Determine df_1 and df_2.
3. Find F_α.

4. Calculate R^2.
5. Calculate the test statistic F.
6. Make a decision.

Table 10.11 *F* Test for the Regression of Education on Siblings, Age, Father's Education, and Mother's Education

1. $\alpha = .05$
2. $df_1 = 4$
   $df_2 = 875 - 4 - 1 = 870$
3. $F_{.05} = 2.38$
4. $R^2 = .299$
5. $F = \dfrac{.299 \, / \, 4}{(1 - .299) \, / \, 870} = 92.771$
6. $(F = 92.771) > (F_{.05} = 2.38)$, $\therefore$ reject H_0
   Conclude $R^2 > 0$
   $p = .000$

greater than the critical F, and thus we can easily reject the null hypothesis that $R^2 = 0$. In other words, we can reject the hypothesis that none of the Xs are related to Y. As a result, we can conclude that $R^2 > 0$, which means that one or more independent variables are related to Y.

Notice that the calculated F in Table 10.11 is slightly less than the value reported in Table 10.8 ($F = 92.798$) because $R^2 = .299$ is rounded to three decimal places. If, however, we use the values of the regression SS and the residual SS from Table 10.8 in the alternate formula for F (Equation 10.19), we get

$$F = \frac{\Sigma(\hat{Y} - \overline{Y})^2/df_1}{\Sigma(Y - \hat{Y})^2/df_2} = \frac{2224.343/4}{5213.424/870} = 92.798$$

This value is equivalent to that from Table 10.8.

t Tests Versus *F* Tests

Why do we need both t tests for partial slopes and F tests for the fit of the entire regression model? Consider our example of education regressed on age, siblings, father's education, and mother's education. The t tests for the partial slopes for all four independent variables were found to be significant at $p < .05$. The significant F test, however, allowed us to conclude only that at least one of the independent variables was significant. In this case, the t tests told us much more than the F test, but at least there

	Significant F	Nonsignificant F
One or more significant t_i	1. Unambiguous	2. Type I errors
No significant t_i	3. Multicollinearity	4. Unambiguous

FIGURE 10.6 Typology of Results for t Tests Versus F Tests

was no contradiction between the two types of tests. This will also be true if only one, two, or three of the t tests is significant. This case falls in cell 1 of Figure 10.6.

There are also many instances in social research where none of the t tests for the partial slopes are significant and the F test is also not significant. This may not be a very satisfactory result to the investigator, but the t tests and the F test are consistent with one another (cell 4, Figure 10.6).

Another possible outcome will appear to be implausible at first glance. The t tests for all the independent variables may be nonsignificant. Thus, you cannot conclude that any of the Xs has a non-zero association with the dependent variable. Surprisingly, however, it is possible that the F test may be *significant!* This would mean that at least one of the independent variables has a non-zero slope, which clearly contradicts the results of the t tests. How can this happen? High multicollinearity between the Xs may cause this result. If each X is highly correlated with the other Xs (i.e., high multi-collinearity, or low tolerance), the SEIF will inflate the standard errors of each X and, consequently, none of the t tests may be significant. This result leaves the researcher in an ambiguous but understandable situation. The F test tells the investigator that some-thing is going on (at least one of the $\beta_i \neq 0$), but the t tests cannot reliably separate the effects of one or more variables from the others. Thus, this is a logical but, thankfully, relatively infrequent outcome (cell 3, Figure 10.6).

The last possible outcome is that F is not significant but one or more of the t tests are significant. The F test implies that none of the Xs are significant, which is contra-dicted by one or more of the t tests. The explanation for this ambiguous result is not as clear as for the previous case. One possibility, however, is that there is a Type I error for one or more of the Xs. This becomes more and more possible as k increases. In general, the greater the number of Xs that are included in the equation, the greater the proba-bility that one of them may be significant at the specified α level by chance. Thus, it may be the case that the F test for R^2 may be correct in that none of the independent variables is related to Y and that there has been one or more Type I errors among the multiple t tests (cell 2, Figure 10.6).

The fourth possibility has led to the proposal (e.g., Cohen and Cohen, 1983) that the F test be used to protect the t tests. This is called a **protected t test.** The proce-dure calls for conducting the F test first. If the F test is significant, we can conclude that one or more of the independent variables is related to the dependent variable. We would then examine the individual t tests to learn which one(s) does indeed have non-zero partial slopes. If, on the other hand, the initial F test is not significant, we should con-clude that none of the partial slopes differ from zero. Consequently, we should not con-duct (or we should ignore) the t tests of the partial slopes. Restraining ourselves in this way will give us some protection against incorrectly rejecting the null hypothesis for one or more independent variables. It should be noted, however, that not all investigators accept the validity of the protected t test. If investigators are not willing to use this protection, they should at least entertain the possibility that a t test may have led to a Type I error when the F test is not significant.

SUMMARY

Multiple regression is the regression of a single dependent variable on two or more independent variables. The slope for each X is a partial slope that indicates the change in Y per unit increase in X with the other Xs held constant. The way that variables are held constant is by computing residuals for each X that is uncorrelated with the other Xs. The bivariate slope for Y regressed on the residuals of X equals the partial slope for that X. The coefficient of multiple determination for the multiple regression equation equals the regression sum-of-squares divided by the total sum-of-squares of Y, which is the same as in the bivariate case. The multiple correlation equals the correlation between Y and the predicted values of Y; it is always positive. Standardized partial slopes are computed in the same way as the unstandardized slopes, except that they are based on standardized z scores instead of raw X scores. Although the standardized slopes do not have maximum and minimum values, they can be used to compare the effect of one X with that of another X.

We can conduct hypothesis tests and construct confidence intervals for partial slopes in basically the same ways as we did for bivariate slopes. There are two differences, however. The formula for the standard error of the partial slope differs from that of the bivariate slope in that it contains the tolerance of X in the denominator. Tolerance equals 1 minus the squared multiple correlation (i.e., the multicollinearity) between one X and the other Xs in the equation. Thus, it indicates the proportion of the variance of each X that is independent of the other Xs. The greater the tolerance, the smaller the standard error of the partial slope will be. The degrees of freedom of t for the partial slope equals n minus the number of Xs minus 1 (i.e., $n - k - 1$). An additional kind of hypothesis test is used in multiple regression, namely, the F test for the coefficient of multiple determination (R-squared). The F test is a one-tailed test of the null hypothesis that R-squared equals 0, or equivalently, that all the partial slopes equal 0. The F test can be used as a protected t test in which the F test must be significant before the t tests of the partial slopes are conducted. The logic behind the protected t test is that some t tests may be significant by chance when the F test is not significant.

KEY TERMS

multiple regression model	multiple correlation	standard error inflation factor
partial regression slope	standardized partial slope	
residualized X_1	standard error of partial slope	F statistic
standard error of estimate		F test
coefficient of multiple determination	multicollinearity	protected t test
	tolerance of X_i	

SPSS PROCEDURES

We will use the 1998 GSS data to regress the education of women 30 or more years old on age, number of siblings, father's education, and mother's education. To select only those cases that are female and 30 or more years of age, click on *Data* and *Select Cases* to open the *Select Cases* window. Click on *If condition is satisfied* and the *If* button, and in the *Select Cases: If* window enter "sex = 0 and age >= 30" in the box. The code for females is 0, and >= means greater than or equal to. After you have clicked on *Continue* and *OK*, only females who are 30 or more years of age will be available for analysis.

To execute the regression analysis, click on *Analyze, Regression,* and *Linear* to get the *Linear Regression* window. Move the variable *educ* to the *Dependent* box, and move the variables *age, sibs, paeduc,* and *maeduc* to the *Independent* box. To get additional statistics to those generated by the default selections, click on the *Statistics* button to open the *Linear Regression: Statistics* window. In addition to *Estimates* and *Model fit,* which are already checked, click on *Confidence intervals* and *Collinearity diagnostics*. After clicking on *Continue* and *OK,* you will find tables for Model Fit, ANOVA, Coefficients, and Collinearity Diagnostics in the output window. The regression statistics in Table 10.8 are the same as those shown in the output window. The regression mean square and residual mean square found in the ANOVA table are used to calculate the F statistic according to Equation 10.19. Note that the slopes in the coefficients table for SIBS and AGE are listed as $-8.188E\text{-}02$ and $-1.233E\text{-}02$, respectively. "E-02" is scientific notation that means we are to move the decimal point two places to the left, which gives $-.08188$ and $-.01233$ as the slopes. The tolerance values $(1 - R_i^2)$ listed under Collinearity Statistics in the coefficients table are used in the denominator of the formula for the standard error of the partial regression slope. The VIF statistic located next to tolerance stands for variance inflation factor, which is $1/(1 - R_i^2)$. The square root of VIF equals SEIF, which means standard error inflation factor. You may ignore the collinearity diagnostics table.

Model Summary

Model	R	R Square	Adjusted R Square	Std. Error of the Estimate
1	.547[a]	.299	.296	2.448

a. Predictors: (Constant), MAEDUC, SIBS, AGE, PAEDUC

ANOVA[b]

Model		Sum of Squares	df	Mean Square	F	Sig.
1	Regression	2224.343	4	556.086	92.798	.000[a]
	Residual	5213.424	870	5.992		
	Total	7437.767	874			

a. Predictors: (Constant), MAEDUC, SIBS, AGE, PAEDUC

b. Dependent Variable: EDUC

Coefficients^a

Model		Unstandardized Coefficients		Standardized Coefficients	t	Sig.	95% Confidence Interval for B		Collinearity Statistics	
		B	Std. Error	Beta			Lower Bound	Upper Bound	Tolerance	VIF
1	(Constant)	10.170	.521		19.528	.000	9.148	11.193		
	SIBS	-8.188E-02	.025	-.095	-3.271	.001	-.131	-.033	.949	1.05
	AGE	-1.233E-02	.006	-.062	-2.037	.042	-.024	.000	.878	1.14
	PAEDUC	.243	.027	.347	9.171	.000	.191	.295	.564	1.77
	MAEDUC	.160	.031	.195	5.153	.000	.099	.221	.564	1.77

a. Dependent Variable: EDUC

EXERCISES

1. Is the partial slope for X_1 greater than 0, less than 0, or 0? Is the partial slope for X_2 greater than 0, less than 0, or 0?

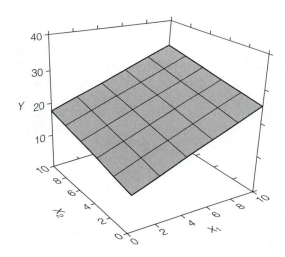

2. Is the partial slope for X_1 greater than 0, less than 0, or 0? Is the partial slope for X_2 greater than 0, less than 0, or 0?

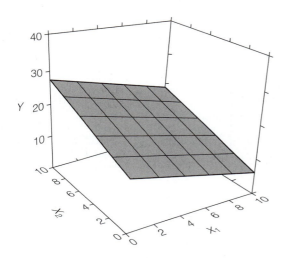

3. In the following table, let IQ be X_1, ED be X_2, and SEI be Y. Use the residualization method to find b_1, b_2, and a for the multiple regression equation.

Case Summaries

	IQ	ED	SEI
1	95	8	32
2	75	10	70
3	105	12	58
4	135	14	86
5	115	16	44

4. In the following table, let SIBS be X_1, EDUC be X_2, and AGEKDBRN be Y. Use the residualization method to find b_1, b_2, and a for the multiple regression equation.

Case Summaries

	SIBS	EDUC	AGEKDBRN
1	2.00	8.00	14.00
2	.00	10.00	27.00
3	3.00	12.00	23.00
4	6.00	14.00	27.00
5	4.00	16.00	24.00

5. The following tables report the regression of homicide rate (homicides per 100,000) (*rathoms*) on percentage of children in single-parent households (*single*) and the percentage of the population below the poverty level (*poverty*) for the 50 states. Interpret the unstandardized slopes, the intercept, the standardized slopes, the coefficient of multiple determination, and the multiple correlation.

Coefficients[a]

	Unstandardized Coefficients	Standardized Coefficients
	B	Beta
(Constant)	-10.380	
SINGLE	.754	.790
POVERTY	.079	.079

a. Dependent Variable: RATHOMS

Model Summary

Model	R	R Square	Std. Error of the Estimate
1	.833[a]	.693	2.2519

a. Predictors: (Constant), POVERTY, SINGLE

6. For a sample of 30 countries circa 1990, the following tables report the regression of female life expectancy in years (*felifex*) on the death rate (deaths per 1,000 persons) (*deathrt*) and the birthrate (births per 1,000 persons) (*birthrt*). Interpret the unstandardized slopes, the intercept, the standardized slopes, the coefficient of multiple determination, and the multiple correlation.

Coefficients[a]

	Unstandardized Coefficients	Standardized Coefficients
	B	Beta
(Constant)	96.607	
BIRTHRT	-.589	-.684
DEATHRT	-1.056	-.399

a. Dependent Variable: FELIFEXP

Model Summary

Model	R	R Square	Std. Error of the Estimate
1	.976[a]	.953	2.801

a. Predictors: (Constant), DEATHRT, BIRTHRT

7. For 516 women aged 40 or more from GSS 2000, the following tables report the regression of number of children (*childs*) on education (*educ*), age at birth of first child (*agekdbrn*), ideal number of children (*chldidel*), and age.

Model Summary

Model	R	R Square	Adjusted R Square	Std. Error of the Estimate
1	a	?	.122	1.366

a. Predictors: (Constant), AGE, CHLDIDEL, AGEKDBRN, EDUC

ANOVA[b]

Model		Sum of Squares	df	Mean Square	F	Sig.
1	Regression	141.282	?	?	?	?[a]
	Residual	953.105	?	?		
	Total	1094.388	515			

a. Predictors: (Constant), AGE, CHLDIDEL, AGEKDBRN, EDUC

b. Dependent Variable: CHILDS

Coefficients[a]

Model		Unstandardized Coefficients		Standardized Coefficients	t	Sig.	Collinearity Statistics	
		B	Std. Error	Beta			Tolerance	VIF
1	(Constant)	3.991	.469		8.501	.000		
	EDUC	-.057	?	-.109	?	?	.823	1.215
	AGEKDBRN	-.068	?	-.247	?	?	.859	1.164
	CHLDIDEL	.093	.037	.104	2.511	.012	.991	1.009
	AGE	.012	.004	.111	2.609	.009	.941	1.062

a. Dependent Variable: CHILDS

a. Calculate the values of the missing statistics (?).

b. What are the degrees of freedom and the critical t for the two-tailed tests of the partial slopes at $\alpha = .05$?

c. What are the critical value of F at $\alpha = .05$ and the decision/conclusion for the F test?

8. From the GSS 2000 for 301 males working full time, the following tables report the regression of occupational status (*sei*) on education (*educ*), vocabulary score (*wordsum*), and father's occupational status (*pasei*).

Model Summary

Model	R	R Square	Adjusted R Square	Std. Error of the Estimate
1	[a]	?	.442	14.5080

a. Predictors: (Constant), PASEI, WORDSUM, EDUC

ANOVA[b]

Model		Sum of Squares	df	Mean Square	F	Sig.
1	Regression	50715.629	?	?	?	?[a]
	Residual	62512.989	?	?		
	Total	113228.6	300			

a. Predictors: (Constant), PASEI, WORDSUM, EDUC

b. Dependent Variable: SEI

Coefficients[a]

Model		Unstandardized Coefficients		Standardized Coefficients	t	Sig.	Collinearity Statistics	
		B	Std. Error	Beta			Tolerance	VIF
1	(Constant)	-16.074	4.507		-3.567	.000		
	EDUC	3.991	?	.553	?	?	.646	1.548
	WORDSUM	1.134	?	.123	?	?	.713	1.403
	PASEI	.095	.048	.093	1.984	.048	.848	1.179

a. Dependent Variable: SEI

    a. Calculate the values of the missing statistics (?).
    b. What are the degrees of freedom and the critical t for the two-tailed tests of the partial slopes at $\alpha = .05$?
    c. What are the critical value of F at $\alpha = .05$ and the decision/conclusion for the F test?

9. Use the method of residualization to compute partial slopes with SPSS. First, using GSS 2000, regress *tvhours* on *educ* and *wordsum*. Second, select cases that do not have missing values on *tvhours, educ,* and *wordsum* by entering "not missing (*tvhours*) and not missing (*educ*) and not missing (*wordsum*)" in the *Select Cases: If* window. Third, regress *educ* on *wordsum* and save the residuals (click on *Save* in the *Linear Regression* window and then click on *Unstandardized* under *Residuals*). The residuals for *educ* will be saved in the data editor as a new variable named, for example, *res_1,* which you may rename to *educres,* for example, if you like. Fourth, regress *tvhours* on *res_1* (or, e.g., *educres*). Compare the slope in the fourth step to the slope you computed for *educ* in the first step; they should be equal. To get the partial slope for *wordsum,* repeat the third and fourth steps by regressing *wordsum on educ,* saving the residuals, and regressing *tvhours* on the residuals of *wordsum.*

10. In GSS 2000, *sexprtyr* is constructed to equal the number of sex partners since age 18 divided by (age − 17). Thus, it is a sex-partner rate—that is, the number of sex partners per year. Using GSS 2000 (*gss2000b*), use SPSS to regress *sexprtyr* on *educ, sei, wordsum,* and *age.* What decision and conclusion do you make on the basis of the F test? What decision and conclusion do you make on the basis of the t test for each partial slope?

11. Using GSS 2000 (*gss2000b*), regress *wordsum* on *educ, age, paeduc, maeduc,* and *sibs.* What are your decision and conclusion on the basis of the F test? What decision and conclusion do you make on the basis of the t test for each partial slope? Which variables are the most closely associated with vocabulary score?

11

Nominal Independent Variables

WEST COAST STORY

"Mother, a new boy in school named Kazuo asked me out, and I don't know what to do."

"What makes you so unsure, Maria? Is there something wrong with him?"

"Well, Mother, yes and no. He seems nice. And he's smart. But he's Japanese-American."

"So, what's wrong with that, Maria?"

"I've never known any boys from that group before. He speaks English with an accent. Aren't Japanese-Americans a lot different from us?"

"Well, his family might speak Japanese at home, like we speak Spanish when we go to Grandmother's. I think it's good to preserve our cultural heritages."

"It's not that simple, Mother. My girlfriends will make fun of me, and the boys will be mad. They'll say, 'Aren't Chicano boys good enough for you anymore, Maria?' I don't want to be unpopular."

"Maria, this is America! Everyone is supposed to be treated as equals in our country. It's a melting pot where ethnic groups learn to mix with one another and enjoy each other's cultures. And deep down inside, we're all pretty much alike."

"I suppose you're right, Mother. Thanks for talking with me about it. I'll think about what you said."

Is Maria's mother right? She paints a very utopian picture of American diversity. Different ethnic groups intermix and share cultural traditions, while at the same time, each group preserves its own heritage—its roots. Out of this intermingling emerges a richer, more complex American culture shared by all people who can reach across traditional differences to readily identify with one another. And, because deep down inside we are all pretty much the same, once we are able to get past our cultural differences, we can recognize our common humanity.

How can we go about studying how similar or different ethnic groups are? In the preceding chapters we assumed that the variables included in the regression equation were measured at the interval or ratio level. To compute sums, deviations, and the amount that Y changes per unit increase in X, we must have measures of variables that indicate how much one case differs from another on each of the variables in the regression analysis. When investigators want to analyze differences between groups on a dependent variable, however, the independent variable representing the groups is often a nominal variable. Nominal variables do not provide information about the intervals between the groups that belong to its categories. Nominal variables simply group units into mutually exclusive categories, where each category is defined by some attribute shared by all its members. Examples of nominal variables are gender (male, female), marital status (married, divorced, widowed, and never married), and ethnicity (e.g., African-, British-, German-, Irish-, Italian-, Mexican-, and Native Americans). For variables such as these, there is no measurement system that specifies how high or low members of one group are positioned in relation to those of other groups.

Nevertheless, nominal variables may still contain information that helps to explain variance in Y. Before we can use regression analysis to extract such potentially valuable information, however, one or more appropriately coded variables must be created for each nominal variable. These new variables are then included in the regression equation to represent the nominal variable. Although the coding scheme for these variables is arbitrary in some respects, some types of coding result in more meaningful interpretations of the regression coefficients. *Dummy variable coding* is the most common type used in the social sciences.[1] We will use dummy variable coding in this chapter.

DICHOTOMOUS NOMINAL VARIABLES

We will begin with the simplest type of nominal variable, one that contains only two groups (i.e., a dichotomous variable). In this case, a single dummy coded variable can be used to represent the two groups. We will consider the interpretation of the slope and intercept of the dummy variable regression and the special meaning of the t test for the slope of the dummy variable.

---

1. Other useful types are effects coding and contrast coding (McClendon, 1994).

Best-Fitting Line

When X is dichotomous, there will be a distribution of Y for group A and a distribution of Y for group B (Figure 11.1). These are called *conditional distributions*. The distribution of Y is conditional on whether X is group A or group B. What will be the best predictions of Y for group A and for group B according to the least-squares criterion? You will remember that the single predicted value for any distribution that minimizes the sum of squared deviations is the mean of the distribution. Therefore, the prediction of $\bar{Y}_A$ for all cases in group A will minimize the sum of squared residuals for group A, $[\Sigma(Y_A - \bar{Y}_A)^2]$, and the prediction of $\bar{Y}_B$ for all cases in group B will minimize the sum of squared residuals for group B, $[\Sigma(Y_B - \bar{Y}_B)^2]$. Consequently, *the least-squares regression line will pass through the two* **conditional means** *to minimize the total sum of squared residuals,* as shown in Figure 11.1.

Notice that no numeric values are shown for groups A and B on the horizontal axis in Figure 11.1. In terms of finding the best-fitting line, how we score A and B is arbitrary. Whatever scores we assign to A and B (which we can refer to as X_A and X_B), the least-squares line will pass through the coordinates $(X_A, \bar{Y}_A)$ and $(X_B, \bar{Y}_B)$. The slope of the line is given by

$$b = \frac{\bar{Y}_B - \bar{Y}_A}{X_B - X_A} = \frac{\bar{Y}_A - \bar{Y}_B}{X_A - X_B} \tag{11.1}$$

Equation 11.1 indicates that we can find the best-fitting line simply by dividing the difference between the means of Y by the difference between the scores we assign to groups A and B. We will see, however, that a certain scoring of X will give particularly meaningful interpretations of the slope as well as of the intercept.

Dummy Coding

Dummy variable coding is a method of scoring a dichotomous independent variable that provides readily interpretable regression coefficients. In coding a **dummy variable,** one group is given a score of 1 and the other a score of 0. The group scored 0 is called the **reference group.** The two alternative ways to assign the dummy codes to groups A and B are shown in Table 11.1. Group A can be coded 1 or group B can be coded 1. We cannot include both X_1 and X_2 in a regression equation because

$$X_1 = 1 - X_2$$

If we know a case's score on one variable, we can always determine its score on the other. Thus, the two variables have a perfect correlation of $r = -1.0$. Because this is true, regression analysis cannot estimate separate slopes for each variable. A regression program will allow only one of the two variables to be entered. Therefore, we have to choose between them.

How do we choose? To a great extent, it doesn't matter much which coding we adopt. It is more or less arbitrary. One criterion, however, is to assign 1 to the group that has or is expected to have the higher mean on Y. As a result, the slope will be positive, as we shall see. Positive slopes are somewhat easier to interpret than negative slopes.

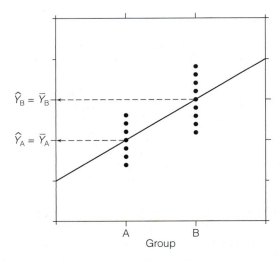

Table 11.1 Two Possible Dummy Codings for a Dichotomous X

Group	X_1	X_2
A	1	0
B	0	1

FIGURE 11.1 Least-Squares Line for Y and a Dichotomous Nominal X

When Y is regressed on a dummy coded X, how do we interpret the slope and intercept? Let

$$Y_0 = \text{scores on } Y \text{ when } X = 0$$

$$Y_1 = \text{scores on } Y \text{ when } X = 1$$

If we enter Y_0 and Y_1 as well as the dummy coding into Equation 11.1, we get

$$b = \frac{\overline{Y}_1 - \overline{Y}_0}{1 - 0} = \overline{Y}_1 - \overline{Y}_0 \tag{11.2}$$

The **slope of a dummy variable** *equals the mean of Y for the group coded 1 on X minus the mean of Y for the group coded 0 on X.* The least-squares slope is equal to a difference of means. Because a slope equals the change in Y per unit increase in X, and because a unit increase in X corresponds to the difference between groups A and B, the slope for a dummy X indicates how much the mean of Y changes when X increases from 0 to 1. This change is illustrated in Figure 11.2. The sign of b indicates whether the group scored 1 has a higher mean (+ slope) or a lower mean (− slope) than the group scored 0. Figure 11.2 also shows that when $X = 0$, the predicted value of Y is the mean of the group coded 0. In other words, *the* **intercept of a dummy variable equation** *equals the mean of Y for the group scored 0* (i.e., the reference group):

$$a = \overline{Y}_0 \tag{11.3}$$

The prediction of Y for the group coded 1 equals

$$a + b = \overline{Y}_0 + (\overline{Y}_1 - \overline{Y}_0) = \overline{Y}_1$$

That is, the mean of the group scored 1 equals the intercept plus the slope.

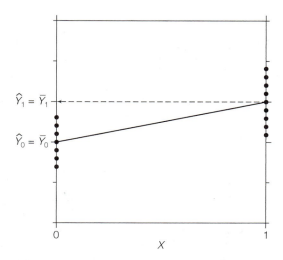

FIGURE 11.2 Least-Squares Line for Y and a Dummy Coded X

As an example of dummy variables, we will regress age at first marriage on gender. Because we expect males to be older than females when they marry for the first time, we will code the dummy variable for gender as shown in Table 11.2.

Table 11.3 shows the means of age at first marriage for males and females from the 1994 GSS. The relevant population consists of all persons 18 years of age or older in 1994 who had ever been married. Males were 2 years older, on average, than females when they first got married.

The two means are all that we need to determine the slope and intercept of the regression equation. Using Equations 11.2 and 11.3, we find that the slope and intercept are

$$b = \overline{Y}_1 - \overline{Y}_0 = 23.93 - 21.82 = 2.11$$

$$a = \overline{Y}_0 = 21.82$$

We can then write the regression equation as

$$\hat{Y} = 21.82 + 2.11X$$

The intercept is the mean of the gender that is coded 0 (females). Females who had ever been married as of 1994 had a mean age at first marriage of 21.82 years. The slope indicates that males ($X = 1$) were 2.11 years older than females ($X = 0$) when they first married. The age of males can be determined from the regression equation as

$$\overline{Y}_{\text{males}} = a + b = 21.82 + 2.11 = 23.92$$

If 1 had been assigned to females and 0 to males, the regression equation would have been

$$\hat{Y} = 23.92 - 2.11X$$

Table 11.2 Dummy Coding of Gender

Gender	X
Males	1
Females	0

Table 11.3 Age at First Marriage by Gender, GSS 1994

Gender	n	Mean	SS	SD
Female	720	21.82	16621.887	4.805
Male	469	23.93	10431.817	4.719
Total	1189	22.65	28319.239	4.880

The intercept would indicate that males ($X = 0$) were 23.92 years of age at first marriage. The slope would indicate that females ($X = 1$) were 2.11 years younger than males at first marriage.

Goodness-of-Fit

Measures of the strength of association for the dummy variable X can be determined just as they are in any bivariate regression. You will remember that one definition of the coefficient of determination is

$$CD = \frac{\text{total SS} - \text{residual SS}}{\text{total SS}}$$

How do we find the residual sum-of-squares? Based on the fact that $\bar{Y}_A$ and $\bar{Y}_B$ are the best predictions for groups A and B, respectively, the residual sum-of-squares equals

$$\text{Residual SS} = \Sigma(Y_A - \bar{Y}_A)^2 + \Sigma(Y_B - \bar{Y}_B)^2 = SS_A + SS_B$$

For the example of gender and age at first marriage (Table 11.3), the residual sum-of-squares is

$$\text{Residual SS} = 16621.887 + 10431.817 = 27053.704$$

The total sum-of-squares listed in Table 11.3 is 28319.239. The coefficient of determination is

$$CD = \frac{28319.239 - 27053.704}{28319.239} = \frac{1265.535}{28319.239} = .0447$$

Gender explains 4.5 percent of the variance in age at first marriage.

In the bivariate case, the standardized slope equals

$$B = \sqrt{CD} = \sqrt{.0447} = .2114$$

Normally, the standardized slope indicates the fraction of a standard deviation that Y changes when X increases by 1 standard deviation. However, is it possible for a dummy X to increase by 1 standard deviation? Can the two groups that compose a dummy variable differ by exactly 1 standard deviation? Before we can answer this question, we must consider what the standard deviation of a dummy variable represents.

The mean of a dummy variable, such as gender, equals

$$\bar{X} = \frac{\Sigma X}{n} = \frac{1 + 1 + \cdots + 1 + 0 + 0 + \cdots + 0}{n} = \frac{n_1}{n} = p_1$$

The sum of dummy scores $\{1, 0\}$ equals the number of cases coded 1 (n_1). Thus, *the* **mean of a dummy variable** *equals* p_1, *the proportion of the cases scored 1.* Using the number of males and the sample size from Table 11.3, the mean for the dummy coded gender variable is

$$\overline{X} = \frac{n_{\text{males}}}{n} = \frac{469}{1189} = .394$$

Thus, the mean of X indicates that .394, or 39.4 percent, of the cases are male. The mean of a dummy variable is always between 0 and 1:

$$\text{If } X = \{0, 1\}: 0 < \overline{X} < 1$$

The mean cannot be exactly 1 or 0 because that would indicate that all cases were 1s or 0s. If this were true, X would have no variance.

The **standard deviation of a dummy variable** (without proof) equals

$$s_X = \sqrt{p_1(1 - p_1)} = \sqrt{p_1 p_0}$$

where p_1 is the proportion of cases coded 1 and $1 - p_1$ is the proportion not coded 1, which is the proportion coded 0 (p_0). This quantity is the same one you would get if you used the general formula for a standard deviation ($s = \sqrt{\Sigma(X - \overline{X})^2/n}$). The standard deviation of gender is thus

$$s = \sqrt{(.394)(1 - .394)} = \sqrt{(.394)(.606)} = .489$$

Because the standard deviation of a dummy variable is the square root of the product of two proportions, it, like the mean, must always be between 0 and 1:

$$\text{If } X = \{0, 1\}: \ 0 < s < 1$$

Now we return to the question of whether the two groups forming a dummy variable can differ by 1 standard deviation. The answer is no. They differ by $1 - 0 = 1$. But the value of a standard deviation of a dummy variable is always less than 1. Therefore, the usual interpretation of a standardized slope is somewhat artificial in the case of a dummy variable. We can still use B, however, to measure the strength of association by considering how close it is to 0 or to either 1 or -1. The closer it is to 1 or -1, the stronger the association.

t Test for the Slope

Hypothesis tests for the slopes of dummy variables are carried out exactly the same way as tests for the slopes of interval/ratio variables. There is one important difference in the interpretation of these tests, however. The most common null hypothesis for the t test is that the population slope equals 0. Because the slope of a dummy variable equals the difference of means between two groups, the hypothesis $\beta = 0$ is equivalent to the hypothesis that the difference of means equals 0. Thus, the null and alternative two-tailed hypotheses are

X in General	*Dummy X*
$H_0: \beta = 0$	$H_0: \mu_1 = \mu_0$
$H_1: \beta \neq 0$	$H_1: \mu_1 \neq \mu_0$

**Table 11.4 *t* Test for the Slope
of Age at First Marriage Regressed
on Gender**

1. $H_0: \mu_1 = \mu_0$
   $H_1: \mu_1 \neq \mu_0$
2. $\alpha = .05$
3. $df = n - 2 = 1189 - 2 = 1187$
4. $t_{\alpha/2,df} = t_{.025,1187} = 1.962$
5. $b = 2.11$
6. $s_b = \dfrac{\hat{s}_{Y-\hat{Y}}}{s_x\sqrt{n}} = \dfrac{4.774}{.489\sqrt{1189}} = .283$
7. $t = \dfrac{b - \beta_0}{s_b} = \dfrac{2.11 - 0}{.283} = 7.456$
8. $(|t| = 7.456) > (t_{.025} = 1.962), \therefore$ reject H_0
   $(t = 7.456) > (t_{.025} = 1.962), \therefore \mu_1 > \mu_0$
   $p < .00005$

If the null hypothesis is rejected, you conclude that the two groups have different means of Y. Other than this special interpretation of the hypotheses when X is a dummy variable, the usual formulas and eight-step t test for the bivariate slope apply.

Let's return to the example of age at first marriage regressed on gender (male = 1, female = 0). The eight-step **t test for a dummy variable slope** is shown in Table 11.4. We will conduct a two-tailed test of the null hypothesis that males and females have equal means of age at first marriage ($\mu_1 = \mu_0$). We adopt the conventional $\alpha = .05$. For $df = 1187$, we determine from Table A.2 that the critical value of t is 1.962. The sample slope of $b = 2.11$ from the previous section indicates that on the average males are 2.11 years older than females at first marriage.

Before executing step 6, we need to calculate the standard error of estimate to use in the formula for the standard error of the slope. In the prior section we determined that the residual sum-of-squares of age at first marriage equals 27053.704. Entering this quantity into the usual formula for the standard error of estimate for a bivariate regression gives

$$\hat{s}_{Y-\hat{Y}} = \sqrt{\frac{\Sigma(Y - \hat{Y})^2}{n - 2}} = \sqrt{\frac{27053.704}{1187}} = \sqrt{22.792} = 4.774$$

We also need the standard deviation of the dummy X, which, from the previous section, equals $s = \sqrt{p_1 p_0} = .489$. Now we can compute the standard error of b in step 6 and t in step 7, the latter of which far exceeds the critical t. Thus, we reject the null hypothesis and conclude that the mean male age at first marriage is greater than the mean female age.

*t Test for Difference of Means

We saw in the previous section that the t test for the slope of a dummy variable is equivalent to a test of whether the two groups represented by the dummy variable have equal or unequal means. This **t test for a difference of means** also exists as a specialized test separate from regression analysis, one with its own formulas and notation. There are special computer programs for carrying out this test. We will describe this popular test and show its equivalence to the regression-based test for a difference of means.

The following test assumes that each group has the same variance. This assumption is equivalent to the homoskedasticity assumption in regression.[2] The symbol for the dependent variable in this test is X. There is no Y variable. The null and alternative hypotheses for a two-tailed t test for equal group means are

$$H_0: \mu_1 = \mu_2$$

$$H_1: \mu_1 \neq \mu_2$$

where μ_1 is the population mean of X in group 1 and μ_2 is the population mean of X in group 2.

The sample statistic that we desire to test is

$$\text{Sample statistic} = \overline{X}_1 - \overline{X}_2$$

which is the difference in sample means between groups 1 and 2. Just as the sample statistics $\overline{X}$ and b have standard errors, $\overline{X}_1 - \overline{X}_2$ also has a standard error. The estimated standard error is

$$s_{\overline{X}_1 - \overline{X}_2} = \sqrt{\frac{n_1 s_1^2 + n_2 s_2^2}{n_1 + n_2 - 2} \left(\frac{1}{n_1} + \frac{1}{n_2} \right)} \tag{11.4}$$

In this equation, n_1 and n_2 are the sample sizes of groups 1 and 2, respectively, and s_1^2 and s_2^2 are the sample variances of the two groups. Equation 11.4 is known as a *pooled variance* estimate of the standard error.

The t statistic for testing the null hypothesis is

$$t = \frac{(\overline{X}_1 - \overline{X}_2) - (\mu_1 - \mu_2)}{s_{\overline{X}_1 - \overline{X}_2}} = \frac{\overline{X}_1 - \overline{X}_2}{s_{\overline{X}_1 - \overline{X}_2}} \tag{11.5}$$

This equation is equivalent in form to that of any t, namely, the sample statistic minus the population parameter with the difference divided by the estimated standard error. Because the null hypothesis states that $\mu_1 - \mu_2 = 0$, $\mu_1 - \mu_2$ is omitted from the numerator.

Now let's use the data on gender differences in age at first marriage (Table 11.3) to test the null hypothesis that males and females have equal mean ages at first marriage. X equals age at first marriage. Let

$$\text{Males} = \text{group 1}$$

$$\text{Females} = \text{group 2}$$

Using the sums-of-squares from Table 11.3, the variances for males and females are

$$s_1^2 = \frac{\Sigma(X_1 - \overline{X}_1)^2}{n_1} = \frac{10431.817}{469} = 22.243$$

$$s_2^2 = \frac{\Sigma(X_2 - \overline{X}_2)^2}{n_2} = \frac{16621.887}{720} = 23.086$$

---

2. The SPSS T-Test procedure provides a test of the assumption of equal variances and an alternative t test for a difference of means that does not assume equal variances. See the example output in the SPSS section.

Substituting n_1, n_2, s_1^2, and s_2^2 into Equation 11.4, we find that the standard error of the difference of means is

$$s_{\bar{X}_1-\bar{X}_2} = \sqrt{\frac{(469)(22.243) + (720)(23.086)}{469 + 720 - 2}\left(\frac{1}{469} + \frac{1}{720}\right)}$$

$$= \sqrt{\frac{27053.887}{1187}(.00213 + .00139)}$$

$$= \sqrt{(22.792)(.00352)}$$

$$= \sqrt{.08023}$$

$$= .283$$

Notice that $s_{\bar{X}_1-\bar{X}_2} = s_b$ where b is the dummy variable slope for age at first marriage regressed on gender (Table 11.4).

Now we can enter the standard error, the male and female means, and $\mu_1 - \mu_2 = 0$ into Equation 11.5 to get t

$$t = \frac{23.93 - 21.82}{.283} = \frac{2.11}{.283} = 7.456$$

This value equals the t in Table 11.4 for the slope of the dummy variable. Because the degrees of freedom are $df = n - 2$ for both t tests, both tests lead to exactly the same conclusion: Females marry at a younger mean age than males.

★Paired–Sample t Test

The t test for the difference of means between two matched samples, such as husbands and wives, or between two repeated measures on each case, was covered in Chapter 6 as a test for the mean of the paired difference scores.

★t Test for Difference of Proportions

Closely related to the t test for a difference of means is the **t test for a difference of proportions.**[3] The null and alternative hypotheses for a two-tailed t test for equal group proportions are

$$H_0: \pi_1 = \pi_2$$

$$H_1: \pi_1 \neq \pi_2$$

where π_1 is the population proportion in group 1 having some attribute, and π_2 is the population proportion in group 2 having the attribute. The null hypothesis implies that each group has the same variance because the variance of a dichotomous variable equals $\pi(1 - \pi)$.

---

3. This test is slightly better than a t test for a difference of means for a dummy coded dependent variable with a pooled-variance estimate of the standard error (Blalock, 1979, p. 233). The latter test can be executed in regression analysis by regressing a dummy coded dependent variable on a dummy coded independent variable. A test for a difference of proportions can also be conducted with the Pearson chi-square statistic for a two-by-two contingency table. The chi-square test is covered in Chapter 13. The t and χ^2 test statistics usually produce very similar results, but they are not identical tests.

Table 11.5 Proportion Remarried by Gender, 50 and Older, 1998 GSS

Gender	n	p
1 Male	294	.3299
2 Females	457	.2013
Total	751	.2517

The sample statistic that we desire to test is

$$\text{Sample statistic} = p_1 - p_2$$

which is the difference in sample proportions between groups 1 and 2. The estimated standard error of $p_1 - p_2$ is

$$s_{p_1-p_2} = \sqrt{p(1-p)\left(\frac{1}{n_1} + \frac{1}{n_2}\right)} \tag{11.6}$$

In this equation, n_1 and n_2 are the sample sizes of groups 1 and 2, respectively, and p is the overall proportion of cases having the specified attribute in the sample.

The t statistic for testing the null hypothesis is

$$t = \frac{(p_1 - p_2) - (\pi_1 - \pi_2)}{s_{p_1-p_2}} = \frac{p_1 - p_1}{s_{p_1-p_2}} \tag{11.7}$$

where $\pi_1 - \pi_2 = 0$ under the null hypothesis.

To illustrate this test, we will address the question of whether men or women are more likely to remarry after a divorce. The null hypothesis says that the proportion of men who remarry after divorce is equal to the proportion of women who remarry. The GSS asks currently married and widowed persons whether they were ever divorced. Those who said yes are counted as remarried persons.[4] Table 11.5 shows the proportions of males and females who are remarried. The test statistic is

$$p_1 - p_2 = .3299 - .2013 = .1286$$

This result indicates that in the sample, about 13 percent more divorced males remarry than females.

The standard error for the gender difference in proportion remarried is

$$s_{p_1-p_2} = \sqrt{.2517(1 - .2517)\left(\frac{1}{294} + \frac{1}{457}\right)} = \sqrt{(.1883)(.0056)} = \sqrt{.00105} = .0324$$

The test statistic divided by its standard error gives the t statistic

$$t = \frac{.1286}{.0324} = 3.96$$

---

4. The GSS did not ask currently divorced persons whether they had been divorced more than once. Thus, there is a slight undercount of remarried persons.

The critical $t = 1.963$ for $df = 751 - 2 = 749$ and $\alpha = .05$. The calculated t is greater than the critical t. Thus, we reject the null hypothesis and conclude that divorced men are more likely than divorced women to remarry.

POLYTOMOUS NOMINAL VARIABLES

When a nominal variable has more than two categories, we can no longer use a single dummy variable to represent all of the groups. We also cannot use a single variable coded with three or more scores because regression programs would treat the numeric codes as if they represented an interval variable. For example, if there are four groups, the scores 1, 2, 3, and 4 might be assigned to groups A, B, C, and D, respectively. These codes would be treated as interval scores where D is one unit higher than C, which is one unit higher than B, which is one unit higher than A. This would not be valid because the groups are not higher or lower than one another. They are just different from one another.

When there are more than two nominal categories, no coding system will allow us to use a single variable to represent these groups in a regression analysis. Instead, we have to use more than one specially coded variable. We will see how to do this with dummy coded variables.

Dummy Coding

How many dummy variables are required to represent all of the groups in a nominal variable? Let g be the number of groups (categories) in a nominal variable. Then g is the maximum number of distinct dummy variables that can be created. Table 11.6 shows the four distinct dummy variables that exist for $g = 4$ groups.

In this coding scheme, each group receives a code of 1 on one and only one variable. Codes of 0 are assigned to each group on all the variables other than the one on which it has a code of 1. Each dummy variable contrasts the group coded 1 with the three groups coded 0. The dummy variable makes no distinction among the groups coded 0. Thus, each variable contrasts two groups, the cases coded 1 versus the cases not coded 1. For example, X_3 compares C and not-C.

The dummy coding scheme creates the following sum of the g dummy variables:

$$X_1 + X_2 + X_3 + X_4 = 1$$

For every case, the sum of the g dummy variables equals 1. Because of this fact, if we know the scores of a case on $g - 1$ of the variables, we can determine its score on the remaining variable. For example, the score of X_4 is determined as follows:

$$X_4 = 1 - X_1 - X_2 - X_3$$

$$= 1 - 1 - 0 - 0 = 0 \quad \text{(for members of group A)}$$

$$= 1 - 0 - 1 - 0 = 0 \quad \text{(for members of group B)}$$

$$= 1 - 0 - 0 - 1 = 0 \quad \text{(for members of group C)}$$

$$= 1 - 0 - 0 - 0 = 1 \quad \text{(for members of group D)}$$

Table 11.6 Four Dummy Variables for a Nominal Variable with $g = 4$ Groups

Group	X_1	X_2	X_3	X_4
A	1	0	0	0
B	0	1	0	0
C	0	0	1	0
D	0	0	0	1

The preceding results show that X_4 is a perfect linear function of the other three dummy variables. Therefore, $R^2 = 1$ for X_4 as the dependent variable and the other three dummies as predictors. In other words, there is perfect multicollinearity among the four variables. If we residualized one dummy variable on the other three dummies, the residual variance would equal 0. Thus, there would be no residual variance with which to determine a partial regression slope as we did in Chapter 10. Consequently, we cannot use all four dummy variables in the regression equation.

The preceding discussion leads to the conclusion that one of the g dummies must be left out of the equation. In general, $g - 1$ is the **maximum number of dummy variables** *that can be used to represent a nominal variable.* The group represented by the dummy variable that will be omitted will not, however, be left out of the analysis. It will serve as a **reference group,** as we shall see. The dummy variable to be omitted depends on which group is of most interest as the reference group. The goodness-of-fit of the model (i.e., R^2) will be the same, no matter which of the g dummies is omitted. Statistically speaking, it is arbitrary which variable is left out.

Let's assume X_4 is excluded from the regression equation. The equation becomes

$$\hat{Y} = a + b_1 X_1 + b_2 X_2 + b_3 X_3$$

What will be the values of the slopes and intercept? In the case of a single dummy variable, the predicted value of Y for every case equals the mean of Y for the group to which the case belongs. These are the conditional means of Y. They minimize the sum of squared residuals. The same is true when the nominal variable comprises three or more groups. The solution for the regression coefficients is the one that leads to predicting the conditional mean of each case's group. The conditional means of Y when $g = 4$ are

$$\overline{Y}_A = \text{mean of } Y \text{ for group A}$$

$$\overline{Y}_B = \text{mean of } Y \text{ for group B}$$

$$\overline{Y}_C = \text{mean of } Y \text{ for group C}$$

$$\overline{Y}_D = \text{mean of } Y \text{ for group D}$$

Because the least-squares solution predicts the conditional mean of Y for the group to which each person belongs, we can interpret each of the regression coefficients by substituting the scores of each group into the equation and simplifying as follows:

$$\hat{Y}_A = \overline{Y}_A = a + b_1(1) + b_2(0) + b_3(0) = a + b_1$$

$$\hat{Y}_B = \overline{Y}_B = a + b_1(0) + b_2(1) + b_3(0) = a + b_2$$

$$\hat{Y}_C = \overline{Y}_C = a + b_1(0) + b_2(0) + b_3(1) = a + b_3$$

$$\hat{Y}_D = \overline{Y}_D = a + b_1(0) + b_2(0) + b_3(0) = a$$

Look at the prediction for group D first. Group D has a score of 0 on each of the three dummy variables. It is called the reference group. Because the intercept is the predicted value for cases with scores of 0 on all variables, a is the predicted value for group D. The predicted value that minimizes the sum of squared residuals for group D is the mean of Y of the group. Therefore, *the* **intercept of the dummy variable equation** *equals the mean of the reference group* (group D). The predicted value for group A, which is the mean of group A, equals the intercept plus the slope for the variable that represents A (i.e., X_1). The mean of group B equals the intercept plus the slope for X_2. And the mean of group C equals the intercept plus the slope for X_3.

Given the preceding, if we subtract the mean of group D, the reference group, from the mean of each of the other groups, we get

$$\overline{Y}_A - \overline{Y}_D = (a + b_1) - a = b_1$$

$$\overline{Y}_B - \overline{Y}_D = (a + b_2) - a = b_2$$

$$\overline{Y}_C - \overline{Y}_D = (a + b_3) - a = b_3$$

This result indicates that *the* **slope for each dummy variable** *equals the mean of the group coded 1 minus the mean of the reference group.* Thus, each slope is equal to a difference of means, and the intercept equals the mean of the reference group. Let

$$\overline{Y}_0 = \text{mean of } Y \text{ of reference group}$$

$$\overline{Y}_i = \text{mean of } Y \text{ of group } i$$

Then, in general,

$$a = \overline{Y}_0$$

$$b_i = \overline{Y}_i - \overline{Y}_0$$

Note that the slope b_i does not equal the difference of means between the group coded 1 on the ith variable and all the groups coded 0 on that variable. For example, b_1 does not equal the difference between the mean of group A and the mean of groups B, C, and D combined. Instead, it is the difference of means between groups A and D only. Why? Consider the meaning of a partial slope, which, for example, is the change in Y per unit increase in X_1 with X_2 and X_3 held constant. The only way X_2 and X_3 can be held constant when X_1 increases by one unit is for X_2 and X_3 to be held constant at 0. In that case, a unit increase in X_1 must refer to a change from group D to group A.

The interpretations of the partial regression coefficients for two or more dummy variables are very similar to those of a bivariate regression equation with a dummy independent variable. We can think of the group coded 0 on the single dummy variable as a reference group because that group equals 0 on all of the Xs. Thus, whether we are using a single dummy variable to stand for a dichotomous nominal variable or we are using two or more dummies to stand for a polytomous nominal variable, the intercept equals the mean of the reference group. Moreover, in either case, the slope of a dummy variable indicates the difference of means between the group coded 1 on the dummy variable and the reference group.

We will now use ethnicity as an example of dummy variable coding of a polytomous nominal variable. Respondents to the GSS were asked: "From which countries or parts of the world did your ancestors come?" If a respondent named more than one country, they were asked: "Which of these countries do you feel closer to?" Respondents were assigned a single ethnic code if they named only one country or chose one that they felt closer to. For our analysis we will use the seven largest ethnic groups of persons with a single ethnic code. Persons who named England, Wales, or Scotland were combined to form a British-American ethnic group. The seven ethnic groups and the seven distinct dummy variables are shown in Table 11.7.

Which dummy variable should we leave out of the regression analysis? We have chosen to omit X_7, the dummy variable representing British-Americans. Thus, British-Americans will be the reference group. The rationale for this choice is that the English were the first Europeans to emigrate to America and their language is the chief medium of communication in the United States. It would thus be of interest to compare British-Americans with later immigrants and with Native Americans. Our regression equation will be

$$\hat{Y} = a + b_1X_1 + b_2X_2 + b_3X_3 + b_4X_4 + b_5X_5 + b_6X_6$$

The dependent variable in our regression analysis will be respondents' ages at the birth of their first child. The population will be women with one or more children. Thus, we are interested in any ethnic group differences in the age at which a woman bears her first child. Table 11.8 shows the ethnic-group means of age at first birth of 643 mothers in the 1998 GSS who belonged to one of the seven listed ethnic groups. The group with the youngest mean age at first birth in this sample is Mexican-American, and the group with the oldest mean age is Italian-American.

Because $a = \bar{Y}_0$ and $b_i = \bar{Y}_i - \bar{Y}_0$, we can determine the intercept and slopes of the regression equation from the means in Table 11.8 as follows:

$$a = \bar{Y}_0 = 23.8$$
$$b_1 = \bar{Y}_1 - \bar{Y}_0 = 20.4 - 23.8 = -3.4$$
$$b_2 = \bar{Y}_2 - \bar{Y}_0 = 23.3 - 23.8 = -.5$$
$$b_3 = \bar{Y}_3 - \bar{Y}_0 = 23.1 - 23.8 = -.7$$
$$b_4 = \bar{Y}_4 - \bar{Y}_0 = 24.1 - 23.8 = .3$$
$$b_5 = \bar{Y}_5 - \bar{Y}_0 = 20.0 - 23.8 = -3.8$$
$$b_6 = \bar{Y}_6 - \bar{Y}_0 = 20.6 - 23.8 = -3.2$$

Table 11.7 Dummy Variables for $g = 7$ Ethnic Groups

Group	X_1	X_2	X_3	X_4	X_5	X_6	X_7
African-American	1	0	0	0	0	0	0
German-American	0	1	0	0	0	0	0
Irish-American	0	0	1	0	0	0	0
Italian-American	0	0	0	1	0	0	0
Mexican-American	0	0	0	0	1	0	0
Native American	0	0	0	0	0	1	0
British-American	0	0	0	0	0	0	1

Table 11.8 Age at Birth of First Child by Ethnic Group, Women, 1998 GSS

Group	Mean	n	SS	St. Dev.
1. African-American	20.4	101	1919.703	4.360
2. German-American	23.3	172	3367.041	4.424
3. Irish-American	23.1	100	2485.000	4.985
4. Italian-American	24.1	47	1034.468	4.691
5. Mexican-American	20.0	30	284.000	3.077
6. Native American	20.6	53	667.019	3.548
7. British-American	23.8	140	2994.536	4.625
Total	22.6	643	14094.479	4.682

Thus, the regression equation is

$$\hat{Y} = 23.8 - 3.4X_1 - .5X_2 - .7X_3 + .3X_4 - 3.8X_5 - 3.2X_6$$

These coefficients indicate that the mean age at first birth of British–American women is 23.8 years, the average African–American has her first child 3.4 years earlier than the average British–American, the average German–American has her first child .5 year earlier than the average British–American, the average Irish–American has her first child .7 year earlier than the average British–American, the average Italian–American has her first child .3 year later than the average British–American, the average Mexican–American has her first child 3.8 years earlier than the average British–American, and the average Native American has her first child 3.2 years earlier than the average British–American, respectively. Remember, these are sample coefficients and they have not yet been tested to see if they are significantly different from 0.

The t test for the partial slope of a dummy variable is the same as that for a partial slope in general. The null and alternative hypotheses for a partial slope are

	X in General	Dummy X
	$H_0: \beta_i = 0$	$H_0: \mu_i = \mu_0$
	$H_1: \beta_i \neq 0$	$H_1: \mu_i \neq \mu_0$

The null hypothesis states that the mean of group i equals the mean of the reference group. For age at first birth regressed on the ethnic dummy variables, the null hypothesis when $i = 3$, for example, is

$$H_0: \mu_3 = \mu_0, \text{ i.e., } \mu_{\text{Irish-Americans}} = \mu_{\text{British-Americans}}$$

$$H_1: \mu_3 \neq \mu_0, \text{ i.e., } \mu_{\text{Irish-Americans}} \neq \mu_{\text{British-Americans}}$$

The null hypothesis specifies that the mean for Irish-Americans (represented by X_3) equals the mean for British-Americans (the reference group).

The standard errors and t scores for partial dummy variables are computed the same way as for partial slopes in general. You will remember that the standard error of the partial slope is

$$s_{b_i} = \frac{\hat{s}_{Y-\hat{Y}}}{s_i \sqrt{n} \sqrt{(1 - R_i^2)}}$$

This formula indicates the four quantities that we need to calculate the standard error of b_3. We know that the total $n = 643$ (Table 11.8). Using n_3 in Table 11.8, we calculate the mean of the Irish-American dummy variable (X_3), which is then used to compute the standard deviation of that variable:

$$\overline{X}_3 = p_3 = \frac{100}{643} = .1555$$

$$s_3 = \sqrt{p_3(1 - p_3)} = \sqrt{.1555(1 - .1555)} = .3624$$

To determine the standard error of estimate, we need to compute the sum of squared residuals. Because the sum of squared residuals for group i equals $\Sigma(Y - \overline{Y}_i)^2$, the total sum of squared residuals equals the sum of $\Sigma(Y - Y_i)^2$ across all groups (Table 11.8):

$$\Sigma(Y - \hat{Y})^2$$
$$= \text{residual SS} = \Sigma(Y - \overline{Y}_1)^2 + \Sigma(Y - \overline{Y}_2)^2 + \Sigma(Y - \overline{Y}_3)^2 + \cdots + \Sigma(Y - \overline{Y}_7)^2$$
$$= 1919.703 + 3367.041 + 2485.000 + 1034.468 + 284.000$$
$$\quad + 667.019 + 2994.536$$
$$= 12751.767$$

Now, the standard error of estimate equals

$$\hat{s}_{Y-\hat{Y}} = \sqrt{\frac{\Sigma(Y - \hat{Y})^2}{n - k - 1}} = \sqrt{\frac{12751.767}{636}} = 4.478$$

The final quantity needed for the standard error of b_3 is $1 - R_i^2$—that is, the tolerance. You will remember that the tolerance indicates the proportion of the variance of X_i that is independent of the other Xs. This statistic cannot be determined from Table 11.8. It is given in the last column of Table 11.9 (labeled TOL), which contains the computer output of regressing age at first birth on the six ethnic variables. The six tolerances are fairly high, indicating that the standard errors of the dummy variable slopes are not inflated a great deal by the multicollinearity among the dummy variables. With the tolerance for the Irish-American variable, we can now compute its standard error:

$$s_{b_3} = \frac{\hat{s}_{Y-\hat{Y}}}{s_3 \sqrt{n} \sqrt{(1 - R_3^2)}} = \frac{4.478}{.3624 \sqrt{643} \sqrt{.691}} = .586$$

Table 11.9 Regression of Age at Birth of First Child on Ethnicity, Females, 1998 GSS

Variables	b	SE	B	t	p	TOL
African-American	−3.445	.585	−.268	−5.893	.000	.689
German-American	−.537	.510	−.051	−1.053	.293	.613
Irish-American	−.721	.586	−.056	−1.231	.219	.691
Italian-American	.285	.755	.016	.377	.706	.808
Mexican-American	−3.821	.901	−.172	−4.242	.000	.864
Native American	−3.255	.722	−.191	−4.508	.000	.791
Intercept	23.821	.378		62.947	.000	
Source	**SS**	**df**	**MS**	**F**	**p**	
Regression	1342.713	6	223.785	11.161	.000	
Residual	12751.766	636	20.050			
Total	14094.479	642				
	R	**R-Squared**	**SE of Est.**			
Goodness-of-fit	.309	.095	4.478			

This corresponds to the value shown in Table 11.9.

The t score for X_3 equals

$$t = \frac{b_3}{s_{b_3}} = \frac{-.721}{.586} = -1.23$$

The critical $t = 1.965$ for $\alpha = .05$ and $df = 636$. Because the absolute value of the calculated t is less than the critical t, we cannot reject the null hypothesis. Therefore, we cannot conclude that Irish- and British-American women have different mean ages at the birth of their first child. The level of significance of b_3 is $p = .219$.

Notice that the slopes and intercept in Table 11.9 are equal to those we calculated from the means in Table 11.8. The absolute value of t for three slopes (b_1, b_5, and b_6) is greater than the critical t. Therefore, we reject the null hypothesis for these slopes. We conclude that African-Americans, Mexican-Americans, and Native Americans have significantly lower mean ages at first birth than British-Americans.

Goodness-of-Fit

The coefficient of multiple determination (CMD) can be used to measure the goodness-of-fit of the multiple dummy-variable equation. One formula for the CMD is the regression sum-of-squares divided by the total sum-of-squares. Because the predicted value of Y for each case is the mean of Y for the group to which the case belongs, the regression SS equals

$$\text{Regression SS} = \Sigma n_i (\bar{Y}_i - \bar{Y})^2 \qquad (11.8)$$

The product of the number of cases in the group and the squared difference between the group mean and the total mean is summed across all groups. We will insert the group means from Table 11.8 into Equation 11.8 to compute the regression SS:

Regression SS

$$= n_1(\bar{Y}_1 - \bar{Y})^2 + n_2(\bar{Y}_2 - \bar{Y})^2 + n_3(\bar{Y}_3 - \bar{Y})^2 + n_4(\bar{Y}_4 - \bar{Y})^2$$
$$+ \cdots + n_7(\bar{Y}_7 - \bar{Y})^2$$
$$= 101(20.4 - 22.6)^2 + 172(23.3 - 22.6)^2 + 100(23.1 - 22.6)^2$$
$$+ 47(24.1 - 22.6)^2 + 30(20.0 - 22.6)^2 + 53(20.6 - 22.6)^2$$
$$+ 140(23.8 - 22.6)^2$$
$$= 101(-2.2)^2 + 172(.7)^2 + 100(.5)^2 + 47(1.5)^2 + 30(-2.6)^2 + 53(-2.0)^2$$
$$+ 140(1.2)^2$$
$$= 1342.71$$

This value equals that shown in Table 11.9. Now, the CMD (R^2) and the multiple correlation (R) are

$$R^2 = \frac{\text{regression SS}}{\text{total SS}} = \frac{1342.71}{14094.479} = .0953$$

$$R = \sqrt{R^2} = \sqrt{.0953} = .3087$$

R^2 indicates that about 10 percent of the variance in women's age at first birth is related to the ethnic groups to which they belong. Even if we use the multiple R, which is always greater than R^2, to judge the goodness-of-fit, the differences in age at first birth within ethnic groups are much greater than the differences between groups. Thus, the groups are more alike than they are different, at least for this dependent variable.

The most common null and alternative hypotheses with respect to the overall group differences are

X in General	*Dummy Xs*
$H_0: R^2 = 0$	$H_0: \mu_1 = \mu_2 = \cdots = \mu_g$
$H_1: R^2 \neq 0$	$H_1: \text{some } \mu_i \neq \mu_j$

The null hypothesis that CMD $= 0$ is equivalent to the hypothesis that each group has the same mean. If each group has the same mean, all the group means must equal the total mean. The alternative hypothesis states that some pairs of groups have different means. In other words, at least one group has a different mean from the others.

To test the null hypothesis, we use the same F test that was introduced in Chapter 10.

$$F = \frac{R^2/k}{(1 - R^2)/(n - k - 1)} = \frac{.0953/6}{(1 - .0953)/(643 - 6 - 1)} = 11.166$$

The numerator R^2 is divided by $df_1 = k = g - 1 = 7 - 1 = 6$, the number of dummy variables in the equation. For $\alpha = .05$, $df_1 = 6$, and $df_2 = 636$, the critical value of F is 2.02 (Table A.3a). Thus, we reject the null hypothesis that all seven ethnic groups have equal mean ages at first birth in favor of the alternative hypothesis that at least one group has a different mean from the others.

The t tests of the six slopes that are shown in Table 11.9 provide tests of six pairs of mean differences. They show that African-Americans, Mexican-Americans, and Native Americans have different (lower) means than British-Americans. There are many more pairs of means that an investigator might want to test. There are a total of $g(g - 1)/2 =$

$7 \cdot 6/2 = 21$ pairs of group means. Whenever you want to test more than the $g - 1$ pairs that are provided by the multiple regression equation, you run a greater risk of making a Type I error. Then you need to use more conservative t tests for the differences of means. Although they are beyond the scope of this text, they are readily available in texts devoting more attention to the analysis of variance (e.g., Keppel, 1991), which is introduced in the section after the next one.

Inclusion of Another *X* with the Dummy Variables

If we add another X to the equation containing the dummy variables, the interpretation of the intercept and dummy variable slopes changes somewhat. For purposes of this discussion, the new X may be an interval/ratio variable, or it may even be an additional dummy variable(s) representing another nominal variable.

In the new equation, the intercept equals the predicted value of Y when all the variables are equal to 0, including the new X. The value of a no longer equals the mean of the reference group; instead, *a is the predicted value for members of the reference group who have a score of 0 on the new X*. The slopes no longer equal the difference between the mean of the group coded 1 and the mean of the reference group, because the new X is now being held constant when the groups are contrasted. The partial slope now equals the predicted difference between the group coded 1 and the reference group for persons who do not differ on the new X. So, *the new slopes represent adjusted mean differences between groups that take into account group differences in the new X* that may in part account for the group differences in Y.

To illustrate, we will add education (years of school completed) to the six ethnic dummy variables as predictors of women's age at first birth. We expect that women who get more education will wait longer to have children than women with less education. Because ethnic groups are likely to differ in levels of education, such group differences in education may account for, or partially account for, the group differences in age at first birth.

The results of regressing age at first birth on the ethnic dummy variables and education are shown in Table 11.10. The slope for education is positive and statistically significant ($p < .001$). The greater the education, the older the woman is at the birth of her first child. More precisely, the partial slope indicates that for each additional year of school completed, age at first birth is delayed by .531 year, controlling for ethnic group. The addition of education caused R^2 to increase from .095 to .184 (R increased from .309 to .429).

What was the effect of controlling for education on the intercept and slopes of the dummy variables? The intercept is now 16.46 years, which is the predicted age at first birth of British-American women with 0 years of education. While it appears to be a plausible value, it may not be valid because there is only one woman with 0 years of education and only ten with fewer than 6 years of education (not shown). The slopes for the African-American, Mexican-American, and Native American variables are negative and statistically significant ($p < .01$). Thus, even with education held constant, women from these ethnic groups have their first child at a younger age than British-Americans. If we compare the slope for Mexican-Americans in Table 11.10 with that in Table 11.9, we see that the slope decreased from about -3.8 without controlling for education to about -2.3 with education controlled, a decrease of 1.5 years. This means that 1.5 years of the mean difference between Mexican- and British-Americans is due

Table 11.10 Regression of Age at First Birth on Ethnicity and Education, Females, 1998 GSS

Variables	b	SE	B	t	p	TOL
African-American	−2.529	.568	−.197	−4.451	.000	.661
German-American	−.014	.490	−.001	−.028	.977	.600
Irish-American	−.066	.565	−.005	−.117	.907	.678
Italian-American	.749	.721	.042	1.038	.299	.801
Mexican-American	−2.296	.878	−.103	−2.616	.009	.824
Native American	−2.461	.695	−.145	−3.542	.000	.774
Education	.531	.064	.307	8.263	.000	.932
Intercept	16.460	.963		17.093	.000	
Source	**SS**	**df**	**MS**	**F**	**p**	
Regression	2589.210	7	369.887	20.374	.000	
Residual	11491.932	633	18.155			
Total	14081.142	640				
	R	**R-Squares**	**SE of Est.**			
Goodness-of-fit	.429	.184	4.26			

to the differing educational attainments of the two ethnic groups. Thus, a considerable amount of the mean difference is due to educational differences, but there are still statistically significant ethnic differences in age at first birth with education controlled. The slopes for African-Americans and Native Americans decreased by about .9 and .8 year, respectively, as a result of controlling for education.

*One-Way Analysis of Variance

The F test for a regression equation with $g - 1$ dummy variables is equivalent to a popular statistical technique called **one-way analysis of variance.** Analysis of variance is often used in experimental research in which subjects are randomly assigned to g treatment groups. In a medical experiment, for example, subjects may be assigned to three groups: two groups that receive different dosages of a new drug that is being tested and a third group that receives a placebo (a harmless pill that doesn't contain any drugs). Analysis of variance, however, can be applied to any variable consisting of two or more groups.

We will illustrate analysis of variance with the seven ethnic groups used in the multiple regression analysis containing six dummy variables. The dependent variable in this example will be the score on a ten-word English vocabulary test included in the 1998 GSS. The score equals the number of words matched with the correct meaning. Table 11.11 shows the mean vocabulary score for each group, the number of cases in each group, and the sum-of-squares and standard deviation of each group. These statistics summarize the scores of the subsample of all males and females who were administered the test. British-Americans have the highest mean and Mexican-Americans have the lowest mean. (All of the vocabulary words and the definitions were given in English.)

Table 11.11 Vocabulary Score by Ethnic Group, 1998 GSS

Group	Mean	n	SS	SD
1. African-American	5.14	87	302.344	1.864
2. German-American	6.39	195	636.406	1.807
3. Irish-American	6.33	131	556.876	2.062
4. Italian-American	6.35	63	210.318	1.827
5. Mexican-American	4.78	23	119.918	2.283
6. Native American	5.34	53	165.888	1.769
7. British-American	7.18	139	564.490	2.015
Total	6.24	691	2882.776	2.043

We want to test the hypothesis that the different group means in Table 11.11 might have occurred by chance when there are no group differences in the population. The null and alternative hypotheses are of the same form as they were in the F test for the dummy variable regression equation:

$$H_0: \mu_1 = \mu_2 = \mu_3 = \mu_4 = \mu_5 = \mu_6 = \mu_7$$

$$H_1: \text{some } \mu_i \neq \mu_j$$

To test the null hypothesis, the analysis of variance breaks down the total sum-of-squares (SS) into between-group SS and within-group SS as follows:

$$\text{Total SS} = \text{between-group SS} + \text{within-group SS}$$

The between-group SS equals the sum of the squared deviations of the group means from the total mean with each squared deviation weighted by the number of cases in that group:

Between SS
$$= \Sigma n_i (\overline{Y}_i - \overline{Y})^2$$
$$= n_1 (\overline{Y}_1 - \overline{Y})^2 + n_2 (\overline{Y}_2 - \overline{Y})^2 + n_3 (\overline{Y}_3 - \overline{Y})^2 + \cdots + n_g (\overline{Y}_g - \overline{Y})^2 \qquad (11.9)$$

where $n_i (\overline{Y}_i - \overline{Y})^2$ is the number of cases in group i times the squared deviation of the mean of group i from the total mean. This is equivalent to summing the squared deviations of each case's group mean from the total mean. For the vocabulary score means in Table 11.11, the between-group SS is

Between SS
$$= 87(5.14 - 6.24)^2 + 195(6.39 - 6.24)^2 + 131(6.33 - 6.24)^2$$
$$\quad + 63(6.35 - 6.24)^2 + 23(4.78 - 6.24)^2 + 53(5.34 - 6.24)^2$$
$$\quad + 139(7.18 - 6.24)^2$$
$$= 87(-1.10)^2 + 195(.15)^2 + 131(.09)^2 + 63(.11)^2 + 23(-1.46)^2$$
$$\quad + 53(-.90)^2 + 139(.94)^2$$
$$= 326.258$$

Table 11.12 Analysis of Variance of Vocabulary Score by Ethnic Group, 1998 GSS

Source	SS	df	MS	F	p
Between groups	326.258	6	54.376	14.551	.000
Within groups	2556.240	684	3.737		
Total	2882.498	690			

The within-group SS equals the sum, across all groups, of each group's sum of squared deviations around its mean (i.e., the SS within each group):

$$\text{Within SS} = \Sigma(Y - \bar{Y}_1)^2 + \Sigma(Y - \bar{Y}_2)^2 + \Sigma(Y - \bar{Y}_3)^2$$
$$+ \cdots + \Sigma(Y - \bar{Y}_g)^2 \tag{11.10}$$

where, for example, $\Sigma(Y - \bar{Y}_2)^2$ is the sum-of-squares for group 2. For the vocabulary score data in Table 11.11, the within-group SS is

$$\begin{aligned}\text{Within SS} &= 302.344 + 636.406 + 556.876 + 210.318 + 119.918 + 165.888 \\ &\quad + 564.490 \\ &= 2556.240\end{aligned}$$

If we sum the within-group SS and the between-group SS, we get the total SS:

$$\text{Total SS} = 326.258 + 2556.240 = 2882.498$$

This total equals that given in Table 11.11 except for a little rounding error in the calculations.

The within and between sum-of-squares are used to create the F statistic:

$$F = \frac{\text{between SS}/(g - 1)}{\text{within SS}/(n - g)} = \frac{\text{between mean square}}{\text{within mean square}} \tag{11.11}$$

where $g - 1 = df_1$ and $n - g = df_2$. When we enter the between SS, within SS, df_1, and df_2 into Equation 11.11, we get

$$F = \frac{326.258/(7 - 1)}{2556.240/(691 - 7)} = \frac{54.376}{3.737} = 14.55$$

For $\alpha = .05$, $df_1 = 6$, and $df_2 = 684$, we find in Table A.3b that the critical F is 2.02. Because the calculated F is greater than the critical F, we reject the null hypothesis and conclude that at least one ethnic group has a different mean vocabulary score from the others. If we want to test all pairs of groups to see which ones were significantly different, we have to use a special t test for a difference of means that protects against the increased risk of a Type I error resulting from conducting $g(g - 1)/2 = (7 \cdot 6)/2 = 21$ t tests (e.g., Keppel, 1991). This test, however, is beyond the scope of this text. The results of the F test are shown in Table 11.12, a typical analysis of variance table.

We stated at the beginning of this section that the analysis of variance F test is identical to the F test for a regression equation with $g - 1$ dummy variables. You probably noticed the similarity between the material in this section and that covered earlier for dummy variables. In fact, the between-group and within-group sums-of-squares are equal to the regression and residual sums-of-squares, respectively:

$$\text{Between SS} = \text{regression SS}$$

$$\text{Within SS} = \text{residual SS}$$

The formula for the F statistic used in analysis of variance (Equation 11.11) is also equal to and similar in appearance to the formula for F given in Chapter 10 (Equation 10.19):

$$F = \frac{\text{between SS}/(g-1)}{\text{within SS}/(n-g)} = \frac{\text{between mean square}}{\text{within mean square}}$$

$$= \frac{\Sigma(\hat{Y}-\bar{Y})^2/df_1}{\Sigma(Y-\hat{Y})^2/df_2} = \frac{\text{regression mean square}}{\text{residual mean square}}$$

where $df_1 = k = g - 1$ and $df_2 = n - k - 1 = n - g$. Thus, the difference between analysis of variance and regression analysis with dummy variables is entirely cosmetic because it consists of nothing more than different names, notation, and software.

SUMMARY

In this chapter we learned how to conduct a regression analysis for a nominal independent variable consisting of two or more groups/categories. Because regression analysis requires at least an interval level of measurement, it is necessary to create one or more special variables to represent the groups composing the nominal variable. In this chapter we used dummy variables on which one group is given a score of 1 and the other group(s) is given a score of 0. If the number of groups is g, we use $g - 1$ dummy variables to fully represent the groups composing the nominal variable. Thus, one group, called the reference group, has a score of 0 on all the dummy variables. Because the best prediction of Y for each group is that group's mean on Y, the intercept of the equation equals the mean of the reference group and the slope for each dummy variable equals the mean of the group coded 1 minus the mean of the reference group. Consequently, the usual t test for the slope of each variable is a test for the difference of means between these two groups. Similarly, the usual F test for the coefficient of multiple determination (R^2) is a test of the null hypothesis that all of the groups composing the nominal variable have equal means on Y. When another X is included in the equation along with the dummy variable(s), there is a slight change in the meaning of the intercept and slope(s). Optional sections described how dummy variable regression produces equivalent results to t tests of differences of means and one-way analysis of variance.

KEY TERMS

conditional mean

dummy variable

reference group

slope of a dummy variable

intercept of a dummy variable equation

mean of a dummy variable

standard deviation of a dummy variable

t test of a dummy variable slope

*t test for difference of means

*t test for difference of proportions

maximum number of dummy variables

*one-way analysis of variance

SPSS PROCEDURES

To conduct a regression analysis with a nominal independent variable, we must first create one or more dummy variables to represent the nominal variable. We will illustrate this procedure with the ethnicity variable, named *ethnic,* from the 1998 GSS that is used in this chapter. The following SPSS frequency distribution for *ethnic* lists 42 groups, each with its own numeric code. Only the eight largest groups (Africa, England and Wales, Germany, Ireland, Italy, Mexico, Scotland, and Native America) are used in the regression analyses. For these analyses, dummy variables were created for Africa, Germany, Ireland, Italy, Mexico, and Native America. Britain (England, Wales, and Scotland) formed the reference group.

We will illustrate how to create a dummy variable for Germany. Click on *Transform, Recode,* and *Into Different Variables* to open the *Recode into Different Variables* window. Move *ethnic* into the box under *Numeric Variable → Output Variable.* In the *Name* box in the *Output Variable* panel, type *german,* which will be the name of the dummy variable that is being created. When you click on *Change, german* will be moved into the following box along with *ethnic* under *Numeric Variable → Output Variable.* Now, click on the button *Old and New Values* to open the window *Recode into Different Variables: Old and New Values.* In the *Old Value* panel, click on *Value* and type 11, which is the numeric code for Germany, in its box. In the *New Value* panel, click on *Value,* type 1 in its box, and click on the *Add* button to move "11 → 1" into the box *Old → New.* You have now assigned the value of 1 to German-Americans on the *german* dummy variable. To assign 0 to the other groups on the dummy variable, click on *All other values* in the panel *Old Values,* click on *Value* in the *New Value* panel, type 0 in its box, and click on the *Add* button to move "Else → 0" into the box *Old → New.* To assign a missing value to the dummy variable *german* if *ethnic* is missing, click on *System- or user-missing* in the *Old Values* panel, click on *System-missing* in the *New Values* panel, and click on *Add* to move MISSING → SYSMISS into the following *Old → New* box. After you have clicked on *Continue* and *OK,* the dummy variable named *german* will appear in the data editor window.

ETHNIC

		Frequency	Percent	Valid Percent	Cumulative Percent
Valid	1 AFRICA	214	7.6	9.6	9.6
	2 AUSTRIA	5	.2	.2	9.8
	3 FRENCH CANADA	44	1.6	2.0	11.8
	4 OTHER CANADA	13	.5	.6	12.3
	5 CHINA	23	.8	1.0	13.4
	6 CZECHOSLOVAKIA	25	.9	1.1	14.5
	7 DENMARK	12	.4	.5	15.0
	8 ENGLAND & WALES	263	9.3	11.8	26.8
	9 FINLAND	3	.1	.1	26.9
	10 FRANCE	56	2.0	2.5	29.4
	11 GERMANY	406	14.3	18.1	47.5
	12 GREECE	9	.3	.4	47.9
	13 HUNGARY	15	.5	.7	48.6
	14 IRELAND	269	9.5	12.0	60.6
	15 ITALY	134	4.7	6.0	66.6
	16 JAPAN	5	.2	.2	66.8
	17 MEXICO	74	2.6	3.3	70.2
	18 NETHERLANDS	34	1.2	1.5	71.7
	19 NORWAY	34	1.2	1.5	73.2
	20 PHILIPPINES	19	.7	.8	74.0
	21 POLAND	65	2.3	2.9	76.9
	22 PUERTO RICO	35	1.2	1.6	78.5
	23 RUSSIA	19	.7	.8	79.4
	24 SCOTLAND	69	2.4	3.1	82.4
	25 SPAIN	16	.6	.7	83.2
	26 SWEDEN	34	1.2	1.5	84.7
	27 SWITZERLAND	5	.2	.2	84.9
	28 WEST INDIES	1	.0	.0	84.9
	29 OTHER	28	1.0	1.3	86.2
	30 AMERICAN INDIAN	110	3.9	4.9	91.1
	31 INDIA	10	.4	.4	91.6
	32 PORTUGAL	9	.3	.4	92.0
	33 LITHUANIA	5	.2	.2	92.2
	34 YUGOSLAVIA	14	.5	.6	92.8
	35 RUMANIA	3	.1	.1	92.9
	36 BELGIUM	5	.2	.2	93.2
	37 ARABIC	3	.1	.1	93.3
	38 OTHER SPANISH	32	1.1	1.4	94.7
	39 NON-SPAN WINDIES	19	.7	.8	95.6
	40 OTHER ASIAN	13	.5	.6	96.2
	41 OTHER EUROPEAN	14	.5	.6	96.8
	97 AMERICAN ONLY	72	2.5	3.2	100.0
	Total	2238	79.0	100.0	
Missing	0 UNCODEABLE	581	20.5		
	99 NA	13	.5		
	Total	594	21.0		
Total		2832	100.0		

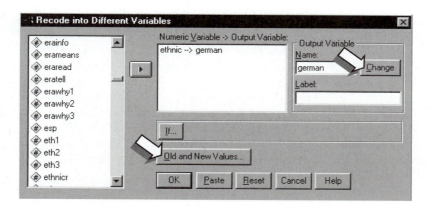

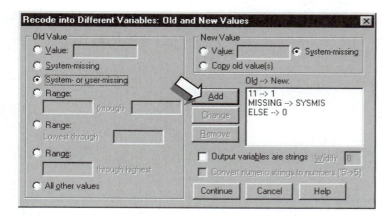

Repeat this procedure for the other five ethnic groups for which we want to create dummy variables. We do not have to create a dummy variable for British-Americans because they have been chosen as the reference group. For each group, type a new variable name (e.g., *mexican*) in the *Output Variable* box in the first Recode window and type the appropriate numeric code for the group (e.g., 17) in the *Old Value* box of the second Recode window.

Before conducting the regression analysis for ethnic differences in women's age at birth of their first child, we must select only those cases that are female and that belong to the eight largest ethnic groups. Click on *Data* and *Select Cases* to open the *Select Cases* window and then click on *If condition is satisfied* and the *If* button. In the *Select Cases: If* window, enter "sex = 0 and (ethnic = 1 or ethnic = 8 or ethnic = 11 or ethnic = 14 or ethnic = 15 or ethnic = 17 or ethnic = 24 or ethnic = 30)" in the box. After you have clicked on *Continue* and *OK*, only female members of those groups will be available for analysis. You can now run the regression procedure with the six ethnic dummy variables as independent variables and *agekdbrn* as the dependent variable.

*We can use SPSS's one-way analysis of variance procedure to conduct the same test. First, recode the *ethnic* variable to assign Scotland the same numeric code as England and Wales (*Old Value* = 24 and *New Value* = 8) and place the recoded variable in a new variable *ethnicr*. Then, select cases that belong to one of the eight largest ethnic groups and that are female. [In the *Select Cases: If* window, enter "(ethnic = 1 or ethnic = 8 or ethnic = 11 or ethnic = 14 or ethnic = 15 or ethnic = 17 or ethnic = 24 or ethnic = 30) and sex = 0" in the box.] Now, click on *Analyze, Compare Means,* and *One-Way ANOVA*. In the *One-Way ANOVA* window, move *agekdbrn* to the *Dependent List* and move *ethnicr* to the following *Factor* box. After clicking *OK,* you can verify that the *F* test shown in the output window in the following table is identical to the *F* test for the regression analysis shown in this chapter.

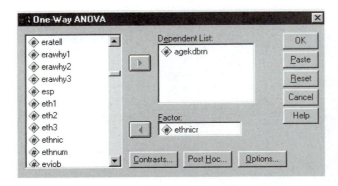

ANOVA

AGEKDBRN

	Sum of Squares	df	Mean Square	F	Sig.
Between Groups	1342.713	6	223.785	11.161	.000
Within Groups	12751.766	636	20.050		
Total	14094.479	642			

*We can use the independent-samples *t* test to test for a gender difference in age of first marriage in the 1994 GSS. Click on *Analyze, Compare Means,* and *Independent-Samples T Test.* Move the variable *agewed* into the *Test Variable(s)* box and move the variable *sex* into the following *Grouping Variable* box. Click on *Define Groups* and enter 1 in the *Group 1* box and enter 0 in the *Group 2* box (*sex* has been recoded to 1 = males and 0 = females). After clicking *Continue* and *OK,* the Group Statistics table and the Independent Samples Test table will appear in the following output window. The *t* test for "Equal variances assumed" is identical to the *t* test for the slope of *agewed* regressed on the dummy variable for gender that is shown in this chapter.

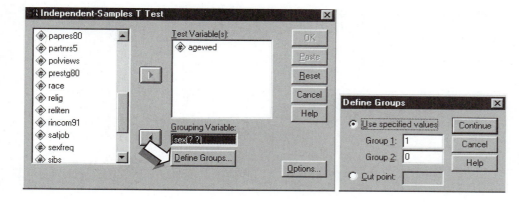

Group Statistics

	SEX	N	Mean	Std. Deviation	Std. Error Mean
AGEWED	MALE	469	23.93	4.721	.218
	FEMALE	720	21.82	4.808	.179

Independent Samples Test

		Levene's Test for Equality of Variances		t-test for Equality of Means						
									95% Confidence Interval of the Difference	
		F	Sig.	t	df	Sig. (2-tailed)	Mean Difference	Std. Error Difference	Lower	Upper
AGEWED	Equal variances assumed	.083	.773	7.452	1187	.000	2.11	.283	1.555	2.667
	Equal variances not assumed			7.480	1013.002	.000	2.11	.282	1.557	2.665

EXERCISES

1. The following graph shows the regression line for a state's homicide rate (homicides per 100,000) circa 1993 regressed on a dummy variable for region, where 1 is the South (AL, AR, DE, FL, GA, KY, LA, MD, MS, NC, OK, SC, TN, TX, VA, WV) and 0 is all other states, which we will call the North. Find the mean for the South, the mean for the North, the intercept, and the slope.

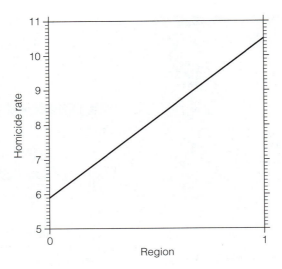

2. For 778 women aged 40 or older in the GSS 2000, the graph shows the regression of age at birth of first child (*agekdbrn*) on a dummy variable for race, where 1 is Black and 0 is White (25 women of other races were omitted). Find the mean for Whites, the mean for Blacks, the intercept, and the slope.

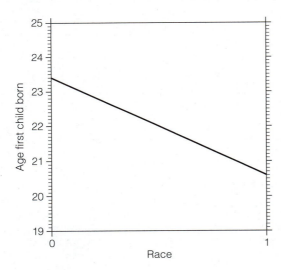

3. For full-time workers in GSS 2000, the table reports the mean hours worked last week (*hrs1*) by whether the person was self-employed or was employed by someone else (*wrkslf*). If *wrkslf* was coded as a dummy variable with SOMEONE ELSE coded 1 and SELF-EMPLOYED coded 0:

a. What would be the mean and standard deviation of *wrkself?*

b. What would be the intercept and slope of *hrs1* regressed on *wrkslf?*

HRS1

WRKSLF	Mean	N	Std. Deviation
SELF-EMPLOYED	50.66	186	15.178
SOMEONE ELSE	44.83	1303	9.504
Total	45.55	1489	10.555

4. For 50 states circa 1993, the following table reports the mean percentage of children living in single-parent households (*single*) by region (dummy coded).

a. What are the mean and standard deviation of *region?*

b. What are the intercept and slope of *single* regressed on *region?*

SINGLE

REGION	Mean	N	Std. Deviation
0 North	20.59	34	3.456
1 South	25.44	16	3.669
Total	22.14	50	4.170

5. Assume you are investigating how the type of students' residence affects their satisfaction with college life. Let there be four types of residence: Greek house, dormitory, parent's house, and apartment. Construct a table showing how you would code dummy variables to represent each type of residence in a regression analysis, assuming that dormitory residence is the reference group.

6. Assume you are investigating how type of degree affects post-graduation earnings. Let there be six degree-granting schools at a university: liberal arts, business, fine arts, engineering, nursing, and education. Choose one school as a reference group and construct a table showing how you would code dummy variables to represent type of degree in a regression analysis.

7. For all respondents in the GSS 2000 who ever had a child, both males and females, the table reports the mean age at birth of first child (*agekdbrn*) by the variable religious preference (*relig5*), which was constructed by recoding all preferences other than the first four into the category Other. What would be the intercept and slopes for the regression of *agekdbrn* on the dummy variables for religious preference, with Protestant as the reference group?

AGEKDBRN

RELIG5	Mean	N	Std. Deviation
Protestant	23.24	1158	5.116
Catholic	24.35	456	5.412
Jewish	26.53	36	5.023
None	23.54	222	5.745
Other	23.71	91	4.996
Total	23.62	1963	5.283

8. For all respondents in the GSS 2000 who ever had a child, both males and females, the following table reports the mean age at birth of first child (*agekdbrn*) by the variable *bible,* which indicates whether the respondent feels the Bible is

a. The actual word of God and is to be taken literally.
b. The inspired word of God but not everything is to be taken literally.
c. An ancient book of fables, legends, history, and moral precepts recorded by man.

AGEKDBRN

BIBLE	Mean	N	Std. Deviation
WORD OF GOD	22.70	607	5.001
INSPIRED WORD	24.09	747	5.299
BOOK OF FABLES	24.27	207	5.729
Total	23.57	1561	5.289

What would be the intercept and slopes for the regression of *agekdbrn* on the dummy variables for *bible,* with Word of God as the reference group?

9. For the GSS 2000, the following tables report the regression of *sexprtyr* on religious preference as represented by four dummy variables, with Protestant as the reference group.

ANOVA[b]

Model		Sum of Squares	df	Mean Square	F	Sig.
1	Regression	43.463	4	10.866	6.640	.000[a]
	Residual	3166.259	1935	1.636		
	Total	3209.722	1939			

[a.] Predictors: (Constant), OTHERRLG, JEWISH, NONE, CATHOLIC
[b.] Dependent Variable: SEXPRTYR

Coefficients[a]

Model		Unstandardized Coefficients		Standardized Coefficients	t	Sig.
		B	Std. Error	Beta		
1	(Constant)	.473	.039		12.048	.000
	CATHOLIC	-.013	.072	-.004	-.183	.855
	JEWISH	.053	.199	.006	.265	.791
	NONE	.413	.086	.112	4.786	.000
	OTHER	.224	.130	.040	1.723	.085

a. Dependent Variable: SEXPRTYR

a. Interpret the intercept and the four partial slopes.
b. What is the two-tailed null hypothesis for each dummy variable?
c. Using t tests at $\alpha = .05$, what decision/conclusion can you make about each variable?
d. What are the null and alternative hypotheses for the F test at $\alpha = .05$, and what decision should you reach?

10. For the GSS 2000, the following tables report the regression of *sexprtyr* on feelings about the Bible as represented by two dummy variables, with Word of God as the reference group.

ANOVA[b]

Model		Sum of Squares	df	Mean Square	F	Sig.
1	Regression	17.806	2	8.903	6.307	.002[a]
	Residual	2222.099	1574	1.412		
	Total	2239.906	1576			

a. Predictors: (Constant), BKFABLES, INSPIRWD
b. Dependent Variable: SEXPRTYR

Coefficients[a]

Model		Unstandardized Coefficients		Standardized Coefficients	t	Sig.
		B	Std. Error	Beta		
1	(Constant)	.394	.053		7.401	.000
	INSPIRWD	.140	.068	.059	2.072	.038
	BKFABLES	.317	.090	.100	3.517	.000

a. Dependent Variable: SEXPRTYR

a. Interpret the intercept and the two partial slopes.
b. What are the null hypothesis and the alternative hypothesis for each dummy variable?
c. Using t tests at $\alpha = .05$, what decision/conclusion can you make about each variable?
d. What are the null and alternative hypotheses for the F test at $\alpha = .05$, and what decision/conclusion should you reach?

11. Use GSS 2000 (*gss2000a* or *gss2000b*) to study differences in sex partners per year by marital status. Use SPSS to create dummy variables for marital status (*marital*), using married as the reference group. Regress *sexprtyr* on the dummy variables representing marital status. Using two-tailed tests with $\alpha = .05$, what decision can you make for each dummy variable? Interpret the partial slopes of those variables that are significant. Now add *age* to the equation and run the regression again. Using a two-tailed test with $\alpha = .05$, what decision can you make for *age,* and how do you interpret the slope for *age* if it is significant? Describe whether controlling for *age* changes the hypothesis tests and the slopes of the dummy variables for marital status. Why does controlling for age result in these changes, if any?

12. Use GSS 2000 (*gss2000a* or *gss2000b*) to study differences in age at birth of first child by marital status for women aged 40 or more. Use SPSS to create dummy variables for marital status (*marital*), using married as the reference group. Regress *agekdbrn* on the dummy variables representing marital status. Using two-tailed tests with $\alpha = .05$, what decision do you make for each dummy variable? Interpret the partial slopes of those variables that are significant. Now, add *age* to the equation and run the regression again. Using a two-tailed test with $\alpha = .05$, what decision do you make for *age,* and how do you interpret the slope for *age* if it is significant? Describe whether controlling for *age* changes the hypothesis tests and the slopes of the dummy variables for marital status. Why does controlling for age result in these changes, if any?

12

Ordinal Dependent Variables

AN A IS AN A IS AN A

"Professor Normal, I came to ask you to introduce a proposal in the faculty senate to improve the grading system here at East Central State University," said Professor Perfect.

"I'm all ears. But what's wrong with our grading system?" asked Professor Normal.

"Grade inflation! Rampant grade inflation!" exclaimed Professor Perfect. "I've been studying the grade distributions in my department. There are lots of courses where most of the students get an A or a B, and there are even some where the majority get an A! A friend of mine has found the same thing in her department. I knew some of my colleagues were easy graders, but I didn't know it was this bad. It devalues an A, and it isn't fair to students who earn A grades in courses with normal grade distributions. I think it is time for the faculty to do something about it," Professor Perfect declared.

"I'm not surprised to hear that. Grade inflation is a national trend. But what can be done about it? You can't tell instructors how many A's and B's to give. It would violate their academic freedom," replied Professor Normal.

"I agree. But there is a way to stop inflation without telling instructors how to grade. And it is based on sound statistical principles. The letter grade that instructors give is an ordinal variable: an A is higher than a B, a B is higher than a C, etc.

The college then assigns grade points to the letter grades to compute a grade point average (GPA) for each student; for example, an A equals 4 points, a B equals 3 points, and so on. By computing the mean of the grade points, the college is treating the grade points as if they stood for an interval variable. This is statistically invalid. It also counts an A in an easy course just as high as an A in a hard course. An A is an A is an A. This is unfair. However, there is a statistically valid way to assign numbers to ordinal letter grades, which is called rank scoring the grades. Without going into great detail, the greater the number of students who receive an A, the lower will be their average rank and, thus, the lower will be the number assigned to an A in that course. Consequently, an A in an easy course will receive fewer points than an A in a harder course and, possibly, even fewer points than a B in a harder course. After standardizing the rank score by dividing by the number of students in the class, the ranks can be averaged together for all of a student's courses to get a grade rank average (GRA). What do you think?" asked Professor Perfect.

Professor Normal scratched his head. "I don't know. I'm not a statistician. What you should do, Professor Perfect, is write all this out in clear and simple terms that can be understood by faculty members who have no statistical training. I would be happy to read that document, and then we can talk again about taking the proposal to the senate. I have to tell you though, in all honesty, I am not very optimistic that your proposal will get much support," concluded Professor Normal.

Professor Perfect claims that letter grades are an ordinal variable, not an interval variable. Is she correct? Aren't instructors supposed to grade in such a way that grades are an equal-interval variable? Shouldn't the difference between an A and a B, for example, be the same as the difference between a B and a C in terms of student performance? If so, is this the way grades are actually assigned? If not, what are the intervals between grades? And if these intervals can be identified, won't they vary from course to course and from professor to professor? In the end, do we know anything more than the fact that in any given course, an A is higher than a B, which is higher than a C, and so forth? Aren't grades just an ordinal variable for which the computation of a grade point average is invalid?

   Intriguing and important as this topic may be, we do not intend to resolve the question of the correct level of measurement of letter grades. The discussion between Professors Perfect and Normal does introduce the topic of how to appropriately score an ordinal variable. If we have an ordinal variable, or think we have an ordinal variable, how should we assign scores to the different attributes so we can measure its association with another variable and test hypotheses about that association? In this chapter, we will explore how to measure and test associations involving an ordinal variable.

RANK SCORING

We are all familiar with the use of ranks to describe differences in performance. Whether in horse races, track and field events, spelling bees, dog shows, baking contests, or most other competitions, the best performance is called *first place,* the next best is *second place,* the one after that is *third place,* and so on. The associated numerical ranks are shown in Table 12.1. In this table, *n* is the number of units in the competition. Many competitions don't

Table 12.1 Everyday Rank Scoring

Performance	Rank
First place	1
Second place	2
Third place	3
Fourth place	4
⋮	⋮
Last place	n

Table 12.2 Statistical Rank Scoring

Attribute	Rank
Lowest	1
Second lowest	2
Third lowest	3
Fourth lowest	4
⋮	⋮
Highest	n

Table 12.3 "Normal" Grade Distribution

Grade	f
F	1
D	2
C	3
B	2
A	1
Total	9

assign ranks all the way up to n. Instead, they typically focus on only the top three performances. For purposes of statistical analysis, however, we need to assign each unit a rank.

In statistical analyses, unlike everyday rankings, we will seldom, if ever, be dealing with formal competitions in which units are assigned first place, second place, third place, and so on. Instead, we will be dealing with ordinal variables consisting of attributes that are ranked from highest to lowest. Also, rather than assigning the highest attribute the rank of 1, we will give it the rank of n, and the lowest attribute will be given the rank of 1, as shown in Table 12.2. The reason for assigning higher ranks to higher attributes is to facilitate the interpretation of measures of association between ordinal variables and other variables. A positive association means that higher-ranked attributes on the ordinal variable are associated with higher attributes on another variable.

Tables 12.1 and 12.2 suggest that no two units have exactly the same attribute—that is, there are no ties. When this is the case, each unit will have a distinct rank. When there are no ties, there will be n distinct ranks. It is usually the case in the social sciences, however, that ordinal variables may have many ties. For example, if an ordinal variable has $g = 5$ attributes and there are $n = 100$ cases, there will be an average of $n/g = 100/5 = 20$ cases per attribute. To illustrate, let's examine the following "normal" grade distribution in a class of $n = 9$ students as shown in Table 12.3. Two students are tied for a B, three students for a C, and two students for a D, with an average of $9/5 = 1.8$ students per grade. We can assign ranks to the A and F students in a straightforward manner because there is only one student with each grade. The A, or highest attribute, gets the rank of $n = 9$. The F, or lowest attribute, receives the rank of 1. How should we assign ranks to the tied students?

Table 12.4 shows how ranks are computed for the other grades. First, the individual grades are listed in ascending order as shown in the grade column. It is arbitrary which student's grade is listed first among those who are tied at a particular grade. Then, ranks from 1 to 9 are assigned to the nine grades in the untied-rank column. Next, the mean of the untied ranks for each letter grade is computed. It is interesting to note that in the case of a subset of adjacent ranks, the mean of the ranks equals the median. Last, the average rank for each grade is assigned to each student with that grade, as shown in the rank column.

What is the logic behind this method of scoring ranks? The two students with a B, for example, are not just tied for the second highest rank. They are tied for the second and third highest ranks, which are 8 and 7, respectively. Therefore, we take the average of these two ranks and assign each B student a rank of 7.5. The three C students who

Table 12.4 Ranks for Letter Grades with Ties

Grade	Untied Rank	Average Untied Rank		Rank
F	1	1 / 1 =	1.0	1.0
D	2	(2 + 3) / 2 =	2.5	2.5
D	3			2.5
C	4			5.0
C	5	(4 + 5 + 6) / 3 = 5.0		5.0
C	6			5.0
B	7	(7 + 8) / 2 =	7.5	7.5
B	8			7.5
A	9	9 / 1 =	9.0	9.0

tied for the fourth, fifth, and sixth highest ranks, which are 6, 5, and 4, respectively, are assigned the average of these three ranks, which is 5.0.

Notice that the intervals between the ranks are not equal. For example, the difference between an A and a B is 1.5 ranks, whereas the difference between a B and a C is 2.5 ranks. Unequal intervals are almost always found with this type of rank scoring. Only when there is an equal number of cases with each attribute (i.e., grade) will the ranks have equal intervals.

There is an easier way to determine the average rank of each attribute than the method shown in Table 12.4. The formula for this method of **rank scoring** is

$$R_i = cf_B + \frac{f_i + 1}{2} \tag{12.1}$$

where

$$R_i = \text{average rank of attribute } i$$

$$cf_B = \text{cumulative frequency below attribute } i$$

$$f_i = \text{frequency of attribute } i$$

In this formula, cf_B is equivalent to the next lower untied rank below attribute i. The next term, $(f_i + 1)/2$, indicates how much we must add to the next lower rank to reach the median of the untied ranks for attribute i. The method is shown in Table 12.5. The use of Equation 12.1 becomes more efficient as sample size increases.

The mean of the ranks is

$$\overline{R} = \frac{\Sigma f R_i}{n} = \frac{(1 \cdot 1.0 + 2 \cdot 2.5 + 3 \cdot 5.0 + 2 \cdot 7.5 + 1 \cdot 9.0)}{9} = \frac{45.0}{9} = 5.0$$

Because this normal grade distribution is symmetrical, the mean is located at the center of the distribution. A quicker way to compute the **mean rank** for any variable is

$$\overline{R} = \frac{n + 1}{2} = \frac{10}{2} = 5$$

Table 12.5 Average Ranks for "Normal" Grade Distribution

Grade	f	cf	$cf_B + (f_i + 1)/2 = R_i$
F	1	1	$0 + (1 + 1)/2 = 1.0$
D	2	3	$1 + (2 + 1)/2 = 2.5$
C	3	6	$3 + (3 + 1)/2 = 5.0$
B	2	8	$6 + (2 + 1)/2 = 7.5$
A	1	9	$8 + (1 + 1)/2 = 9.0$

Table 12.6 Average Ranks for Upward Skewed Grade Distribution

Grade	f	cf	$cf_B + (f_i + 1)/2 = R_i$
C	2	2	$0 + (2 + 1)/2 = 1.5$
B	4	6	$2 + (4 + 1)/2 = 4.5$
A	3	9	$6 + (3 + 1)/2 = 8.0$

Table 12.7 Ranks of Grades in Normal and Skewed Distributions

Grade	Normal Distribution	Skewed Distribution
F	1.0	. . .
D	2.5	. . .
C	5.0	1.5
B	7.5	4.5
A	9.0	8.0

Now, let's look at a different grade distribution to see how it changes the ranks assigned to letter grades. A skewed distribution is shown in Table 12.6. This distribution contains more A's than C's, and B is the modal grade. There are no D's or F's. The average ranks for the three letter grades are again computed using Equation 12.1. The mean of the ranks is $(n + 1)/2 = (9 + 1)/2 = 5$, the same as the mean of the normal distribution.

The ranks of the letter grades in the normal and skewed distributions are displayed in Table 12.7. As you can see, the shape of the distribution can have a dramatic effect on the ranks of the attributes. An A in the skewed distribution counts less than it would in the normal distribution. A grade of B in the skewed distribution ranks lower than even a C in the normal distribution. And the C in the skewed distribution gets fewer points than the D in the normal distribution. If ranks are used to assign grade points, it is not true that "an A is an A is an A."

ASSOCIATION BETWEEN
TWO ORDINAL VARIABLES

If we want to measure the association between two ordinal variables, we can use the ranks for each variable to conduct a regression analysis in the usual fashion. The only difference between a regression analysis performed on two rank-scored variables and that for two interval variables is that we must interpret the regression coefficients for the ordinal variables in terms of the ranks. The linear **rank-regression equation** is

$$\hat{R}_Y = a + bR_X \tag{12.2}$$

where

$$R_X = \text{rank of } X$$
$$\hat{R}_Y = \text{predicted rank of } Y$$
$$a = \hat{R}_Y \text{ when } R_X \text{ equals } 0$$
$$b = \text{rank change in } Y \text{ per rank increase in } X$$

The intercept a is interpreted as the predicted rank of Y when the rank of X equals 0. Because the smallest rank for any variable is 1, 0 is out of the range of R_X, and thus the intercept is not, strictly speaking, meaningful. We can adjust for this bias by subtracting 1 from all the ranks. After subtracting, 0 will be the minimum possible rank and the intercept will theoretically be meaningful. Aside from the fact that, traditionally, ranks do not have a value of 0, there is a second trait about ranks that usually negates the value of such an adjustment. If there is more than one case that has the lowest-ranked attribute, no case will have a rank of 1. Because there are usually many ties for each ordinal position on realistic ordinal variables, the minimum possible rank will seldom, if ever, be assigned. As a result, it is best not to interpret the intercept, but instead, simply to view it as a constant that must be added to the product bR_X to give accurate predictions.

The **rank–regression slope** *is the change in the rank of Y per rank increase in X.* That is, if the rank of X increases by 1, the slope indicates the number of ranks that Y is expected to increase or decrease, depending on the sign of b. This is a simple and meaningful interpretation because when both variables are rank scored, each variable has the same measurement scale. Thus, it is valid to size up the amount R_Y changes per unit increase in R_X despite the fact that b is an unstandardized slope. For reasons to be described shortly, its range will usually fall between -1 and 1.

To illustrate regression for rank-scored ordinal variables, we will examine the association between how strongly Democratic or Republican a person is and how conservative or liberal the person is. The General Social Survey asked respondents to pick one of seven political party labels that best described them. These categories ranged from strong Democrat to strong Republican. The seven party labels and the frequency distribution for this ordinal variable are shown in the first panel of Table 12.8. People are more likely to label themselves Democrats than Republicans. Respondents were also asked to choose one of seven labels that described their political views. These labels ranged from extremely liberal to extremely conservative. The labels and the frequencies for this variable are shown in the second panel of Table 12.8. Although respondents were more likely to label themselves conservative than liberal, they were most likely to say they were moderate in their political views. A comparison of the shapes of the distributions in Figure 12.1 suggests that there is greater dispersion in political party identification than in political views. Because there is no special measure of dispersion for

Table 12.8 Political Party and Political Views, GSS 1998

	f	Pct	cf	Rank
DEMOCRAT–REPUBLICAN				
Strong Democrat	362	13.8	362	181.5
Not very strong Democrat	570	21.7	932	647.5
Independent, near Democrat	342	13.0	1274	1103.5
Independent	411	15.7	1685	1480.0
Independent, near Republican	233	8.9	1918	1802.0
Not very strong Republican	470	17.9	2388	2153.5
Strong Republican	234	8.9	2622	2505.5
Total	2622	100.0		
LIBERAL–CONSERVATIVE				
Extremely liberal	64	2.4	64	32.5
Liberal	345	13.2	409	237.0
Slightly liberal	343	13.1	752	581.0
Moderate	960	36.6	1712	1232.5
Slightly conservative	423	16.1	2135	1924.0
Conservative	403	15.4	2538	2337.0
Extremely conservative	84	3.2	2622	2580.5
Total	2622	100.0		

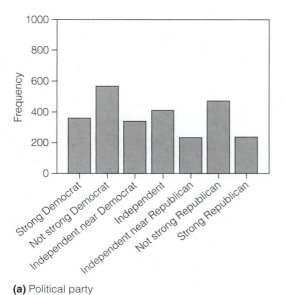

(a) Political party

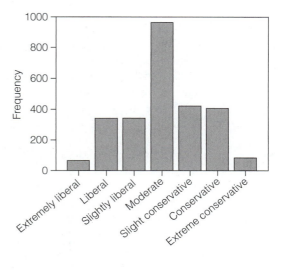

(b) Political views

FIGURE 12.1 Political Party and Political Views, GSS 1998

ordinal variables, we must use the index of qualitative variation (IQV) to measure dispersion in these variables (Chapter 4). The values of IQV are

$$\text{IQV} = \frac{\sum f_i f_j}{\dfrac{g(g-1)}{2}\left(\dfrac{n}{g}\right)^2} = \frac{2901555}{\dfrac{7(7-1)}{2}\left(\dfrac{2622}{7}\right)^2} = .985 \ (\text{political party})$$

$$\text{IQV} = \frac{\sum f_i f_j}{\dfrac{g(g-1)}{2}\left(\dfrac{n}{g}\right)^2} = \frac{2682060}{\dfrac{7(7-1)}{2}\left(\dfrac{2622}{7}\right)^2} = .910 \ (\text{political views})$$

These results confirm that the variation in party identification is greater than that for political views.

Before we can execute the rank-regression analysis, we must first assign ranks to each attribute on both variables. To assign ranks, we must decide which pole of the response scale is the highest attribute. For example, should strong Republican or strong Democrat be designated as the highest party attribute? When it comes to party identification and political views, the decisions are arbitrary. For this analysis, strong Republican and extremely conservative have been designated as the highest attributes. Thus, we can think of the variables as representing how Republican and how conservative a person happens to be. Equation 12.1 can be used to compute the ranks. The rank for the independent category of the political-party variable is

$$R_{\text{Independent}} = cf_B + \frac{f_i + 1}{2} = 1274 + \frac{411 + 1}{2} = 1274 + 206 = 1480$$

The results of regressing party-identification rank on political-views rank are shown in Table 12.9. First, look at the means and standard deviations. The means of both variables are equal because both variables have 2,622 observations and thus the means equal $2623/2 = 1311.5$. There is no simple formula for the standard deviation of the ranks. The greater the n, however, the greater the standard rank deviation will be. In this example, the standard rank deviation of party identification is slightly greater than that of political views. Two ordinal variables usually have very similar standard rank deviations when n is the same for both variables because both variables will have the same maximum possible range of ranks, which is 1 to n.

We see that the slope for party identification regressed on political views is .392, which indicates that an increase of one rank in political views is associated with .392 of a rank increase in party identification. The more conservative a person is (or the less liberal), the greater the degree to which the person will identify with the Republican Party. The standardized slope indicates that party identification becomes more Republican by .385 of a standard rank deviation per standard rank deviation increase in conservative political views. Because this is a bivariate regression, *the* **standardized rank slope,** *or the Pearson correlation for ranks, ranges from −1 to 1.* Thus, the linear relationship between ranks is .385 of its maximum positive value. There is a substantial relationship between liberal–conservative views and party identification, but the magnitude of the correlation suggests that there are lots of conservatively oriented people who are not Republicans and many

Table 12.9 Regression of Party Identification Rank on Political Views Rank, GSS 1998

Variables	Mean	St. Dev.	n		
Party identification	1311.5	746.945	2622		
Political views	1311.5	733.272	2622		
Variables	**b**	**SE**	**B**	**t**	**p**
Political views	.392	.018	.385	21.327	.0000
Intercept	797.687	27.602		28.899	.0000

liberally oriented people who are not Democrats. The unstandardized slope is nearly identical to the standardized slope because the standard rank deviations of the variables are so similar. In general, this is the case. Thus, the unstandardized slope in a rank regression will be a good indicator of the strength of association.

The intercept indicates that a person with a rank of 0 on political views would be predicted to have a rank of about 798 out of 2,622 on party identification. As discussed earlier, because there is no rank of 0 and because the closest rank to 0 is 32.5 for those who are extremely liberal, the intercept is not a meaningful coefficient.

The standard error and t score of a rank-regression slope are calculated the same way as a regular slope. The value of t indicates that the slope is 21.372 times as large as its standard error. The two-tailed level of significance (p) is less than .00005. Thus, we reject the null hypothesis that $\beta = 0$ in the population.

We have seen that a regression analysis of ranks is conducted in an identical fashion to that for interval scores. The only difference is that when we interpret the coefficients, we must remember to do so in terms of ranks instead of interval scores.

The standardized rank slope, or the Pearson correlation, for two rank-scored variables equals the Spearman correlation, a nonparametric measure of association for two ordinal variables that is described in the following optional section.

*Spearman's Rank Correlation

The **Spearman rank correlation** is a frequently used measure of association between ordinal variables. This correlation is defined as

$$r_S = \text{correlation between } R_Y \text{ and } R_X \qquad (12.3)$$

where

$$r_S = \text{Spearman rank correlation}$$
$$R_Y = \text{ranks of } Y$$
$$R_X = \text{ranks of } X$$

Because r_S is an ordinary correlation computed on ranks, it has a range of values from -1 to 1. The closer r_S is to -1 or 1, the stronger the association between ranks.

The test statistic for evaluating the null hypothesis that the population correlation equals 0 (i.e., $\rho_S = 0$) is

$$t = r_S \sqrt{\frac{n-2}{1-r_S^2}} \qquad (12.4)$$

When n is not small (i.e., $n \geq 10$), this test statistic is compared with a critical value from the t distribution for $df = n - 2$ and the selected α to determine whether to reject the null hypothesis. If n is small, exact probability tables for r_S may be found in Siegel (1956).

Because the ordinary correlation equals the standardized slope, *the values of r_S and t (Equation 12.4) are equivalent to the standardized slope and the computed t from the regression of R_Y on R_X.* Thus, the bivariate regression analysis discussed in the previous section gives results that are equivalent to those obtained with the Spearman rank correlation.

*Other Measures of Ordinal Association

A family of additional correlations for ordinal variables uses similar ways of measuring association. These correlation coefficients are Goodman and Kruskal's **gamma,** Somers's d_{YX} and d_{XY}, and Kendall's **tau-b.** Instead of using each case as the unit of calculation, as does Spearman's correlation, these four measures use each distinct pair of cases as the unit of calculation. For each pair of cases that are not tied on either X or Y, we determine whether the case with the higher rank on X has a higher rank on Y (concordant pair) or a lower rank on Y (discordant pair). If there are more concordant pairs than discordant pairs, the correlation is positive. If there are more discordant than concordant pairs, the association is negative. If the numbers of concordant and discordant pairs are equal, there is no association.

These measures differ in the way they handle pairs of cases with the same rank on one or both variables (i.e., ties). Gamma excludes all pairs of tied cases from its computation. Pairs that are tied on Y but not X are used in d_{YX}, and pairs that are tied on X but not Y are used in d_{XY}. The latter two measures are asymmetric correlations because their value depends on whether Y is the dependent variable (d_{YX}) or X is the dependent variable (d_{XY}). Tau-b includes all tied pairs of cases. Because of the differing ways that tied pairs are handled, gamma is greater than or equal to the other three measures, and tau-b is an "average" of the two ds (tau-$b = \sqrt{d_{YX}d_{XY}}$). Despite these differences, all of them are standardized to range from -1 to 1, although tau-b cannot equal its maximum absolute value unless there are equal numbers of ranked attributes on X and Y. Table 12.10 shows the association between party identification and political views as described by each of the four measures and by Spearman's correlation.

Because of the uncertainty surrounding the best way to handle tied pairs of cases, it is difficult to say which of these measures is best. There is also little agreement on the relative superiority of these measures in comparison with Spearman's rank correlation. This text uses Spearman's correlation, or the equivalent standardized rank slope, because of its continuity with regression analysis. When we are testing the null hypothesis of no association, *the observed level of significance (p) of Spearman's correlation will closely resemble the results for gamma, tau, and d.*

**Table 12.10 Ordinal Correlation
Coefficients for Party Identification
and Political Views, GSS 1998**

Spearman correlation	.385
Gamma	.383
Somers's d_{YX} (Y = party identification)	.326
Kendall's tau-b	.313
Somers's d_{XY} (Y = party identification)	.301
n	2622

ASSOCIATION BETWEEN AN ORDINAL AND A NOMINAL VARIABLE

If the independent variable is nominal, its attributes are not ordered and thus we cannot legitimately compute ranks for its categories, which means we cannot use the regression and correlation methods described in the previous sections. Instead, we have to use one or more dummy variables to represent the nominal variable and include them as independent variables in a regression equation where the dependent variable is the rank-scored ordinal variable. If the *dependent* variable is nominal, we have to use the cross-tabulation techniques described in Chapter 13 in place of ordinary regression analysis.

A Dichotomous Nominal Variable

In the case of a nominal independent variable with just two categories, we can regress a rank-scored dependent variable on a single dummy variable. The **regression equation for a rank-scored Y and a dummy X** and the meaning of the coefficients are given by

$$\hat{R}_Y = a + bX$$

where

$$\hat{R}_Y = \text{predicted rank of } Y$$

$$X = \{1, 0\} \text{ for groups A and B}$$

$$a = \overline{R}_{Y_B}$$

$$b = \overline{R}_{Y_A} - \overline{R}_{Y_B}$$

The regression intercept equals the mean rank of Y for the group coded 0 on the dummy variable X (i.e., group B). *The regression slope equals the mean rank of Y for the group coded 1 on X minus the mean rank of Y for the group coded 0 on X (i.e., group B).* Thus, the slope is a difference of mean ranks between the two groups. The usual two-tailed null and alternative hypotheses for the slope are

$$H_0: \mu_{R_A} = \mu_{R_B}$$

$$H_1: \mu_{R_A} \neq \mu_{R_B}$$

Thus, *the t test of the slope is a **t test for a difference of mean ranks**.*

**Table 12.11 Dummy
Coded Gender**

Gender	X
Male	1
Female	0

**Table 12.12 Regression of Party-Identification Rank on
Gender, GSS 1998**

Variables	Mean	St. Dev.	n		
Party identification	1311.500	746.945	2622		
Gender	.439	.496	2622		
Variables	b	SE	B	t	p
Gender	154.284	29.250	.103	5.275	.0000
Intercept	1243.773	19.380		64.178	.0000

To illustrate this technique, we will test for a gender difference in political-party identification. Who tends to be more Republican, males or females? Party identification is scored again in accordance with the ranks shown in Table 12.8, where the higher the rank is, the greater the identification with the Republican Party. The dummy variable is coded as shown in Table 12.11.

The results of the regression are shown in Table 12.12. Because the number of cases is the same as in the previous regression, the mean rank for party identification is again 1311.5. The mean for the gender variable indicates that .439 (43.9%) of the respondents are male. The intercept (1243.773) indicates the mean rank of females (the group coded 0), which is less than the mean for males and females combined (1311.5). This means that females are closer to the Democratic end of the ordinal scale than to the Republican end. The slope speaks directly to the male-female difference. It is positive, which indicates that males (the group coded 1) are more Republican than females. The absolute value of the slope tells us that the mean rank for males is 154.3 ranks higher than the mean rank for females.

Is the gender difference in party identification likely to have occurred by chance? Using $\alpha = .05$ and $df = n - 2 = 2620$, we find from Table A.2 that the critical t is 1.961. Because the calculated t (5.275) is much larger than the critical t, we reject the null hypothesis and conclude that males are more Republican than females. The p value indicates that the level of significance is less than .00005.

The level of significance indicates that the gender difference in party is extremely unlikely to have occurred by chance. Is the difference of 154.3 ranks a large difference? The standardized slope, or the Pearson correlation, is only .103. Thus, males are only slightly more likely to be Republicans than are females.

The t test for an equation with a single dummy variable is very nearly equal to the nonparametric Mann-Whitney-Wilcoxon test, which is covered in an optional section later in this chapter.

Rank Scoring the Dichotomous Variable

An alternate way of scoring the gender variable is instructive. Any dichotomous variable can be considered to be an ordinal variable if we select one attribute to be the high attribute. For example, we can arbitrarily specify males as high. If we do so, we can think of the gender variable as indicating whether a person is male or not male. Then we can determine the ranks of the attributes in the usual fashion by applying Equation 12.1 to a frequency distribution for the gender variable, which results in a rank of 2,047 for

Table 12.13 Regression of Party-Identification Rank on Gender Rank, GSS 1998

Variables	b	SE	B	t	p
Gender	.118	.022	.103	5.275	.0000
Intercept	1157.158	32.664		35.426	.0000

males and 736 for females. If we regress the party-identification variable on the rank-scored gender variable, we get the results shown in Table 12.13.

The intercept indicates the predicted rank on the party variable for a person with a rank of 0 on the gender variable. As discussed earlier, because no one has a rank of 0, this is not a meaningful prediction. The slope means that a one-rank increase in gender is related to an increase of .118 rank in political party. This increase is not meaningful because the two attributes of the gender variable (male and female) differ by $2047 - 736 = 1311$ ranks, not 1 rank. A comparison of Table 12.13 with Table 12.12 shows that the standardized slope, the calculated t, and the p value are the same in both regressions. Thus, dummy scoring and rank scoring lead to the same conclusions with respect to the strength of the relationship and the hypothesis test. In fact, when there are only two attributes, any scoring will lead to identical results for these statistics. The reason for preferring the dummy coding is that the unstandardized slope and the intercept are meaningful statistics under such scoring.

*Mann–Whitney–Wilcoxon Test

The **Mann–Whitney test** and the **Wilcoxon test** are identical tests for a difference of ranks between two groups. The objective of these tests is the same as the t test for the regression slope described in the previous section, namely, to determine if the mean rank for an ordinal dependent variable differs between two groups. The Mann–Whitney test statistic, or U, is defined in one of two alternative ways:

$$U = n_1 n_2 + \frac{n_1(n_1 + 1)}{2} - \Sigma R_1$$

$$\text{(12.5)}$$

$$U' = n_1 n_2 + \frac{n_2(n_2 + 1)}{2} - \Sigma R_2$$

where

$$n_1 = \text{sample size in group 1}$$

$$n_2 = \text{sample size in group 2}$$

$$\Sigma R_1 = \text{sum of ranks in group 1}$$

$$\Sigma R_2 = \text{sum of ranks in group 2}$$

First, ranks are computed for groups 1 and 2 combined ($n = n_1 + n_2$). Then, either the sum of the ranks in group 1 (ΣR_1) or the sum of the ranks in group 2 (ΣR_2) is computed, whichever is most convenient. It is not necessary to compute both sums because we need to compute only U or U'. As we will see, using U or U' leads to the same conclusion.

To illustrate, we will use the same data used in the previous section, namely, the variables political–party identification and gender. The mean rank and sum of ranks for females and males are given in Table 12.14. The mean ranks are the same as in the previous section and indicate, again, that males have a higher rank on the average, meaning that they are more closely identified with the Republican Party.

According to Equation 12.5, all we need to compute U or U' is the number of females, the number of males, and the sum of ranks for females or for males. Plugging these numbers into Equation 12.5 gives

$$U = (1471)(1151) + \frac{1471(1471 + 1)}{2} - 1829590 = 946187$$

$$U' = (1471)(1151) + \frac{1151(1151 + 1)}{2} - 1609163 = 746934$$

For small samples (neither n_1 nor $n_2 \geq 20$), tables of exact values of U can be used to test the null hypothesis of no difference between groups (Siegel, 1956). As the sample sizes increase, however, the distribution of U becomes closer and closer to the normal distribution. Thus, the normal distribution is typically used to test the null hypothesis. Before comparing U to the appropriate critical value from the normal distribution, we must standardize U with the following formula:

$$Z_U = \frac{U - \mu_U}{\sigma_U} \qquad (12.6)$$

where

μ_U = the mean of the sampling distribution of U

σ_U = the standard error of the sampling distribution of U

The mean of the sampling distribution of U is

$$\mu_U = \frac{n_1 n_2}{2} \qquad (12.7)$$

which is the mean of U if the usual null hypothesis of no difference between groups is true. For our example, the mean is

$$\mu_U = \frac{(1471)(1151)}{2} = \frac{1693121}{2} = 846560.5$$

Notice that this mean is exactly halfway between U and U'.

The standard error of U is

$$\sigma_U = \sqrt{\left(\frac{n_1 n_2}{n(n-1)}\right)\left(\frac{n^3 - n}{12} - \Sigma T_i\right)} \qquad (12.8)$$

where

$$T_i = \frac{f_i^3 - f_i}{12}$$

f_i = frequency of attribute i

Table 12.14 Political Party Ranks by Gender, GSS 1998

Gender	n	Mean Rank	Sum of Ranks
1 Female	1471	1243.77	1829590
2 Male	1151	1398.06	1609163

Table 12.15 ΣT for Political Party Identification, GSS 1998

	f	$(f^3 - f)/12$ =	T
Strong Democrat	362	$(362^3 - 362)/12 =$	3953130.5
Not very strong Democrat	570	$(570^3 - 570)/12 =$	15432702.5
Independent, near Democrat	342	$(342^3 - 342)/12 =$	3333445.5
Independent	411	$(411^3 - 411)/12 =$	5785510.0
Independent, near Republican	233	$(233^3 - 233)/12 =$	1054092.0
Not very strong Republican	470	$(470^3 - 470)/12 =$	8651877.5
Strong Republican	234	$(234^3 - 234)/12 =$	1067722.5
Total	2622		39278480.5

ΣT_i is a correction for ties. For each attribute, we subtract its frequency from the cube of its frequency and then divide by 12 to get T for attribute i. ΣT_i is then the sum of T_i across all attributes. The computation of ΣT_i is illustrated in Table 12.15 for the political-party variable. We can now enter n_1, n_2, n, and ΣT_i into Equation 12.8 to get the standard error of U:

$$\sigma_U = \sqrt{\left(\frac{(1471)(1151)}{2622(2622 - 1)}\right)\left(\frac{2622^3 - 2622}{12} - 39278480.5\right)}$$

$$= \sqrt{360410982.3} = 18984.49$$

With the standard error and mean of U in hand, we use Equation 12.6 to standardize both U and U', as follows:

$$Z = \frac{U - \mu_U}{\sigma_U} = \frac{946187 - 846560.5}{18984.49} = \frac{99626.5}{18984.49} = 5.248$$

$$Z' = \frac{U' - \mu_U}{\sigma_U} = \frac{746934 - 846560.5}{18984.49} = \frac{-99626.5}{18984.49} = -5.248$$

The absolute values of Z and Z' are equal because U is the same distance above the mean as U' is below the mean. The null hypothesis of no difference between groups and the alternative hypothesis are

$$H_0: \mu_{R_1} = \mu_{R_2}$$

$$H_1: \mu_{R_1} \neq \mu_{R_2}$$

If $\alpha = .05$, the critical value of Z is 1.96 (Table A.1). Comparing the absolute value of the computed Z or Z' with its critical value leads to the rejection of the null hypothesis and the conclusion that males identify with the Republican Party more strongly than females do.

In the previous section we used regression analysis with a dummy independent variable and a rank-scored dependent variable to conduct a t test of the same null hypothesis that we have just tested with the Mann-Whitney U. Are the regression t test and the Mann-Whitney U test different tests or equivalent tests? The following formula shows that that the t statistic (called t_R) is an exact function of the standardized U:

$$t_R = \frac{Z_U}{\sqrt{\dfrac{n-1}{n-2} - \dfrac{Z_U^2}{n-2}}} \tag{12.9}$$

where

$$t_R = t \text{ test for difference in mean ranks}$$

$$Z_U = \text{standardized Mann-Whitney } U$$

If we enter Z and n into Equation 12.9, we get

$$t_R = \frac{Z_U}{\sqrt{\dfrac{n-1}{n-2} - \dfrac{Z_U^2}{n-2}}} = \frac{5.248}{\sqrt{\dfrac{2622-1}{2622-2} - \dfrac{5.248^2}{2622-2}}} = \frac{5.248}{.9948} = 5.275$$

This is the same value we got for t in the regression of the rank-scored party-identification variable on the dummy gender variable (Table 12.12). The sampling distributions of t_R and Z_U are approximated by the t and normal distributions, respectively, that are used for critical values in hypothesis tests. Because of this, the p values for the two statistics are slightly different. They are similar enough, however, that the Mann-Whitney test can be performed using t_R as the test statistic if desired (Conover and Iman, 1981), which is what we did in the previous section.

A Polytomous Nominal Variable

In the case of a nominal independent variable with three or more groups, we can use a **rank-scored Y regressed on multiple dummy Xs.** The number of dummy variables will equal $g - 1$, where g is the number of groups/attributes in the nominal variable. If there are $g = 4$ groups, the regression equation is

$$\hat{R}_Y = a + b_1 X_1 + b_2 X_2 + b_3 X_3$$

where

$$\hat{R}_Y = \text{predicted rank of } Y$$

$$X_1 = \{1, 0, 0, 0\} \text{ for groups A, B, C, and D}$$

$$X_2 = \{0, 1, 0, 0\} \text{ for groups A, B, C, and D}$$

$$X_3 = \{0, 0, 1, 0\} \text{ for groups A, B, C, and D}$$

$$a = \overline{R}_{Y_D}$$

$$b_1 = \overline{R}_{Y_A} - \overline{R}_{Y_D}$$

$$b_2 = \overline{R}_{Y_B} - \overline{R}_{Y_D}$$

$$b_3 = \overline{R}_{Y_C} - \overline{R}_{Y_D}$$

Group D, which is coded 0 on each of the three dummy variables, is the reference group. The intercept equals the mean rank of group D, and *the slope of each dummy variable equals the difference between the mean rank of the group coded 1 on the variable and the mean rank of the reference group* (group D).

To illustrate, we again use the ethnicity variable that was introduced in Chapter 11. Table 12.16 shows how the seven major ethnic groups are coded on six dummy variables. British-Americans are coded 0 on all six variables, and thus they are the reference group.

The dependent variable will again be political-party identification. Table 12.17 shows the frequency distribution for the seven party-identification categories. There are 1,509 cases with valid party-identification scores for the seven major ethnic groups, which is fewer than the 2,622 cases for which ranks were computed in Table 12.8. Therefore, it is necessary to compute again the ranks using Equation 12.1. Because of the large difference in n and the slightly different percentage distribution, the ranks in Table 12.17 differ greatly from those in Table 12.8.

We will first test whether there is any relationship between ethnicity and party identification. The null and alternative hypotheses are

$$H_0\colon \mu_{R_1} = \mu_{R_2} = \mu_{R_3} = \cdots = \mu_{R_7}$$

$$H_1\colon \text{some } \mu_{R_i} \neq \mu_{R_j}$$

The null hypothesis states that all seven groups have equal mean ranks for party identification. The alternative hypothesis says that at least one group has a different mean rank from the other groups.

To evaluate these hypotheses, we regress the rank-scored party-identification variable on the six ethnic dummy variables. Table 12.18 shows the results of the regression analysis, including the **F test for a difference of mean ranks.** The critical value of F for $df_1 = 6$, $df_2 = 1502$, and $\alpha = .05$ is 2.10 (Table A.3a). Because the computed F (32.526) is greater than the critical value of F, we reject the null hypothesis and conclude that at least one of the groups has a different mean from the others. Thus, there is a relationship between ethnicity and party identification. The multiple correlation of .339 ($R^2 = .115$) indicates these ethnic differences in politics are not inconsequential. At the same time, the fact that R is closer to 0 than to 1 (and $1 - R^2 = .885$) indicates that the ethnic differences between groups are not as great as the differences in party identification among persons within the same group. The groups are more alike than they are different.

Now we can examine some specific differences between ethnic groups by comparing each group with the reference group. First, the slope for the African-American variable indicates that the average rank of African-Americans is 432.832 less than that of British-Americans, which the intercept indicates is 868.555. Because strong Republicans have the highest rank, this slope means that African-Americans are not as Republican

Table 12.16 Dummy Variables for
$g = 7$ Ethnic Groups

Group	X_1	X_2	X_3	X_4	X_5	X_6
1. African-American	1	0	0	0	0	0
2. German-American	0	1	0	0	0	0
3. Irish-American	0	0	1	0	0	0
4. Italian-American	0	0	0	1	0	0
5. Mexican-American	0	0	0	0	1	0
6. Native American	0	0	0	0	0	1
7. British-American	0	0	0	0	0	0

Table 12.17 Political Party Identification, GSS 1998

	f	Pct	cf	Rank
Strong Democrat	222	14.7	222	111.5
Not very strong Democrat	322	21.3	544	383.5
Independent, near Democrat	181	12.0	725	635.0
Independent	224	14.8	949	837.5
Independent, near Republican	136	9.0	1085	1017.5
Not very strong Republican	277	18.4	1362	1224.0
Strong Republican	147	9.7	1509	1436.0
Total	1509	100.0		

Table 12.18 Regression of Party-Identification Rank on Ethnicity, GSS 1998

Variables	b	SE	B	t	p
African-American	−432.832	35.742	−.350	−12.110	.000
German-American	−13.064	30.346	−.013	−.431	.667
Irish-American	−86.090	33.627	−.076	−2.560	.011
Italian-American	−174.969	41.734	−.115	−4.193	.000
Mexican-American	−253.055	52.512	−.126	−4.819	.000
Native American	−90.593	45.349	−.054	−1.998	.046
Intercept	868.555	22.489		38.621	.000
Source	**SS**	**df**	**MS**	**F**	**p**
Regression	32079078.324	6	5346513.054	32.526	.000
Residual	246893659.676	1502	164376.604		
Total	278972738.000	1508			
	R	**R-Squared**	**SE of Est.**		
Goodness-of-fit	.339	.115	405.434		

as British-Americans. Is this difference statistically significant? For $df = 1502$ and $\alpha = .05$, the critical $t = 1.962$ (Table A.2), which is less than the absolute value of the computed $t = -12.11$. Therefore, we easily reject a null hypothesis that African- and British-Americans have equal mean ranks.

The slopes of all the other ethnic variables are also negative, indicating that each group is less Republican, or more Democratic, than British-Americans. For $\alpha = .05$, the t test for each of these slopes except German-Americans is significant. Thus, we can conclude that five ethnic groups (African-, Irish-, Italian-, Mexican-, and Native Americans) identify more closely with the Democratic Party than do British-Americans. A comparison of the standardized slopes (B) indicates that African-Americans are the most different, followed by Italian- and Mexican-Americans and then by Irish- and Native Americans.

The statistical analysis of the relationship between an ordinal dependent variable and a polytomous nominal variable is identical to that for an interval dependent variable and a polytomous nominal variable that was covered in Chapter 11. Because the ordinal variable *is* scored for ranks, there is a difference in the way the results of the regression analyses are interpreted. In the case of ordinal dependent variables, we must always be careful to interpret the slopes for the dummy variables as differences between *mean ranks* rather than differences between means on an interval scale.

The F test for an equation with two or more dummy variables is basically equivalent to the nonparametric Kruskal–Wallis test covered in the following optional section.

★Kruskal–Wallis Test

The **Kruskal–Wallis test** looks for differences between three or more groups in their ranks on a dependent variable. The objective of this test is the same as that of the F test for the regression equation described in the previous section, namely, to determine whether the mean rank of an ordinal dependent variable differs between three or more groups. The Kruskal–Wallis test statistic, or H, is

$$H = \frac{\Sigma((\Sigma R_i)^2/n_i) - n(n + 1)^2/4}{(\Sigma(\Sigma R_i^2) - n(n + 1)^2/4)/(n - 1)} \tag{12.10}$$

where

$$n_i = \text{number of cases in group } i$$

$$\Sigma R_i = \text{sum of ranks in group } i$$

$$\Sigma R_i^2 = \text{sum of squared ranks in group } i$$

The ranks of the attributes of the dependent ordinal variable are computed in the same method used throughout this chapter (Equation 12.1). The computation of the H statistic in Equation 12.10 can be broken down into the following eight steps:

1. Compute the sum of the ranks of the ordinal variable in each category of the nominal variable: ΣR_i.

2. Square the sums of ranks computed in step 1 and divide by the number of cases in each category: $(\Sigma R_i)^2/n_i$.

3. Sum the quantities computed in step 2 across all categories of the nominal variable: $\Sigma((\Sigma R_i)^2/n_i)$.

4. Subtract $n(n + 1)^2/4$ from the quantity computed in step 3 to get the numerator of the H statistic: $\Sigma((\Sigma R_i)^2/n_i) - n(n + 1)^2/4$.

5. Compute the sum of the *squared ranks* of the ordinal variable in each category of the nominal variable: ΣR_i^2.

6. Sum the quantities computed in step 5 across all categories of the nominal variable: $\Sigma(\Sigma R_i^2)$.

7. Subtract $n(n + 1)^2/4$ from the quantity computed in step 6 and divide the difference by $n - 1$ to get the denominator of the H statistic: $(\Sigma(\Sigma R_i^2) - n(n + 1)^2/4)/(n - 1)$.

8. Divide the numerator from step 4 by the denominator from step 7 to get H.

The Kruskal-Wallis test will be illustrated with the data on political-party identification and ethnic group membership used in the previous section. The ranks of the party-identification attributes are given in Table 12.17. Table 12.19 shows the calculation of certain key quantities needed for the H statistic. Column 2 gives the sum of party ranks in each ethnic group (step 1). Column 3 contains the squared sum of ranks divided by the number of people in each group (step 2). The sum of column 3 gives the quantity desired in step 3. Column 4 lists the sum of squared ranks for each group (step 5) and the sum of column 4 equals the quantity specified in step 6.

We now enter the sums of columns 3 and 4 into Equation 12.10, along with $n(n + 1)^2/4$ and $n - 1$. Steps 4, 7, and 8 are calculated in the process of computing H as follows:

$$
\begin{aligned}
H &= \frac{892246803.3 - 1509(1509 + 1)^2/4}{(1139140463.0 - 1509(1509 + 1)^2/4)/(1509 - 1)} \\[2mm]
&= \frac{892246803.3 - 860167725}{(1139140463.0 - 860167725)/(1509 - 1)} \\[2mm]
&= \frac{32079078.3}{278972738/1508} \\[2mm]
&= \frac{32079078.3}{184995.2} \\[2mm]
&= 173.405
\end{aligned}
$$

H is used to test the same hypotheses specified in the previous section, which are

$$H_0: \mu_{R_1} = \mu_{R_2} = \mu_{R_3} = \cdots = \mu_{R_7}$$

$$H_1: \text{some } \mu_{R_i} \neq \mu_{R_j}$$

H is approximately distributed as *chi-square* (χ^2) with $df = k - 1$. The chi-square distribution will be described in Chapter 13. Critical values of chi-square may be found in Table A.4. For $\alpha = .05$ and $df = 7 - 1 = 6$, $\chi^2 = 12.59$. Because the computed H (173.405) is greater than the critical chi-square, we reject the null hypothesis that the mean party ranks of all ethnic groups are equal.

In the previous section we used regression analysis with six dummy independent variables and a rank-scored dependent variable to conduct an F test of the same null hypothesis that we have just tested with the Kruskal-Wallis H. Are the regression F test

Table 12.19 Sum of Ranks and Sum of Squared Ranks for Political Party by Ethnic Group, 1998 GSS

	n_i	ΣR_i	$(\Sigma R_i)^2/n_i$	ΣR_i^2
African-Americans	213	92809.0	40439016.3427	64423878.00
British-Americans	325	282280.5	245176248.2469	306433574.25
German-Americans	396	338774.5	289818590.5309	355146603.75
Irish-Americans	263	205788.5	161022459.0580	211469149.75
Italian-Americans	133	92247.0	63981270.7444	82500690.50
Mexican-Americans	73	44931.5	27655338.2500	35310406.25
Native Americans	106	82464.0	64153880.1509	83856160.50
Total	1509		892246803.3239	1139140463.00

and the Kruskal–Wallis H test different tests or equivalent tests? The following formula shows that the F statistic for ranks (called F_R) is an exact function of H:

$$F_R = \frac{H/(k-1)}{(n-1-H)/(n-k)} \tag{12.11}$$

where

$$F_R = \text{the } F \text{ test for group differences in mean ranks}$$

$$H = \text{the Kruskal–Wallis } H \text{ statistic}$$

$$k = \text{the number of groups}$$

Substituting H, k, and n into Equation 12.11 gives

$$F_R = \frac{H/(k-1)}{(n-1-H)/(n-k)} = \frac{173.405/(7-1)}{(1509-1-173.405)/(1509-7)} = 32.526$$

This value is the same one we got for F in the dummy variable regression test (Table 12.18). The sampling distributions of F_R and H are approximated by the F and chi-square distributions, respectively, which are used for critical values in hypothesis tests. Because the two test statistics are compared to different sampling distributions, the p values for the two statistics will be slightly different. Iman and Davenport (1976) show that the F approximation is preferred to the chi-square approximation in most cases.

SUMMARY

This chapter describes how to measure and test associations involving an ordinal dependent variable. The first step is to assign rank scores to the ordinal variable. To do this, we list the cases in ascending order according to their attributes. The case with the highest attribute receives the rank of n, and the lowest case is ranked 1. Cases that have the same attribute (i.e., tied cases) receive a rank score equal to the mean of the ranks for which they are tied. After ordinal variables have been rank scored, we can use them in regression analyses in the same way that we use interval variables.

When a dependent ordinal variable that has been rank scored is regressed on an independent ordinal variable that has been rank scored, the unstandardized slope indicates the rank change in the dependent variable per rank increase in the independent variable. The null hypothesis for this slope is tested with the usual t test. The unstandardized slope is approximately equal to the standardized slope because the standard deviations of two rank-scored variables are approximately equal. The standardized slope, or the Pearson correlation, for two rank-scored variables equals the Spearman correlation, a nonparametric measure of association for two ordinal variables.

When the independent variable is a nominal variable, we regress the rank-scored dependent variable on one or more dummy variables. The slope for each dummy variable indicates the average rank difference between the group coded 1 on the dummy variable and the reference group. The F statistic for the equation provides a test of the null hypothesis that all groups have the same mean rank. The t test for an equation with a single dummy variable is nearly equivalent to the nonparametric Mann-Whitney test and the Wilcoxon test, and the F test for an equation with two or more dummy variables is nearly equivalent to the nonparametric Kruskal-Wallis test.

KEY TERMS

rank scoring

mean rank

rank-regression equation

rank-regression slope

standardized rank slope

★Spearman rank correlation

★gamma, d, and tau-b

regression equation for a rank-scored Y and a dummy X

t test for a difference of mean ranks

★Mann-Whitney test

★Wilcoxon test

rank-scored Y regressed on multiple dummy Xs

F test for a difference of mean ranks

★Kruskal-Wallis test

SPSS PROCEDURES

Before we can use ordinal variables in regression analysis, we must compute ranks for the attributes of the variables. When we regress political party identification (*partyid*) on political ideology (*polviews*), we want to assign ranks to only those cases that do not have missing values on either variable. To select only those cases that do not have missing values, click on *Data* and *Select Cases* to open the *Select Cases* window, where you will click on *If condition is satisfied* and the *If* button. In the *Select Cases: If* window, enter "not missing (*partyid*) and not missing (*polviews*)" in the box, click on *Continue,* and click on *OK*. To compute ranks for these variables now, click on *Transform* and *Rank Cases,* move the variable (e.g., *partyid*) into the *Variable(s)* box of the following *Rank Cases* window, and click *OK*. SPSS will create a new variable containing the ranks and give it a name (*rpartyid*). To see the rank scores for each of the seven attributes, run *Frequencies* for the new variable shown after the *Rank Cases* window. After you have computed ranks for *polviews,* you can run the regression procedure for the two rank-scored variables to get the statistics shown in Table 12.9.

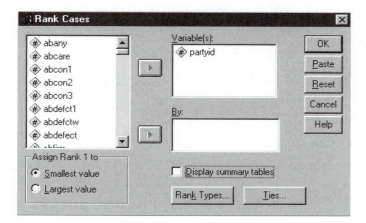

RPARTYID

		Frequency	Percent	Valid Percent	Percent
Valid	181.500	362	13.8	13.8	13.8
	647.500	570	21.7	21.7	35.5
	1103.500	342	13.0	13.0	48.6
	1480.000	411	15.7	15.7	64.3
	1802.000	233	8.9	8.9	73.2
	2153.500	470	17.9	17.9	91.1
	2505.500	234	8.9	8.9	100.0
	Total	2622	100.0	100.0	

★You can compute the Spearman correlation by clicking on *Analyze, Correlate,* and *Bivariate.* In the *Bivariate Correlations* window, move the variables *partyid* and *polviews* into the *Variables* box, click on *Spearman* in the panel *Correlation Coefficients,* and click on *OK.*

★To conduct the Mann-Whitney test, click on *Analyze, Nonparametric,* and *2 Independent Samples.* Move the ordinal dependent variable (*partyid*) into the box *Test Variable List,* move the dichotomous nominal variable (*sex*) into the *Grouping Variable* box (you must also click on *Define Groups* to specify the values for each group), click on *Mann-Whitney U,* and click on *OK.*

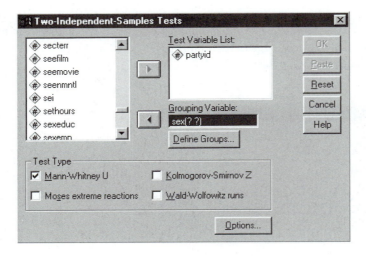

Mann-Whitney Test

Ranks

	SEX	N	Mean Rank	Sum of Ranks
PARTYID	female	1471	1243.77	1829590.00
	MALE	1151	1398.06	1609163.00
	Total	2622		

Test Statistics^a

	PARTYID
Mann-Whitney U	746934.000
Wilcoxon W	1829590.000
Z	-5.248
Asymp. Sig. (2-tailed)	.000

a. Grouping Variable: SEX

★To conduct the Kruskal–Wallis test for party identification by ethnic group, first re-code the *ethnic* variable to assign Scotland the same numeric code as England and Wales (*Old Value* = 24; *New Value* = 8) and place the recoded variable in a new variable *ethnicr.* Then, select cases that belong to one of the eight largest ethnic groups. (In the *Select Cases: If* window, enter "ethnic = 1 or ethnic = 8 or ethnic = 11 or ethnic = 14 or ethnic = 15 or ethnic = 17 or ethnic = 24 or ethnic = 30" in the box.) Now, click on *Analyze, Nonparametric,* and *K Independent Samples,* move the ordinal dependent variable (*partyid*) into the box *Test Variable List,* move the polytomous nominal variable (*ethnicr*) into the *Grouping Variable* box (you must also click on *Define Range* to specify the range of values for the groups—that is, 1–30), click on *Kruskal-Wallis H,* and click on *OK.*

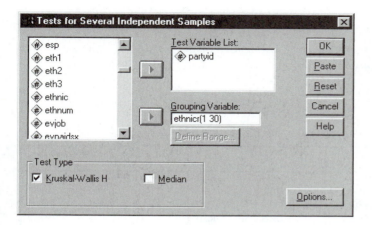

Kruskal-Wallis Test

Ranks

	ethnic summary	N	Mean Rank
RANK of PARTYID	African	213	435.72
	England, Scotland, Wales	325	868.56
	Germany	396	855.49
	Ireland	263	782.47
	Italy	133	693.59
	Mexico	73	615.50
	American Indian	106	777.96
	Total	1509	

Test Statistics[a,b]

	RANK of PARTYID
Chi-Square	173.405
df	6
Asymp. Sig.	.000

a. Kruskal Wallis Test

b. Grouping Variable: ethnic summary

EXERCISES

1. The GSS 2000 asked whether it was wrong for a man and a woman to have sexual relations before marriage (*premarsx*). The following table gives the frequency distribution for the response alternatives. Treating not wrong at all as the highest attribute, calculate the ranks. What is the mean rank?

PREMARSX -- SEX BEFORE MARRIAGE

	Frequency	Percent	Cumulative Frequency	Cumulative Percent
ALWAYS WRONG	490	27.9	490	27.9
ALMST ALWAYS WRG	153	8.7	643	36.7
SOMETIMES WRONG	374	21.3	1017	58.0
NOT WRONG AT ALL	737	42.0	1754	100.0
Total	1754	100.0		

2. The GSS 2000 asked, "How often do you attend religious services?" (*attend*). The following table gives the frequency distribution for the response alternatives for those respondents who also answered *premarsx* in Exercise 1. Calculate the ranks. What is the mean rank?

ATTEND -- HOW OFTEN R ATTENDS RELIGIOUS SERVICES

	Frequency	Percent	Cumulative Frequency	Cumulative Percent
NEVER	386	22.0	386	22.0
LT ONCE A YEAR	133	7.6	519	29.6
ONCE A YEAR	220	12.5	739	42.1
SEVRL TIMES A YR	233	13.3	972	55.4
ONCE A MONTH	128	7.3	1100	62.7
2-3X A MONTH	148	8.4	1248	71.2
NRLY EVERY WEEK	78	4.4	1326	75.6
EVERY WEEK	302	17.2	1628	92.8
MORE THN ONCE WK	126	7.2	1754	100.0
Total	1754	100.0		

3. The following table reports the regression of the ranks of *premarsx* (see Exercise 1) on the ranks of religious-service attendance (*attend*, see Exercise 2). Interpret the unstandardized and standardized slopes and the *t* test.

Coefficients[a]

	Unstandardized Coefficients		Standardized Coefficients		
	B	Std. Error	Beta	t	Sig.
(Constant)	1216.694	21.102		57.657	.000
RATTEND RANK of ATTEND	-.387	.021	-.404	-18.507	.000

a. Dependent Variable: RPREMARS RANK of PREMARSX

4. The GSS 2000 asked how often respondents had sex during the past year (*sexfreq:* not at all, once or twice, once a month, 2 to 3 times a month, weekly, 2 to 3 times per week, 4+ times per week) and how often they went to a bar or tavern (*socbar:* never, once a year, several times a year, once a month, several times a month, several times a week, almost daily). The following table reports the ranks of *sexfreq* regressed on the ranks of *socbar.* Interpret the unstandardized and standardized slopes and the *t* test.

Coefficients[a]

	Unstandardized Coefficients		Standardized Coefficients		
	B	Std. Error	Beta	t	Sig.
(Constant)	564.956	22.484		25.127	.000
RSOCBAR RANK of SOCBAR	.254	.026	.243	9.721	.000

a. Dependent Variable: RSEXFREQ RANK of SEXFREQ

5. For the GSS 2000, the ranks of frequency of sex in the past year (*rsexfreq,* see Exercise 4) were regressed on gender (*sex:* males = 1, females = 0). The regression equation is RSEXFREQ = 711.0 + 105.2SEX, R-squared = .015, p < .0005. Interpret the intercept, slope, and coefficient of determination.

6. For the GSS 2000, the ranks of frequency of attendance of religious services (*rattend,* see Exercise 2) are regressed on religious preference as represented by four dummy variables, with Protestant as the reference group. Interpret the *t* tests for the dummy variables, the intercept, the unstandardized slopes for those variables that are significant at $\alpha = .05$, and the coefficient of multiple determination.

Model Summary

R	R Square	Std. Error of the Estimate
.415[a]	.172	456.192411

a. Predictors: (Constant), OTHERRLG, JEWISH, NONE, CATHOLIC

Coefficients[a]

	Unstandardized Coefficients		Standardized Coefficients		
	B	Std. Error	Beta	t	Sig.
(Constant)	987.795	14.747		66.985	.000
CATHOLIC	-45.381	26.790	-.039	-1.694	.090
JEWISH	-359.620	76.434	-.103	-4.705	.000
NONE	-593.937	32.051	-.419	-18.531	.000
OTHERRLG	-99.215	51.084	-.043	-1.942	.052

a. Dependent Variable: RATTEND RANK of ATTEND

7. Respondents in the GSS 2000 (*gss2000a* or *gss2000b*) were asked whether they thought their family income was far below average, below average, average, above average, or far above average, compared with American families in general (*finrela*). They were also asked in which social class they belong—the lower class, the working class, the middle class, or the upper class (*class*). Use SPSS to rank each of these variables. Then regress the ranks of *class* on the ranks of *finrela* to see how relative financial status affects subjective class identification. Interpret the unstandardized and standardized slopes and the *t* test.

8. Respondents in the GSS 2000 (*gss2000a* or *gss2000b*) were asked if they agreed or disagreed that government should reduce income differences in the country (*goveqinc:* strongly agree, agree, neither agree nor disagree, disagree, or strongly disagree). Rank *goveqinc* giving strongly agree the highest rank. Regress the ranks of *goveqinc* on region of residence as represented by dummy variables for Northeast, South, and West, with Midwest as the reference group (Northeast = New England + Middle Atlantic; South = South Atlantic + East South Central + West South Central; West = Mountain + Pacific; Midwest = East North Central + West North Central). Interpret the *t* tests for the dummy variables, the intercept, the unstandardized slopes for those variables that are significant at $\alpha = .05$, and the coefficient of multiple determination.

13

Nominal Dependent Variables

AN UNEXPECTED CONTINGENCY

Professor Smart announced, "I'm going to divide the class into four groups. Each group will make a class presentation and write a joint paper on a topic to be assigned. I think it will be easier to get the members of the groups working together if they have something in common on which to build. Therefore, I'm going to assign you to groups based on your answers to a couple of questions about yourself. Please write your answers to the following questions on a piece of paper. One, would you rather go to a big party or hang out with a few friends? . . . (pause) . . . Two, when you go out with friends, do you usually go to the same places, or do you like to try new places as much as possible?"

After collecting their papers, Professor Smart quickly tallied their answers. To Smart's surprise and pleasure, the answers of the twenty students split evenly on both questions. On the first question, ten said they preferred big parties and ten said they preferred to hang out with a few friends, and on the second question, ten said they went to the same places and ten said they tried different places. The professor thought, "That ought to give us approximately equal-sized groups. There will probably be about five students in each group."

Professor Smart announced, "Now, I'll direct you to the corner of the room where your group will form. Will those of you who prefer a big party go to the left side of the room and those who prefer hanging out with a few friends go to the right side." Ten students went to the left side and ten moved to the right. "OK, those who said they usually go out to the same places go to the front of your side of the room, and those who like to try new places go to the back of your side of the room." Ten headed for the front and ten to the back.

After everyone reached their corners, Brandy exclaimed from the left-front corner, "It's not fair! I'm the only one in this group. I'll have to do all the work!" Then Phillip quickly called from the right-rear corner, "And I'm the only one in this group! This isn't right." Professor Smart, caught off guard by the result of his experiment, exclaimed, "OK, OK. This is not what I expected. We'll have to make some adjustments."

What went wrong with Professor Smart's experiment? Can you draw a diagram of the room divided into four quadrants and enter the number of students in each quadrant/group? The technical term for the diagram is a crosstabular table. The table is a display of the crosstabulation of the responses to the two questions that the professor asked. (See Figure 13.1.)

In statistical terminology, Professor Smart unconsciously expected a null hypothesis of no association between the two questions to be true. It turned out that there was a strong association between the students' answers to the questions. There are two explanations for what was wrong with Smart's expectation. One possibility is that the professor erroneously overlooked the fact that there would be a true association between the questions. For example, students may prefer large parties because they hope to make new acquaintances, and they may also like to try different places when they go out because they enjoy new experiences. A second possibility is that the two questions were associated merely by chance. This possibility is somewhat plausible in a relatively small class like the professor's. In this chapter we will learn how to measure the strength of association in a crosstabulation and how to evaluate whether that association is real or likely to have occurred by chance.

VISUALIZING ASSOCIATIONS

When the dependent variable is a nominal variable, we cannot order the attributes from high to low and thus we cannot compute ranks for the attributes. Consequently, we are not able to use regression analysis as we know it to study the association between an independent variable (whether nominal, ordinal, or interval) and a nominal dependent variable.[1] The method we will use instead is called **crosstabulation.** Crosstabulation involves crossing one variable with another, which consists of tabulating the distribution of the nominal dependent variable within each category of the independent variable. We will first examine scatterplots to visualize the association, and then we will use crosstabular tables to read the association more precisely.

---

1. Regression-like methods may be used for nominal dependent variables. An introduction to one such method, *logistic regression,* can be found in Agresti and Finlay (1997). Logistic regression, however, is sufficiently different from the least-squares regression covered in this book that it is beyond the scope of an introductory text such as this one.

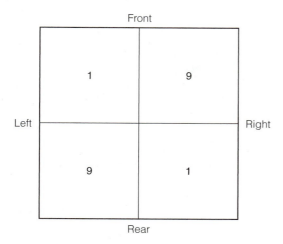

FIGURE 13.1 An Unexpected Contingency

Jittered Scatterplots

Scatterplots involving one or two nominal variables are made the same way they are for interval variables. We will start with a simple example of two dichotomous variables. Let the two attributes of each variable be assigned the codes 1 and 2. There are four possible pairs of coordinates for two dichotomous variables: (1, 1), (1, 2), (2, 1), and (2, 2). Figure 13.2a shows a scatterplot for hypothetical data with at least one case for each possible pair of coordinates. The scatterplot is divided into four quadrants by inserting dashed lines midway between the two attributes of X and Y. The insertion points do not represent the means of the variables, as was the case in Chapter 7. The markers for the cases with the same pair of attributes will be stacked on top of one another so that we will see only one marker for each pair of coordinates. Thus, we will not be able to see the joint distribution underlying the scatterplot. A solution to the problem is to create a **jittered scatterplot** by adding a random number from a uniform distribution with a mean of 0 to the numeric codes assigned to each attribute of X and Y. This technique was introduced in Chapter 7 and is shown in Figure 13.2b.

After jittering the data, we can see that two pairs of coordinates, (1, 2) and (2, 1), have many more cases than the other two pairs of coordinates, (1, 1) and (2, 2). This joint distribution indicates that cases with attribute 1 on X are more likely to have attribute 2 than attribute 1 on Y, whereas cases with attribute 2 on X are more likely to have attribute 1 than attribute 2 on Y. That is, when X is 1, Y tends to be 2, but when X is 2, Y tends to be 1. Thus, there is an association between X and Y. Based on the way we interpreted scatterplots in Chapter 7, we are inclined to say that as X increases, Y decreases, which indicates a negative association. Remember, however, that we are assuming that Y is a nominal variable, and thus we cannot speak of increases or decreases in Y. This means that it is not valid to describe associations as positive or negative. Instead, we must describe how the distribution of Y differs across the attributes of X. To repeat the example in Figure 13.2, when X is 1, Y tends to be 2, but when X is 2, Y tends to be 1.

Figure 13.3a shows an association between X and Y that is just the opposite of the one in Figure 13.2b. When X is 1, the mode of Y is 1, but when X is 2, the mode of Y is 2. For

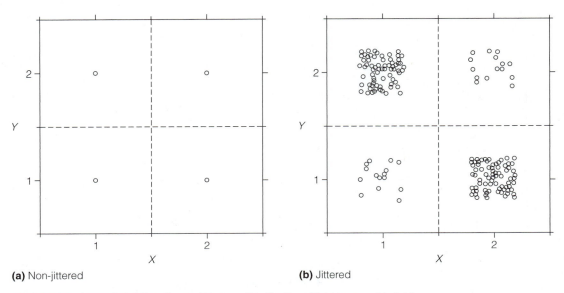

(a) Non-jittered   **(b)** Jittered

FIGURE 13.2 Jittered and Non-jittered Scatterplots for Two Dichotomous Variables

any given pattern of association, it is important to be able to recognize what the scatterplot would look like if the total distributions of X and Y were unchanged but the joint distribution showed no association between X and Y. Figure 13.3b displays a scatterplot in which the total distributions of X and Y are the same as in Figure 13.3a—that is, half the cases have attribute 1 on both X and Y (not shown)—but the joint distribution shows no association. When X is 1, Y is equally likely to be 1 or 2, and when X is 2, Y is also equally likely to be 1 or 2. That is, the distribution of Y is bimodal both when X is 1 and when X is 2.

Figure 13.3c illustrates an additional type of association. For both attributes of X, the mode of Y is 1. That is, Y tends to be 1 whether X is 1 or X is 2. This statement sounds like a description of no association. When X is 1, however, Y is only slightly more likely to be 1 than 2, but when X is 2, Y is much more likely to be 1 than 2. In other words, the probability that Y is 1 is greater when X is 2 than when X is 1. Figure 13.3d shows a scatterplot in which the total distributions of X and Y are the same as in Figure 13.3c— that is, 50 percent of the cases have attribute 1 on X and 70 percent have attribute 1 on Y (not shown). In Figure 13.3d, however, there is no association between X and Y; the mode of Y is 1 for both attributes of X, but there is no difference in the tendency for Y to be 1. That is, the concentration of cases at coordinates $(1, 1)$ is the same as it is at $(2, 1)$.

Why is it important to recognize the pattern of no association that corresponds to any given pattern of association? The answer is that the pattern of no association will serve as a reference for comparison with the pattern of association. The more the pattern of association differs from the pattern of no association, the less likely it will be that the association may have resulted by chance. This will be important when we learn how to compute a test of the null hypothesis that there is no association between X and Y.

To illustrate the association between two dichotomous variables with real data, we will examine racial differences in attitudes toward capital punishment. Who do you

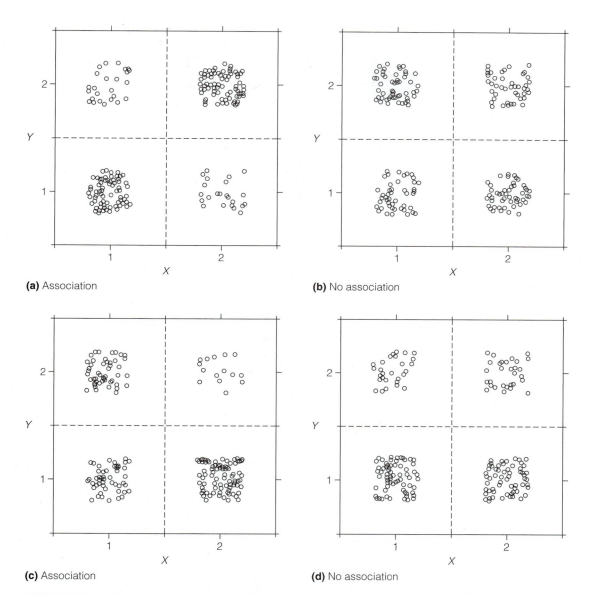

FIGURE 13.3 Two Patterns of Association and Corresponding Patterns of No Association

think will be more likely to favor the death penalty, Blacks or Whites? Respondents to the 1998 GSS were asked, "Do you favor or oppose the death penalty for persons convicted of murder?" Figure 13.4 shows the jittered scatterplot for race (1 = White, 2 = Black) and capital punishment (1 = favor, 2 = oppose) for 2,424 respondents. What is the nature, if any, of the racial difference in attitudes?

We can see by the density of the data points in the four quadrants that there are many more Whites than Blacks. Looking at the two quadrants for Black respondents, it

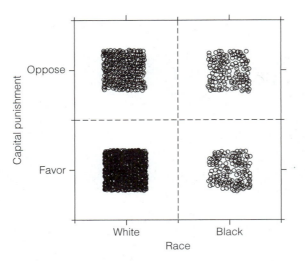

FIGURE 13.4 Capital Punishment Attitude and Race, 1998 GSS

appears that the numbers of Blacks who favor and oppose the death penalty are about equal. The distribution of attitudes is bimodal for Blacks. For Whites, on the other hand, the modal attitude is clearly in favor of capital punishment. Thus, Whites are more likely to favor the death penalty than Blacks.

We will now examine the association between nominal variables with more than two categories. The dependent variable is the candidate for whom respondents reported voting in the 1992 presidential race: William Clinton, George Bush, or Ross Perot. The very small numbers of persons voting for other candidates are excluded from the analysis. The independent variable is marital status (married, widowed, divorced/separated, never married). Because there are four categories for marital status and three for votes, there are $4 \cdot 3 = 12$ pairs of coordinates in the scatterplot for the two variables. The jittered scatterplot is shown in Figure 13.5. Does the distribution of votes for president differ by marital status?

We can see that for each marital status, the number of votes for Perot is less than the number for either Bush or Clinton. Thus, there are no obvious differences in the vote for Perot across marital statuses. When we compare the votes for Bush and Clinton, however, there are apparent differences by marital status. The number of votes for Bush and Clinton appear to be about equal among married people. Among the other three statuses, however, we can see that there are more Clinton voters than Bush voters. Finally, Clinton's advantage over Bush seems to be the largest among never-married persons. In conclusion, Clinton did best among people who had never been married, next best among those who once were married (divorced/separated and widowed), and poorest among married voters.

In scatterplots of nominal variables, the order in which attributes are placed along the horizontal and vertical axes is arbitrary. As a consequence, we cannot describe associations as being positive or negative. Instead, we have to describe how the distribution

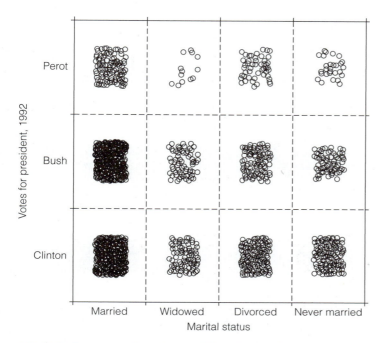

FIGURE 13.5 Votes for President in 1992 by Marital Status, 1998 GSS

of the dependent variable differs between different categories of the independent variable.

READING ASSOCIATIONS

The crosstabulation of the independent and dependent variables provides the numerical information needed to read the association between the variables. Crosstabulations also supply the data needed for inferential tests of association and for standardized measures of association. In this section we will learn how to construct crosstabular tables, including percentage tables, and how to read the tables to understand the nature of any association that may be present.

Constructing Crosstabular Tables

We will illustrate the construction of a crosstabular table with a random sample of twenty cases from the 1998 General Social Survey. The example is limited to twenty cases to simplify the crosstabulations. The variables to be crosstabulated are race and death penalty attitude, which we examined in Figure 13.4. The racial and attitudinal attributes of each case are shown in Table 13.1.

Table 13.1 Race and Opinion on Death Penalty for Random Sample of 20 Cases from 1998 GSS

Race	Death Penalty
Black	Favor
White	Oppose
White	Favor
White	Favor
Black	Favor
White	Favor
Black	Favor
White	Favor
White	Favor
Black	Oppose
White	Favor
White	Favor
White	Favor
White	Oppose
White	Favor
Black	Oppose
White	Favor
White	Favor
White	Oppose
White	Favor

The construction of the crosstabular table can be broken down into five steps (Box 13.1). The first step is to lay out the skeleton, or structure, of the table. Because there are two attributes for each variable, there will be $2 \cdot 2 = 4$ cells in the table. Each cell represents a specific pair of attributes of the independent and dependent variables: that is, (White, oppose), (White, favor), (Black, oppose), and (Black, favor). The column label and row label associated with the cell give the attribute pair for any cell. Which attributes should be placed in each row and in each column? The attributes of the independent variable should form the columns of the table and those of the dependent variable should form the rows. Thus, in our example, the racial attributes (White, Black) define the columns and the death penalty attributes (favor, oppose) are the rows. It doesn't matter, however, which column represents White and which Black, or which row represents favor and which oppose, because these are nominal variables.

The second step is to tally the joint attributes for each cell or, in other words, count the number of cases that go in each cell. This step involves going through the data matrix (Table 13.1) case by case, noting the attribute pair for each case and placing a tally mark in the appropriate cell of the crosstabular table. For example, because the first case in Table 13.1 is Black and in favor of the death penalty, a tally is placed in the upper right-

---

BOX 13.1   Building Crosstabular Tables

1. *Construct the skeleton of the table.* The columns are the attributes of the independent variable, and the rows are the attributes of the dependent variable.

Death penalty	Race	
	White	Black
Favor		
Oppose		

2. *Tally the joint frequency for each cell.* For each case in the raw data matrix (Table 13.1) that is White and approves of capital punishment, for example, one tally mark is placed in the upper left-hand cell.

Death penalty	Race	
	White	Black
Favor	IIIII IIIII II	III
Oppose	III	II

3. *Convert the tallies to numbers.* These are the crosstabulations.

Death penalty	Race	
	White	Black
Favor	12	3
Oppose	3	2

4. *Sum the rows and columns.* These are the marginal distributions. We now have the complete crosstabular table.

Death penalty	Race		Total
	White	Black	
Favor	12	3	15
Oppose	3	2	5
Total	15	5	20

5. *Compute the column percentages.* Divide the joint frequencies and row marginals by the number of cases in their corresponding columns. This is percentaging in the direction of the independent variable. We now have the percentage crosstabular table.

Death penalty	Race		Total
	White	Black	
Favor	$\frac{12}{15}100 = 80.0\%$	$\frac{3}{5}100 = 60.0\%$	$\frac{15}{20}100 = 75.0\%$
Oppose	$\frac{3}{15}100 = 20.0\%$	$\frac{2}{5}100 = 40.0\%$	$\frac{5}{20}100 = 25.0\%$
Total	100.0% (15)	100.0% (5)	100.0% (20)

hand cell. Because the second case is White and opposed, a tally is placed in the lower left-hand cell. The last case is White and in favor, and thus a tally is made in the upper left-hand cell. When each case has been tallied, the tallies shown in step 2 are obtained.

Step 3 is simply to convert the tallies to numbers. These are the joint frequencies. Whereas a simple frequency, like the one found in a frequency distribution, indicates the number of cases with some specific attribute on a single variable, *a* **joint frequency** *indicates the number of cases with a specific pair of attributes on two variables.* We can read from the joint frequencies in step 3 in the table that there are 12 Whites who are in favor, 3 Whites who are opposed, 3 Blacks who are in favor, and 2 Blacks who are opposed.

The fourth step is to compute the row and column totals. The row totals are the sums of the joint frequencies in each row. These sums are placed in a Total column at the right of the table. This column shows that $12 + 3 = 15$ people are in favor of the death penalty and $3 + 2 = 5$ people are opposed. The column totals are the sums of the joint frequencies in each column. The Total row at the bottom of the table indicates there are 15 Whites and 5 Blacks. The row and column totals are simple frequency distributions for the dependent and independent variables, respectively. They are called **marginal frequencies,** or marginal distributions, because they are located in the margins of the crosstabular table. Finally, the sum of either the row marginals or the column marginals gives the total number of cases in the table ($n = 20$).

The fifth step is to convert joint frequencies into column percentages. Because, in general, the numbers of cases with each attribute of the independent variable are not equal, it is difficult to compare the joint frequencies in one column with those in another. The solution is to convert all joint frequencies to column percentages. *A* **column percentage** *equals a joint frequency divided by the marginal frequency in its column with the result multiplied by 100.*

$$P_c = \frac{f_{ij}}{f_j} \, 100 \qquad\qquad (13.1)$$

where

$$P_c = \text{column percentage}$$

$$f_{ij} = \text{joint frequency in row } i \text{ and column } j$$

$$f_j = \text{marginal frequency in column } j$$

For example, the column percentage for the 12 Whites who are in favor of the death penalty is $(12 \div 15)100 = 80\%$. The sum of the percentages in each column equals 100 percent. *A column percentage indicates the percentage of all persons with a given attribute on the independent variable that have a certain attribute on the dependent variable.* For example, 40 percent of all the Blacks in the sample are opposed to the death penalty. The row marginals are also converted to column percentages by dividing by the total number of cases in the table (i.e., 20) and multiplying by 100.

Reading a Crosstabular Table

We have learned how to construct crosstabular tables and percentage crosstabular tables. Now we will learn how to read any association that may be in the table. Table 13.2a gives the joint frequencies and Table 13.2b gives the column percentages for the crosstabulation of death-penalty attitude by race from the 1998 GSS. The column percentages are used to read the association.

Each column of joint frequencies and each column of percentages within the margins of a crosstabular table is a **conditional distribution** of the dependent variable.

Table 13.2 Crosstabulation of Death-Penalty Attitude by Race, 1998 GSS

(a) Joint Frequencies				(b) Column Percentages			
Death penalty	Race		Total	Death penalty	Race		Total
	White	Black			White	Black	
Favor	1616	179	1795	Favor	78.3%	49.9%	74.1%
Oppose	449	180	629	Oppose	21.7%	50.1%	25.9%
Total	2065	359	2424	Total	100.0% (2065)	100.0% (359)	100.0% (2424)

Table 13.3 Death-Penalty Attitudes for Whites and Blacks

Death penalty	White	Death penalty	Black
Favor	78.3%	Favor	49.9%
Oppose	21.7%	Oppose	50.1%
Total	100.0% (2065)	Total	100.0% (359)

Thus, *a conditional distribution is the distribution of Y for cases with a specific attribute of X.* The conditional distribution of death-penalty attitudes for Whites is shown in Table 13.3a, and the conditional distribution of death-penalty attitudes for Blacks is shown in Table 13.3b.

If there is an association between X and Y, people who are different on X will tend to be different on Y. In terms of conditional distributions, *the* **criterion for an association** *between X and Y is that at least one conditional distribution of Y is different from the others.* To determine whether there is an association in Table 13.2, for example, we ask the question shown in Figure 13.6. Because Blacks and Whites clearly have different conditional distributions, we can conclude that race and death-penalty attitude are associated.

Having recognized that the two variables are associated, we can describe the nature of the association. To do this, we pick a row of the crosstabular table and compare the percentages across that row. That is, *to* **read an association,** *we compare how different are the likelihoods of having a given attribute of Y between people with different attributes of X.* In Table 13.2, for example, we can read across the row for those who favor the death penalty. One way of putting into words what we read in that row is to say that *78.3 percent of Whites favor the death penalty in comparison with 49.9 percent of Blacks.* A less precise way of making the same statement is to say that *Whites are more likely to favor the death*

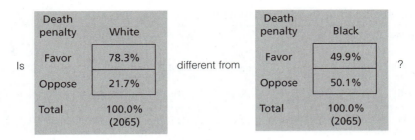

FIGURE 13.6 Criterion for an Association between X (e.g., Race) and Y (e.g., Death-Penalty Attitude)

penalty than Blacks. Neither of these statements directly indicates how much more likely Whites are to favor the death penalty. To describe this difference, we must subtract one percentage from the other to get the difference (D):

$$D = 78.3\% - 49.9\% = 28.4\%$$

Based on this result, we can say that *Whites are 28.4 percent more likely to favor the death penalty than Blacks.*

The following three statements are logically equivalent to the preceding three statements and thus are also legitimate interpretations of the association between race and death-penalty attitude:

- 49.9 percent of Blacks favor the death penalty in comparison with 78.3 percent of Whites.
- Blacks are less likely to favor the death penalty than Whites.
- Blacks are 28.4 percent less likely to favor the death penalty than Whites.

Instead of reading the association off the top row of the table, we can also read it off the bottom row (oppose death penalty). The percentage difference on the bottom row is

$$D = 21.7\% - 50.1\% = -28.4\%$$

Notice that the bottom-row D is the negative of the top-row D:

$$D_{\text{bottom row}} = -D_{\text{top row}} \tag{13.2}$$

This will always be true for a dichotomous dependent variable because each bottom-row percentage equals 100 minus the top-row percentage:

$$P_{\text{bottom row}} = 100 - P_{\text{top row}} \tag{13.3}$$

The two sets of equivalent bottom-row descriptions of the association are

- 21.7 percent of Whites oppose the death penalty in comparison with 50.1 percent of Blacks.
- Whites are less likely to oppose the death penalty than Blacks.
- Whites are 28.4 percent less likely to oppose the death penalty than Blacks.

and

- 50.1 percent of Blacks oppose the death penalty in comparison with 21.7 percent of Whites.
- Blacks are more likely to oppose the death penalty than Whites.
- Blacks are 28.4 percent more likely to oppose the death penalty than Whites.

We have derived four sets of descriptions of the association in a column–percentage crosstabular table. Any one of the sets is legitimate, and because they are all logically equivalent, there is no need to use more than one. Also, there is no need to use more than one statement in a set. The one that provides the most information is the first statement, but the one that most directly and precisely speaks to the association is the third statement (e.g., Blacks are 28.4 percent more likely to oppose the death penalty than Whites.).

Next we will look at a larger crosstabular table. Table 13.4 shows the crosstabulation of vote in the 1992 presidential election by marital status. This table has three rows and four columns (a 3 $\times$ 4 table), which makes reading the association considerably more complex than it is in a 2 $\times$ 2 table. First, because there are four columns, there are now six Ds in each row. Once we know any three Ds, however, we can compute the other three (proof not shown). Thus, if we pick the three most informative Ds, we can describe the differences in any given row with maximum efficiency. Second, because there are three rows, a D in one row does not equal the negative of a D in another row for the same columns as it does in a 2 $\times$ 2 table. Instead, each D equals the negative of the sum of the other two Ds. For example,

$$D_{\text{bottom row}} = -(D_{\text{top row}} + D_{\text{middle row}}) \qquad (13.4)$$

As a consequence, once we have described the association in any two rows, it is not necessary to do so in the third row, mathematically speaking.

Let's read the top row of Table 13.4 first, which corresponds to the candidate (Clinton) who received the greatest percentage of votes (48.0%) from the GSS respondents. We will read the difference between the marital status that gave Clinton the greatest percentage of votes and the status that gave him the second greatest percentage, the difference between the second highest status and the third highest, and the difference between the third highest and the fourth highest. *Never-married persons were 6.1 percent more likely to vote for Clinton than widowed persons, who were 1.5 percent more likely to vote for Clinton than divorced/separated persons, who were 10.7 percent more likely to vote for Clinton than married persons.* Because there is very little difference between widowed and divorced/ separated persons, this row indicates that never-married persons gave Clinton the most votes and married persons gave him the fewest votes, with those who had once been married (widowed and divorced/separated) falling between the other two groups.

We used three Ds to describe the association in the top row by comparing the highest percentage to the second highest, the second highest to the third highest, and the third highest to the fourth highest. It is legitimate to pick our comparisons this way because the attributes of the nominal independent variable are not ranked. Next, if we are primarily interested in discovering who voted for the two major party candidates, we will read the middle row (Bush) with the same method we used to read the top row. *Married persons were 5.8 percent more likely to vote for Bush than widowed persons, who were*

Table 13.4 Crosstabulation of 1992 Presidential Vote by Marital Status, 1998 GSS

Presidential vote 1992	Married		Widowed		Divorced/ separated		Never married		Total	
					Marital Status					
Clinton	374	41.6%	107	53.8%	173	52.3%	163	59.9%	817	48.0%
Bush	409	45.5%	79	39.7%	105	31.7%	79	29.0%	672	39.5%
Perot	116	12.9%	13	6.5%	53	16.0%	30	11.0%	212	12.5%
Total	899	100.0%	199	100.0%	331	100.0%	272	100.0%	1701	100.0%

8.0 percent more likely to vote for Bush than divorced/separated persons, who were 2.7 percent more likely to vote for Bush than never-married persons. The association in the middle row is pretty much the reverse of that in the top row. Married persons gave Bush the most votes, never-married persons gave him the fewest votes, and widowed and divorced/separated persons were in between. The difference between the widowed and divorced/separated statuses, however, were not the reverse of that in the top row. Widowed persons were more likely to vote for both Clinton and Bush than divorced/separated persons, which must be because the latter status must have voted more for the third party candidate (Perot) than the former status. We can use the Ds for these groups from the top two rows to compute their D on the bottom row, as follows:

$$D_{\text{bottom row}} = -(D_{\text{top row}} + D_{\text{middle row}}) = -(1.5 + 8.0) = -9.5$$

This result indicates that widowed persons were 9.5 percent less likely to vote for Perot than divorced/separated persons. It also illustrates that it is not mathematically necessary to read the third row once two rows have been read. We will do so, however, to describe explicitly the entire association in the table. *Divorced/separated persons were 3.1 percent more likely to vote for Perot than married persons, who were 1.9 percent more likely to vote for Perot than never-married persons, who were 4.5 percent more likely to vote for Perot than widowed persons.*

We used nine Ds to describe the relationship between marital status and presidential vote compared to only one D to describe the association between race and gun control attitude. The greater the number of rows and the greater the number of columns in the table, the greater the number of Ds that are needed to describe the association. As you can see, the description can get rather lengthy, even in tables as small as the 3 × 4 table we analyzed.

Although we have outlined a method for describing the association in crosstabulations of two nominal variables, there is no generally accepted formula for doing this. Some investigators pick out just a few of the largest Ds to describe the highlights of the association, leaving the percentages in the table to speak for themselves with respect to the remaining differences in the table. On the other hand, the focus of the study may dictate which Ds are emphasized. For example, if the objective of the study is to identify sources of support for third-party candidates, the investigator may stress the differences in the row corresponding to a third-party candidate (e.g., Perot).

TESTING ASSOCIATIONS

We have learned how to both visualize and read associations in crosstabular tables. We must remember that any association we observe in our sample data may have occurred by chance due to random sampling error. Thus, we need an inferential test to evaluate the hypothesis that there is no association in the population from which our sample was selected. In this section we will learn how to conduct the Pearson chi-square test of independence, or no association. The logic of the chi-square test is identical to the t and F tests that we used in regression analysis. The difference lies in the details of the chi-square statistic and its sampling distribution.

Pearson Chi-Square Statistic

Chi-square, pronounced "kai-square" and symbolized by the Greek χ^2, is a statistic for testing the null hypothesis of no association in the *population*. To understand the chi-square statistic we must construct a model of what the crosstabulation would look like if there were no association in the *sample*. We will then compare the observed crosstabulation to this model crosstabulation, and if the difference is sufficiently large, we will reject the null hypothesis. So, the first thing we need to do is specify a model of a crosstabulation with no association.

When we were learning to read crosstabular tables, we defined an association as existing when at least one conditional *percentage* distribution of Y is different from the other conditional *percentage* distributions. **The criterion for no association** is that all of the conditional percentage *distributions will be identical.* Thus, we want to determine which joint frequencies would produce identical conditional percentage distributions of Y.

To find these joint frequencies, we must first specify what the identical conditional percentage distributions are. For the example of race and attitude on capital punishment, Table 13.5 shows the conditional percentage distributions for our model of no association. It shows that in both conditional distributions (i.e., Blacks and Whites), 74.1 percent favor the death penalty and 25.9 percent oppose it. Thus, there is no association between race and death-penalty attitude. But, why have we chosen a 74.1/25.9 split instead of a 70/30 split, a 50/50 split, or any other split? The answer is that we have chosen the observed row marginal percentages (Table 13.2) to be the conditional percentage distributions. We have also chosen the observed column marginals (Table 13.2) to be the number of persons in each conditional distribution. These choices were made because the observed marginal distributions are the distributions that are most likely to exist in the population, even if there is no association in the population. That is, our sample distributions for race and death-penalty attitude are our best estimates of the population distributions, whether or not there is an association between the two variables.

To compute the Pearson chi-square statistic, we need to know the joint frequencies that are expected in each cell of the crosstabular table if the conditional percentage distributions were equal. These **expected frequencies** are

$$f_e = p_r n_c \tag{13.5}$$

Table 13.5 Model of No Association Between Race and Death Penalty Attitude, 1998 GSS

Death penalty	Race White	Black	Total
Favor	74.1%	74.1%	74.1%
Oppose	25.9%	25.9%	25.9%
Total	100.0% (2065)	100.0% (359)	100.0% (2424)

Table 13.6 Expected Frequencies by Equation 13.5

Death penalty	Race White	Black
Favor	$.741 \cdot 2065 = 1530.2$	$.741 \cdot 359 = 266.0$
Oppose	$.259 \cdot 2065 = 534.8$	$.259 \cdot 359 = 93.0$
Total	2065	359

where

$$f_e = \text{expected frequency}$$

$$p_r = \text{row proportion}$$

$$n_c = \text{column total}$$

which means that if there is no association between X and Y, the expected frequency in any cell of the crosstabular table equals the proportion of all cases that are in the corresponding row times the number of cases that are in the corresponding column. Note that it is the row proportion, not the row percent, that is used.

Table 13.6 gives the calculation of the expected frequencies for our example. To calculate the expected frequency for Whites who favor the death penalty, multiply the proportion of all the people who favor the death penalty (row proportion = .741) by the total number of Whites (column total = 2065). The result means that we would expect 1530.2 cases to be Whites who favor the death penalty if there is no association between the two variables. Expected frequencies may contain fractions. Notice that the sum of the expected frequencies in each column equals the observed column total.

Although Equation 13.5 best expresses the logic behind the calculation of expected frequencies, the accuracy of the resulting expected frequency may be less than desired because the proportion is rounded before multiplying. For better accuracy, we can re-express Equation 13.5 as follows:

Table 13.7 Expected Frequencies by Equation 13.6

Death penalty	Race		Total
	White	Black	
Favor	$\dfrac{1795 \cdot 2065}{2424} = \dfrac{3706675}{2424} = 1529.2$	$\dfrac{1795 \cdot 359}{2424} = \dfrac{644405}{2424} = 265.8$	1795
Oppose	$\dfrac{629 \cdot 2065}{2424} = \dfrac{1298885}{2424} = 535.8$	$\dfrac{629 \cdot 359}{2424} = \dfrac{225811}{2424} = 93.2$	629
Total	2065	359	2424

$$f_e = p_r n_c = \frac{n_r}{n} n_c = \frac{n_r n_c}{n} \tag{13.6}$$

where

$$f_e = \text{expected frequency}$$

$$n_r = \text{row total}$$

$$n_c = \text{column total}$$

$$n = \text{sample size}$$

This equation indicates that the expected frequency is obtained by multiplying the row total by the column total and dividing the product by the total sample size. There is no rounding until the calculation is completed. The calculation of the expected frequencies by Equation 13.6 is given in Table 13.7. The expected frequency for Blacks who oppose the death penalty, for example, equals the product of the row total (629) and the column total (359) divided by the total sample size (2,424). The product is a whole number, so there is no rounding until the final result is determined. The two expected frequencies for Whites differ from their counterparts in Table 13.6 by 1 and −1; the differences for Blacks are only .2 and −.2. Notice that the sum of the expected frequencies in each column equals the observed column marginals, and the sum in each row equals the row marginals.

Now that we have the joint frequencies that would be expected if there were no association, we can evaluate the hypothesis of no association by creating a statistic that is a function of the difference between the observed frequencies and the expected frequencies. The Pearson **chi-square statistic** is such a function:

$$\chi^2 = \Sigma \frac{(f_o - f_e)^2}{f_e} \tag{13.7}$$

where

$$\chi^2 = \text{Pearson chi-square}$$

$$f_o = \text{observed frequency}$$

$$f_e = \text{expected frequency}$$

Table 13.8 Calculation of Pearson Chi-Square for Race and Death Penalty Attitude

Race	Death Penalty	f_o	f_e	$f_o - f_e$	$(f_o - f_e)^2$	$(f_o - f_e)^2/f_e$
White	Favor	1616	1529.2	86.8	7534.24	4.927
Black	Favor	179	265.8	−86.8	7534.24	28.346
White	Oppose	449	535.8	−86.8	7534.24	14.062
Black	Oppose	180	93.2	86.8	7534.24	80.839
Total		2424	2424.0	.0		$\chi^2 = 128.174$

The summation is done across all the cells in the crosstabular table. In each cell the squared difference between the observed frequency and the expected frequency is divided by the expected frequency. The resulting quotients are then summed to give chi-square. For any given sample size, the larger the value of chi-square, the more the observed frequencies depart from those that would be expected if there were no association. The larger chi-square is, the less likely it is that the observed association could have occurred by chance.

The calculation of chi-square is illustrated in Table 13.8. The observed frequencies are from Table 13.2a, and the expected frequencies are from Table 13.7. The observed and expected frequencies both sum to the sample size ($n = 2424$). The differences between the observed and expected frequencies ($f_o - f_e$) sum to 0,[2] in much the same way that deviation scores and regression residuals sum to 0. We can get around this problem by squaring the differences, just as we square deviation scores when we calculate the variance and standard deviation. The final step before summing is to divide the squared differences by the expected frequencies. We do this because cells with large expected frequencies tend to have larger differences between the observed and expected frequencies. Dividing by the expected frequency somewhat offsets this tendency and allows all cells, whether large or small, to have a more equal opportunity of influencing the value of chi-square. In Table 13.8, the two smallest cells (those for which race is Black) have the largest value of ($f_o - f_e)^2/f_e$. The final step is to sum ($f_o - f_e)^2/f_e$ to get chi-square (128.174).

Table 13.9 shows the calculation of chi-square for the crosstabulation of vote in 1992 by marital status. The observed frequencies are from Table 13.4, and the expected frequencies can be calculated from the marginals of Table 13.4. The sum of the differences ($f_o - f_e$) is slightly different from 0 because of rounding error.[3] Notice that the two

---

2. Although the absolute value of the difference is the same for each cell (86.8), this result occurs only when there are two rows and two columns, forming a 2 × 2 table. When there are two rows and more than two columns, the differences in each column have the same absolute value. When there are two columns and more than two rows, the differences in each row have the same absolute value. If there are three or more rows and columns, none of the absolute values of the differences will be equal.

3. Because there are three or more rows and columns, none of the absolute values of the differences between the observed and expected frequencies are the same.

Table 13.9 Calculation of Pearson Chi-Square for Marital Status and Vote in 1992

Marital Status	Vote 1992	f_o	f_e	$f_o - f_e$	$(f_o - f_e)^2$	$(f_o - f_e)^2/f_e$
Married	Clinton	374	431.8	−57.8	3340.84	7.7370
Widowed	Clinton	107	95.6	11.4	129.96	1.3594
Divorced	Clinton	173	159.0	14.0	196.00	1.2327
Never married	Clinton	163	130.6	32.4	1049.76	8.0380
Married	Bush	409	355.2	53.8	2894.44	8.1488
Widowed	Bush	79	78.6	.4	.16	.0020
Divorced	Bush	105	130.8	−25.8	665.64	5.0890
Never married	Bush	79	107.5	−28.5	812.25	7.5558
Married	Perot	116	112.0	4.0	16.00	.1429
Widowed	Perot	13	24.8	−11.8	139.24	5.6145
Divorced	Perot	53	41.3	11.7	136.89	3.3145
Never married	Perot	30	33.9	−3.9	15.21	.4487
Total		1701	1701.0	−.1		$\chi^2 = 48.6833$

largest values of $(f_o - f_e)^2$ are in cells that include married persons, but after dividing by f_e, which is higher for the two cells than for any others, the contribution to chi-square of these cells is more or less equal to that of the never-married-Clinton and the never-married-Bush cells. When $(f_o - f_e)^2/f_e$ is summed, we get a chi-square of 48.68.

We have calculated chi-square for two crosstabulation tables. Can we conclude anything directly from these two values? The answer is no. Just as we had to compare t and F to their distributions in order to evaluate them, we must compare χ^2 to the chi-square distribution to determine what proportion of the distribution is this large or larger. We also cannot compare one χ^2 to the other to evaluate the relative strength of the associations because, as we shall see, χ^2 is directly affected by the size of the table (the numbers of rows and columns) and the number of cases in the table. Thus, we cannot use the fact that χ^2 is larger for the crosstabulation of race and capital-punishment attitude than for marital status and 1992 vote to conclude that the former association is stronger than the latter.

Chi-Square Distribution

Just as the t and F distributions are families of distributions with different but related shapes for different degrees of freedom, so too is the chi-square distribution. The **degrees of freedom of the χ^2** statistic are

$$df = (r - 1)(c - 1) \tag{13.8}$$

where

$$r = \text{number of rows}$$

$$c = \text{number of columns}$$

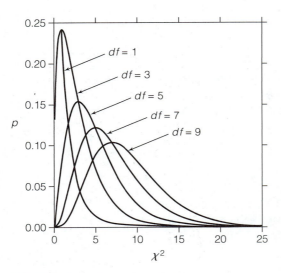

FIGURE 13.7 Chi-Square Distributions

The greater the number of rows and the greater the number of columns, the greater will be the degrees of freedom. The 2×2 table for race and death penalty has $(2 - 1)(2 - 1) = 1$ degree of freedom and the 3×4 table for marital status and vote in 1992 has $(3 - 1)(4 - 1) = 6$ degrees of freedom. Thus, larger tables (i.e., more rows and columns) have greater degrees of freedom.

The degrees of freedom are important because as the degrees of freedom increase, the value of the chi-square statistic tends to increase. Figure 13.7 shows chi-square distributions for five different degrees of freedom. You can see that each of the chi-square distributions is skewed to the right, but the greater the degrees of freedom, the less skewed the distribution. It is also obvious that the mode of the distribution shifts to the right as the degrees of freedom increase. Thus, there are fewer small values of chi-square and more large values of chi-square as the degrees of freedom increase, which means that we cannot compare the chi-square for a large table (large df) with that for a small table (small df).

Table A.4 gives values of chi-square for selected degrees of freedom and areas (A) under the distribution. For example, when $df = 1$, .10 of the distribution is greater than 2.706 and .05 of the distribution is greater than 3.841. When $df = 6$, on the other hand, .10 of the distribution is greater than 10.645 and .05 of the distribution is greater than 12.592. Figure 13.8 shows the upper 5 percent (A = .05) of the chi-square distribution for 1 and 6 degrees of freedom. The 95th percentiles for these degrees of freedom are 3.841 and 12.592. If we were conducting a chi-square test at $\alpha = .05$, these percentiles would be our critical values of chi-square. Notice that for the .05 column in Table A.4, as the degrees of freedom increase, the critical value of chi-square increases. Thus, the larger the crosstabulation table, the greater will be the value of chi-square that is needed to reject the null hypothesis. The same is true for each column in Table A.4.

What should we conclude about the associations between race and capital punishment (Table 13.2) and between marital status and presidential vote in 1992 (Table 13.4)? The computed values of chi-square for these crosstabulations are 128.17 and

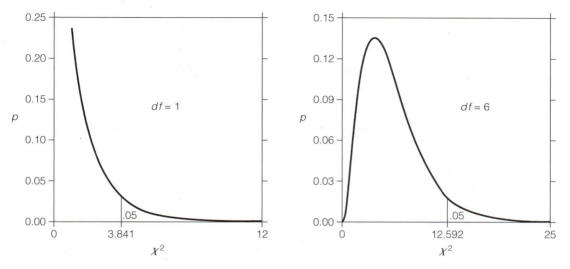

FIGURE 13.8 Critical Values of Chi-Square for $\alpha = .05$

48.68, respectively. Because both computed values are greater than the critical values of 3.841 and 12.592, respectively, we reject the null hypothesis for each pair of variables and conclude that race is related to attitude on capital punishment and that marital status is related to vote for president in 1992.

Chi-Square Test

We will now outline the chi-square test of no association in a crosstabular table. The logic of the test is identical to that of the t and F tests described in Chapters 9 and 10. Only the details differ. The null and alternative **hypotheses for the chi-square test** are

$$H_0: \text{all } Ds = 0$$

$$H_1: \text{at least one } D \neq 0$$

You will remember that D is the difference between two column percentages in some row. In a 2×2 table there is only one independent D because $D_{\text{bottom row}} = -D_{\text{top row}}$. As the number of rows and/or columns increases, the number of independent Ds increases. If all Ds equal 0 (the null hypothesis), there is no association in the table. If at least one D is not 0, there is an association between X and Y.

The t test for the regression slope can be one-tailed or two-tailed, and the null hypothesis can state that the slope is 0 or some value other than 0. In other words, there can be different specifications of the null and alternative hypotheses in different t tests. The chi-square test, however, is like the F test in that the null and alternative hypotheses are always the same. Therefore, we don't have to state these hypotheses each time we conduct the chi-square test, although we have to understand what they are.

Box 13.2 outlines the seven-step chi-square test. The first step is to specify the α level that will be used. The second step is to determine the degrees of freedom according to Equation 13.8. In step 3, the critical value of chi-square associated with α and df

---

BOX 13.2   Chi-Square Test

1. Specify α.
2. Determine degrees of freedom.
3. Find $\chi^2_{\alpha, df}$.
4. Construct a crosstabular table.

5. Calculate the expected frequencies.
6. Calculate χ^2.
7. Make decision and draw conclusion.

Table 13.10 Chi-Square Test for Crosstabulation of 1992 Presidential Vote by Gender of Married Persons

1. $\alpha = .05$
2. $df = (r - 1)(c - 1) = (3 - 1)(2 - 1) = 2$
3. $\chi^2_{\alpha, df} = \chi^2_{.05,2} = 5.991$

4. $f_o =$

213	161
198	211
53	63

(See Table 13.11.)

5. $f_e =$

193.0	181.0
211.1	197.9
59.9	56.1

(See Table 13.12.)

6. $\chi^2 = 7.606$ (See Table 13.13.)
7. $(\chi^2 = 7.606) > (\chi^2_{.05,2} = 5.991)$, $\therefore$ reject H_0
   $p = .023$

---

is found in Table A.4. Step 4 consists of constructing the crosstabular table according to the directions in Box 13.1. The fifth step is to use Equation 13.6 to calculate the expected frequencies that would exist if there were no association. Step 6 is to calculate the chi-square test statistic with Equation 13.7. The seventh and final step is to decide whether to reject the null hypothesis of no association by comparing the computed chi-square with the critical chi-square.

We will illustrate the chi-square test by looking at the association between gender and vote for president in 1992 (Clinton, Bush, Perot) among then married persons. The independent variable is gender, which has two attributes. Therefore, we will be making a 3×2 crosstabulation. The seven-step test is shown in Table 13.10. In step 1 we choose the usual .05 level for α. In step 2 we calculate the degrees of freedom to be $(3 - 1)(2 - 1) = 2 \cdot 1 = 2$. The third step is to look under A = .05 at $df = 2$ in Table A.4 to find $\chi^2 = 5.991$, the critical value of chi-square. The crosstabular table is constructed in step 4 and is shown in Table 13.11. You will remember that married persons were more likely to vote for Bush than Clinton, which is shown in the row marginals

Table 13.11 Presidential Vote by Gender, Married Persons, 1998 GSS

Presidential vote 1992	Gender Female		Male		Total	
Clinton	213	45.9%	161	37.0%	374	41.6%
Bush	198	42.7%	211	48.5%	409	45.5%
Perot	53	11.4%	63	14.5%	116	12.9%
Total	464	100.0%	435	100.0%	899	100.0%

Table 13.12 Expected Frequencies for Gender and 1992 Vote

Presidential vote 1992	Gender Female	Male	Total
Clinton	$\dfrac{374 \cdot 464}{899} = 193.0$	$\dfrac{374 \cdot 435}{899} = 181.0$	374
Bush	$\dfrac{409 \cdot 464}{899} = 211.1$	$\dfrac{409 \cdot 435}{899} = 197.9$	409
Perot	$\dfrac{116 \cdot 464}{899} = 59.9$	$\dfrac{116 \cdot 435}{899} = 56.1$	116
Total	464	435	899

of the table. The table indicates that females are more likely to vote for Clinton than are males ($D = 8.9\%$), whereas males are more likely to vote for Bush and Perot than are females. In step 5 we calculate the expected frequencies that would occur if there were no association between gender and vote (see Table 13.12). The sixth step is to use the observed and expected frequencies to calculate a chi-square of 7.606, as shown in Table 13.13. In the seventh and final step we note that the computed chi-square is greater than the critical chi-square and thus we reject the null hypothesis and conclude that females were more likely to vote for Clinton than males. We also note from computer output (not shown) that the observed level of significance is $p = .027$, indicating that the probability of incorrectly rejecting the null hypothesis is only .027.

Small Sample Tests

When conducting the chi-square test, we compare the computed χ^2 statistic ($\Sigma(f_o - f_e)^2/f_e$) with a critical value from the chi-square distribution for the specified α (e.g., Figure 13.7). The distribution of the computed χ^2 statistic, however, is only approximately the same as the theoretical chi-square distribution. The larger the sample size,

Table 13.13 Calculation of Pearson Chi-Square for Gender and 1992 Vote

Marital Status	Vote 1992	f_o	f_e	$f_o - f_e$	$(f_o - f_e)^2$	$(f_o - f_e)^2/f_e$
Female	Clinton	213	193.0	20.0	400.00	2.0725
Male	Clinton	161	181.0	−20.0	400.00	2.2099
Female	Bush	198	211.1	−13.1	171.61	.8129
Male	Bush	211	197.9	13.1	171.61	.8672
Female	Perot	53	59.9	−6.9	47.61	.7948
Male	Perot	63	56.1	6.9	47.61	.8487
Total		899	899.0	.0		$\chi^2 = 7.6060$

the closer the correspondence between the computed χ^2 statistic and the theoretical chi-square distribution. Thus, when the sample size is large, the level of significance (p) of the computed χ^2 will be more accurate and the decision to reject or not reject the null hypothesis is more likely to be made correctly. When sample size is small, on the other hand, we will be more likely to make an incorrect decision about whether to reject the null hypothesis. Therefore, when sample size is too small, we should not use the chi-square test.

How small can n become without invalidating the chi-square test? It turns out that it is not n itself that we must consider, but the sizes of the expected frequencies. Sample size and the expected frequencies, however, are not unrelated. The smaller the sample size, the smaller the expected frequencies will be. A frequently used guideline is that expected frequencies less than 5 are problematic for the chi-square test. More specifically, if any expected frequency in a 2 × 2 table (i.e., $df = 1$) is less than 5, the chi-square test should not be used. If the table is larger than 2 × 2 (i.e., $df > 1$), some experts say that it is all right to use the test if only a small percentage of the expected frequencies are less than 5 (e.g., 20% or less).

What are our options if there are too many small expected frequencies to use the chi-square test? One alternative is to combine some columns and/or some rows to create a smaller table (i.e., fewer degrees of freedom). For example, if marital status is one of the variables, it may help to combine divorced and separated persons into a single category if the expected frequencies for one group (e.g., separated) are less than 5. In general, we normally combine our smaller categories in an attempt to make expected frequencies greater than 5. If we have a 2 × 2 table, however, we cannot combine any categories because that would result in a table with only one row or one column, and thus there would be no variance for one of the variables.

If we have a 2 × 2 table with expected frequencies less than 5, we can use **Fisher's exact test** in place of chi-square test. Because this test is very complex and difficult to compute, we will not describe it here. Suffice it to say that some statistical programs (e.g., SPSS) provide this test automatically when you request a chi-square test for a 2 × 2 table. Fisher's exact test produces a level of significance (p) that is compared to your α to reach a decision about the null hypothesis. It is also possible to calculate Fisher's exact test for tables with df greater than 1. Special software packages may be required for this purpose; for example, you may purchase an exact test module in SPSS that includes Fisher's test for tables larger than 2×2.

MEASURING ASSOCIATIONS

We have learned how to read associations in crosstabular tables and how to test for *any* association in the population. In this section we will learn how to measure the strength of association in a crosstabulation. Although there are numerous measures of association for nominal variables, just as there are for ordinal variables (Chapter 12), we will emphasize Cramér's *V*, a measure that is based on the chi-square statistic. Other measures will be described briefly and compared with Cramér's *V*.

Cramér's *V*

Because the chi-square statistic is a function of how much the joint frequencies depart from those that would be expected if there were no association, it is potentially useful in constructing a measure of strength of association. Chi-square, however, cannot itself serve as a measure of association because its magnitude is affected not only by the strength of association in the crosstabulation but also by the sample size and the number of rows and columns in the table. We do not want a measure of association to be affected by the latter two factors because we want to be able to use our measure to compare the strength of association in tables that differ in sample size and/or number of rows and columns. Thus, we will examine exactly how sample size and table size affect chi-square so that we can standardize chi-square by controlling for these extraneous factors.

In general, the larger the sample size *n*, the larger chi-square tends to be. This fact is important when we use chi-square to test the null hypothesis because the larger the *n*, the less likely it is that the observed association in the sample could have occurred by chance. The specific way that **n affects chi-square** is that *if two tables are identical in all ways but n, then chi-square in the second table will be n_2/n_1 times as large as chi-square in the first table:*

$$\chi_2^2 = \frac{n_2}{n_1} \chi_1^2, \text{ all other things being equal} \tag{13.9}$$

Box 13.3 illustrates the effect of *n* on chi-square. Both tables in this box are of the same size—that is, they are 2×2 tables. The association is also the same in both tables, as measured by the difference in percentages in a row, i.e., $D = 80\% - 20\% = 60\%$. The sample size in the second table ($n_2 = 100$), however, is twice as large as that in the first table ($n_1 = 50$). In each table, each row and column has the same number of cases, and thus the expected frequency is the same for every cell of the table. The tables were designed this way to simplify the calculations, not because this is a necessary criterion for Equation 13.9. The calculated χ^2 in the first table is 18 compared to 36 in the second table. Thus, doubling the sample size while maintaining all other things causes a doubling of chi-square, as Equation 13.9 specifies. The larger χ^2 in Table 2 in Box 13.3 means that we would have a smaller probability of incorrectly rejecting the null hypothesis in Table 2, but it does not mean that there is a stronger association in Table 2.

BOX 13.3 The Effect of Doubling Sampling Size on Chi-Square

Table 1

20	80%	5	20%	25
5	20%	20	80%	25
25	100%	25	100%	50

Table 2

40	80%	10	20%	50
10	20%	40	80%	50
50	100%	50	100%	100

$$f_e = \frac{25 \cdot 25}{50} = 12.5 \text{ per cell}$$

$$f_e = \frac{50 \cdot 50}{100} = 25 \text{ per cell}$$

$$\chi_1^2 = 2\left(\frac{(20 - 12.5)^2}{12.5}\right) + 2\left(\frac{(5 - 12.5)^2}{12.5}\right) = 18$$

$$\chi_2^2 = 2\left(\frac{(40 - 25)^2}{25}\right) + 2\left(\frac{(10 - 25)^2}{25}\right) = 36$$

$$\chi_2^2 = \frac{n_2}{n_1} \chi_1^2 = \frac{100}{50} 18 = 2 \cdot 18 = 36$$

Next, we will examine the **effect of table size on chi-square.** Let k be the minimum of the number of rows minus 1 or the number of columns minus 1:

$$k = \min(r - 1, c - 1)$$

If, for example, $r = 3$ and $c = 4$, then $k = 3 - 1 = 2$. Don't confuse k with $df = (r - 1)(c - 1)$. As k increases, chi-square gets larger, all other things being equal. We may refer to r and c as the *dimensions* of the table. *As the minimum dimension of the table increases, chi-square increases.* More precisely, the relationship is

$$\chi_2^2 = \frac{k_2}{k_1} \chi_1^2, \text{ all other things being equal} \tag{13.10}$$

To illustrate, Box 13.4 shows two tables that both have perfect associations, the strongest possible association. In Table 1, all of the cases in column 1 are in row 1, whereas all of the cases in column 2 are in row 2. Thus, the two conditional distributions could not be more different, which means that $D = 100\%$, a perfect association. Table 2 in Box 13.4, a 3×3 table, also has a perfect association because the three conditional distributions could not be more different from what is shown. $D = 100\%$ for two Ds in each row of Table 2, a perfect association, which means that both tables have equally strong associations. In addition to having equal associations, both tables also have the same number of cases, $n_1 = n_2 = 90$. The tables differ only in k, the minimum of $(r - 1)$ and $(c - 1)$. Because $r = c$ in both tables, k equals either $r - 1$ or $c - 1$. Thus, $k_1 = 2 - 1 = 1$ and $k_2 = 3 - 1 = 2$. After computing chi-square for both tables, we can see that chi-square in Table 2 ($\chi^2 = 180$) is twice as great as that in Table 1 ($\chi^2 = 90$) because $k_2 / k_1 = 2$.

BOX 13.4 Effect of the Number of Rows and Columns on Chi-Square

Table 1

45	0	45
0	45	45
45	45	90

Table 2

30	0	0	30
0	30	0	30
0	0	30	30
30	30	30	90

$$f_e = \frac{45 \cdot 45}{90} \doteq 22.5 \text{ per cell}$$

$$\chi_1^2 = 2\left(\frac{(45 - 22.5)^2}{22.5}\right) + 2\left(\frac{(0 - 22.5)^2}{22.5}\right) = 90$$

$$f_e = \frac{30 \cdot 30}{90} = 10 \text{ per cell}$$

$$\chi_2^2 = 3\left(\frac{(30 - 10)^2}{10}\right) + 6\left(\frac{(0 - 10)^2}{10}\right) = 180$$

$$\chi_2^2 = \frac{k_2}{k_1} \chi_1^2 = \frac{2}{1} 90 = 180$$

We have seen how n and k affect the value of chi-square. Now, what is the maximum value for chi-square in any table? The maximum χ^2 occurs when there is a perfect association in the table, but it can be specified by merely knowing n and k. The **maximum value of chi-square** is

$$\text{Max } \chi^2 = nk \qquad (13.11)$$

For Tables 1 and 2 in Box 13.4, the maximum chi-squares are

$$\text{Max } \chi_1^2 = n_1 k_1 = 90 \cdot 1 = 90$$

$$\text{Max } \chi_2^2 = n_2 k_2 = 90 \cdot 2 = 180$$

These are the same values as the ones computed in Box 13.4. Notice that in a 2×2 table (Table 1), the maximum chi-square equals n because $k = 1$. We can generalize further by saying that whenever $k = 1$, maximum $\chi^2 = n$, whether it is a 2×3 table, a 4×2 table, or any other table where one dimension (r or c) is 2.

Now that we know what the maximum value of chi-square will be for any table, we can standardize χ^2 to range from 0 to 1 by dividing by the maximum chi-square. After dividing, we also take the square root of the quotient because chi-square is based on squared differences between observed and expected frequencies [i.e., $(f_o - f_e)^2$]. The name of this statistic is **Cramér's V:**

$$\text{Cramér's } V = \sqrt{\frac{\chi^2}{nk}} \qquad (13.12)$$

Table 13.14 Strength of Association in Different Sized Tables as Measured by Cramér's V

	Section 1 Death penalty approval by race	Section 2 1992 vote by marital status	Section 3 1992 vote by gender
$r \times c$	2×2	3×4	3×2
n	2424	1701	899
χ^2	128.174	48.683	7.606
V	.230	.120	.092

where

$$k = \min(r - 1, c - 1)$$

Because we are dividing the computed chi-square by the maximum possible chi-square, V must range from 0 to 1:

$$\text{If there is no association, } \chi^2 = 0 \text{ and } V = \sqrt{\frac{0}{nk}} = 0$$

$$\text{If there is a perfect association, } \chi^2 = nk \text{ and } V = \sqrt{\frac{nk}{nk}} = 1$$

Now we can use Cramér's V to determine for which pair of variables the association is stronger: race and death-penalty attitude, marital status and 1992 presidential vote, or gender and 1992 presidential vote. The relevant statistics from our previous analyses are given in Table 13.14. Notice that each association is based on a different sample size and each table has different dimensions (i.e., $r \times c$). Thus, the chi-squares cannot be compared.

Cramér's V for each section in Table 13.14 is

$$V_1 = \sqrt{\frac{\chi^2}{nk}} = \sqrt{\frac{128.174}{2424 \cdot 1}} = \sqrt{.0529} = .230$$

$$V_2 = \sqrt{\frac{\chi^2}{nk}} = \sqrt{\frac{48.683}{1701 \cdot 2}} = \sqrt{.0143} = .120$$

$$V_3 = \sqrt{\frac{\chi^2}{nk}} = \sqrt{\frac{7.606}{899 \cdot 1}} = \sqrt{.0085} = .092$$

The Vs indicate that the association between race and death-penalty attitude is almost twice as strong ($.23/.12 = 1.92$) as that between marital status and 1992 presidential vote, and it is two and one-half times as strong ($.23/.092 = 2.5$) as the association between gender and the 1992 presidential vote of married persons. Thus, the strongest association is between race and death-penalty attitude.

How do we interpret the values of V? Unlike measures of strength of association for interval variables, the values of V have no clear meaning. For example, V does not indicate the proportion of variance in Y that is explained by X or the standard deviation change in Y per standard deviation increase in X, as the coefficient of determination and the standardized slope do, respectively. The only meaning that V has is that it indicates what proportion of a perfect association is the observed association. For example, $V = .23$ indicates that the association between race and death-penalty attitude is .23 as strong as a perfect association.

*Other Measures of Association

A number of additional measures of association for crosstabulations are available in computer programs that you may encounter in the research literature. Therefore, it is useful to look at some of their properties and to compare them with Cramér's V.

*Chi-Square-Based Measures

In addition to Cramér's V, there are several **other chi-squared measures of association.** Pearson's *phi* coefficient is

$$\phi = \sqrt{\frac{\chi^2}{n}}$$

Because χ^2 is divided by n, which is the maximum value of χ^2 in tables where $k = \min(r - 1, c - 1) = 1$, phi is standardized to range from 0 to 1 when $k = 1$. Because $V = \sqrt{\chi^2/n}$ when $k = 1$, $\phi = V$ in such tables. When $k > 1$, phi is no longer standardized because n is no longer the maximum value of χ^2 and thus phi may be greater than 1. For this reason, $\phi > V$ when $k > 1$. In sum, because phi is not standardized, it should not be used to measure the strength of association.

A second measure is Pearson's coefficient of contingency, or C:

$$C = \sqrt{\frac{\chi^2}{\chi^2 + n}}$$

Because the denominator under the radical is always greater than the numerator, $(\chi^2 + n) > \chi^2$, C can never reach 1. Because its maximum value depends on the size of the table, C is not standardized.

Another measure is Tschuprow's T:

$$T = \sqrt{\frac{\chi^2}{n\sqrt{df}}}$$

where $df = (r - 1)(c - 1)$, the degrees of freedom of chi-square. In a square table where $r = c$, $\sqrt{df} = \min(r - 1, c - 1)$ and $T = V$. Thus, T is standardized in square tables. When $r \neq c$, however, $\sqrt{df} > \min(r - 1, c - 1)$ and $T < V$, which also means that T can never reach the value of 1 in tables that are not square. Thus, in general, T is not standardized.

In sum, *because ϕ, C, and T are not standardized for all crosstabular tables, Cramér's V is the preferred chi-square-based measure of association.*

*Proportional Reduction in Error Measures

We can illustrate what is meant by proportional reduction in error (PRE) measures by considering the familiar coefficient of determination (CD) for interval variables. Based on Equation 8.17, the coefficient of determination is

$$CD = \frac{\text{total variance} - \text{residual variance}}{\text{total variance}}$$

We may think of the total variance as the total amount of errors we would make if we used the mean of Y as the predicted value for every case. The residual variance is the amount of error that remains after using X to predict Y. Because the residual variance is always less than the total variance, the numerator of CD indicates the reduction in error achieved by using X to predict Y. After dividing by the total variance to get CD, we have the proportional reduction in error that results from predicting Y from X. For example, if CD = .30, it means that errors are reduced by 30 percent by using X to predict Y. Although we are used to interpreting CD as the *proportion of explained variance*, this interpretation is identical to a PRE interpretation because

$$\text{Explained variance} = \text{total variance} - \text{residual variance}$$

$$= \text{total errors} - \text{residual errors} \qquad (13.13)$$

$$= \text{reduction in errors}$$

There are two **PRE measures of association** for nominal variables that have been developed by Goodman and Kruskal. The first is named *lambda* (λ). Lambda uses the mode of Y as the best prediction that can be made for every case without using X. Thus, total errors equal the number of cases that are not in the modal category of Y. The best prediction of Y based on X is to predict the conditional mode of Y within each category of X. The errors within each category of X equal the number of cases in that category that are not at the conditional mode of Y. The within-category errors are summed across all categories to get the residual errors. The total errors and residual errors are then used to compute lambda as follows:

$$\lambda = \frac{\text{total errors} - \text{residual errors}}{\text{total errors}}$$

We will illustrate the statistic with the data on gender and 1992 vote of married persons shown in Table 13.15. The modal category of Y is Bush, so Bush is the prediction for all persons when we are not using gender. This will result in 374 + 116 = 490 total errors, the number that did not vote for Bush. Now, when we use gender to predict vote, we will predict that all females vote for Clinton because Clinton is the mode in the female distribution of votes. That prediction will result in 198 + 53 = 251 errors for females. The mode for males is Bush, and that prediction will result in 161 + 63 = 224 errors for males. The sum of the errors for males and female is 251 + 224 = 475, which is the residual error.

Entering the total and residual errors into the formula gives

$$\lambda = \frac{490 - 475}{490} = \frac{15}{490} = .031$$

Table 13.15 Crosstabulation of Married Persons' Presidential Votes in 1992 by Gender, 1998 GSS

Vote 1992	Female	Male	Total
Clinton	213	161	374
Bush	198	211	409
Perot	53	63	116
Total	464	435	899

Lambda indicates that the proportional reduction in errors achieved by using gender to predict vote is .031, or gender reduces the prediction errors by 3.1 percent.

Lambda ranges from 0 to 1 for any size of table. It also has the advantage of having a PRE interpretation. The value of lambda, unlike that of CD, is asymmetrical; that is, $\lambda_{YX} \neq \lambda_{XY}$. For example, if we used vote to predict gender, which is nonsense, $\lambda = [435 - (161 + 198 + 53)]/435 = .051$. *The disadvantage of lambda is that if each category of X has the same conditional mode of Y, lambda will equal 0.* This result will occur because lambda will predict the same attribute of Y for every attribute of X. This is undesirable because there may be large percentage differences in the modal row of the table indicating that the conditional distributions of Y are unequal, as illustrated here:

90%	60%
10%	40%

$$D = 30\%$$

$$\lambda = 0$$

Because of this characteristic, use of lambda is not recommended even though it has a desirable PRE interpretation.

The second PRE measure is Goodman and Kruskal's *tau* (τ). Because tau is more complex to describe and compute than lambda, it will not be explained here. Suffice it to say that it uses a method of counting prediction errors that is quite different from lambda. Like lambda, tau ranges from 0 to 1 and is an asymmetric measure. *Unlike lambda, tau does not necessarily equal 0 when the conditional modes of Y are the same for each category of X.* This makes tau an attractive measure of association.

When considering the use of a PRE measure versus Cramér's V, it is important to realize that the strength of association indicated by a PRE measure is not comparable to that of V. In a 2 × 2 table ($df = 1$), tau equals the square of V:

$$\tau = V^2 \text{ when } df = 1$$

This means that $V > \tau$ in a 2 × 2 table, which points up the fact that PRE measures will be less than measures based on chi-square. This is analogous to the fact that CD is always less than the correlation coefficient because CD $= r^2$. Thus, PRE measures are like coefficients of determination, as we have already stressed, and chi-square measures are like correlations. Chi-square measures, however, do not indicate the direction of correlation because they are best applied to nominal variables for which direction, or order, is not a valid concept.

SUMMARY

In this chapter we learned how to analyze the association between a nominal dependent variable and another variable. We first examined the association in a jittered scatterplot in order to visualize it. It is necessary to jitter the data points (i.e., add or subtract a small random number from the numeric code for each variable) because there is usually a big stack of cases at each pair of coordinates in the plot. When looking for an association in the scatterplot, we do not look for linear patterns as we did in scatterplots for interval variables. Instead, we must compare the scatter of Y for each category of X with that for the other categories to see which conditional distributions are different from one another.

We next outlined some principles of reading the association from a table. This process involves crosstabulating the attributes of the two variables to determine the joint frequencies—that is, the number of cases for each pair of attributes. These joint frequencies are placed in a crosstabular table where the attributes of the independent variable form the columns of the table and the attributes of the dependent variable form the rows. The sum of joint frequencies in each column gives the column marginals, or the distribution of the independent variable. The sum across each row gives the row marginals, or the distribution of the dependent variable. Finally, each joint frequency is expressed as a percentage of the number of cases in its column. Now the table is ready to be analyzed and read. To read the table we compare the percentage distributions in each column—that is, the conditional percentage distributions of Y. If all the conditional distributions are the same, there is no association. If any conditional percentage distribution is different from the others, there is at least some association. The more these distributions differ, the stronger the association. To facilitate this comparison, we compare percentages across a given row. The differences in percentages between each pair of columns in a row, the D statistics, indicate how strong the association is. In a 2×2 table, there is only one independent D to read. The more rows and columns the table has, the more independent Ds there are and thus the more complex it is to read the association.

We next learned how to conduct a chi-square test of the null hypothesis of no association. The chi-square statistic is a function of the difference between the observed joint frequency and the joint frequency that would be expected if there were no association. The squared difference divided by the expected frequency is summed across all cells to get the computed chi-square. After specifying α and the degrees of freedom, the critical value of chi-square can be determined from a table of selected values from the chi-square distributions. There is a different distribution for each degree of freedom. If the computed chi-square is greater than or equal to the critical chi-square, the null hypothesis is rejected and we conclude that the two variables are associated.

Last, we used the computed chi-square to create a measure of the strength of association. Chi-square itself cannot serve as a measure of the strength of association because its magnitude is a positive function of the degrees of freedom and the number of cases in the table. The degrees of freedom and sample size determine the maximum value of chi-square for a given table. By dividing chi-square by its maximum possible value and taking the square root of the quotient, we create a statistic that ranges from 0 to 1. The statistic is Cramér's V, and it indicates the proportion of the maximum possible association that the observed association represents.

KEY TERMS

crosstabulation

jittered scatterplots

joint frequency

marginal frequency

column percentage

conditional distribution

criterion for an association

reading an association

criterion for no association

expected frequency

chi-square statistic

degrees of freedom of chi-square

hypotheses for chi-square test

Fisher's exact test

effect of n on chi-square

effect of table size on chi-square

maximum value of chi-square

Cramér's V

★other chi-squared measures of association

★PRE measures of association

SPSS PROCEDURES

We will construct a crosstabular table, compute chi-square, and compute several measures of association for the association between race (*race*) and attitude on capital punishment (*cappun*). First, we will select only those cases that are Black or White on the race variable. Click on *Data* and *Select Cases* to open the Select Cases window, where you will click on *If condition is satisfied* and the *If* button. In the *Select Cases: If* window, enter "race = 1 or race = 2" in the box, click on *Continue*, and click on *OK*. To conduct the crosstabulation analysis, click on *Analyze, Descriptive Statistics,* and *Crosstabs* to open the following *Crosstabs* window. Move the dependent variable *cappun* into the *Row(s)* box and move *race* into the *Column(s)* box. Click on the *Cells* button, and in the following *Crosstabs: Cell Display* window, click on *Observed Counts, Expected Counts,* and *Column Percentages*. Click on *Continue* to return to the *Crosstabs* window, where you will click on the *Statistics* button to open the following *Crosstabs: Statistics* window. Click on *Chi-square, Contingency coefficient, Phi and Cramér's V,* and *Lambda,* and then click on *Continue* and *OK* to produce the crosstabular table and statistics shown after the windows.

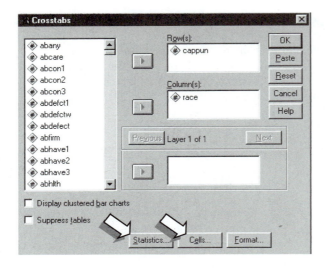

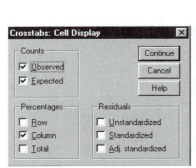

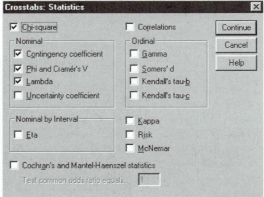

CAPPUN * RACE Crosstabulation

			RACE		
			WHITE	BLACK	Total
CAPPUN	FAVOR	Count	1616	179	1795
		Expected Count	1529.2	265.8	1795.0
		% within RACE	78.3%	49.9%	74.1%
	OPPOSE	Count	449	180	629
		Expected Count	535.8	93.2	629.0
		% within RACE	21.7%	50.1%	25.9%
Total		Count	2065	359	2424
		Expected Count	2065.0	359.0	2424.0
		% within RACE	100.0%	100.0%	100.0%

Chi-Square Tests

	Value	df	Asymp. Sig. (2-sided)	Exact Sig. (2-sided)	Exact Sig. (1-sided)
Pearson Chi-Square	128.335[b]	1	.000		
Continuity Correction[a]	126.861	1	.000		
Likelihood Ratio	115.271	1	.000		
Fisher's Exact Test				.000	.000
Linear-by-Linear Association	128.282	1	.000		
N of Valid Cases	2424				

a. Computed only for a 2x2 table

b. 0 cells (.0%) have expected count less than 5. The minimum expected count is 93.16.

Directional Measures

			Value	Asymp. Std. Error[a]	Approx. T[b]	Approx. Sig.
Nominal by Nominal	Lambda	Symmetric	.001	.019	.053	.958
		CAPPUN Dependent	.002	.030	.053	.958
		RACE Dependent	.000	.000	.[c]	.[c]
	Goodman and Kruskal tau	CAPPUN Dependent	.053	.010		.000[d]
		RACE Dependent	.053	.010		.000[d]

a. Not assuming the null hypothesis.

b. Using the asymptotic standard error assuming the null hypothesis.

c. Cannot be computed because the asymptotic standard error equals zero.

d. Based on chi-square approximation

Symmetric Measures

		Value	Approx. Sig.
Nominal by Nominal	Phi	.230	.000
	Cramer's V	.230	.000
	Contingency Coefficient	.224	.000
N of Valid Cases		2424	

a. Not assuming the null hypothesis.

b. Using the asymptotic standard error assuming the null hypothesis.

EXERCISES

1. GSS 2000 asked respondents if it should be possible for a pregnant woman to obtain a legal abortion if the woman wants it for any reason (*abany*). Is there an association between abortion attitude and gender in the following jittered scatterplot? Explain.

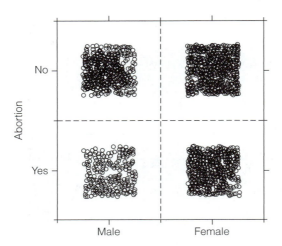

2. GSS 2000 asked whether the respondent feels the Bible is (a) "The actual word of God and is to be taken literally"; (b) "The inspired word of God, but not everything is to be taken literally"; or (c) "An ancient book of fables, legends, history, and moral precepts recorded by man." Is there an association between abortion attitude (Exercise 1) and interpretation of the Bible in the following jittered scatterplot? Explain.

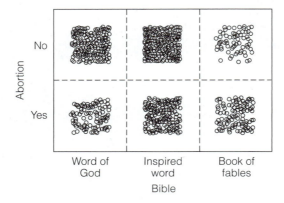

3. GSS 2000 asked if respondents agreed or disagreed with the statement that it is better
if the man is the achiever outside the home and the woman takes care of the home
and family (*fefam2*). The following table reports a random sample of 20 cases from
GSS 2000 with valid values on *fefam2* (1 = agree, 2 = disagree) and the variable *abany*
(1 = yes, 2 = no) described in Exercise 1. Construct crosstabular tables for the joint
frequencies and the column percentages. Let *abany* be the row variable. Interpret the
association.

	1	2	3	4	5	6	7	8	9	10	11	12	13	14	15	16	17	18	19	20
ABANY ABORTION IF WOMAN WANTS FOR ANY REASON	2	1	1	2	1	2	1	2	2	2	2	2	2	2	1	1	1	2	1	
FEFAM2 better for man to work and woman tend home, recode	2	2	2	2	1	1	2	1	2	2	1	1	2	2	1	2	2	1	1	1

4. The following table reports a random sample of 20 cases from GSS 2000 with valid
values on *pres96* (1 = Clinton, 2 = Dole, 3 = Perot) and *bible* (1 = word of God,
2 = inspired word, 3 = book of fables) described in Exercise 2. Construct crosstabular
tables for the joint frequencies and the column percentages. Let *pres96* be the row
variable. Interpret the association.

	1	2	3	4	5	6	7	8	9	10	11	12	13	14	15	16	17	18	19	20
PRES96 VOTE FOR CLINTON, DOLE, PEROT	1	2	1	2	1	1	2	1	1	3	1	1	3	1	3	2	3	2	1	2
BIBLE FEELINGS ABOUT THE BIBLE	2	2	2	2	1	2	2	2	1	1	1	3	1	2	3	1	1	1	3	2

5. Conduct an hypothesis test at $\alpha = .05$ of the association between *abany* and *fefam2* using the crosstabulation table from GSS 2000.

	FEFAM2 better for man to work and woman tend home, recode		
	1 Agree	2 Disagree	Total
ABANY ABORTION 1 YES IF WOMAN WANTS	108	217	325
FOR ANY REASON 2 NO	241	274	515
Total	349	491	840

a. What are the degrees of freedom and the critical $\chi^2_{.05,df}$?
b. Calculate the expected frequencies and the chi-square statistic.
c. State your decision and conclusion.

6. Conduct a hypothesis test at $\alpha = .05$ of the association between *pres96* and *bible* using the crosstabulation table from GSS 2000.

		BIBLE FEELINGS ABOUT THE BIBLE			
		WORD OF GOD	INSPIRED WORD	BOOK OF FABLES	Total
PRES96 VOTE FOR CLINTON, DOLE, PEROT	CLINTON	247	342	149	738
	DOLE	143	210	39	392
	PEROT	42	105	38	185
Total		432	657	226	1315

a. What are the degrees of freedom and the critical $\chi^2_{.05,df}$?
b. Calculate the expected frequencies and the chi-square statistic.
c. State your decision and conclusion.

7. a. Calculate Cramér's V for *abany* and *fefam2* in Exercise 5.
   b. Exercise 2 shows a scatterplot of *abany* and *bible*. In the crosstabulation of these variables, $n = 1255$ and $\chi^2 = 101.0$. Calculate Cramér's V.
   c. Which is more strongly related to *abany, fefam2* or *bible*? Why?

8. The following tables report the column percentages for the crosstabulation of *pres96* by *relig5* and the crosstabulation of *pres96* by *bible*.

	RELIG5				
PRES96	Protestant	Catholic	Jewish	None	Other
CLINTON	51.2%	58.3%	80.5%	67.5%	71.9%
DOLE	35.6%	29.0%	14.6%	16.0%	15.6%
PEROT	13.2%	12.7%	4.9%	16.5%	12.5%
	100.0%	100.0%	100.0%	100.0%	100.0%

$n = 1630, \chi^2 = 49.933$

	BIBLE		
PRES96	WORD OF GOD	INSPIRED WORD	BOOK OF FABLES
CLINTON	57.2%	52.1%	65.9%
DOLE	33.1%	32.0%	17.3%
PEROT	9.7%	16.0%	16.8%
Total	100.0%	100.0%	100.0%

$n = 1315, \chi^2 = 29.156$

a. Interpret the association in each table.
b. Calculate Cramér's V for each table.
c. Is *relig5* or *bible* more strongly related to *pres96*? Why?

9. The GSS 2000 (*gss2000a* or *gss2000b*) asked if respondents would favor or oppose a law that would require a person to obtain a police permit before he or she could buy a gun (*gunlaw*). Use SPSS to crosstabulate *gunlaw* by *sex* and to obtain the appropriate statistics to evaluate the association.

a. Interpret the association between *gunlaw* and *sex*.
b. What decision/conclusion would you make on a test of the null hypothesis at $\alpha = .05$?
c. Evaluate the strength of association between the two variables.

10. The GSS 2000 (*gss2000a* or *gss2000b*) asked respondents, "If some people in your community suggested that a book against churches and religion should be taken out of your public library, would you favor removing this book, or not?" (*libath*). GSS 2000 also asked: "Generally speaking, would you say that most people can be trusted or that you can't be too careful in dealing with people?" (*trust*). Use SPSS to crosstabulate *libath* by *trust* and to obtain the appropriate statistics to evaluate the association.

a. Interpret the association between *libath* and *trust*.
b. What decision/conclusion would you make on a test of the null hypothesis at $\alpha = .05$?
c. Evaluate the strength of association between the two variables.

11. GSS 2000 asked Jewish people if they consider themselves Orthodox, Conservative, Reform, or none of these (*jew*). Using GSS 2000 (*gss2000a*), use SPSS to crosstabulate *bible* (excluding 4 = other) with *jew* (make *jew* the column variable) and to obtain the appropriate statistics to evaluate the association.

a. Interpret the association between *jew* and *bible* in the table.
b. Evaluate the strength of association in the table.
c. Is it valid to use the chi-square statistic to test the null hypothesis?

Appendix A

Statistical Tables

Table A.1 Areas (A) of the Standard Normal Distribution from $Z \rightarrow \infty$ and $-Z \rightarrow -\infty$

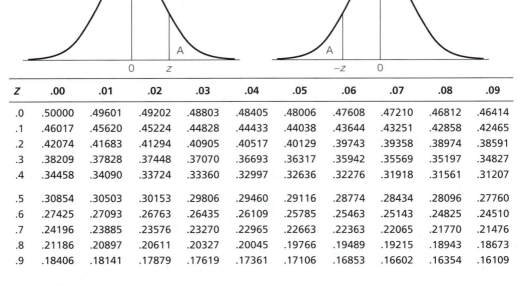

Z	.00	.01	.02	.03	.04	.05	.06	.07	.08	.09
.0	.50000	.49601	.49202	.48803	.48405	.48006	.47608	.47210	.46812	.46414
.1	.46017	.45620	.45224	.44828	.44433	.44038	.43644	.43251	.42858	.42465
.2	.42074	.41683	.41294	.40905	.40517	.40129	.39743	.39358	.38974	.38591
.3	.38209	.37828	.37448	.37070	.36693	.36317	.35942	.35569	.35197	.34827
.4	.34458	.34090	.33724	.33360	.32997	.32636	.32276	.31918	.31561	.31207
.5	.30854	.30503	.30153	.29806	.29460	.29116	.28774	.28434	.28096	.27760
.6	.27425	.27093	.26763	.26435	.26109	.25785	.25463	.25143	.24825	.24510
.7	.24196	.23885	.23576	.23270	.22965	.22663	.22363	.22065	.21770	.21476
.8	.21186	.20897	.20611	.20327	.20045	.19766	.19489	.19215	.18943	.18673
.9	.18406	.18141	.17879	.17619	.17361	.17106	.16853	.16602	.16354	.16109

Table A.1 *(continued)*

Z	.00	.01	.02	.03	.04	.05	.06	.07	.08	.09
1.0	.15866	.15625	.15386	.15151	.14917	.14686	.14457	.14231	.14007	.13786
1.1	.13567	.13350	.13136	.12924	.12714	.12507	.12302	.12100	.11900	.11702
1.2	.11507	.11314	.11123	.10935	.10749	.10565	.10383	.10204	.10027	.09853
1.3	.09680	.09510	.09342	.09176	.09012	.08851	.08691	.08534	.08379	.08226
1.4	.08076	.07927	.07780	.07636	.07493	.07353	.07215	.07078	.06944	.06811
1.5	.06681	.06552	.06426	.06301	.06178	.06057	.05938	.05821	.05705	.05592
1.6	.05480	.05370	.05262	.05155	.05050	.04947	.04846	.04746	.04648	.04551
1.7	.04457	.04363	.04272	.04182	.04093	.04006	.03920	.03836	.03754	.03673
1.8	.03593	.03515	.03438	.03362	.03288	.03216	.03144	.03074	.03005	.02938
1.9	.02872	.02807	.02743	.02680	.02619	.02559	.02500	.02442	.02385	.02330
2.0	.02275	.02222	.02169	.02118	.02068	.02018	.01970	.01923	.01876	.01831
2.1	.01786	.01743	.01700	.01659	.01618	.01578	.01539	.01500	.01463	.01426
2.2	.01390	.01355	.01321	.01287	.01255	.01222	.01191	.01160	.01130	.01101
2.3	.01072	.01044	.01017	.00990	.00964	.00939	.00914	.00889	.00866	.00842
2.4	.00820	.00798	.00776	.00755	.00734	.00714	.00695	.00676	.00657	.00639
2.5	.00621	.00604	.00587	.00570	.00554	.00539	.00523	.00508	.00494	.00480
2.6	.00466	.00453	.00440	.00427	.00415	.00402	.00391	.00379	.00368	.00357
2.7	.00347	.00336	.00326	.00317	.00307	.00298	.00289	.00280	.00272	.00264
2.8	.00256	.00248	.00240	.00233	.00226	.00219	.00212	.00205	.00199	.00193
2.9	.00187	.00181	.00175	.00169	.00164	.00159	.00154	.00149	.00144	.00139
3.0	.00135	.00131	.00126	.00122	.00118	.00114	.00111	.00107	.00104	.00100
3.1	.00097	.00094	.00090	.00087	.00084	.00082	.00079	.00076	.00074	.00071
3.2	.00069	.00066	.00064	.00062	.00060	.00058	.00056	.00054	.00052	.00050
3.3	.00048	.00047	.00045	.00043	.00042	.00040	.00039	.00038	.00036	.00035
3.4	.00034	.00032	.00031	.00030	.00029	.00028	.00027	.00026	.00025	.00024
3.5	.00023	.00022	.00022	.00021	.00020	.00019	.00019	.00018	.00017	.00017
3.6	.00016	.00015	.00015	.00014	.00014	.00013	.00013	.00012	.00012	.00011
3.7	.00011	.00010	.00010	.00010	.00009	.00009	.00008	.00008	.00008	.00008
3.8	.00007	.00007	.00007	.00006	.00006	.00006	.00006	.00005	.00005	.00005
3.9	.00005	.00005	.00004	.00004	.00004	.00004	.00004	.00004	.00003	.00003

Entries were computed with SPSS's cumulative normal distribution function CDF.NORMAL(q,mean,stddev).

Table A.2 Values of $t_{A,df}$ and $-t_{A,df}$ for Selected Areas (A) of the t Distributions

df	.125	.100	.075	.050	.025	.010	.005	.001	.0005	.00005
1	2.414	3.078	4.165	6.314	12.706	31.821	63.657	318.309	636.619	6366.215
2	1.604	1.886	2.282	2.920	4.303	6.965	9.925	22.327	31.599	99.992
3	1.423	1.638	1.924	2.353	3.182	4.541	5.841	10.215	12.924	28.000
4	1.344	1.533	1.778	2.132	2.776	3.747	4.604	7.173	8.610	15.544
5	1.301	1.476	1.699	2.015	2.571	3.365	4.032	5.893	6.869	11.178
6	1.273	1.440	1.650	1.943	2.447	3.143	3.707	5.208	5.959	9.082
7	1.254	1.415	1.617	1.895	2.365	2.998	3.499	4.785	5.408	7.885
8	1.240	1.397	1.592	1.860	2.306	2.896	3.355	4.501	5.041	7.120
9	1.230	1.383	1.574	1.833	2.262	2.821	3.250	4.297	4.781	6.594
10	1.221	1.372	1.559	1.812	2.228	2.764	3.169	4.144	4.587	6.211
11	1.214	1.363	1.548	1.796	2.201	2.718	3.106	4.025	4.437	5.921
12	1.209	1.356	1.538	1.782	2.179	2.681	3.055	3.930	4.318	5.694
13	1.204	1.350	1.530	1.771	2.160	2.650	3.012	3.852	4.221	5.513
14	1.200	1.345	1.523	1.761	2.145	2.624	2.977	3.787	4.140	5.363
15	1.197	1.341	1.517	1.753	2.131	2.602	2.947	3.733	4.073	5.239
16	1.194	1.337	1.512	1.746	2.120	2.583	2.921	3.686	4.015	5.134
17	1.191	1.333	1.508	1.740	2.110	2.567	2.898	3.646	3.965	5.044
18	1.189	1.330	1.504	1.734	2.101	2.552	2.878	3.610	3.922	4.966
19	1.187	1.328	1.500	1.729	2.093	2.539	2.861	3.579	3.883	4.897
20	1.185	1.325	1.497	1.725	2.086	2.528	2.845	3.552	3.850	4.837
25	1.178	1.316	1.485	1.708	2.060	2.485	2.787	3.450	3.725	4.619
30	1.173	1.310	1.477	1.697	2.042	2.457	2.750	3.385	3.646	4.482
35	1.170	1.306	1.472	1.690	2.030	2.438	2.724	3.340	3.591	4.389
40	1.167	1.303	1.468	1.684	2.021	2.423	2.704	3.307	3.551	4.321
45	1.165	1.301	1.465	1.679	2.014	2.412	2.690	3.281	3.520	4.269
50	1.164	1.299	1.462	1.676	2.009	2.403	2.678	3.261	3.496	4.228
60	1.162	1.296	1.458	1.671	2.000	2.390	2.660	3.232	3.460	4.169
70	1.160	1.294	1.456	1.667	1.994	2.381	2.648	3.211	3.435	4.127
80	1.159	1.292	1.453	1.664	1.990	2.374	2.639	3.195	3.416	4.096
90	1.158	1.291	1.452	1.662	1.987	2.368	2.632	3.183	3.402	4.072
100	1.157	1.290	1.451	1.660	1.984	2.364	2.626	3.174	3.390	4.053
150	1.155	1.287	1.447	1.655	1.976	2.351	2.609	3.145	3.357	3.998

Table A.2 *(continued)*

df	.125	.100	.075	.050	.025	.010	.005	.001	.0005	.00005
200	1.154	1.286	1.445	1.653	1.972	2.345	2.601	3.131	3.340	3.970
300	1.153	1.284	1.443	1.650	1.968	2.339	2.592	3.118	3.323	3.944
400	1.152	1.284	1.442	1.649	1.966	2.336	2.588	3.111	3.315	3.930
500	1.152	1.283	1.442	1.648	1.965	2.334	2.586	3.107	3.310	3.922
750	1.151	1.283	1.441	1.647	1.963	2.331	2.582	3.101	3.304	3.912
1000	1.151	1.282	1.441	1.646	1.962	2.330	2.581	3.098	3.300	3.906
1500	1.151	1.282	1.440	1.646	1.962	2.329	2.579	3.096	3.297	3.901
∞	1.150	1.282	1.440	1.645	1.960	2.326	2.576	3.090	3.291	3.891

Entries were computed with SPSS's inverse t distribution function IDF.T(p,df).

Table A.3a Values of F_A from the F Distributions for $A = .05$

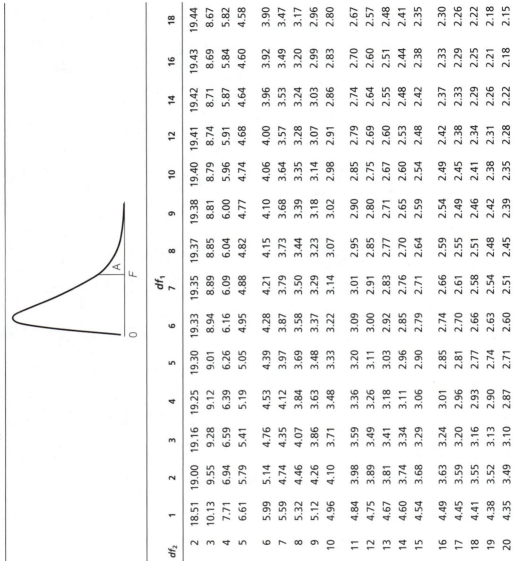

df_2	1	2	3	4	5	6	7	8	9	10	12	14	16	18
2	18.51	19.00	19.16	19.25	19.30	19.33	19.35	19.37	19.38	19.40	19.41	19.42	19.43	19.44
3	10.13	9.55	9.28	9.12	9.01	8.94	8.89	8.85	8.81	8.79	8.74	8.71	8.69	8.67
4	7.71	6.94	6.59	6.39	6.26	6.16	6.09	6.04	6.00	5.96	5.91	5.87	5.84	5.82
5	6.61	5.79	5.41	5.19	5.05	4.95	4.88	4.82	4.77	4.74	4.68	4.64	4.60	4.58
6	5.99	5.14	4.76	4.53	4.39	4.28	4.21	4.15	4.10	4.06	4.00	3.96	3.92	3.90
7	5.59	4.74	4.35	4.12	3.97	3.87	3.79	3.73	3.68	3.64	3.57	3.53	3.49	3.47
8	5.32	4.46	4.07	3.84	3.69	3.58	3.50	3.44	3.39	3.35	3.28	3.24	3.20	3.17
9	5.12	4.26	3.86	3.63	3.48	3.37	3.29	3.23	3.18	3.14	3.07	3.03	2.99	2.96
10	4.96	4.10	3.71	3.48	3.33	3.22	3.14	3.07	3.02	2.98	2.91	2.86	2.83	2.80
11	4.84	3.98	3.59	3.36	3.20	3.09	3.01	2.95	2.90	2.85	2.79	2.74	2.70	2.67
12	4.75	3.89	3.49	3.26	3.11	3.00	2.91	2.85	2.80	2.75	2.69	2.64	2.60	2.57
13	4.67	3.81	3.41	3.18	3.03	2.92	2.83	2.77	2.71	2.67	2.60	2.55	2.51	2.48
14	4.60	3.74	3.34	3.11	2.96	2.85	2.76	2.70	2.65	2.60	2.53	2.48	2.44	2.41
15	4.54	3.68	3.29	3.06	2.90	2.79	2.71	2.64	2.59	2.54	2.48	2.42	2.38	2.35
16	4.49	3.63	3.24	3.01	2.85	2.74	2.66	2.59	2.54	2.49	2.42	2.37	2.33	2.30
17	4.45	3.59	3.20	2.96	2.81	2.70	2.61	2.55	2.49	2.45	2.38	2.33	2.29	2.26
18	4.41	3.55	3.16	2.93	2.77	2.66	2.58	2.51	2.46	2.41	2.34	2.29	2.25	2.22
19	4.38	3.52	3.13	2.90	2.74	2.63	2.54	2.48	2.42	2.38	2.31	2.26	2.21	2.18
20	4.35	3.49	3.10	2.87	2.71	2.60	2.51	2.45	2.39	2.35	2.28	2.22	2.18	2.15

df2														
25	2.04	2.07	2.11	2.16	2.24	2.28	2.34	2.40	2.49	2.60	2.76	2.99	3.39	4.24
30	1.96	1.99	2.04	2.09	2.16	2.21	2.27	2.33	2.42	2.53	2.69	2.92	3.32	4.17
35	1.91	1.94	1.99	2.04	2.11	2.16	2.22	2.29	2.37	2.49	2.64	2.87	3.27	4.12
40	1.87	1.90	1.95	2.00	2.08	2.12	2.18	2.25	2.34	2.45	2.61	2.84	3.23	4.08
45	1.84	1.87	1.92	1.97	2.05	2.10	2.15	2.22	2.31	2.42	2.58	2.81	3.20	4.06
50	1.81	1.85	1.89	1.95	2.03	2.07	2.13	2.20	2.29	2.40	2.56	2.79	3.18	4.03
60	1.78	1.82	1.86	1.92	1.99	2.04	2.10	2.17	2.25	2.37	2.53	2.76	3.15	4.00
70	1.75	1.79	1.84	1.89	1.97	2.02	2.07	2.14	2.23	2.35	2.50	2.74	3.13	3.98
80	1.73	1.77	1.82	1.88	1.95	2.00	2.06	2.13	2.21	2.33	2.49	2.72	3.11	3.96
90	1.72	1.76	1.80	1.86	1.94	1.99	2.04	2.11	2.20	2.32	2.47	2.71	3.10	3.95
100	1.71	1.75	1.79	1.85	1.93	1.97	2.03	2.10	2.19	2.31	2.46	2.70	3.09	3.94
150	1.67	1.71	1.76	1.82	1.89	1.94	2.00	2.07	2.16	2.27	2.43	2.66	3.06	3.90
200	1.66	1.69	1.74	1.80	1.88	1.93	1.98	2.06	2.14	2.26	2.42	2.65	3.04	3.89
300	1.64	1.68	1.72	1.78	1.86	1.91	1.97	2.04	2.13	2.24	2.40	2.63	3.03	3.87
400	1.63	1.67	1.72	1.78	1.85	1.90	1.96	2.03	2.12	2.24	2.39	2.63	3.02	3.86
500	1.62	1.66	1.71	1.77	1.85	1.90	1.96	2.03	2.12	2.23	2.39	2.62	3.01	3.86
750	1.62	1.66	1.70	1.77	1.84	1.89	1.95	2.02	2.11	2.23	2.38	2.62	3.01	3.85
1000	1.61	1.65	1.70	1.76	1.84	1.89	1.95	2.02	2.11	2.22	2.38	2.61	3.00	3.85
1500	1.61	1.65	1.70	1.76	1.84	1.89	1.94	2.02	2.10	2.22	2.38	2.61	3.00	3.85
∞	1.60	1.64	1.69	1.75	1.83	1.88	1.94	2.01	2.10	2.21	2.37	2.60.	3.00	3.84

Entries were computed with SPSS's inverse F distribution function IDF.F(p,df1,df2).

Table A.3b Values of F_A from the F Distributions for A = .01

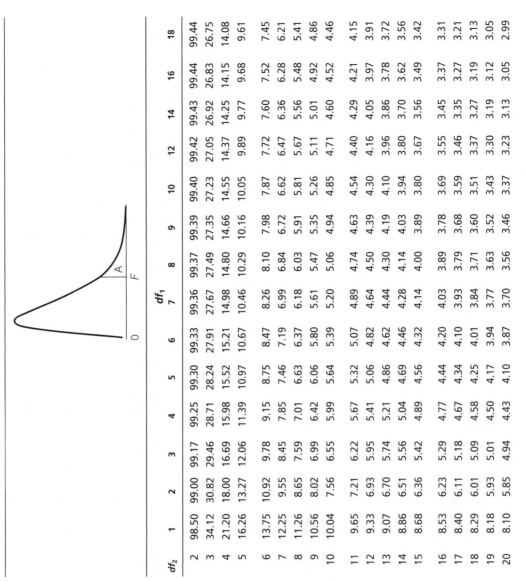

df_2	1	2	3	4	5	6	7	8	9	10	12	14	16	18
2	98.50	99.00	99.17	99.25	99.30	99.33	99.36	99.37	99.39	99.40	99.42	99.43	99.44	99.44
3	34.12	30.82	29.46	28.71	28.24	27.91	27.67	27.49	27.35	27.23	27.05	26.92	26.83	26.75
4	21.20	18.00	16.69	15.98	15.52	15.21	14.98	14.80	14.66	14.55	14.37	14.25	14.15	14.08
5	16.26	13.27	12.06	11.39	10.97	10.67	10.46	10.29	10.16	10.05	9.89	9.77	9.68	9.61
6	13.75	10.92	9.78	9.15	8.75	8.47	8.26	8.10	7.98	7.87	7.72	7.60	7.52	7.45
7	12.25	9.55	8.45	7.85	7.46	7.19	6.99	6.84	6.72	6.62	6.47	6.36	6.28	6.21
8	11.26	8.65	7.59	7.01	6.63	6.37	6.18	6.03	5.91	5.81	5.67	5.56	5.48	5.41
9	10.56	8.02	6.99	6.42	6.06	5.80	5.61	5.47	5.35	5.26	5.11	5.01	4.92	4.86
10	10.04	7.56	6.55	5.99	5.64	5.39	5.20	5.06	4.94	4.85	4.71	4.60	4.52	4.46
11	9.65	7.21	6.22	5.67	5.32	5.07	4.89	4.74	4.63	4.54	4.40	4.29	4.21	4.15
12	9.33	6.93	5.95	5.41	5.06	4.82	4.64	4.50	4.39	4.30	4.16	4.05	3.97	3.91
13	9.07	6.70	5.74	5.21	4.86	4.62	4.44	4.30	4.19	4.10	3.96	3.86	3.78	3.72
14	8.86	6.51	5.56	5.04	4.69	4.46	4.28	4.14	4.03	3.94	3.80	3.70	3.62	3.56
15	8.68	6.36	5.42	4.89	4.56	4.32	4.14	4.00	3.89	3.80	3.67	3.56	3.49	3.42
16	8.53	6.23	5.29	4.77	4.44	4.20	4.03	3.89	3.78	3.69	3.55	3.45	3.37	3.31
17	8.40	6.11	5.18	4.67	4.34	4.10	3.93	3.79	3.68	3.59	3.46	3.35	3.27	3.21
18	8.29	6.01	5.09	4.58	4.25	4.01	3.84	3.71	3.60	3.51	3.37	3.27	3.19	3.13
19	8.18	5.93	5.01	4.50	4.17	3.94	3.77	3.63	3.52	3.43	3.30	3.19	3.12	3.05
20	8.10	5.85	4.94	4.43	4.10	3.87	3.70	3.56	3.46	3.37	3.23	3.13	3.05	2.99

df_1

25	7.77	5.57	4.68	4.18	3.85	3.63	3.46	3.32	3.22	3.13	2.99	2.89	2.81	2.75
30	7.56	5.39	4.51	4.02	3.70	3.47	3.30	3.17	3.07	2.98	2.84	2.74	2.66	2.60
35	7.42	5.27	4.40	3.91	3.59	3.37	3.20	3.07	2.96	2.88	2.74	2.64	2.56	2.50
40	7.31	5.18	4.31	3.83	3.51	3.29	3.12	2.99	2.89	2.80	2.66	2.56	2.48	2.42
45	7.23	5.11	4.25	3.77	3.45	3.23	3.07	2.94	2.83	2.74	2.61	2.51	2.43	2.36
50	7.17	5.06	4.20	3.72	3.41	3.19	3.02	2.89	2.78	2.70	2.56	2.46	2.38	2.32
60	7.08	4.98	4.13	3.65	3.34	3.12	2.95	2.82	2.72	2.63	2.50	2.39	2.31	2.25
70	7.01	4.92	4.07	3.60	3.29	3.07	2.91	2.78	2.67	2.59	2.45	2.35	2.27	2.20
80	6.96	4.88	4.04	3.56	3.26	3.04	2.87	2.74	2.64	2.55	2.42	2.31	2.23	2.17
90	6.93	4.85	4.01	3.53	3.23	3.01	2.84	2.72	2.61	2.52	2.39	2.29	2.21	2.14
100	6.90	4.82	3.98	3.51	3.21	2.99	2.82	2.69	2.59	2.50	2.37	2.27	2.19	2.12
150	6.81	4.75	3.91	3.45	3.14	2.92	2.76	2.63	2.53	2.44	2.31	2.20	2.12	2.06
200	6.76	4.71	3.88	3.41	3.11	2.89	2.73	2.60	2.50	2.41	2.27	2.17	2.09	2.03
300	6.72	4.68	3.85	3.38	3.08	2.86	2.70	2.57	2.47	2.38	2.24	2.14	2.06	1.99
400	6.70	4.66	3.83	3.37	3.06	2.85	2.68	2.56	2.45	2.37	2.23	2.13	2.05	1.98
500	6.69	4.65	3.82	3.36	3.05	2.84	2.68	2.55	2.44	2.36	2.22	2.12	2.04	1.97
750	6.67	4.63	3.81	3.34	3.04	2.83	2.66	2.53	2.43	2.34	2.21	2.11	2.02	1.96
1000	6.66	4.63	3.80	3.34	3.04	2.82	2.66	2.53	2.43	2.34	2.20	2.10	2.02	1.95
1500	6.65	4.62	3.79	3.33	3.03	2.81	2.65	2.52	2.42	2.33	2.20	2.09	2.01	1.95
∞	6.63	4.61	3.78	3.32	3.02	2.80	2.64	2.51	2.41	2.32	2.18	2.08	2.00	1.93

Entries were computed with SPSS's inverse F distribution function IDF.F(p,df1,df2).

Table A.3c Values of F_A from the F Distributions for $A = .001$

df_2	1	2	3	4	5	6	7	8	9	10	12	14	16	18
2	998.50	999.00	999.17	999.25	999.30	999.33	999.36	999.37	999.39	999.40	999.42	999.43	999.44	999.44
3	167.03	148.50	141.11	137.10	134.58	132.85	131.58	130.62	129.86	129.25	128.32	127.64	127.14	126.74
4	74.14	61.25	56.18	53.44	51.71	50.53	49.66	49.00	48.47	48.05	47.41	46.95	46.60	46.32
5	47.18	37.12	33.20	31.09	29.75	28.83	28.16	27.65	27.24	26.92	26.42	26.06	25.78	25.57
6	35.51	27.00	23.70	21.92	20.80	20.03	19.46	19.03	18.69	18.41	17.99	17.68	17.45	17.27
7	29.25	21.69	18.77	17.20	16.21	15.52	15.02	14.63	14.33	14.08	13.71	13.43	13.23	13.06
8	25.41	18.49	15.83	14.39	13.48	12.86	12.40	12.05	11.77	11.54	11.19	10.94	10.75	10.60
9	22.86	16.39	13.90	12.56	11.71	11.13	10.70	10.37	10.11	9.89	9.57	9.33	9.15	9.01
10	21.04	14.91	12.55	11.28	10.48	9.93	9.52	9.20	8.96	8.75	8.45	8.22	8.05	7.91
11	19.69	13.81	11.56	10.35	9.58	9.05	8.66	8.35	8.12	7.92	7.63	7.41	7.24	7.11
12	18.64	12.97	10.80	9.63	8.89	8.38	8.00	7.71	7.48	7.29	7.00	6.79	6.63	6.51
13	17.82	12.31	10.21	9.07	8.35	7.86	7.49	7.21	6.98	6.80	6.52	6.31	6.16	6.03
14	17.14	11.78	9.73	8.62	7.92	7.44	7.08	6.80	6.58	6.40	6.13	5.93	5.78	5.66
15	16.59	11.34	9.34	8.25	7.57	7.09	6.74	6.47	6.26	6.08	5.81	5.62	5.46	5.35
16	16.12	10.97	9.01	7.94	7.27	6.80	6.46	6.19	5.98	5.81	5.55	5.35	5.20	5.09
17	15.72	10.66	8.73	7.68	7.02	6.56	6.22	5.96	5.75	5.58	5.32	5.13	4.99	4.87
18	15.38	10.39	8.49	7.46	6.81	6.35	6.02	5.76	5.56	5.39	5.13	4.94	4.80	4.68
19	15.08	10.16	8.28	7.27	6.62	6.18	5.85	5.59	5.39	5.22	4.97	4.78	4.64	4.52
20	14.82	9.95	8.10	7.10	6.46	6.02	5.69	5.44	5.24	5.08	4.82	4.64	4.49	4.38

25	13.88	9.22	7.45	6.49	5.89	5.46	5.15	4.91	4.71	4.56	4.31	4.13	3.99	3.88
30	13.29	8.77	7.05	6.12	5.53	5.12	4.82	4.58	4.39	4.24	4.00	3.82	3.69	3.58
35	12.90	8.47	6.79	5.88	5.30	4.89	4.59	4.36	4.18	4.03	3.79	3.62	3.48	3.38
40	12.61	8.25	6.59	5.70	5.13	4.73	4.44	4.21	4.02	3.87	3.64	3.47	3.34	3.23
45	12.39	8.09	6.45	5.56	5.00	4.61	4.32	4.09	3.91	3.76	3.53	3.36	3.23	3.12
50	12.22	7.96	6.34	5.46	4.90	4.51	4.22	4.00	3.82	3.67	3.44	3.27	3.14	3.04
60	11.97	7.77	6.17	5.31	4.76	4.37	4.09	3.86	3.69	3.54	3.32	3.15	3.02	2.91
70	11.80	7.64	6.06	5.20	4.66	4.28	3.99	3.77	3.60	3.45	3.23	3.06	2.93	2.83
80	11.67	7.54	5.97	5.12	4.58	4.20	3.92	3.70	3.53	3.39	3.16	3.00	2.87	2.76
90	11.57	7.47	5.91	5.06	4.53	4.15	3.87	3.65	3.48	3.34	3.11	2.95	2.82	2.71
100	11.50	7.41	5.86	5.02	4.48	4.11	3.83	3.61	3.44	3.30	3.07	2.91	2.78	2.68
150	11.27	7.24	5.71	4.88	4.35	3.98	3.71	3.49	3.32	3.18	2.96	2.80	2.67	2.56
200	11.15	7.15	5.63	4.81	4.29	3.92	3.65	3.43	3.26	3.12	2.90	2.74	2.61	2.51
300	11.04	7.07	5.56	4.75	4.22	3.86	3.59	3.38	3.21	3.07	2.85	2.69	2.56	2.46
400	10.99	7.03	5.53	4.71	4.19	3.83	3.56	3.35	3.18	3.04	2.82	2.66	2.53	2.43
500	10.96	7.00	5.51	4.69	4.18	3.81	3.54	3.33	3.16	3.02	2.81	2.64	2.52	2.41
750	10.91	6.97	5.48	4.67	4.15	3.79	3.52	3.31	3.14	3.00	2.78	2.62	2.49	2.39
1000	10.89	6.96	5.46	4.65	4.14	3.78	3.51	3.30	3.13	2.99	2.77	2.61	2.48	2.38
1500	10.87	6.94	5.45	4.64	4.13	3.77	3.50	3.29	3.12	2.98	2.76	2.60	2.47	2.37
∞	10.83	6.91	5.42	4.62	4.10	3.74	3.47	3.27	3.10	2.96	2.74	2.58	2.45	2.35

Entries were computed with SPSS's inverse F distribution function IDF.F(p,df1,df2).

Table A.4 Values of $\chi^2_{A,df}$ for Selected Areas (A) of the Chi-Square Distribution

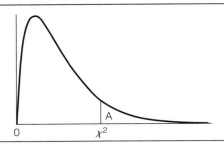

df	.25	.20	.15	.10	A .05	.01	.005	.001	.0005	.0001
1	1.323	1.642	2.072	2.706	3.841	6.635	7.879	10.828	12.116	15.137
2	2.773	3.219	3.794	4.605	5.991	9.210	10.597	13.816	15.202	18.421
3	4.108	4.642	5.317	6.251	7.815	11.345	12.838	16.266	17.730	21.108
4	5.385	5.989	6.745	7.779	9.488	13.277	14.860	18.467	19.997	23.513
5	6.626	7.289	8.115	9.236	11.070	15.086	16.750	20.515	22.105	25.745
6	7.841	8.558	9.446	10.645	12.592	16.812	18.548	22.458	24.103	27.856
7	9.037	9.803	10.748	12.017	14.067	18.475	20.278	24.322	26.018	29.878
8	10.219	11.030	12.027	13.362	15.507	20.090	21.955	26.124	27.868	31.828
9	11.389	12.242	13.288	14.684	16.919	21.666	23.589	27.877	29.666	33.720
10	12.549	13.442	14.534	15.987	18.307	23.209	25.188	29.588	31.420	35.564
11	13.701	14.631	15.767	17.275	19.675	24.725	26.757	31.264	33.137	37.367
12	14.845	15.812	16.989	18.549	21.026	26.217	28.300	32.909	34.821	39.134
13	15.984	16.985	18.202	19.812	22.362	27.688	29.819	34.528	36.478	40.871
14	17.117	18.151	19.406	21.064	23.685	29.141	31.319	36.123	38.109	42.579
15	18.245	19.311	20.603	22.307	24.996	30.578	32.801	37.697	39.719	44.263
16	19.369	20.465	21.793	23.542	26.296	32.000	34.267	39.252	41.308	45.925
17	20.489	21.615	22.977	24.769	27.587	33.409	35.718	40.790	42.879	47.566
18	21.605	22.760	24.155	25.989	28.869	34.805	37.156	42.312	44.434	49.189
19	22.718	23.900	25.329	27.204	30.144	36.191	38.582	43.820	45.973	50.795
20	23.828	25.038	26.498	28.412	31.410	37.566	39.997	45.315	47.498	52.386
21	24.935	26.171	27.662	29.615	32.671	38.932	41.401	46.797	49.011	53.962
22	26.039	27.301	28.822	30.813	33.924	40.289	42.796	48.268	50.511	55.525
23	27.141	28.429	29.979	32.007	35.172	41.638	44.181	49.728	52.000	57.075
24	28.241	29.553	31.132	33.196	36.415	42.980	45.559	51.179	53.479	58.613
25	29.339	30.675	32.282	34.382	37.652	44.314	46.928	52.620	54.947	60.140
26	30.435	31.795	33.429	35.563	38.885	45.642	48.290	54.052	56.407	61.657
27	31.528	32.912	34.574	36.741	40.113	46.963	49.645	55.476	57.858	63.164
28	32.620	34.027	35.715	37.916	41.337	48.278	50.993	56.892	59.300	64.662
29	33.711	35.139	36.854	39.087	42.557	49.588	52.336	58.301	60.735	66.152
30	34.800	36.250	37.990	40.256	43.773	50.892	53.672	59.703	62.162	67.633

Table A.4 *(continued)*

df	.25	.20	.15	.10	A .05	.01	.005	.001	.0005	.0001
31	35.887	37.359	39.124	41.422	44.985	52.191	55.003	61.098	63.582	69.106
32	36.973	38.466	40.256	42.585	46.194	53.486	56.328	62.487	64.995	70.571
33	38.058	39.572	41.386	43.745	47.400	54.776	57.648	63.870	66.403	72.030
34	39.141	40.676	42.514	44.903	48.602	56.061	58.964	65.247	67.803	73.481
35	40.223	41.778	43.640	46.059	49.802	57.342	60.275	66.619	69.199	74.926
36	41.304	42.879	44.764	47.212	50.998	58.619	61.581	67.985	70.588	76.365
37	42.383	43.978	45.886	48.363	52.192	59.893	62.883	69.346	71.972	77.798
38	43.462	45.076	47.007	49.513	53.384	61.162	64.181	70.703	73.351	79.225
39	44.539	46.173	48.126	50.660	54.572	62.428	65.476	72.055	74.725	80.646
40	45.616	47.269	49.244	51.805	55.758	63.691	66.766	73.402	76.095	82.062

Entries were computed with SPSS's inverse chi-square distribution function IDF.CHISQ(p,df).

Appendix B

Answers to Exercises

Chapter 1: Statistics and Social Science

1. a. variable   b. unit of analysis   c. $3,549   **3.** a. ordinal   b. nominal
c. interval   **5.** reliable   **7.** no   **9.** descriptive statistic   **11.** no

Chapter 2: Distributions of Variable Attributes

1. a.

Region	f	Pct
Northeast	6	20.0
Midwest	7	23.3
South	11	36.7
West	6	20.0
Total	30	100.0

b. 11 out of 30 persons are from the South.
c. 20 out of every 100 persons are from the West.
d. A cumulative distribution is not valid because region is a nominal variable.
3. Ratio = 1248/389 = 3.2. For every remarried person, there are 3.2 married persons who have never been divorced or separated.

5. a.

Response	cPct
Always wrong	28.0
Almost always wrong	36.8
Sometimes wrong	58.2
Not wrong at all	100.0

b. 58.2 percent believe premarital sex is at least sometimes wrong.
c. Ratio = 749/502 = 1.5. There are 1.5 persons who believe premarital sex is not wrong at all for each person who believes it is always wrong.

7.

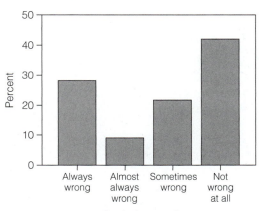

9.

GPA	f	Pct
1.745–1.995	2	5.6
1.995–2.245	2	5.6
2.245–2.495	4	11.1
2.495–2.745	1	2.8
2.745–2.995	12	33.3
2.995–3.245	6	16.7
3.245–3.495	2	5.6
3.495–3.745	4	11.1
3.745–3.995	3	8.3
Total	36	100.0

11. a.

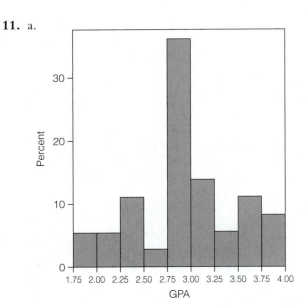

b.

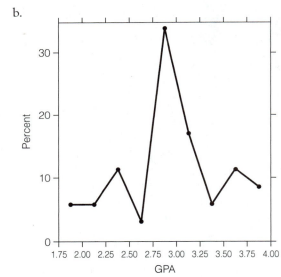

Chapter 3: Central Tendency

1. a. 9   b. 9.5   c. 10   **3.** right-skewed   **5.** a. 12.5   b. 13.8   c. 16.0
7. a. 10.5   b. 62.7   **9.** a. 15 years   b. −.71 year   c. If the case with a survival time of 15 years were removed from the distribution, the mean would decline by
.71 year.   **11.** a. The mode would stay the same.   b. The median would stay the
same.   c. The mean would increase.

Chapter 4: Dispersion

1. a. 214   b. 42.8   c. 6.542   d. 1.308   **3.** The standard amount that the number of female sex partners deviates from the mean for females is 6.54 partners. **5.** a. $-.611$, $-.611$, $-.459$, $-.306$, 1.987   b. .00   c. 1.00   d. Yes, because the mean and standard deviation of z scores are always 0 and 1, respectively.   **7.** The woman who had 18 sex partners is more standard deviations from the mean because the absolute value of her z score is larger.   **9.** a. 1.453   b. right-skewed c. .185   d. more peaked than normal   **11.** a. .865   b. The dispersion is .865 of the maximum possible dispersion.

Chapter 5: Sampling Error

1. The population consists of households in Boise, Idaho, with listed telephone numbers.   **3.** $30, 27, 21, 18, 22, 37, 01, 13$ **5.** a. 21.5   b. 22.6   c. 1.1   **7.** a. 3.3 b. 47.6   **9.** a. 22.7%   b. 77.3%   c. 11.6%   d. 4.55%   **11.** a. .995; it can be ignored.   b. 1.74, 1.23, .87, .71   c. It should be reduced by 50%. Yes d. .71 without FPC, .67 with FPC.

Chapter 6: Inferences about the Mean

1. a. .173
   b. 1.962
   c. 3.30–3.68

3. 1. $H_0: \mu = 12.3; H_1: \mu \neq 12.3$
    2. $\alpha = .05$
    3. $df = 1309$
    4. $t_{.025} = 1.962$

5. $\overline{X} = 31.5$
6. $s_{\overline{X}} = .482$
7. $t = 39.83$
8. reject H_0, conclude $\mu > 12.3, p < .0005$

5. 1. $H_0: \mu \geq 50.0; H_1: \mu < 50.0$
    2. $\alpha = .05$
    3. $df = 1309$
    4. $t_{.05} = 1.646$

5. $\overline{X} = 31.5$
6. $s_{\overline{X}} = .482$
7. $t = -38.38$
8. reject H_0, conclude $\mu < 50.0, p < .0005$

7. 1. $H_0: \mu_D = .0; H_1: \mu_D \neq .0$
    2. $\alpha = .05$
    3. $df = 1804$
    4. $t_{.025} = 1.961$

5. $\overline{D} = -.139$
6. $s_{\overline{X}} = .076$
7. $t = -1.83$
8. do not reject $H_0, p = .072$

9. 1. $H_0: \pi = .5; H_1: \pi \neq .5$
    2. $\alpha = .10$
    3. $Z_{.05} = 1.646$
    4. $p = .508$

5. $\sigma_p = .014$
6. $t = .57$
7. do not reject $H_0, p = .569$

Chapter 7: Associations

1. *emailjit:*   2.8   2.8 $-.4$ 5.3 $-.4$ 4.6
   wwwjit: 2.1   5.4   2.1   3.5   2.3   3.8

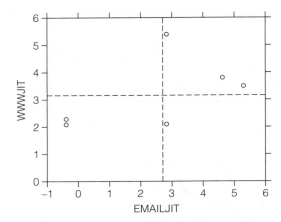

3. a. .9
   b. −.5
   c. .2
5. a. 62
   b. 12.4
   c. There is a positive association, meaning that wives with above average gender-role attitudes tend to have husbands who are above average, and wives who are below average tend to have husbands who are below average.

Chapter 8: Bivariate Regression Analysis

1. $b = 1.0, a = -11.7$
3. $b = .8, a = -10$
5. a. $b = .725, a = -7.60$. There is an increase of .725 homicide per 100,000 for each percentage increase in children living in single-parent households. The predicted homicides per 100,000 is −7.60 when the percentage of children in single-parent households is 0.
   b. $CD = .748$. The percentage of children in single-parent households explains 74.8 percent of the variance in homicides per 100,000.
   c. $B = .865$. Homicides per 100,000 increase by .865 of a standard deviation per standard deviation increase in the percentage of children living in single-parent households.
   d. $\hat{s}_{Y-\hat{Y}} = 2.605$. The estimated standard deviation of the residuals in the population is 2.605 homicides per 100,000.
7. $R = .830$, R-square $= .688$, standard error of the estimate $= 2.245$, constant $= -10.207, B = .792, \beta = .830$.
9. a. For each year of increase in father's education, son's education increases by .33 year.
   b. 2.14, −.56, and −3.26 years
   c. $b = .632, a = 2.34$

Chapter 9: Inferences about the Regression Slope

1. Standard error $= .0012$, $t = -8.701$, Sig $< .00005$

3. $.281 - .371$

5. 1. $H_0: \beta = 0; H_1: \beta \neq 0$ 5. $b = .0478$

     2. $\alpha = .05$ 6. $s_b = .020$

     3. $df = 432$ 7. $t = 2.383$

     4. $t_{.05} = 1.966$ 8. reject H_0, conclude $\beta > 0$, $.02 > p > .01$

7. a. $b - \beta = .722$ b. 95%CI $= 3.551 - 6.273$. Yes   c. 75%CI $= 4.133 - 5.691$. No

Chapter 10: Multiple Regression

1. $b_1 > 0$, $b_2 > 0$

3. $b_1 = .2$, $b_2 = 1$, $a = 25$

5. b_1: Homicides increase by .754 per 100,000 for each percent increase in children in single-parent households, with percentage below the poverty level held constant.

     b_2: Homicides increase by .079 per 100,000 for each percent increase in poverty, with percentage of children in single-parent households held constant.

     a: The predicted number of homicides per 100,000 is -10.38 when percentage in poverty equals 0 and percentage of children in single-parent households equals 0.

     B_1: Homicides per 100,000 increase by .79 of a standard deviation per standard deviation increase in percentage of children in single-parent households, with percentage in poverty held constant.

     B_2: Homicides per 100,000 increase by .079 of a standard deviation per standard deviation increase in percentage in poverty, with percentage of children in single-parent households held constant.

     R^2: Percentage of children in single-parent households and percentage in poverty explain 69.3 percent of the variance in homicides per 100,000.

     R: Homicides per 100,000 increase by .833 of a standard deviation per standard deviation increase in predicted homicides.

7. a. R-square $= .129$, regression $df = 4$, residual $df = 511$, regression mean square $= 35.321$, residual mean square $= 1.865$, $F = 18.937$, Sig $< .001$, EDUC Standard error $= .024$, AGEKDBRN Standard error $= .012$, EDUC $t = -2.392$, AGEKDBRN $t = -5.536$, EDUC Sig $< .05$, AGEKDBRN Sig $< .00005$.

     b. $df = 511$, $t_{.025} = 1.965$

     c. $F_{.05} = 2.39$, reject H_0 that $R^2 = 0$ and conclude $R^2 > 0$, or that at least one $b_i \neq 0$.

Chapter 11: Nominal Independent Variables

1. South mean $= 10.5$, North mean $= 5.9$, intercept $= 5.9$, slope $= 4.6$

3. a. mean $= .875$, st. dev. $= .331$

     b. intercept $= 50.66$, slope $= -5.83$

5.

Group	X_1	X_2	X_3
Greek	1	0	0
Parents	0	1	0
Apt.	0	0	1
Dorm	0	0	0

7. AGEKDBRN = 23.24 + 1.11CATHOLIC + 3.29JEWISH + .30NONE + .47OTHER

9. a. The mean number of sex partners per year for Protestants is .473; the mean sex partners per year for Catholics is .013 less than the mean for Protestants; the mean sex partners per year for Jews is .053 more than the mean for Protestants; the mean sex partners per year of no religion is .413 greater than that for Protestants; and the mean sex partners per year for other religions is .224 greater than that for Protestants.

b. The mean sex partners per year for Catholics equals that for Protestants. The mean sex partners per year for Jews equals that for Protestants. The mean sex partners per year for those with no religion equals that for Protestants. The mean sex partners per year for those with other religions equals that for Protestants.

c. Do not reject the null hypothesis of equal Catholic and Protestant means. Do not reject the null hypothesis of equal Jewish and Protestant means. Reject the null hypothesis for nones and Protestants; conclude that the mean sex partners per year for those with no religion is greater than that for Protestants. Do not reject the null hypothesis that other religions and Protestants have equal means.

d. The null hypothesis is that Catholics, Jews, nones, and others have equal mean numbers of sex partners per year; the alternative hypothesis is that at least one religion has a mean number of sex partners different from the others. Reject the null hypothesis that the five religious groups have equal means in favor of the alternative that at least one group has a mean that is different from the others.

Chapter 12: Ordinal Dependent Variables

1. Always wrong = 245.5; almost always wrong = 567; sometimes wrong = 830.5; not wrong at all = 1386. Mean = 877.5.

3. Tolerance of premarital sex decreases by .387 of a rank for each rank increase in frequency of religious service attendance. Tolerance of premarital sex decreases by .404 of a standard rank deviation per standard rank deviation increase in attendance at religious services. Reject the null hypothesis that the rank slope equals 0 at $p < .00005$ and conclude that the slope is negative.

5. The mean rank of sex frequency of females is 711.0. The mean rank of sex frequency of males is 105.2 greater than that of females. Gender explains 1.5 percent of the variance in the ranks of sex frequency.

Chapter 13: Nominal Dependent Variables

1. No. Males and females are both more likely to say no than yes to abortion for any reason, and the proportion responding no appears to be about the same for each gender.

3.

	FEFAM2 better for man to work and woman tend home, recode		
	1 Agree	2 Disagree	Total
ABANY ABORTION 1 YES IF WOMAN WANTS FOR ANY REASON	3	5	8
	33.3%	45.5%	40.0%
2 NO	6	6	12
	66.7%	54.5%	60.0%
Total	9	11	20
	100.0%	100.0%	100.0%

People who disagree with the idea that it is better for the man to be the achiever outside the home are 12.2 percent more likely to say that a pregnant woman should be able to have a legal abortion if she wants one for any reason.

5. a. $df = 1, \chi^2 = 3.841$

b.

Expected Count

		FEFAM2	
		Agree	Disagree
ABANY	YES	135.0	190.0
	NO	214.0	301.0

chi-square $= 15.1$

c. Reject the null hypothesis of no association and conclude that people who believe that it is better for the man to be the achiever outside the home are more likely to say that a pregnant woman should not be able to have a legal abortion.

7. a. .134

b. .284

c. *bible.* Cramér's V is larger for *abany* and *bible* than for *abany* and *fefam2*.

References

Agresti, Alan, and Finlay, Barbara. 1997. *Statistical Methods for the Social Sciences,* 3rd Edition. Upper Saddle River, NJ: Prentice Hall.

Allison, Paul D. 2001. *Missing Data.* Thousand Oaks, CA: Sage.

Blalock, Hubert M., Jr. 1979. *Social Statistics,* 2nd Edition. New York: McGraw-Hill.

Cohen, Jacob, and Cohen, Patricia. 1983. *Applied Multiple Regression/Correlation Analysis for the Behavioral Sciences,* 2nd Edition. Hillsdale, NJ: Lawrence Erlbaum.

Conover, W. J. 1980. *Practical Nonparametric Statistics,* 2nd Edition. New York: Wiley.

Conover, W. J., and Iman, Ronald L. 1981. "Rank Transformations as a Bridge Between Parametric and Nonparametric Statistics." *The American Statistician,* 35:124–129.

Fox, John. 1997. *Applied Regression Analysis, Linear Models, and Related Methods.* Thousand Oaks, CA: Sage.

Galton, Francis. 1877. "Typical Laws of Heredity." *Nature* 15:492–495, 512–514, 532–533.

Galton, Francis. 1886. "Regression towards Mediocrity in Hereditary Stature." *Journal of the Anthropological Institute* 15:246–263.

Huff, Darrell. 1954. *How to Lie with Statistics.* New York: W. W. Norton.

Iman, Ronald L., and Davenport, James M. 1976. "Approximations to the Exact Distribution of the Kruskal-Wallis Test Statistic." *Communications in Statistics,* Series A, 5:1335–1348.

Judd, Charles M., and McClelland, Gary H. 1989. *Data Analysis: A Model-Comparison Approach.* San Diego: Harcourt Brace Jovanovich.

Keppel, Geoffrey. 1991. *Design and Analysis: A Researcher's Handbook,* 3rd Edition. Englewood Cliffs, NJ: Prentice Hall.

Loether, Herman J., and McTavish, Donald G. 1976. *Descriptive and Inferential Statistics: An Introduction.* Boston: Allyn & Bacon.

McClendon, McKee J. 1994. *Multiple Regression and Causal Analysis.* Itasca, IL: F.E. Peacock. Reissued in 2002 by Waveland Press, Inc., Prospect Heights, IL.

Radloff, Lenore E. 1977. "The CES-D Scale: A Self-Report Depression Scale for Research in the General Population." *Applied Psychological Measurement,* 1:385–401.

Siegel, Sidney. 1956. *Nonparametric Statistics for the Behavioral Sciences.* New York: Mc-Graw-Hill.

Index